Advances in Experimental Medicine and Biology

Volume 876

More information about this series at http://www.springer.com/series/5584

Clare E. Elwell • Terence S. Leung •
David K. Harrison

Editors

Oxygen Transport to Tissue XXXVII

 Springer

OPEN

Editors
Clare E. Elwell
Department of Medical Physics
  & Biomedical Engineering
University College London
London
United Kingdom

Terence S. Leung
Department of Medical Physics & Biomedical
  Engineering
University College London
London
United Kingdom

David K. Harrison
Microvascular Measurements
St. Lorenzen, Italy

ISSN 0065-2598   ISSN 2214-8019   (electronic)
Advances in Experimental Medicine and Biology
ISBN 978-1-4939-3022-7   ISBN 978-1-4939-3023-4   (eBook)
DOI 10.1007/978-1-4939-3023-4

Library of Congress Control Number: 2015958657

Springer New York Heidelberg Dordrecht London

Printed on acid-free paper

Springer Science+Business Media LLC New York is part of Springer Science+Business Media (www.springer.com)

Preface

The 42nd Annual Meeting of International Society on Oxygen Transport to Tissue was held at University College London (UCL) in the UK between 28th June and 3rd July 2014. This volume contains the papers submitted, reviewed and accepted from presentations at that meeting.

Clare Elwell, the President of ISOTT 2014, and her organising committee were particularly keen to represent a broad range of oxygen-related scientific topics at the meeting and to include contributions from researchers new to ISOTT. This resulted in a total of 146 submitted abstracts and attendance of 192 participants including 40 students.

UCL has a long tradition of multidisciplinary research in basic and clinical sciences and this was reflected in the contributions from local keynote speakers Nick Lane and Monty Mythen who between them took us from the origin of life to the heights of Everest. Keynote talks were also provided by Sir Moncada Salvador (University of Manchester, UK), Carsten Lundby (University of Zurich, Switzerland), Elizabeth Hillman (Columbia University, USA) and Can Ince (Erasmus University of Rotterdam, Holland). Submitted contributions were given as oral and poster presentations in sessions on cellular hypoxia, blood substitutes and oxygen therapeutics, oxygen transport in neonatal and adult critical care medicine, cancer metabolism, altitude, muscle oxygenation, multimodal imaging techniques, brain oxygenation, optical techniques for oxygen measurement and mathematical modelling of oxygen transport. The 2014 Kovach Lecture was delivered by Edwin Nemoto. Lunchtime sessions were presented on new advances in optical neuroimaging and the future of blood substitutes. We were also pleased to welcome Harriet Tuckey, who spoke to us about the scientific contribution of her father Griffin Pugh, to the 1953 Everest expedition.

The conference was supported by generous sponsorship from Hitachi High Tech, Hamamatsu Photonics, NanoBlood LLC, Seahorse Bioscience, Baker Ruskinn, Gowerlabs, Moor Instruments, Swinco AG/Secret Training and Oxford Optronics.

The Annual General Meeting agreed to establish the "John Severinghaus Lecture" in honour and recognition of the significant contributions that John

Severinghaus has and continues to make to ISOTT and the field of oxygen transport. Some ISOTT members were lucky to share a memorable dinner with John and his colleagues at the Royal Society of Medicine during which extensive and stimulating discussions were held about those who discovered oxygen and many other topics.

This exciting scientific programme was matched by social events reflecting the vibrant central London location. ISOTT2014 took place on the UCL Bloomsbury campus in the heart of London providing excellent local venues for social events allowing scientific interaction in an informal atmosphere. The Welcome Reception was hosted in the UCL Cloisters complete with string quartet. On Sunday evening, we honoured UCL's spiritual father with a Great British Pub evening in the Jeremy Bentham pub. On Monday evening, we viewed a wide and varied range of zoological specimens at the local Grant Museum before enjoying a barbeque in the main UCL quad. The highlight social event was the Thames River Cruise Banquet which provided spectacular views of the city as well as music and dancing.

The meeting concluded with the Awards Ceremony during which the Melvin H. Knisely award was presented to David Highton, the Dietrich Lübbers award to Felix Scholkmann and the Britton Chance award to Anne Riemann. Bruley travel awards were presented to Nassim Nasseri, Stefan Kleiser, Shinsuke Nirengi, Marie-Aline Neveu, Geraldine Depreter, Ben Jones and Takuya Osawa.

London, UK   Clare Elwell

Acknowledgements

As President of the 2014 Conference of the International Society on Oxygen Transport to Tissue, held from 28th June to 3rd July at University College London, UK, I would like to gratefully acknowledge the support of our sponsors:

University College London

Hitachi High-Technologies

Hamamatsu Photonics KK

BAKER RUSKINN

Baker Ruskinn

Seahorse Bioscience

Swiss Innovative Concepts

Moor Instruments

Gowerlabs

NanoBlood LLC

Oxford Optronix

Panel of Reviewers

Reviewer	Affiliation
James Bassingthwaighte	University of Washington, USA
Duane Bruley	Synthesizer, Inc., Ellicott City, USA
Alexander Caicedo	Catholic University of Leuven, Belgium
Chris Cooper	University of Essex, Colchester, UK
Victor Darley-Usmar	University of Alabama at Birmingham, USA
Clare Elwell	University College London, UK
Malou Friederich-Persson	University of Uppsala, Sweden
Howard Halpern	University of Chicago, USA
David Harrison	Microvascular Measurements, St. Lorenzen, Italy
David Highton	University College London, UK
Elizabeth Hillman	Columbia University, USA
Can Ince	University of Amsterdam, The Netherlands
Benjamin Jones	University of Essex, Colchester, UK
Peter Keipert	Keipert Corp. Consulting, San Diego, USA
Royatoro Kime	Tokyo Medical University, Japan
Stefan Kleiser	University Hospital Zurich, Switzerland
Joseph LaManna	Case Western Reserve University, Cleveland, USA
Terence Leung	University College London, UK
Lin Li	University of Pennsylvania, USA
Qingming Luo	Huazhong University of Science and Technology, China
Arnulf Mayer	University Medical Centre, Mainz, Germany
Mark Muthalib	University of Montpellier, France
Nassim Nasseri	University Hospital of Zurich, Switzerland
Edwin Nemoto	University of New Mexico, USA
Marie-Aline Neveu	Catholic University of Leuven, Belgium
Takuya Osawa	Japan Institute of Sports Sciences, Tokyo, Japan
Patrik Persson	University of Uppsala, Sweden
Kaoru Sakatani	Nihon University School of Medicine, Japan

(continued)

Reviewer	Affiliation
Hitomi Sakai	Nara Medical University, Japan
Felix Scholkmann	University of Zurich, Switzerland
Akitoshi Seiyama	Kyoto University, Japan
John Severinghaus	University of California San Francisco, USA
Justin Skownow	The Children's Hospital at Westmead, Australia
André Steimers	University of Applied Sciences Koblenz, Remagen, Germany
Harold Swartz	EPR Center at Darmouth, Hanover, USA
Ilias Tachtsidis	University College London, UK
Eiji Takahashi	Saga University, Japan
Oliver Thews	University of Halle, Germany
Clare Thorn	University of Exeter Medical School, UK
Sabine Van Huffel	Catholic University of Leuven, Belgium
Koen van Rossem	GSK, Belgium
Peter Vaupel	University Medical Centre, Mainz, Germany
Peter Wagner	University of California San Diego, USA
Michael Wilson	University of Essex, Colchester, UK
Martin Wolf	University of Zurich, Switzerland
Ursula Wolf	University of Bern, Switzerland
Christopher Wolff	St. Mary's College and Barts, UK
Kui Xu	Case Western Reserve University, Cleveland, USA
Technical Reviewers	
Laraine Visser-Isles	Rotterdam, The Netherlands
Eileen Harrison	St. Lorenzen, Italy

ISOTT participants in front of the UCL Wilkins' Building

Local Organizing Committee

Clare Elwell
Paolo Demaria
Andy O'Reilly
Ilias Tachtsidis
Terence Leung
Chris Cooper
David Highton
Tharindi Hapuarcachchi
Gemma Bale
Luke Dunne

Scientific Organizing Committee

Chris Cooper (Essex, UK)
Clare Elwell (London, UK)
Paul Okunieff (Gainesville, USA)
Fredrik Palm (Uppsala, Sweden)
Kaoru Sakatani (Tokyo, Japan)
Ilias Tachtsidis (London, UK)
Martin Wolf (Zurich, Switzerland)

ISOTT Officers and Executive Committee

President
Name: Clare Elwell
Country: UK
Telephone: (+44) 20-7679-0270
Fax: (+44) 20-7679-0255
Email: c.elwell@ucl.ac.uk

Past President
Name: Harold M. Swartz
Country: USA
Telephone: (+1) 603-650-1955
Fax: (+1) 603-650-1717
Email: Harold.m.swartz@dartmouth.edu

Presidents-Elect
Name: Qingming Luo
Country: China
Telephone: (+86) 27-8779-2033
Fax: (+86) 27-8779-2034
Email: qluo@mail.hust.edu.cn
Email: qluo@mail.hust.edu.cn

Name: Lin Li
Country: USA
Telephone: (+1) 215-898-1805
Fax: (+1) 215-573-2113
Email: linli@mail.med.upenn.edu

Secretary
Name: Oliver Thews
Country: Germany
Telephone: (+49) 345-557-4048
Fax: (+49) 345-557-4019
Email: oliver.thews@medizin.uni-halle

Treasurer
Name: Peter E. Keipert
Country: USA
Telephone: (+1) 858-450-2445
Fax: (+1) 858-450-2499
Email: pkeipert@sangart.com

Chairman, Knisely Award Committee
Name: Duane F. Bruley
Country: USA
Telephone: (+1) 410-313-9939
Fax: F: (+1) 410-313-9939
Email: bruley33@verizon.net

Executive Committee
Chris Cooper (UK)
Jerry D. Glickson (USA)
Howard Halpern (USA)
Terence Leung (UK)
Rammohan Maikala (USA)
Kazuto Masamoto (Japan)
Masaomi Nangaku (Japan)
Frederik Palm (Sweden)

ISOTT Award Winners

The Melvin H. Knisely Award

The Melvin H. Knisely Award was established in 1983 to honor Dr. Knisely's accomplishments in the field of the transport of oxygen and other metabolites and anabolites in the human body. Over the years, he has inspired many young investigators and this award is to honor his enthusiasm for assisting and encouraging young scientists and engineers in various disciplines. The award is to acknowledge outstanding young investigators. This award was first presented during the banquet of the 1983 annual conference of ISOTT in Ruston, Louisiana. The award includes a Melvin H. Knisely plaque and a cash prize.

Melvin H. Knisely Award Recipients

1983   Antal G. Hudetz (Hungary)
1984   Andras Eke (Hungary)
1985   Nathan A. Bush (USA)
1986   Karlfried Groebe (Germany)
1987   Isumi Shibuya (Japan)
1988   Kyung A. Kang (Korea/USA)
1989   Sanja Batra (Canada)
1990   Stephen J. Cringle (Australia)
1991   Paul Okunieff (USA)
1992   Hans Degens (The Netherlands)
1993   David A. Benaron (USA)
1994   Koen van Rossem (Belgium)
1995   Clare E. Elwell (UK)
1996   Sergei A. Vinogradov (USA)
1997   Chris Cooper (UK)
1998   Martin Wolf (Switzerland)
1999   Huiping Wu (USA)

2000 Valentina Quaresima (Italy)
2001 Fahmeed Hyder (Bangladesh)
2002 Geofrey De Visscher (Belgium)
2003 Mohammad Nadeem Khan (USA)
2004 Fredrick Palm (Sweden)
2005 Nicholas Lintell (Australia)
2006 –
2007 Ilias Tachtsidis (UK)
2008 Kazuto Masamoto (Japan)
2009 Rossana Occhipinti (USA)
2010 Sebatiano Cicco (Italy)
2011 Mei Zhang (USA)
2012 Takahiro Igarashi (Japan)
2013 Malou Friederich-Persson (Sweden)
2014 David Highton (UK)

The Dietrich W. Lübbers Award

The Dietrich W. Lübbers Award was established in honor of Professor Lübbers's long-standing commitment, interest, and contributions to the problems of oxygen transport to tissue and to the society. This award was first presented in 1994 during the annual conference of ISOTT in Istanbul, Turkey.

Dietrich W. Lübbers Award Recipients

1994 Michael Dubina (Russia)
1995 Philip E. James (UK/USA)
1996 Resit Demit (Germany)
1997 Juan Carlos Chavez (Peru)
1998 Nathan A. Davis (UK)
1999 Paola Pichiule (USA)
2000 Ian Balcer (USA)
2001 Theresa M. Busch (USA)
2002 Link K. Korah (USA)
2003 James J. Lee (USA)
2004 Richard Olson (Sweden)
2005 Charlotte Ives (UK)
2006 Bin Hong (China/USA)
2007 Helga Blockx (Belgium)
2008 Joke Vanderhaegen (Belgium)
2009 Matthew Bell (UK)
2010 Alexander Caicedo Dorado (Belgium)

2011 Malou Friedrich (Sweden)
2012 Maria Papademetriou (UK)
2013 Nannan Sun (China)
2014 Felix Scholkmann (Switzerland)

The Britton Chance Award

The Britton Chance Award was established in honor of Professor Chance's long-standing commitment, interest, and contributions to the science and engineering aspects of oxygen transport to tissue and to the society. This award was first presented in 2004 during the annual conference of ISOTT in Bari, Italy.

Britton Chance Award Recipients

2004 Derek Brown (Switzerland)
2005 James Lee (USA)
2006 Hanzhu Jin (China/USA)
2007 Eric Mellon (USA)
2008 Jianting Wang (USA)
2009 Jessica Spires (USA)
2010 Ivo Trajkovic (Switzerland)
2011 Alexander Caicedo Dorado (Belgium)
2012 Felix Scholkmann (Switzerland)
2013 Tharindi Hapuarachchi (UK)
2014 Anne Riemann (Germany)

The Duane F. Bruley Travel Awards

The Duane F. Bruley Travel Awards were established in 2003 and first presented by ISOTT at the 2004 annual conference in Bari, Italy. This award was created to provide travel funds for student researchers in all aspects of areas of oxygen transport to tissue. The awards signify Dr. Bruley's interest in encouraging and supporting young researchers to maintain the image and quality of research associated with the society. As a cofounder of ISOTT in 1973, Dr. Bruley emphasizes cross-disciplinary research among basic scientists, engineers, medical scientists, and clinicians. His pioneering work constructing mathematical models for oxygen and other anabolite/metabolite transport in the microcirculation, employing computer solutions, was the first to consider system nonlinearities and time dependence, including multidimensional diffusion, convection, and reaction kinetics. It is hoped that receiving the Duane F. Bruley Travel Award will inspire students to excel in their research and will assist in securing future leadership for ISOTT.

The Duane F. Bruley Travel Award Recipients

2004   Helga Blocks (Belgium), Jennifer Caddick (UK), Charlotte Ives (UK), Nicholas Lintell (Australia), Leonardo Mottola (Italy), Samin Rezania (USA/Iran), Ilias Tachtsidis (UK), Liang Tang (USA/China), Iyichi Sonoro (Japan), Antonio Franco (Italy)

2005   Robert Bradley (UK), Harald Oey (Australia), Kathy Hsieh (Australia), Jan Shah (Australia)

2006   Ben Gooch (UK), Ulf Jensen (Germany), Smruta Koppaka (USA), Daya Singh (UK), Martin Tisdall (UK), Bin Wong (USA), and Kui Xu (USA)

2007   Dominique De Smet (Belgium), Thomas Ingram (UK), Nicola Lai (USA), Andrew Pinder (UK), Joke Vanderhaegen (Belgium)

2008   Sebastiano Chicco (Italy)

2009   Lei Gao (UK), Jianting Wang (USA), Obinna Ndubuizu (USA), Joke Vanderhaegen (Belgium)

2010   Zareen Bashir (UK), Tracy Moroz (UK), Mark Muthalib (Australia), Catalina Meßmer (USA), Takashi Eriguchi (Japan), Yoshihiro Murata (Japan), Jack Honeysett (UK), Martin Biallas (Switzerland)

2011   Catherine Hesford (UK), Luke S. Holdsworth (UK), Andreas Metz (Switzerland), Maria D. Papademetriou (UK), Patrik Persson (Sweden), Felix Scholkmann (Switzerland), Kouichi Yoshihara (Japan)

2012   Allann Al-Armaghany (UK), Malou Friederich-Persson (Sweden), Tharindi Hapuarachchi (UK), Benjamin Jones (UK), Rebecca Re (Italy), Yuta Sekiguchi (Japan), Ebba Sivertsson (Sweden), André Steimers (Germany)

2013   Allann Al-Armaghany (UK), Gemma Bale (UK), Alexander Caicedo-Dorado (Belgium), Luke Dunne (UK)

2014   Geraldine De Preter (Belgium), Benjamin Jones (UK), Stefan Kleiser (Switzerland), Nassimsadat Nasseri (Switzerland), Marie-Aline Neveu (Belgium), Shinsuke Nirengi (Japan), Takuya Osawa (Japan)

Kovach Lecture

The Kovach Lecture is presented periodically to honor a career dedicated to oxygenation research. Arisztid Kovach was a world-renowned cardiovascular physiologist and one of the early leaders of ISOTT. This lecture is dedicated to his remarkable scientific and teaching career.

Kovach Lecture Recipients

2011   John Severinghaus
2012   Peter Vaupel
2013   No Recipient
2014   Edwin Nemoto

Contents

Chapter 1
The Most Important Discovery of Science

John W. Severinghaus

Abstract Oxygen has often been called the most important discovery of science. I disagree. Over five centuries, reports by six scientists told of something in air we animals all need. Three reported how to generate it. It acquired many names, finally oxygen. After 8 years of studying it, Lavoisier still couldn't understand its nature. No special date and no scientist should get credit for discovering oxygen. Henry Cavendish discovered how to make inflammable air (H2). When burned, it made water. This was called impossible because water was assumed to be an element. When Lavoisier repeated the Cavendish test on June 24, 1783, he realized it demolished two theories, phlogiston and water as an element, a Kuhnian paradigm shift that finally unlocked his great revolution of chemistry.

Keywords Plagiarism • Cavendish, Henry • Sendivogius, Michael • Discoveries of oxygen • Scheele, Carl Wilhelm

J.W. Severinghaus (✉)
Anaesthesiology and Cardiovascular Research Institute, University of California, San Francisco, CA 94957, USA
e-mail: jwseps@comcast.net

© Springer Science+Business Media, New York 2016   1
C.E. Elwell et al. (eds.), *Oxygen Transport to Tissue XXXVII*, Advances in Experimental Medicine and Biology 876, DOI 10.1007/978-1-4939-3023-4_1

1   Ibn al Nafis 1213–1288

The story of oxygen begins in the thirteenth century with Ibn al Nafis, a Syrian physician and anatomist who became the Egyptian sultan's doctor and chief of Cairo's hospital. In 1250 AD he correctly described the pulmonary circulation and oxygen in a single sentence: "Pulmonary arterial blood passes through invisible pores in the lung, where it mingles with air to form the vital spirit and then passes through the pulmonary vein to reach the left chamber of the heart". Vital spirit was a term used by Galen about 200 AD to refer to red (arterialized) blood.

Nafis clearly was the first to understand lung blood flow, 400 years before William Harvey. This finding was scarcely known in the West until discovered in the Prussian state library in 1924 [1].

Nafis was also the first author to write both SCI FI stories and humorous theological novels.

2   Michael Servetus 1511–1553

Three hundred years later Michael Servetus rediscovered the pulmonary circulation. He was a brilliant polymath, poet, politician, geographer, mapmaker, biblical scholar and doctor who became physician to the archbishop of Lyon, France. Early in Martin Luther's Reformation, Servetus got into trouble by repeatedly writing that doctrines of the trinity and infant baptism were not biblical. Both Catholics and Protestants declared him a heretic. To avoid the Inquisition, he hid in Strasburg under a new name. In 1553 he wrote "Christianity Restored", proposing that the Reformation should return to biblical roots. He sent John Calvin a copy. Here is the cover page.

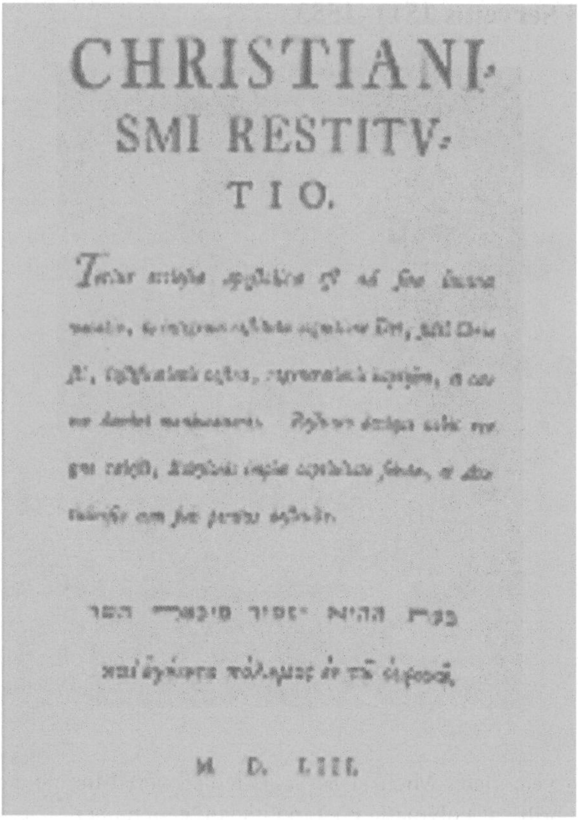
CHRISTIANI-
SMI RESTITV-
TIO.

M. D. LIII.

Here is the cover page. Calvin threatened to have him killed for heresy if he came to Geneva. Inquisition leaders burned him in effigy.

While escaping to Italy, rashly and inexplicably, he snuck into Calvin's church in Geneva. He was recognized, reported, arrested and jailed as a heretic by order of Geneva's government. He was held for over 2 months in prison while Calvin consulted with all the Swiss cantons, and then indicted.

On Oct. 27, 1553, Servetus was burned at the stake in the Champel district of Geneva, with a copy of "Christianity Restored" strapped to his leg.

Calvin ordered the burning of all copies of his book that contained Servetus's only description of the effects of air on blood in the lung. His science report remained unknown for 400 years.

A French historian, August Dide, published a book called "Heretics and Revolutionaries" in 1887. After his election as a Senator in 1900, Dide proposed erecting a monument in Geneva to the most important French heretical martyr, Michael Servetus. In 1907, Rodin's pupil Clothilde Roch finished a sculpture of Servetus in prison. The Calvinist Geneva town council refused to celebrate a heretic. So Dide had it mounted nearby in France, angering the Calvinists. During WWII the pro-Nazi Vichy government melted the monument down because Servetus had been a martyr to freedom of speech.

In 1953, 400 years after his martyrdom, Roland Bainton, Yale professor of ecclesiastical history [2], published the most authoritative Servetus biography, "Hunted Heretic". He had spent most of 20 years studying Servetus' life and works. (His son is an anesthesiologist at UCSF and his daughter married my wife's brother Bill Peck.) Also in 1953 at Yale, John Fulton, professor of physiology, found Servetus's unknown description of the pulmonary circulation in one of the three surviving copies of his heretical book [3, 4]. It is now clear that Servetus had discovered, a century before Sir William Harvey, that blood flows through the lung tissue, excreting waste products into the air and changing blood color.

In 2011, just 3 years ago, on Servetus's 500th birthday, Roch's sculpture was recast and mounted in Geneva near the site of his burning. Even now, Calvinist authorities refused to attend.

3   Michael Sendivogius 1566–1636

The third discoverer of oxygen, Michael Sendivogius, was a famous sixteenth century Polish noble, physician and alchemist. In 1604 he published that air contains what he named the *Secret Food of Life*. He had realized that this "food" was the same gas emitted by heating saltpetre (potassium nitrate). He called it part of the "salt-nitre of the earth". Sendivogius's writings were frequently copied and read for over a century. However, no one grasped the chemical importance of his discovery, not even Boyle and Newton at Oxford 60 years later.

In 1621, a Dutch inventor named Drebbel built the world's first manned submarine. King James I and a horde of viewers watched it travel submerged 10 miles down the Thames from Westminster to Greenwich. Drebbel never revealed the secret of refreshing submarine air but newspapers and others believed that he had learned from Sendivogius in Prague how to make oxygen on board. I doubt that and suspect he pumped in air through snorkles from little surface floats some said they saw.

Forty years later, Robert Boyle in Oxford wrote that he had spoken with a mathematician who had been on Drebbel's submarine. He said that "Drebbel used a chemical liquor to replace the quintessence of air". Although both Boyle and Newton read Sendivogius, they ignored his work, perhaps because they didn't grasp the importance of his chemistry. Or perhaps they thought he was just another fraudulent alchemist.

Sendivogius' "Food of Life" story was only recently rediscovered. I first learned about him a year ago from Martyna Elas, a Polish biochemist. A biography of Sendivogius was published in Warsaw by Roman Bugnai in 1968 [5] and a study of his alchemical pursuits, "Water which does not wet hands; the alchemy of Michael Sendivogius", was published in 1994 by Andrew Szydlo [6]. Szydlo himself was born of Polish parents in England and educated in London's universities. He teaches chemistry at London's Highgate School. Last June, when in London, I invited him to my lecture and a dinner discussion honoring him. He is a dramatic lecturer with demonstrations to fascinated audiences.

4   John Mayow 1641–1679

In Oxford in 1668 Robert Boyle's former student John Mayow, wrote a book about a part of air he named *SPIRITUS NITRO-AEREUS*. He wrote that it is consumed in fire and that we breathe it to provide both body heat and energy. He was the fourth discoverer of oxygen but ignored Sendivogius.

Mayow's own work sank into obscurity when all scientists accepted the false phlogiston theory postulated by Georg Ernst Stahl. This mythical substance, invented to avoid need for air, obstructed reality-based science for nearly a century.

In 1955, Mayow's book about oxygen was rediscovered and published by Donald Proctor at Johns Hopkins University School of Medicine [7].

5   Carl Wilhelm Scheele 1742–1786

The fifth discoverer of oxygen, Carl Wilhelm Scheele, was born in German-speaking Pomerania. Trained in Sweden, he became a very clever apothecary and chemist. In 1768, he published new studies of metal chemistry. In 1770, he became director of Locke Pharmacy in Uppsala. He worked at the university there with famous chemist Torbern Bergman who helped him publish his work in Latin in the journal Nova Acta.

In 1774, Lavoisier sent his new chemistry textbook to the Academy, including a copy for Scheele whose published work he much admired.

Scheele promptly wrote Lavoisier to thank him for the book. He explained how, in 1771, he had generated a strange new gas by heating certain metallic earths. He had named it FIRE AIR because it greatly brightened a candle flame and supported life in mice. Because neither he nor Bergman could relate it to the phlogiston theory, Scheele had delayed publishing. He asked Lavoisier to repeat the experiment and then help him explain it.

By the time Scheele finally decided in 1775 to publish his fire air with all his experimental findings, he had read about Joseph Priestley's 1774 discovery of the same gas, so it was too late to claim its discovery. Scheele wrote that Lavoisier never answered his letter [8]. His book was finally published in 1777 but he could not confirm his claim of discovering fire air. Scientists and historians therefore ignored his oxygen discovery.

Working in Uppsala and later in Köping, Sweden, as an apothecary, Scheele became one of the world's greatest experimental chemists. He discovered seven elements and much other new chemistry. On February 4, 1775, he was elected to membership in the Royal Swedish Academy of Sciences although his discovery of oxygen was unpublished and unknown. This great honor, with the King of Sweden attending, had never been given to an apothecary.

In 1893, French historian Édouard Grimaux was shown Scheele's letter to Lavoisier, never previously seen. He published the text but claimed the letter then had disappeared.

In 1993, to avoid taxation, various Lavoisier artifacts were donated to the French Academy of Science. Among them was Scheele's 1774 letter to Lavoisier [9]. Descendants of Lavoisier' wife had hidden the letter for 219 years because they feared the great chemist may have used Scheele's methods without acknowledgement, inferring guilt of plagiarism.

6   Joseph Priestley 1733–1804

Joseph Priestley
by Rembrandt Peale

Next comes oxygen's sixth discoverer, Unitarian minister Joseph Priestley. Two churches fired him as too radical but he soon became famous while teaching grammar and science at Warrington Academy near Manchester, helping to make it a leading school for dissenters – in fact, the "cradle of Unitarianism" as one scholar called it.

In the 1760s, Benjamin Franklin, while living in England, befriended Priestley and provided books on electricity as well as experimental apparatus. In 1766 Franklin persuaded the Royal Society to elect him to membership at age 33. The next year, Priestley published the most authoritative and popular text on electricity. He was awarded the Royal Society's Copley medal in 1773, partly for discovering how to make cheap soda water from brewery exhaust. Encouraged by Franklin, the Earl of Shelburne offered Priestley a laboratory on his Bowood estate in Wiltshire. On August 1, 1774, Priestley, by heating red mercury calc, generated a new gas that caused a glowing splinter to burst into flame and supported life in a mouse longer than air in a sealed bottle. Still in thrall to phlogiston theory, he described it as DEPHLOGISTICATED AIR [10] and published. He would write that, in October 1774 in Paris, he described to Lavoisier his method of making this new air. He said Lavoisier never acknowledged his help.

7 Antoine Laurent Lavoisier 1743–1794

Lavoisier had become the world's most brilliant chemist by age 30. In the spring of 1775 he began his studies of Priestley's new gas. He named it *principe oxigene* but continued to refer to it as 'vital air'. However, after 8 years of extensively examining its chemistry he was still unable to prove whether vital air was a new element or a compound. And he doubted but still couldn't disprove the phlogiston theory. By 1783 he was stalled by these dilemmas [11].

8   Henry Cavendish 1731–1810

In 1766, Cavendish discovered and published how to make an inflammable gas (H_2) by putting iron filings in strong acid [12]. In the late 1770s, others noted that burning inflammable air caused dew to form on the burner tube's walls. Cavendish analyzed and proved the dew was pure water. His peers in the Royal Society considered his claim of making water to be wrong, since all assumed that water is an element that can't be made.

In 1783, Cavendish' associate, physicist Charles Blagden, persuaded the Royal Society to invite Cavendish to try to explain his 'impossible' finding. Cavendish replied that he had repeated burning inflammable gas with Priestley's air and it again produced water. It remained a mystery. Blagden then visited Lavoisier, asking help to understand the unbelievable Cavendish report. His visit stimulated Lavoisier to test the Cavendish-Priestley observation by burning inflammable air himself.

On June 24, 1783, Lavoisier invited eight chemists to watch him prove that Cavendish was wrong. When water appeared, Lavoisier was stunned! He suddenly realized that the Cavendish observation had revealed a universally accepted error. He declared: "Inflammable air and vital air are elements. Water is *not an element* but a compound made of them".

Lavoisier's insight created the greatest *Kuhnian paradigm shift* [13] in the history of chemistry. It unlocked his great chemical revolution. Cavendish had provided the key.

Lavoisier named inflammable air hydrogen and vital air oxygen. Over the next 6 years Lavoisier demolished phlogiston and revised all chemical theory. In 1788, at age 45 he finished writing his great treatise Elements of Chemistry [14].

In celebration, his 29-year-old wife Marie Anne commissioned a portrait from France's greatest artist of the time, Jacques Louis David. He charged perhaps the highest artistic fee of all time, 7000 livre, about a quarter million US dollars for this now famous huge (2.6 m tall) double portrait of Lavoisier and his wife. It is well worth visiting the Met.

9   What About Lavoisier?

In his Elements of Chemistry Lavoisier falsely wrote: "This species of air was discovered almost at the same time by Mr Priestley, Mr Scheele, and myself", both a bold lie and proof that he had read Scheele's 1774 letter revealing his method of making fire air but failed to acknowledgement it!

Lavoisier was condemned in print by Edmond Genet, later French ambassador to the United States, by Joseph Black, the discoverer of CO_2, and by Priestley himself who accused him of plagiarism [15] for claims of discoveries that actually belonged to others.

Lavoisier was a brilliant polymath, meticulous scientist and the most able chemist of his time. Born into a wealthy family, he became even richer as a despised tax "farmer" in France's *ancien régime* who built a hated wall around Paris to ensure that incoming merchants would pay taxes. As a leading figure in the Academy of Science in 1780 he belittled the work of an aspiring academician, Jean Paul Marat, making a bitter enemy. More than a decade later, during the French Revolution, the now radical revolutionary publicist Marat would repeatedly demand in newspapers that Lavoisier be guillotined as among the worst of the *ancien régime*. At the peak of the Reign of Terror, the Revolutionary Tribunal tried, convicted and beheaded the great chemist all in 1 day, May 8, 1794 [11].

10   What About Henry Cavendish, FRS?

He was part of eight centuries of an immensely wealthy aristocratic family. He was very shy and reclusive. He never married and never sat for a portrait. He built his laboratory and lived most of his life in the home of his father, Lord Charles Cavendish, in Bloomsbury on the corner of Bedford Square beside the British Museum and University College. This plaque is on the wall there.

Cavendish was a natural philosopher, scientist, and a great experimental and theoretical chemist and physicist, distinguished for accuracy and precision in his researches into the composition of atmospheric air, the properties of different gases and the law governing electrical attraction and repulsion. He was first to weight the earth by a method still named the Cavendish experiment. The very famous original and new physics laboratories in Cambridge are named after him.

The Old Cavendish

Credits: Milton Djuric, history editor, Williams College and Martyna Elas, biochemist, Krakow Univ (for Sendivogius).

References

1. West JR (2008) Ibn al-Nafis, the pulmonary circulation and the islamic golden age. J Appl Physiol 105:1877–1880
2. Bainton RH (1953) Hunted heretic: the life and death of Michael Servetus, 1511–1553. Beacon, Boston
3. O'Malley CD (1953) Michael Servetus. American Philosophical Society, Philadelphia, pp 195–208
4. Fulton JF (1953) Michael Servetus, humanist and martyr. Herbert Reichner, New York
5. Bugnaj R (1968) Michael Sendivogius (1566–1636) his life and works, 126. Ossolineum, London/Warsaw
6. Szydlo A (1994) Water that does not wet hands: the Alchemy of Michael Sendivogius. Polish Academy of Science, Warsaw
7. Proctor DF (ed) (1995) A history of breathing physiology, vol 83, Lung biology in health and disease. Marcel Dekker, New York
8. Scheele CW (1777) Chemical treatise on air and fire. Magnus Swederus, Uppsala
9. Grimaux E (1890) Une Iettre inedite de Scheele a Lavoisier. Revue générale des sciences pures et appliquées 1:1–2
10. Priestley J (1775) An account of further discoveries in air. Phil Trans R Soc Lond 65:384–394
11. Guerlac H (1975) Antoine-Laurent Lavoisier. Charles Scribner's Sons, New York
12. Cavendish H (1766) Experiments on factitious air. Part I. Containing experiments on inflammable air. Phil Trans R Soc Lond 56:144–159
13. Kuhn TS (1962) The structure of scientific revolutions. University of Chicago Press, Chicago
14. Lavoisier A (1789) Traité élémentaire de chimie. Chez Cuchet, Paris
15. Schofield RE (2004) The enlightened Joseph Priestley: a study of his life and work from 1773 to 1804. University Park, Penn State University Press, Philadelphia

Part I
Muscle, Oxygen and Exercise

Chapter 2
The Spatial Distribution of Absolute Skeletal Muscle Deoxygenation During Ramp-Incremental Exercise Is Not Influenced by Hypoxia

T. Scott Bowen, Shunsaku Koga, Tatsuro Amano, Narihiko Kondo, and Harry B. Rossiter

Abstract Time-resolved near-infrared spectroscopy (TRS-NIRS) allows absolute quantitation of deoxygenated haemoglobin and myoglobin concentration ([HHb]) in skeletal muscle. We recently showed that the spatial distribution of peak [HHb] within the quadriceps during moderate-intensity cycling is reduced with progressive hypoxia and this is associated with impaired aerobic energy provision. We therefore aimed to determine whether reduced spatial distribution of skeletal muscle [HHb] was associated with impaired aerobic energy transfer during exhaustive ramp-incremental exercise in hypoxia. Seven healthy men performed ramp-incremental cycle exercise (20 W/min) to exhaustion at 3 fractional inspired O_2 concentrations (F_IO_2): 0.21, 0.16, 0.12. Pulmonary O_2 uptake ($\dot{V}O_2$) was measured using a flow meter and gas analyser system. Lactate threshold (LT) was estimated non-invasively. Absolute muscle deoxygenation was quantified by multichannel TRS-NIRS from the *rectus femoris* and *vastus lateralis* (proximal and distal regions). $\dot{V}O_{2peak}$ and LT were progressively reduced ($p < 0.05$) with hypoxia. There was a significant effect ($p < 0.05$) of F_IO_2 on [HHb] at baseline, LT, and

T.S. Bowen (✉)
Department of Internal Medicine and Cardiology, Leipzig University, Heart Center, Strümpellstraße 39, 04289 Leipzig, Germany
e-mail: bows@med.uni-leipzig.de

S. Koga
Applied Physiology Laboratory, Kobe Design University, Kobe, Japan

T. Amano • N. Kondo
Graduate School of Human Development and Environment, Laboratory for Applied Human Physiology, Kobe University, Kobe, Japan

H.B. Rossiter
Division of Respiratory and Critical Care Physiology and Medicine, Rehabilitation Clinical Trials Center, Los Angeles Biomedical Research Institute and Harbor-UCLA Medical Center, Torrance, CA, USA

School of Biomedical Sciences, University of Leeds, Leeds, UK

© Springer Science+Business Media, New York 2016
C.E. Elwell et al. (eds.), *Oxygen Transport to Tissue XXXVII*, Advances in Experimental Medicine and Biology 876, DOI 10.1007/978-1-4939-3023-4_2

peak. However the spatial variance of [HHb] was not different between F_IO_2 conditions. Peak total Hb ([Hb_{tot}]) was significantly reduced between F_IO_2 conditions ($p < 0.001$). There was no association between reductions in the spatial distribution of skeletal muscle [HHb] and indices of aerobic energy transfer during ramp-incremental exercise in hypoxia. While regional [HHb] quantified by TRS-NIRS at exhaustion was greater in hypoxia, the spatial distribution of [HHb] was unaffected. Interestingly, peak [Hb_{tot}] was reduced at the tolerable limit in hypoxia implying a vasodilatory reserve may exist in conditions with reduced F_IO_2.

Keywords Heterogeneity • HHb • Near infrared-spectroscopy • NIRS • Skeletal muscle

1 Introduction

In skeletal muscle during exercise the balance between the rate of oxygen utilisation ($\dot{V}O_2$) and oxygen delivery ($\dot{Q}O_2$) underlies the ability to meet cellular energetic demands through oxidative metabolism [1]. Heterogeneity in muscle fibre oxidative capacity, capillarity, blood flow, and recruitment means that skeletal muscle $\dot{Q}O_2/\dot{V}O_2$ is also widely heterogeneous within and between skeletal muscles [1]. A wide spatial heterogeneity in $\dot{Q}O_2/\dot{V}O_2$ may reflect a beneficial metabolic 'flexibility', i.e. by maintaining muscle regions with high PO_2 and thus high potential for oxidative energy provision and fatigue resistance. Alternatively, wide spatial heterogeneity in $\dot{Q}O_2/\dot{V}O_2$ may reflect a detrimental condition impairing oxidative energy provision in muscle regions that contribute to limiting the system's entire output (and thus limiting exercise tolerance). Whether spatial heterogeneity in $\dot{Q}O_2/\dot{V}O_2$ is a beneficial or detrimental physiological response to exercise remains unclear.

The dynamics of the $\dot{Q}O_2/\dot{V}O_2$ ratio is the predominant variable determining changes in the near-infrared spectroscopy (NIRS) derived deoxyhaemoglobin and myoglobin signal (hereafter termed HHb, for simplicity) during exercise. In humans, a hypoxia-induced slowing of $\dot{V}O_2$ kinetics during moderate intensity exercise is associated with reduced [HHb] heterogeneity [2]. This suggests that a narrow spatial distribution in $\dot{Q}O_2/\dot{V}O_2$ may reflect conditions detrimental to high rates of aerobic energy transfer. While the mechanisms matching regional $\dot{Q}O_2$ to $\dot{V}O_2$ remain equivocal, hypoxia may reduce nitric oxide (NO) bioavailability (a potent vasodilator) and limit the ability to maintain muscle regions with a high $\dot{Q}O_2/\dot{V}O_2$ ratio [3].

It remains unknown whether [HHb] heterogeneity is reduced during maximal exercise, where it has the potential to contribute to the mechanisms limiting exercise tolerance [1]. We therefore determined the association between the spatial distribution of skeletal muscle [HHb] and indices of aerobic energy transfer (pulmonary $\dot{V}O_{2peak}$ and lactate threshold) during ramp-incremental exercise to the limit of tolerance in normoxia and hypoxia. Absolute [HHb] was measured by multi-channel time resolved (TRS)-NIRS. We hypothesized that in hypoxia: (1) altered control of regional microvascular blood flow would cause the spatial distribution of [HHb] to become more uniform; and (2) a reduced metabolic flexibility reflected by low [HHb] heterogeneity would correlate with reduced aerobic energy provision and exercise intolerance.

2   Methods

Seven healthy men (mean ± SD: age, 22 ± 2 years; height, 172 ± 6 cm; weight, 61 ± 6 kg; and $\dot{V}O_{2peak}$, 50 ± 8 ml/kg/min) provided written informed consent, as approved by the Human Subjects Committee of Kobe Design University. Detailed descriptions of all procedures, equipment, and measurements have been previously published [2]. Briefly, a cycle ergometer ramp-incremental (RI) exercise test (20 W/min) was performed at 60 rpm to exhaustion with three inspired fractional oxygen concentrations (F_IO_2): 0.21; 0.16; and 0.12. Each test was performed on a different day in a randomised order. Participants breathed through a mouthpiece connected to a low-resistance, two-way non-rebreathing valve (2700, Hans Rudolph, Shawnee, KS, USA), linked to rubber tubing that supplied humidified air from Douglas bags filled with room air ($F_IO_2 = 0.21$) or room air diluted with N_2 ($F_IO_2 = 0.16$ or 0.12).

Pulmonary $\dot{V}O_2$ was measured breath-by-breath using a flow meter and gas analyser system (Aeromonitor AE-300S; Minato Medical Science, Osaka, Japan). Lactate threshold (LT) was estimated non-invasively from combined ventilatory and gas exchange criteria (e.g. [2]). Deoxygenation of the right quadriceps was quantified by multi-channel TRS-NIRS at three muscle sites: the distal (VL_d) and proximal (VL_p) regions of the *vastus lateralis*, and the mid region of the *rectus femoris* (RF) (TRS-20, Hamamatsu Photonics KK, Hamamatsu, Japan). Each TRS-20 probe (consisting of a detector fixed at 3 cm from the emitter) provided pico-second light pulses at three different wavelengths (760, 795, and 830 nm) to measure (in micromoles) absolute muscle deoxygenation ([deoxy(Hb + Mb)]; here termed [HHb]), oxygenation ([oxy(Hb + Mb)]; here termed [HbO$_2$]), and total haemoglobin concentration ([Hb + Mb]; [Hb$_{tot}$]) at an output frequency of 0.5 Hz. Tissue oxygen saturation ([HbO$_2$]/[Hb$_{tot}$]; S$_t$O$_2$ %) was also calculated. To account for the influence of adipose tissue thickness (ATT) on the absolute haem concentrations, a linear regression was applied to the relationship between resting [Hb$_{tot}$] and ATT ([Hb$_{tot}$] $= -21.4 \cdot$ (ATT) $+ 220$; $r^2 = 0.77$; $P < 0.001$), and data were normalized to an ATT of zero as previously described [2].

Variables at baseline, LT, and peak exercise were determined from the average of 30 s. Data are presented as mean $\pm$ SD. Differences in NIRS variables were assessed by two-way repeated measures ANOVA (F_IO_2 $\times$ muscle region). Other data were assessed by one-way repeated measures ANOVA among F_IO_2 conditions. Post hoc Bonferroni corrected t-tests determined the location of the differences where appropriate. The effect of F_IO_2 on spatial heterogeneity of [HHb] was determined by the coefficient of variation (CV (%): $100 \times$ SD/mean of 3 muscle regions). Significance was accepted at $p < 0.05$. Analyses were completed using SPSS v.16.0 (SPSS Inc., Chicago, IL, USA).

3 Results

Lactate threshold (26 ± 6, 22 ± 5, 20 ± 3 ml/kg/min, at F_IO_2 of 0.21, 0.16, and 0.12, respectively) and pulmonary $\dot{V}O_{2peak}$ (50 ± 8, 45 ± 5, 37 ± 4 ml/kg/min) were progressively reduced with hypoxia (each $p < 0.05$). As expected, the work rate at LT and peak exercise was also reduced (111 ± 29, 101 ± 34, 91 ± 17 and 282 ± 54, 262 ± 45, 221 ± 36 W, respectively; $p < 0.05$). The change in absolute quadriceps deoxygenation in each F_IO_2 condition is presented in Fig. 2.1. There was a significant effect ($p < 0.05$) of F_IO_2 on [HHb] at baseline (57 ± 5, 61 ± 5, 69 ± 6 uM), at LT (67 ± 4, 70 ± 3, 80 ± 5 uM), and at peak (92 ± 8, 97 ± 7, 101 ± 11 uM). Interestingly, peak [Hb$_{tot}$] was also lower in hypoxia: 237 ± 24, 230 ± 23, 227 ± 26 uM ($p < 0.001$; Fig. 2.1), as was S_tO_2: 62 ± 2, 59 ± 2, and 57 ± 3 % ($p < 0.001$; Fig. 2.1). The spatial heterogeneity in [HHb] for a representative subject within and between muscle regions across F_IO_2 conditions is presented in Fig. 2.2. The spatial variance of [HHb] was not different between F_IO_2 conditions: baseline (8 ± 4, 7 ± 5, 9 ± 5 %), LT (9 ± 4, 7 ± 4, 9 ± 4 %), and peak (13 ± 6, 11 ± 4, 13 ± 8 %; Fig. 2.3). Thus, there was no association between [HHb] heterogeneity and indices of aerobic energy transfer in the different F_IO_2 conditions at baseline, LT, or peak power output.

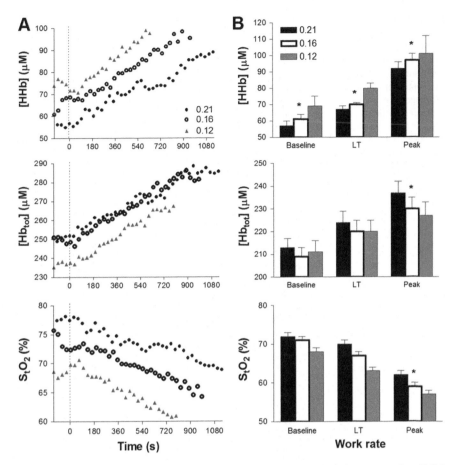

Fig. 2.1 Deoxyhaemoglobin ([HHb]), total haemoglobin ([Hb$_{tot}$], and tissue saturation (S$_t$O$_2$) during ramp-incremental exercise in different F$_I$O$_2$ conditions (0.21, 0.16, 0.12). (**a**) NIRS variables in the *vastus lateralis* (proximal region) from a representative subject; data are a 30 s average. (**b**) Group NIRS variables (mean ± SD) averaged across three muscle regions (*rectus femoris* and proximal and distal regions of the *vastus lateralis*). *p < 0.05, significant effect of F$_I$O$_2$. *Dashed vertical line* represents start of ramp-incremental exercise

T.S. Bowen et al.

Fig. 2.2 Quadriceps deoxygenation ([HHb]) during ramp-incremental exercise in different F_IO_2 conditions (0.21, 0.16, 0.12) in a representative subject. Deoxygenation was measured in the *rectus femoris* (RF) and proximal and distal regions of the *vastus lateralis* (VL_p and VL_d). *Dashed vertical lines* are the start and end of ramp-incremental exercise

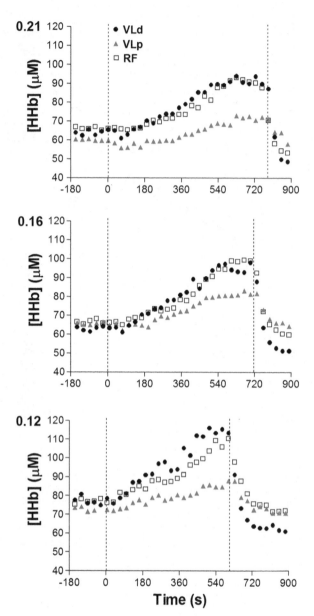

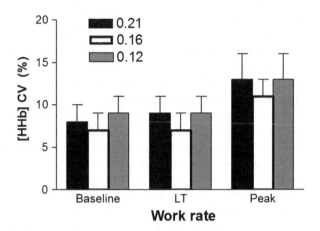

Fig. 2.3 Mean coefficient of variation (CV) in quadriceps deoxygenation ([HHb]) during ramp-incremental exercise in different F_IO_2 conditions (0.21, 0.16, 0.12). CV values were determined from the spatial heterogeneity among the *rectus femoris* and proximal and distal regions of the *vastus lateralis*

4   Discussion

We previously demonstrated that a hypoxia-induced slowing of $\dot{V}O_2$ kinetics during moderate intensity exercise was associated with reduced [HHb] heterogeneity [2]. As such, we hypothesized here that [HHb] spatial distribution would also become more homogenous during RI exercise in hypoxia. This was based on the notion that hypoxia may modulate microvascular blood flow and limit the capacity to maintain heterogeneity in muscle $\dot{Q}O_2/\dot{V}O_2$ during exercise—coincident with reduced metabolic flexibility in hypoxic conditions. To our surprise, however, we found the spatial distribution of [HHb] was unaffected by hypoxia during RI exercise.

Absolute skeletal muscle deoxygenation was progressively increased with the severity of hypoxia during RI exercise, despite progressively reduced peak power output. This may be due to a lower absolute baseline $\dot{Q}O_2$, which requires an increased O_2 extraction during exercise, even if the exercise $\dot{Q}O_2$ increment ($\Delta\dot{Q}O_2/\Delta\dot{V}O_2$) is unaffected [4]; and/or due to a reduced exercise $\Delta\dot{Q}O_2/\Delta\dot{V}O_2$ response in hypoxia. The overall exercise-induced [HHb] increase was not different between conditions (~35 uM), suggesting that a baseline reduction in $\dot{Q}O_2$ in hypoxia may have predominated. Nevertheless, whichever the mechanism, the observed hypoxic modulation in microvascular blood volume (reduced peak $[Hb_{tot}]$) is consistent with an attenuated vasodilation in hypoxia, suggesting that a vasodilatory reserve exists in hypoxia that may limit aerobic energy production.

Hypoxia has been suggested to reduce NO bioavailabilty, which serves as an important metabolite for vasodilation [5]. It therefore remains possible that hypoxia may have limited NO function and thus vasodilation, which could have disrupted the balance between $\dot{Q}O_2$ and $\dot{V}O_2$. Evidence in humans showing NOS inhibition impairs forearm vasodilation during exercise to a greater degree in hypoxia compared to normoxia supports this notion [5]. However, whether hypoxia decreases or increases NO remains controversial [3, 5]. Exercise intensity is also known to alter blood flow spatial distribution in quadriceps muscles [6], with higher compared to lower power outputs associated with a more heterogeneous response. This may be related to an altered contribution of various vasoregulatory metabolites at different exercise intensities, or in different fibre types, and/or an increased distribution of blood flow towards newly recruited muscle fibres [6]. This may also help explain why we found no reduction in the spatial heterogeneity of [HHb] during RI exercise: the expected reduction in [HHb] heterogeneity (based on findings from moderate intensity cycling [2]) were offset by an increased heterogeneity during high intensity exercise recruiting poorly perfused muscle regions [6]. Although CV for [HHb] was not different between F_IO_2 conditions, it is estimated that, with seven subjects, we had power to detect an ~3 % difference in regional deoxygenation CV.

In conclusion, absolute regional skeletal muscle deoxygenation measured by multi-channel TRS-NIRS was greater at rest and at the limit of tolerance in hypoxia compared to normoxia. However, contrary to our hypothesis this was not associated with a reduction in the spatial heterogeneity of [HHb] within and between quadriceps muscles. Of interest, the peak [Hb_{tot}] was reduced at the limit of tolerance in hypoxia compared to normoxia, implicating a vasodilatory reserve exists in hypoxia. This is consistent with the notion that hypoxia reduces exercise vasodilation and modulates the mechanisms linking $\dot{Q}O_2$ and $\dot{V}O_2$ that contribute to reduced metabolic flexibility.

Acknowledgments Support provided by the Medical Research Council UK (studentship to TSB), BBSRC UK (BB/1024798/1; BB/I00162X/1), and The Japan Society for the Promotion of Science, the Ministry of Education, Science, and Culture of Japan (Grant-in-Aid: 22650151 to SK). TSB is a Postdoctoral Research Fellow of the Alexander von Humboldt Foundation.

References

1. Koga S, Rossiter HB, Heinonen I et al (2014) Dynamic heterogeneity of exercising blood flow and O2 utilization. Med Sci Sports Exerc 46:860–876
2. Bowen TS, Rossiter HB, Benson AP et al (2013) Slowed oxygen uptake kinetics in hypoxia correlate with the transient peak and reduced spatial distribution of absolute skeletal muscle deoxygenation. Exp Physiol 98:1585–1596
3. Ferreira LF, Hageman KS, Hahn SA et al (2006) Muscle microvascular oxygenation in chronic heart failure: role of nitric oxide availability. Acta Physiol (Oxf) 188:3–13
4. Benson AP, Grassi B, Rossiter HB (2013) A validated model of oxygen uptake and circulatory dynamic interactions at exercise onset in humans. J Appl Physiol 115:743–755
5. Casey DP, Walker BG, Curry TB et al (2011) Ageing reduces the compensatory vasodilation during hypoxic exercise: the role of nitric oxide. J Physiol 589:1477–1488
6. Heinonen I, Sergey VN, Kemppainen J et al (2007) Role of adenosine in regulating the heterogeneity of skeletal muscle blood flow during exercise in humans. J Appl Physiol 103:2042–2048

Chapter 3
Do Two Tissue Blood Volume Parameters Measured by Different Near-Infrared Spectroscopy Methods Show the Same Dynamics During Incremental Running?

Takuya Osawa, Takuma Arimitsu, and Hideyuki Takahashi

Abstract Both the change in total hemoglobin concentration (cHb), assessed by near-infrared continuous-wave spectroscopy (NIR-CWS), and the normalized tissue hemoglobin index (nTHI), assessed by NIR spatially resolved spectroscopy (NIR-SRS), were used to quantify changes in tissue blood volume. However, it is possible that these parameters may show different changes because of the different measurement systems. The present study aimed to compare changes in cHb and nTHI in working muscles, which were selected for measurement because the parameters changed dynamically. *Methods*: After a standing rest, seven male runners (age 24 ± 3 years, mean $\pm$ S.D.) performed an incremental running exercise test on a treadmill (inclination = 1 %) from 180 to 300 m min^{-1}. During the tests, cHb and nTHI were monitored from the vastus lateralis (VL) and medial gastrocnemius (GM) muscles. These parameters were relatively evaluated from the minimal to maximal values through the test. *Results*: When the exercise began, cHb and nTHI quickly decreased and then gradually increased during running. In comparison with both VL and GM, there was significant interaction between cHb and nTHI. *Conclusions*: The present results suggest that cHb and nTHI in working muscles are not always synchronized, particularly at the onset of exercise and at high intensities. Although cHb was previously used as the change of tissue blood volume, it is implied that tissue blood volume assessed by cHb is overestimated.

Keywords Near-infrared continuous wave spectroscopy • Near-infrared spatially resolved spectroscopy • Tissue blood volume • Working muscle • Running

T. Osawa (✉) • T. Arimitsu • H. Takahashi
Department of Sports Science, Japan Institute of Sports Sciences, 3-15-1 Nishigaoka, Kita-ku, Tokyo 115-0056, Japan
e-mail: takuya.osawa@jpnsport.go.jp

© Springer Science+Business Media, New York 2016
C.E. Elwell et al. (eds.), *Oxygen Transport to Tissue XXXVII*, Advances in Experimental Medicine and Biology 876, DOI 10.1007/978-1-4939-3023-4_3

1   Introduction

The total hemoglobin concentration (cHb) in the tissue is measured by near-infrared continuous-wave spectroscopy (NIR-CWS) and is widely used to quantify changes in tissue blood volume (BV) [1, 2]. Monitoring changes in BV is essential to understand other NIR-CWS parameters, including oxygenated and deoxygenated hemoglobin/myoglobin (oxy-Hb/Mb and deoxy-Hb/Mb, respectively).

Over the previous decade, NIR spatially resolved spectroscopy (NIR-SRS) has been used to study the tissue O_2 index (TOI) in parallel with NIR-CWS [3, 4]. In addition to TOI, NIR-SRS provides a normalized tissue hemoglobin index (nTHI), which is considered to reflect BV, similar to cHb. Previous reports have examined whether cutaneous blood flow and fat layer thickness affects oxy-Hb/Mb, deoxy-Hb/Mb, and TOI [5–9]. However, cHb and nTHI are rarely compared. Since oxy-Hb/Mb monitored by NIR-CWS was influenced by cutaneous blood flow and overestimated during exercise, especially under hot condition, it was be possible that cHb overestimates the change of tissue blood volume, similar to oxy-Hb/Mb. The BV parameters mostly changed in working muscles during exercise [10]. Hence, cutaneous blood flow largely increases during whole-body exercise [9]. The present study aimed to examine whether two tissue blood volume parameters, cHb and nTHI, indicate the same trend during incremental running.

2   Methods

2.1   Subjects

Seven male runners participated in this study (age 24 ± 3 years; height 1.68 ± 0.06 m; body mass 57.5 ± 4.1 kg, mean $\pm$ S.D.). Before participating in the experiment, subjects received an explanation of the study procedures and potential risks, and they then signed an informed consent document. This study was approved by the local ethics committee and all work was performed in accordance with the Declaration of Helsinki.

2.2   Experimental Design

After the subjects were familiarized with the experiments, they performed an incremental running exercise test on a treadmill (inclination = 1 %). Following an initial 2-min standing rest period, the subjects ran at 180 m min^{-1} for 3 min, and the running speed was then increased 10 m min^{-1} every minute until it reached 300 m min^{-1}.

2.3 Measurements

The two BV parameters cHb and nTHI were simultaneously monitored throughout the test from the bellies of the vastus lateralis (VL) and medial gastrocnemius (GM) muscles using the NIRS system (NIRO-200NX, Hamamatsu Photonics, Shizuoka, Japan).

Two fiber optic bundles transmitted the NIR light to the tissues of interest. The NIR light source was three laser diodes that emitted light at different wavelengths (775, 810, and 850 nm), the intensity of which was continuously measured at 2 Hz. The source–detector distances were 2.5 and 3 cm. The probes were housed in a plastic holder and fixed with tape and bandages. The differential pathlength factor was omitted from the measurements in this study because this factor is unknown for VL and GM. The data of cHb and nTHI were averaged at 15 s. The data were normalized to a percentage from rest to the end of exercise (min = 0 %, max = 100 %) because of the large differences in these parameters between subjects.

2.4 Statistical Analyses

The data are represented as the mean $\pm$ S.D. Statistical analyses were performed using the statistical package SPSS for Windows (version 19.0; SPSS, Chicago, IL). For comparisons between cHb and nTHI, a two-way repeated-measures analysis of variance (ANOVA) was used to identify significant differences; a paired t-test was performed for post hoc comparisons when a significant effect was identified. P values <0.05 were considered significant.

3 Results

The raw cHb data in VL and GM (cHb_VL, cHb_GM) and nTHI in VL and GM (nTHI_VL, nTHI_GM) for all subjects are presented in Fig. 3.1. Both cHb_VL and nTHI_VL acutely decreased at the beginning of exercise, gradually increased during exercise, but then decreased as the exercise intensity increased in the most subjects. The maximum cHb_VL, but not nTHI_VL, during exercise was higher than that observed at rest. In contrast, cHb and nTHI were higher in GM than in VL. In addition, cHb_GM and nTHI_GM increased as the running speed increased

Figure 3.2 shows the relative changes (%) in the study parameters. There was a significant interaction between cHb and nTHI in both GM and VL (Fig. 3.2). Moreover, nTHI_VL was significantly higher than cHb_VL from 90 to 120 s (P < 0.05) and tended to be higher from 135 to 225 s (P < 0.1). Similarly, nTHI_GM showed significantly higher values than cHb_GM from 90 to 285 s (P < 0.05).

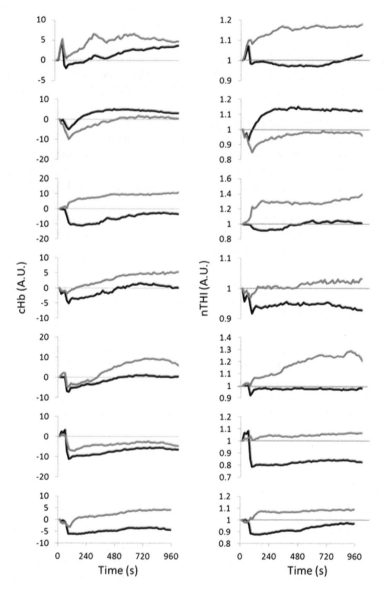

Fig. 3.1 Changes in cHb (*left*) and nTHI (*right*) in VL (*black lines*) and GM (*gray lines*) in all subjects. The *horizon lines* represent their resting values

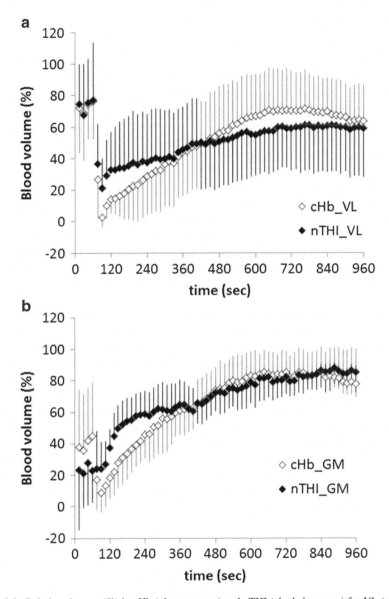

Fig. 3.2 Relative changes (%) in cHb (*clear squares*) and nTHI (*shaded squares*) for VL (**a**) and GM (**b**) from rest to a running exercise performed at 300 m min^{-1}. Values are means $\pm$ S.D. There were significant interactions between cHb and nTHI. Therefore, cHb and nTHI were different as the main effect. *P < 0.05, #P < 0.1

4 Discussion

We compared the two BV parameters assessed by two NIRS systems, NIR-CWS and NIR-SRS, in working muscles during running exercise. The main findings of this study are that changes in these BV parameters showed different dynamics and that, in comparison to relative changes, cHb showed greater dynamics than nTHI during exercise. The parameter of nTHI was thought to be less influenced by increase of skin blood flow and temperature in the measurement part, similar to TOI monitored by NIR-SRS [9]. One of the factors in the greater dynamics of cHb than nTHI might be the change of skin blood flow and tissue temperature. Some previous studies have reported cHb as the change of tissue blood volume, these might overestimate the change of tissue blood volume, especially at the onset of exercise, at high intensity and during over 10-min exercise.

In previous studies, cHb and nTHI were simultaneously monitored at the same part. Robbins et al. [11] observed similar dynamics between the two parameters in a working muscle (lateral gastrocnemius) with whole-body vibration. However, to the best of our knowledge, a comparison between cHb and nTHI has not previously been made.

In this study, the two BV parameters were evaluated in working muscles during dynamic, whole-body, and high-intensity exercise. During exercise, particularly within minutes of beginning the exercise and at higher intensities, BV in working muscles increased because of vasodilation and an increase in capillary recruitment. However, it is possible the BV changes measured with NIR-CWS were influenced by increased cutaneous blood flow and that BV changes evaluated by cHb were overestimated. On the other hand, nTHI may be less influenced by changes in cutaneous blood flow, considering that TOI assessed by NIR-SRS was thought to be only slightly influenced by changes in cutaneous blood flow [9].

In this study, cHb and nTHI were different at rest although this difference was not significant (Fig. 3.2). Considering there were large individual differences between their dynamics, cHb and nTHI were relatively evaluated from the minimal to maximal values throughout the test.

References

1. Ferrari M, Mottola L, Quaresima V (2004) Principles, techniques, and limitations of near infrared spectroscopy. Can J Appl Physiol 29:463–487
2. Hamaoka T, McCully KK, Quaresima V et al (2007) Near-infrared spectroscopy/imaging for monitoring muscle oxygenation and oxidative metabolism in healthy and diseased humans. J Biomed Opt 12:062105
3. Bowen TS, Murgatroyd SR, Cannon DT et al (2011) A raised metabolic rate slows pulmonary O_2 uptake kinetics on transition to moderate-intensity exercise in humans independently of work rate. Exp Physiol 96:1049–1061
4. Osawa T, Kime R, Fujioka M et al (2013) O_2 saturation in the intercostal space during moderate and heavy constant-load exercise. Adv Exp Med Biol 789:143–148

5. Jones AM, Davies RC, Ferreira LF et al (2009) Reply to Quaresima and Ferrari. J Appl Physiol 107:372–373
6. Buono MJ, Miller PW, Hom C et al (2005) Skin blood flow affects in vivo near-infrared spectroscopy measurements in human skeletal muscle. Jpn J Physiol 55:241–244
7. Davis SL, Fadel PJ, Cui J et al (2006) Skin blood flow influences near-infrared spectroscopy-derived measurements of tissue oxygenation during heat stress. J Appl Physiol 100:221–224
8. Quaresima V, Ferrari M (2009) Muscle oxygenation by near-infrared-based tissue oximeters. J Appl Physiol 107:371
9. Tew GA, Ruddock AD, Saxton JM (2010) Skin blood flow differentially affects near-infrared spectroscopy-derived measures of muscle oxygen saturation and blood volume at rest and during dynamic leg exercise. Eur J Appl Physiol 110:1083–1089
10. Osawa T, Kime R, Hamaoka T et al (2011) Attenuation of muscle deoxygenation precedes EMG threshold in normoxia and hypoxia. Med Sci Sports Exerc 43:1406–1413
11. Robbins D, Elwell C, Jimenez A et al (2012) Localised muscle tissue oxygenation during dynamic exercise with whole body vibration. J Sports Sci Med 11:346–351

Chapter 4
Underwater Near-Infrared Spectroscopy: Muscle Oxygen Changes in the Upper and Lower Extremities in Club Level Swimmers and Triathletes

B. Jones and C.E. Cooper

Abstract To date, measurements of oxygen status during swim exercise have focused upon systemic aerobic capacity. The development of a portable, waterproof NIRS device makes possible a local measurement of muscle hemodynamics and oxygenation that could provide a novel insight into the physiological changes that occur during swim exercise. The purpose of this study was to observe changes in muscle oxygenation in the vastus lateralis (VL) and latissimus dorsi (LD) of club level swimmers and triathletes. Ten subjects, five club level swimmers and five club level triathletes (three men and seven women) were used for assessment. Swim group; mean $\pm$ SD = age 21.2 ± 1.6 years; height 170.6 ± 7.5 cm; weight 62.8 ± 6.9 kg; vastus lateralis skin fold 13.8 ± 5.6 mm; latissimus dorsi skin fold 12.6 ± 3.7. Triathlete group; mean $\pm$ SD = age 44.0 ± 10.5 years; height 171.6 ± 7.0 cm; weight 68.6 ± 12.7 kg; vastus lateralis skin fold 11.8 ± 3.5 mm; latissimus dorsi skin fold 11.2 ± 3.1. All subjects completed a maximal 200 m freestyle swim, with the PortaMon, a portable NIR device, attached to the subject's dominant side musculature. ΔTSI % between the vastus lateralis and latissimus dorsi were analysed using either paired (2-tailed) t-tests or Wilcoxon signed rank test. The level of significance for analysis was set at $p < 0.05$. No significant difference ($p = 0.686$) was found in ΔTSI (%) between the VL and LD in club level swimmers. A significant difference ($p = 0.043$) was found in ΔTSI (%) between the VL and LD in club level triathletes. Club level swimmers completed the 200 m freestyle swim significantly faster ($p = 0.04$) than club level triathletes. Club level swimmers use both the upper and lower muscles to a similar extent during a maximal 200 m swim. Club level triathletes predominately use the upper body for propulsion during the same exercise. The data produced by NIRS in this study are the first of their kind and provide insight into muscle oxygenation changes during swim exercise which can indicate the contribution of one muscle compared to another. This also enables a greater understanding of the differences in

B. Jones (✉) • C.E. Cooper
School of Biological Sciences, Centre for Sports and Exercise Science, University of Essex, Colchester CO4 3SQ, UK
e-mail: bjonesa@essex.ac.uk

© Springer Science+Business Media, New York 2016   35
C.E. Elwell et al. (eds.), *Oxygen Transport to Tissue XXXVII*, Advances in Experimental Medicine and Biology 876, DOI 10.1007/978-1-4939-3023-4_4

swimming techniques seen between different cohorts of swimmers and potentially within individual swimmers.

Keywords Underwater • NIRS • Swimming • Muscle oxygenation

1 Introduction

It has been shown that individual swimmers vary considerably with regard to swim technique and style [1]. It has also been reported that swim athletes display specific training adaptations in comparison to land based athletes and triathletes [2]. Specifically, it has been reported that triathletes and swimmers vary in swim velocity and propulsion efficiency [3]. It is therefore of interest to understand the different physiological responses that may occur within different swim athletes. Roels et al. demonstrated that during an incremental swim test, competition swimmers produced significantly higher heart rate maximum, maximal oxygen consumption (VO_2max) and maximal swim velocities in comparison with competition triathletes. Current physiological assessment during swim exercise has predominately focused upon global measurements such as heart rate [4], blood lactate [5] and VO_2max [6]. Systemic measurements of oxygen status have been made possible through the development of specialized snorkels such as the Aqua Trainer Valve Cosmed technologies©, which now allows for breath-by-breath analysis within the pool [7]. It has been suggested that a local measurement of muscle oxygenation during swim exercise would contribute to a multi-modality approach that is perhaps warranted, to enable a more detailed evaluation of swim athletes [8].

Near-infrared spectroscopy (NIRS) has been used successfully in the non-invasive observation of changes in local muscle oxygenation and hemodynamic responses in tissue within a variety of dynamic land based sports including; short track speed skating [9], downhill skiing [10] and sprint running [11] and recently the NIR technique has been utilised with swim athletes [12], albeit during incremental cycle exercise. To the best of our knowledge the concurrent observation of the muscle oxygenation in both the vastus lateralis (VL) and latissimus dorsi (LD) muscles, using NIRS, during underwater swimming has not been carried out. The development of a portable waterproof NIRS device, utilising an optically clear, waterproof, silicone covering, has enabled the provision of a local measurement of muscle oxygenation that could provide novel insight into the physiological changes that occur during swim exercise. Therefore, the aims of the study were: to advance the assessment and utility of an underwater near-infrared spectroscopy device and to monitor the differences (if any) in muscle oxygenation changes within two different cohorts of swim athletes in the upper and lower extremities.

2   Methods

Club level swimmers and triathletes from Essex based (United Kingdom) swim clubs were asked on a voluntary basis to take part in a 200 m maximal swim effort in their club training pool. Ten subjects, five club level swimmers and five club level triathletes, (three men and seven women) volunteered for assessment. Swim group; mean $\pm$ SD = age 21.2 $\pm$ 1.6 years; height 170.6 $\pm$ 7.5 cm; weight 62.8 $\pm$ 6.9 kg; vastus lateralis skin fold 13.8 $\pm$ 5.6 mm; latissimus dorsi skinfold 12.6 $\pm$ 3.7. Triathlete group; mean $\pm$ SD = age 44.0 $\pm$ 10.5 years; height 171.6 $\pm$ 7.0 cm; weight 68.6 $\pm$ 12.7 kg; vastus lateralis skin fold 11.8 $\pm$ 3.5 mm; latissimus dorsi skinfold 11.2 $\pm$ 3.1. All subjects were experienced swim athletes who had been involved in swim training for 11.1 $\pm$ 3.4 years. Subjects were asked not to perform any strenuous activity 24 h prior to the observational assessment. The study received ethical approval and all subjects provided informed consent.

Each subject was asked to perform a maximal 200 m freestyle swim (8 $\times$ 25 m) in a short course swimming pool. Subjects were made to complete a 400 m warm up comprised of 300 m (freestyle) at a self-selected pace and 100 m (freestyle) consisting of 15 m hard and 35 m easy $\times$ 2. Following warm up, subjects passively stood in the shallow end of the swimming pool (waist immersion), with their arms by their sides and their body weight evenly distributed over each leg, for a 2-min period to establish a baseline NIR reading. Subjects then started freestyle swimming without diving from the pool edge and were able to perform regular ('tumble turn') turning motions at the end of the lane. Upon completion subjects passively stood for 3 min at the shallow end of the pool, following the same procedure as for the baseline NIR attainment.

The portable NIRS apparatus (PortaMon, Artinis, Medical Systems, BV, the Netherlands) used in this study was a dual wavelength continuous system, which simultaneously uses the modified Beer-Lambert law and spatially resolved spectroscopy (SRS) methods. Changes in tissue oxyhemoglobin (O_2Hb), deoxyhemoglobin (HHb) and total hemoglobin (tHb) were measured using the difference in absorption characteristics of wavelength at 750 and 850 nm. The tissue hemoglobin saturation index (TSI) (expressed in % and calculated as $[O_2Hb]/([O_2Hb] + [HHb]) \times 100$), was calculated using the SRS method [13]. During testing, the PortaMon stored the data within the device's internal memory capacity. These data were subsequently downloaded onto a personal computer for analysis through an online software program. Data acquisition was sampled at a rate of 10 Hz for analogue-to-digital conversion and subsequent analysis. Baseline and minimum TSI (%) values were calculated as the 30-s average prior to the start of the exercise and the 3-s average surrounding the lowest value during the 200 m swim effort. The ΔTSI was therefore calculated as the baseline value minus the minimum value e.g. TSI Baseline – TSI Min = ΔTSI. The NIRS devices were positioned on the belly of the vastus lateralis muscle, midway between the greater trochanter of the femur and the lateral femoral epicondyle and for the latissimus dorsi muscle, at the midpoint between the mid-axilla and the spinal column. To ensure the optodes and detector did not move relative to the subject's skin, the device was fixed into position using sports waterproof adhesive tape and secured using the subject's own specialist swim apparel.

Descriptive statistics are presented as a mean ± SD unless otherwise stated. Each variable was examined with Kolmogorov-Smirnov normality test. ΔTSI %, between muscle groups and time to swim completion (s) were analysed using either paired (2-tailed) t-tests or Wilcoxon signed rank test when the sample normality test failed as per Bravo and colleagues [14]. The level of significance for analyses was set at $p < 0.05$. All analyses were performed using Graphpad Prism 6 (Graphpad Software, San Diego, CA).

3   Results

Figures 4.1a–e and 4.2a–e show the individual tissue saturation index (TSI %) trends for the vastus lateralis and latissimus dorsi muscle for the club level swimmers and club level triathletes groups, respectively. Club level swimmers completed the 200 m freestyle swim significantly faster (p = 0.04) than the triathlete group. Swim group time; 172.4 ± 14.6 (s), triathlete group time; 232.6 ± 39.6 (s).

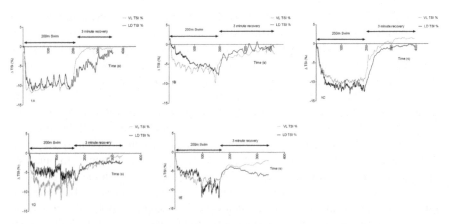

Fig. 4.1 Individual (subject's (**a–e**)) swim group ΔTSI (%) for the vastus lateralis and latissimus dorsi muscles. Group average data show there was no significant difference (p = 0.686) between the extent of desaturation in the upper and lower extremity; VL ΔTSI % = −9.44 ± 1.50, LD ΔTSI % = −9.00 ± 3.27

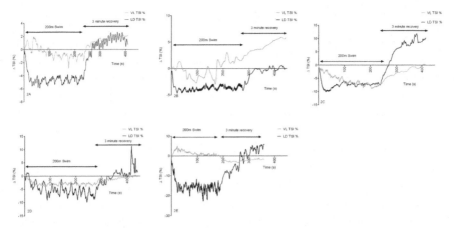

Fig. 4.2 Individual (subject's (**a–e**)) triathlete group ΔTSI (%) for the vastus lateralis and latissimus dorsi muscles. Group average data show there was a significant difference (p=0.043) between the extent of desaturation in the upper and lower extremity; VL ΔTSI %=−3.27±2.74, LD ΔTSI %=−8.79±5.24

4   Conclusions

This pioneering study has demonstrated how, for the first time, a waterproofed NIR device was able to report on muscle oxygen changes at the local level in both the upper and lower body in club level swimmers and triathletes. Furthermore, the data reported in this study show that, during a maximal 200 m swim, club level swimmers displayed no significant difference in ΔTSI (%) in the upper and lower body (p = 0.686), whereas the club level triathletes experienced a significantly greater drop in TSI (%) in the upper body compared with the lower (p = 0.043). These data suggest that club level swimmers use both the upper and lower body muscles to a similar extent during a maximal 200 m swim effort. Club level triathletes, however, predominantly use the upper body for propulsion during the same exercise. A significant difference in swim time (p = 0.04) was seen, with the club level swimmers completing the 200 m swim distance faster than the triathlete group. As previously mentioned, notable differences in swim velocity and propulsion [3] and marked systemic differences [2] have previously been observed between these swim groups. The most likely explanation for the difference in swim time would be related to the contribution of the lower body (legs) to the swim exercise. It has been shown that when the leg kicking action is added to the arm action, a 10 % increment in swim speed can be seen during swim sprint efforts [15]. However, even when told to swim as fast as possible, triathletes used only the upper body, presumably due to the need to spare the leg muscles for the later cycling and running stages of their event. It must be stated that all subjects were asked to swim maximally and 'normally' during the swim task; the differences in swim style were, however, very apparent within this particular level (club) of triathlete and the NIRS data reflected these biomechanical differences. In conclusion, the waterproofed NIR device was able to report upon muscle oxygenation changes in the upper and lower extremities in club level swimmers and triathletes. The ΔTSI % changes indicated that club level swimmers use both the upper and lower muscles to a similar extent during a maximal 200 m swim, whereas there was minimal desaturation in the lower body muscles of the triathletes.

Acknowledgments We like to especially thank all the athletes and swim coaches for their time and effort during this study.

References

1. Aspenes ST, Karlsen T (2012) Exercise-training intervention studies in competitive swimming. Sports Med 42:527–543
2. Roels B, Gimenes AC, Nascimmento RB et al (2005) Specificity of Vo2max and the ventilatory threshold in free swimming and cycle ergometry: comparison between triathletes and swimmers. Br J Sports Med 39:965–968
3. Toussaint HM (1990) Differences in propelling efficiency between competitive and triathlon swimmers. Med Sci Sports Exerc 22:409–415
4. Costa MJ, Bragada JA, Marinho DA et al (2012) Longitudinal interventions in elite swimming: a systematic review based on energetics, biomechanics, and performance. J Strength Cond Res 26:2006–2016
5. Pyne DB, Lee H, Swanwick KM (2001) Monitoring the lactate threshold in world-ranked swimmers. Med Sci Sports Exerc 33:291–297
6. Libicz S, Roels B, Millet GP (2005) Responses to intermittent swimming sets at velocity associated with max. Can J Appl Physiol 30:543–553
7. Reis V, Marinho D, Policarpo F et al (2010) Examining the accumulated oxygen deficit method in front crawl swimming. Int J Sports Med 31:421–427
8. Reis JF, Alves FB, Bruno PM et al (2012) Effects of aerobic fitness on oxygen uptake kinetics in heavy intensity swimming. Eur J Appl Physiol 112:1689–1697
9. Hesford CM, Laing SJ, Cardinale M et al (2012) Asymmetry of quadriceps muscle oxygenation during elite short-track speed skating. Med Sci Sports Exerc 44:501–508
10. Szmedra L, Nioka S, Chance B et al (2001) Hemoglobin/myoglobin oxygen desaturation during Alpine skiing. Med Sci Sports Exerc 33:232
11. Buchheit M, Ufland P, Haydar B et al (2011) Reproducibility and sensitivity of muscle reoxygenation and oxygen uptake recovery kinetics following running exercise in the field. Clin Physiol Funct Imaging 31:337–346
12. Wang B, Tian Q, Zhang Z et al (2012) Comparisons of local and systemic aerobic fitness parameters between finswimmers with different athlete grade levels. Eur J Appl Physiol 112:567–578
13. Suzuki S, Takasaki S, Ozaki T et al (1999) Tissue oxygenation monitor using NIR spatially resolved spectroscopy. Proc SPIE 3597:582–592
14. Bravo DM, Gimenes AC, Nascimento RB et al (2012) Skeletal muscle reoxygenation after high-intensity exercise in mitochondrial myopathy. Eur J Appl Physiol 112:1763–1771
15. Fulton SK, Pyne DB, Burkett B (2009) Quantifying freestyle kick-count and kick-rate patterns in Paralympic swimming. J Sport Sci 27:1455–1461

Chapter 5
Muscle Oxygenation During Running Assessed by Broad Band NIRS

A. Steimers, M. Vafiadou, G. Koukourakis, D. Geraskin, P. Neary, and M. Kohl-Bareis

Abstract We used spatially resolved near-infrared spectroscopy (SRS-NIRS) to assess calf and thigh muscle oxygenation during running on a motor-driven treadmill. Two protocols were used: An incremental speed protocol was performed in 5-min stages, while a pacing paradigm modulated the step frequency (2.3 Hz [S_{Low}]; 3.3 Hz [S_{High}]) during a constant velocity for 2 min each. A SRS-NIRS broadband system was used to measure total haemoglobin concentration and oxygen saturation (SO_2). An accelerometer was placed on the hip joints to measure limb acceleration through the experiment. The data showed that the calf desaturated to a significantly lower level than the thigh. During the pacing protocol, SO_2 was significantly different between the high and low step frequencies. Additionally, physiological data as measured by spirometry were different between the S_{Low} vs. S_{High} pacing trials. Significant differences in VO_2 at the same workload (speed) indicate alterations in mechanical efficiency. These data suggest that SRS broadband NIRS can be used to discern small changes in muscle oxygenation, making this device useful for metabolic exercise studies in addition to spirometry and movement monitoring by accelerometers.

Keywords Muscle oxygenation • Energy expenditure • Broad band NIRS • Spirometry • Accelerometry

1 Introduction

We used spatially resolved, broadband (650–1000 nm) near-infrared spectroscopy (SRS-NIRS) to assess calf and thigh muscle oxygenation during running, and compared these with standard physiology parameters (VO_2, VCO_2, heart rate) and body movement (accelerometry) in healthy volunteers. An incremental speed

A. Steimers (✉) • M. Vafiadou • G. Koukourakis • D. Geraskin • M. Kohl-Bareis
RheinAhrCampus, University of Applied Sciences Koblenz, Remagen, Germany
e-mail: steimers@rheinahrcampus.de

P. Neary
Faculty of Kinesiology and Health Studies, University of Regina, Regina, Canada

© Springer Science+Business Media, New York 2016
C.E. Elwell et al. (eds.), *Oxygen Transport to Tissue XXXVII*, Advances in Experimental Medicine and Biology 876, DOI 10.1007/978-1-4939-3023-4_5

protocol was performed, and a pacing paradigm with a modulated step frequency during a constant velocity.

The purpose of this study was to use NIRS for the study of muscle oxygenation [1], and to evaluate differences between the calf and thigh muscle oxygenation during treadmill running and to correlate the oxygen saturation (SO_2) with whole body physiology (VO_2, VCO_2) and body movement. Because the large intersubject variability of NIRS might be seen as a methodological limitation, we documented the sensitivity of broadband SRS to small changes in physiological function during exercise by using two different pacing (step frequency) protocols. By this, the broader picture of the exercise includes body movement as a measure for mechanical energy expenditure (as measured by hip acceleration), systemic whole body energy consumption (as measured by spirometry and heart rate) and the local energy consumption and expenditure (as expressed in muscle oxygenation parameters). We hypothesised that: (1) the oxygenation of calf and thigh muscle is markedly different during running, and (2) the step frequency does modulate the body movement (acceleration) and thereby the energy consumption as well as metabolic running efficiency and subsequently oxygen saturation. The co-registration of physiological and mechanical parameters is key for interpretation of NIRS data.

2   Methods

Seven subjects (three female; 25 (± 4) years, 65 (± 14) kg, 175 (± 12) cm) were fitted with a 3-axis accelerometer (LIS3L02AS4, ST Microelectronics, CH) fixed on the hip and the mean acceleration <a> was taken as the standard deviation of the absolute values. A spirometry system (Jaeger Oxycon Mobile, Viasys Medical, USA) served to record pulmonary function including oxygen (VO_2), carbon dioxide (VCO_2) consumption and heart rate (HR).

Velocity Paradigm: A continuous incremental protocol was performed with the velocity increased to 6, 8, 10 and 12 km/h every 5 min at zero degrees (horizontal) elevation.

Pacing Paradigm: After running at a self-selected constant speed, each runner was instructed to keep pace with the audible signal set to 2.3 Hz (S_{Low}) and 3.3 Hz (S_{High}) for 2 min each. This pattern was alternated so that three trials each were performed, followed by a rest period.

The spectroscopy system is based on a cooled CCD-camera (PIXIS 512F, Roper Scientific GmbH, Germany) in conjunction with a spectrometer with the aim of high light throughput (transmission) and simultaneous detection of up to six reflectance spectra. To overcome low light levels the design is based on a lens spectrometer with a high f—number (f/# = 1:1.2; f = 58 mm). To optimize both imaging quality and spectral resolution, the lens spectrometer was designed with a ray tracing software (OSLO, Lambda Research Corporation, USA) resulting in a system with micrometer controllable entrance slit, external focus control, optional mechanical shutter in the housing and a holder for standard diffraction gratings. For a slit width of 30 μm, the spectral resolution is limited by the pixel size of the CCD (20 μm), and is about $\Delta\lambda < 1.5$ nm (Full Width Half Maximum) for a grating with 600 lines/mm.

Two monitoring channels (probes) for independent recording of the gastrocnemius (calf) and rectus femoris (thigh) muscles were used. The first NIRS probe was placed in the centre of the gastrocnemius muscle at its maximum circumference, while the second probe was placed along the longitudinal axis of the rectus femoris approximately in the centre of the thigh between the superior border of the patella and the inguinal fold. Skinfold measurements were taken at the site of light interrogation (mean $\pm$ SD for calf: 3.2 ± 0.6 mm; for thigh: 6.8 ± 2.4 mm), with skin and adipose layer thickness being half these values.

Each NIRS probe (channel) used three light detecting fibre bundles of 1 mm diameter (length 6 m) arranged in a line separated by $\Delta\rho = 2.5$ mm at a mean distance $\rho = 30$ mm from the light delivering bundle (3 mm diameter) resulting in six independent recorded spectra. A halogen light bulb (12 V, 50 W) was used in conjunction with condenser optics as the light source. Data acquisition, control of all camera functions and on-line analysis of the spectra was programmed in LabVIEW (National Instruments Inc., USA). Data acquisition rate was set to 0.6 Hz, which is sufficiently low to smooth movement artefacts. Intensity spectra measured on the tissue were converted to attenuation spectra (A) by use of a reference measurement with an integrating sphere. The data analysis of the spectra is based on the spatially-resolved spectroscopy (SRS) approach which calculates tissue absorption spectra $\mu_a(\lambda)$ from the slope of attenuation ΔA with source-detector distance $\Delta\rho$ [2–4].

$$\mu_a(\lambda) = \frac{1}{\mu_s'(\lambda)} \cdot \frac{1}{3} \cdot \left(\ln(10) \frac{\Delta A(\lambda)}{\Delta\rho} - \frac{2}{\rho} \right)^2 \tag{5.1}$$

The transport scattering coefficient $\mu_s'(\lambda)$ is unknown but can be approximated from literature values [2, 5]. For a wavelength range 720–860 nm these absorption coefficient spectra were converted into haemoglobin concentrations for oxyHb and deoxyHb via matrix inversion by assuming that the absorption is dominated by haemoglobin. The total haemoglobin concentration is tHb = oxyHb + deoxyHb. Since the oxygen saturation (SO_2 = oxyHb/tHb) is a relative number it remains unaffected by the absolute magnitude of μ_s'. Throughout the text, we refer to haemoglobin concentration when strictly speaking it is a mixture with a small contribution of myoglobin.

3   Results and Discussion

Figure 5.1 illustrates the mean ($\pm$ SD) response of SO_2, tHb, VO_2, VCO_2, heart rate and mean acceleration to incremental exercise of the velocity paradigm. SO_2 reached a significantly lower level in the calf vs. thigh muscle, while tHb remained relatively similar. The intersubject variability in tHb was high, with the mean tHb higher in the thigh by about 50 % than in the calf. VO_2, VCO_2, HR and acceleration increased approximately linearly with increasing velocity (6 km/h $\leq$ v $\leq$ 12 km/h). The decline in the tissue oxygenation is similar as noted by Quaresima et al. [6], and in fact, they showed that there was approximately a 15 % reduction in calf at 9.6 km/h, while we showed a 15 % SO_2 reduction at 10 km/h. Consistent with previous research findings during incremental exercise, VO_2 and VCO_2 increased in a linear fashion. The mean data showed a strong relationship between SO_2 vs. VO_2 for both the calf ($r^2 = 0.98$) and the thigh ($r^2 = 0.84$), suggesting a good coupling between cellular respiration (i.e. oxidative metabolism) and oxygen consumption.

During the pacing protocol (Fig. 5.2), there was a reduction in saturation at the beginning of exercise (self-selected pace), followed by an undulating decrease and increase during the S_{Low} and S_{High} pacing stages with a significantly lower level in the calf vs. thigh. tHb was significantly different between the calf and thigh for each step frequency, while VO_2, VCO_2, heart rate and acceleration undulated in response to the step frequency.

To further analyse these data changes in measured parameters as a function of the step condition phases (S_{Low} or S_{High}) for the pacing protocol were calculated after subtracting a linear trend (see Fig. 5.3). In all measurement parameters the undulating due to the step frequency is apparent.

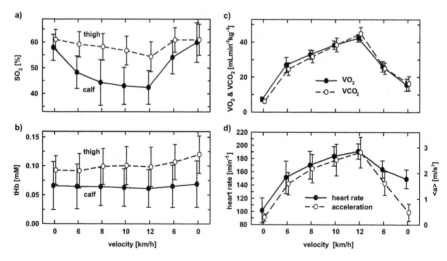

Fig. 5.1   Mean data (n = 7) of (**a**) SO_2, (**b**) total haemoglobin tHb, (**c**) VO_2 and VCO_2 and (**d**) heart rate and mean acceleration of the hip <a> measured during the velocity paradigm (v = 0–12 km/ h)

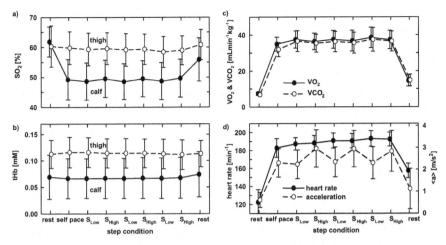

Fig. 5.2 Mean data (n = 7) of (**a**) SO_2, (**b**) total haemoglobin tHb, (**c**) VO_2 and VCO_2 and (**d**) heart rate and mean acceleration of the hip <a> of the pacing protocol as a function of the step condition

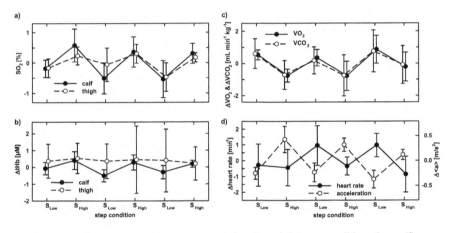

Fig. 5.3 Changes in measured parameters as a function of the step condition phases (S_{Low} or S_{High}) for the pacing protocol, based on the data of Fig. 5.2, after subtracting a linear trend. Values are mean ± SD

The correlation between the parameters for the step frequency conditions S_{Low} and S_{High} after subtraction of a linear trend was $r^2 = 0.46$–0.98. During the pacing protocol, SO_2 was significantly different between the S_{Low} (48.6 and 58.9 % for calf and thigh) vs. S_{High} (49.5 %; 59.3 %) trials for both the calf and thigh muscles, respectively. Whole body oxygen consumption (VO_2) (2563 ± 586 vs. 2503 ± 605 mL min^{-1}) was different for S_{Low} vs. S_{High}.

Our pacing results showed consistent with the velocity data that the calf reached a lower level of desaturation (SO_2) and tHb than the thigh ($p \leq 0.05$), but also showed the undulating changes in SO_2, and all other physiological variables during each step frequency trial.

The haemoglobin concentrations as well as the oxygen saturation are prone to methodological limitations of NIRS in general, e.g. the influence of adipose tissue and uncertainties in the water content and tissue inhomogeneity. However, this affects primarily the absolute values of the calculated parameters, rather than their changes as considered here.

4 Conclusion

We conclude that the less desaturation (SO_2) and less VO_2 during the S_{High} pace vs. S_{Low} pace is related to the lower stride frequency, longer stride length, and increased contact time on the treadmill during the S_{Low} pace protocol. Therefore, based on our results we speculate that at S_{Low} pace, energy cost is greater to try and maintain a longer stride length, and greater 'push-off' phase needed to maintain the S_{Low} pace. Significant differences in VO_2 at the same workload (treadmill speed) would indicate alterations in mechanical efficiency. Previous research has confirmed that stride length and frequency alter the metabolic cost of running.

The results of this study are novel in that we have demonstrated that SRS broadband NIRS can be used to discern small changes (<1 %) in muscle oxygenation during different running paradigms. In addition, this is the first report combining NIRS with both pulmonary gas exchange and accelerometry thus giving indications on the metabolic energy efficiency of the body in response to different modes, durations and intensities of exercise. We found accelerometry (body movement) as an additional tool to complement spirometry and NIRS, with the correlation of the NIRS and physiological parameters being similar. The combination of accelerometry and NIRS oximetry seems especially promising for future work on truly portable systems which can be used in the outdoors rather than during laboratory conditions. Further research is needed to evaluate biomechanical energy based on accelerometers added by gyroscopes with NIRS parameters.

References

1. Wolf M, Ferrari M, Quaresima V (2007) Progress of near-infrared spectroscopy and topography for brain and muscle clinical applications. J Biomed Opt 12(062104):1–14
2. Matcher SJ, Kirkpatrick P, Nahid K et al (1995) Absolute quantification methods in tissue near infrared spectroscopy. Proc SPIE 2389:486–495
3. Soschinski J, Ben Mine L, Geraskin D et al (2007) Cerebral oxygenation monitoring during cardiac bypass surgery in infants with broad band spatially resolved spectroscopy. Proc SPIE 6629(Diffuse Optical Imaging of Tissue): OU1–OU6

4. Geraskin D, Platen P, Franke J et al (2005) Muscle oxygenation during exercise under hypoxic conditions assessed by spatial-resolved broad-band NIR spectroscopy. Proc SPIE 5859:58590L-01-06
5. Matcher SJ, Cope M, Deply DT (1997) In vivo measurements of the wavelength dependence of tissue-scattering coefficients between 760 and 900 nm measured with time-resolved spectroscopy. Appl Opt 36(1):386–396
6. Quaresima V, Komiyamab T, Ferrari M (2002) Differences in oxygen re-saturation of thigh and calf muscles after two treadmill stress tests. Comp Biochem Physiol A 132:67–73

Chapter 6
Regional Differences in Muscle Energy Metabolism in Human Muscle by ^{31}P-Chemical Shift Imaging

Ryotaro Kime, Yasuhisa Kaneko, Yoshinori Hongo, Yusuke Ohno, Ayumi Sakamoto, and Toshihito Katsumura

Abstract Previous studies have reported significant region-dependent differences in the fiber-type composition of human skeletal muscle. It is therefore hypothesized that there is a difference between the deep and superficial parts of muscle energy metabolism during exercise. We hypothesized that the inorganic phosphate (Pi)/phosphocreatine (PCr) ratio of the superficial parts would be higher, compared with the deep parts, as the work rate increases, because the muscle fiber-type composition of the fast-type may be greater in the superficial parts compared with the deep parts. This study used two-dimensional ^{31}Phosphorus Chemical Shift Imaging (^{31}P-$132\#CSI) to detect differences between the deep and superficial parts of the human leg muscles during dynamic knee extension exercise. Six healthy men participated in this study (age 27 ± 1 year, height 169.4 ± 4.1 cm, weight 65.9 ± 8.4 kg). The experiments were carried out with a 1.5-T superconducting magnet with a 5-in. diameter circular surface coil. The subjects performed dynamic one-legged knee extension exercise in the prone position, with the transmit-receive coil placed under the right quadriceps muscles in the magnet. The subjects pulled down an elastic rubber band attached to the ankle at a frequency of 0.25, 0.5 and 1 Hz for 320 s each. The intracellular pH (pHi) was calculated from the median chemical shift of the Pi peak relative to PCr. No significant difference in Pi/PCr was

R. Kime (✉) • T. Katsumura
Department of Sports Medicine for Health Promotion, Tokyo Medical University, Tokyo, Japan
e-mail: kime@tokyo-med.ac.jp

Y. Kaneko
Department of Sports Medicine for Health Promotion, Tokyo Medical University, Tokyo, Japan

Department of Oriental Medicine, Kuratake College of Medical Arts & Sciences, Tokyo, Japan

Y. Hongo • Y. Ohno
Kuratake Medical Clinic, Kuratake College of Medical Arts & Sciences, Saitama, Japan

A. Sakamoto
Department of Oriental Medicine, Kuratake College of Medical Arts & Sciences, Tokyo, Japan

Kuratake Medical Clinic, Kuratake College of Medical Arts & Sciences, Saitama, Japan

© Springer Science+Business Media, New York 2016
C.E. Elwell et al. (eds.), *Oxygen Transport to Tissue XXXVII*, Advances in Experimental Medicine and Biology 876, DOI 10.1007/978-1-4939-3023-4_6

observed between the deep and the superficial parts of the quadriceps muscles at rest. The Pi/PCr of the superficial parts was not significantly increased with increasing work rate. Compared with the superficial areas, the Pi/PCr of the deep parts was significantly higher ($p < 0.05$) at 1 Hz. The pHi showed no significant difference between the two parts. These results suggest that muscle oxidative metabolism is different between deep and superficial parts of quadriceps muscles during dynamic exercise.

Keywords Oxidative metabolism • Magnetic resonance spectroscopy • Dynamic exercise • Quadriceps • pH

1 Introduction

Previous study has reported that there are significant region-dependent differences in the fiber-type composition of human skeletal muscle [1, 2]. It is therefore hypothesized that there is a difference between the deep and superficial parts of muscle energy metabolism during exercise. Chance et al. [3] have reported that high oxidative capacity muscle demonstrated lower inorganic phosphate (Pi)/ phosphocreatine (PCr) at the same relative work rate compared with low oxidative capacity muscle. Recently, creating a localized metabolic map during exercise has been possible using [31]Phosphorus Chemical Shift Imaging (^{31}P-CSI) [4–6]. Therefore, the purpose of this study was to detect metabolic disturbances between the deep and superficial parts of the human leg muscles during dynamic knee extension exercise using two-dimensional ^{31}P-CSI. We hypothesized that the Pi/PCr of the superficial parts would be higher, compared with the deep parts, as the work rate increases, because the muscle fiber-type composition of the fast-type may be higher in the superficial parts compared with the deep parts.

2 Methods

2.1 Subjects

Six healthy men (age: 28 ± 1 year, height: 170.4 ± 4.1 cm, weight: 66.8 ± 7.4 kg) participated in this study. All subjects were briefed about the experimental protocol, and written informed consent was obtained before the experiment. The institutional review board of the Tokyo Medical University approved the research protocol.

2.2 Experimental Design

After receiving written informed consent from the subjects, the experiments were carried out with a 1.5-T superconducting magnet (GE Healthcare, Milwaukee, WI, USA) with a circular surface coil (GE Healthcare, Milwaukee, WI, USA) double-tuned to ^{1}H at 63.5 MHz and ^{31}P at 25.8 MHz. The subjects lay prone in the bore to obtain T_2-weighted ^{1}H images. After completion, regional differences in phosphorus signals were obtained by ^{31}P-CSI. Before the ^{31}P-CSI acquisitions, the magnetic field homogeneity was optimized using the localized water signal from a quadriceps muscles.

2.3 Measurements

A one-pulse 31P-MRS acquisition was carried out with a 5-in. transmit/receive surface coil placed under the knee extensors [rectus femoris (RF), vastus medialis (VM), vastus intermedius (VI), and vastus lateralis (VL)] of the right leg. Spatially resolved acquisition relied on ^{31}P-CSI with 3-cm slice thickness, and a 24-cm^2 field of view. The volume of each voxel was $3 \times 3 \times 3$ cm, or 27 cm^3. A TR of 1000 ms was used, and a two-dimensional ^{31}P metabolite map was generated every 388 s. This acquisition time was determined from pilot measurements to ensure good spectral data from the quadriceps and represents an optimal compromise between signal-to-noise, temporal resolution, and spatial resolution.

^{31}P-CSI post processing was done using Mnova software (Mestrelab Research, Spain). Baseline correction was performed semi-automatically by setting the peak ranges of Pi and PCr as references, and phase correction was applied semi-automatically using Pi and PCr as reference peaks. Supplementary manual phase or baseline correction was performed, if necessary. The peak areas and peak positions of PCr and Pi were fitted in the frequency domain. The intracellular pH (pHi) was calculated from the median chemical shift of the Pi peak relative to PCr [7].

At rest, T_2-weighted ^{1}H images were obtained with the whole body imaging coil with TR = 3200 ms and TE = 95 ms. Scan time for T_2 image was 39 s.

2.4 Exercise Protocol

The subjects performed dynamic one-legged knee extension exercise in the prone position. A broad nonelastic strap over the hips served to stabilize the subject during exercise. A rubber band was attached to the ankle and the subjects exercised by expanding the rubber band at 0.25, 0.5 and 1 Hz for 380 s each. As limited by the scanner bore, a range of motion of knee extension exercise was 0–30°.

2.5   Statistics

The changes in Pi/PCr and pHi during the experiments were analyzed by two-way ANOVA for repeated measurements. The significance level was set to 0.05. Statistics were completed using the Statistical Package for the Social Sciences (SPSS) Statics (IBM, Chicago, IL).

3   Results

Typical examples of ^{31}P-CSI spectra are shown in Fig. 6.1. To compare the metabolic differences between superficial and deep parts, three voxels of superficial parts and three of voxels of deep parts were selected, as all the data from the voxel could be monitored by all subjects. No significant difference in Pi/ PCr was observed between the deep and the superficial parts of the quadriceps muscles at rest (superficial parts; 0.21 ± 0.04, deep parts; 0.18 ± 0.03). At the superficial parts, the Pi/PCr were not significantly changed with increasing work rate. Compared with the superficial parts, the Pi/PCr of the deep parts were significantly higher ($p < 0.05$) at 1 Hz, and the highest peak of Pi/PCr was 0.85 ± 0.08 at D_1. The averaged resting pHi was 7.08 ± 0.02, and the spatial differences of the resting pHi were not observed between every measurement part. Also, the pHi during exercise showed no significant difference between the two parts at all exercise frequencies (superficial parts; 7.02 ± 0.04, deep parts; 6.99 ± 0.03) (Fig. 6.2).

Fig. 6.1 Typical example of ^{31}P-CSI spectra and two-dimensional ^{1}H magnetic resonance image in thigh muscles. Each was $3 \times 3 \times 3$ cm, or 27 cm^3. To compare the metabolic differences between superficial and deep parts, three voxels of superficial parts and three of voxels of deep parts were selected, as all the data from the voxel could be monitored by all subjects. The voxel was numbered from the medial to lateral parts at both superficial and deep parts

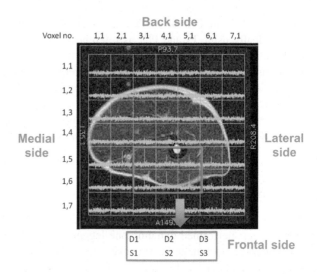

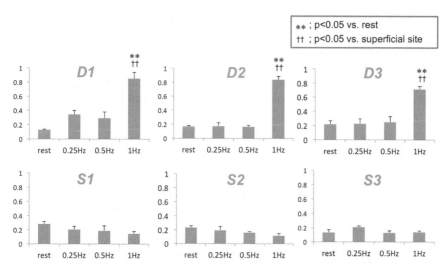

Fig. 6.2 The *upper figures* show the Pi/PCr of deep parts at rest and during dynamic exercise at various work rates. The *bottom figures* show the Pi/PCr of superficial parts at rest and during exercise at various work rates. *Double asterisks* denote significant differences compared with rest below the 0.01 level. *Double hash symbols* denote significant differences compared with superficial parts below the 0.01 level

4   Discussion

The study demonstrated that the Pi/PCr of the superficial parts were not significantly increased with increasing work rate. In comparison, the Pi/PCr of the deep parts were significantly higher at 1 Hz. These results suggest that muscleoxidative metabolism is different between deep and superficial parts of quadriceps muscles during dynamic exercise. Pesta et al. [4] recently reported that PCr changes during exercise were uniform across the quadriceps muscles within sprint-trained, endurance-trained and untrained groups, and the results in the present study are different from the reported data. The reason for the discrepancy is uncertain but may be due to differences in both the type of exercise and exercise time. In almost all previous studies, work rate was increased by altering workload while keeping the rate of contraction steady. Therefore, more fast-twitch fibers may be recruited at higher workloads, and the type of exercise can lead to physiological maximal effort. In contrast, the work rate in the present study was increased by altering the rate of contraction while keeping the workload steady. As this type of exercise reduces the relaxation period of the duty cycle, muscle recruitment may be lower at increasing rate of contraction protocol than increasing workload protocol. In addition, the exercise time was long (388 s) for acquisition of spatially-resolved spectra in each voxel examined, due to limitations in our technique. In the future, we need to shorten the acquisition time by increasing slice thickness. Also, we need to localize the specific muscle signals of the thigh muscles such as rectus femoris and vastus intermedius.

Although muscle recruitment patterns may affect Pi/PCr independently of fiber type composition, we were unable to measure T_2-weighted imaging using f-MRI during exercise. Further investigations are needed to determine the regional differences in muscle energy metabolism and muscle recruitment patterns.

5   Conclusions

We determined the difference between the deep and superficial areas of muscle energy metabolism during dynamic knee extension exercise using ^{31}P- chemical shift imaging. The study demonstrated that the Pi/PCr of the superficial parts were not significantly increased with increasing work rate. In contrast, the Pi/PCr of the deep parts were significantly higher at a higher work rate. These results could reflect (a) regional differences in fiber composition, which was our original hypothesis, or (b) different regional recruitment patterns for this particular form of exercise. These two explanations cannot be discriminated from the present study.

Acknowledgments   The authors are grateful for revision of this manuscript by Andrea Hope. This study was supported in part by Grant-in-Aid for scientific research from the Japan Society for Promotion of Science (24500799) to R. K.

References

1. Johnson MA, Polgar J, Weightman D et al (1973) Data on the distribution of fibre types in thirty-six human muscles. An autopsy study. J Neurol Sci 18(1):111–129
2. Dahmane R, Djordjevic S, Simunic B et al (2005) Spatial fiber type distribution in normal human muscle: histochemical and tensiomyographical evaluation. J Biomech 38 (12):2451–2459
3. Chance B, Leigh JS Jr, Clark BJ et al (1985) Control of oxidative metabolism and oxygen delivery in human skeletal muscle: a steady-state analysis of the work/energy cost transfer function. Proc Natl Acad Sci U S A 82:8384–8388
4. Pesta D, Paschke V, Hoppel F et al (2013) Different metabolic responses during incremental exercise assessed by localized 31P MRS in sprint and endurance athletes and untrained individuals. Int J Sports Med 34(8):669–675
5. Cannon DT, Howe FA, Whipp BJ et al (2013) Muscle metabolism and activation heterogeneity by combined 31P chemical shift and T2 imaging, and pulmonary O2 uptake during incremental knee-extensor exercise. J Appl Physiol 115(6):839–849
6. Forbes SC, Slade JM, Francis RM et al (2009) Comparison of oxidative capacity among leg muscles in humans using gated 31P 2-D chemical shift imaging. NMR Biomed 22 (10):1063–1071
7. Petroff OA, Prichard JW, Behar KL et al (1985) Cerebral intracellular pH by 31P nuclear magnetic resonance spectroscopy. Neurology 35:781–788

Chapter 7
Sex-Related Difference in Muscle Deoxygenation Responses Between Aerobic Capacity-Matched Elderly Men and Women

Shun Takagi, Ryotaro Kime, Masatsugu Niwayama, Takuya Osada, Norio Murase, Shizuo Sakamoto, and Toshihito Katsumura

Abstract Muscle O_2 dynamics during ramp cycling exercise were compared between aerobic capacity-matched elderly men ($n = 8$, age 65 ± 2 years) and women ($n = 8$, age 66 ± 3 years). Muscle O_2 saturation (SmO_2) and relative change in deoxygenated (Δdeoxy-Hb) and total hemoglobin concentration (Δtotal-Hb) were monitored continuously during exercise in the vastus lateralis (VL) and gastrocnemius medialis (GM) by near infrared spatial resolved spectroscopy. SmO_2 was significantly higher during exercise in women than in men in VL, but not in GM. In VL, Δdeoxy-Hb and Δtotal-Hb were significantly higher in men than in women, especially during high intensity exercise. However, no significant difference was observed in Δdeoxy-Hb or Δtotal-Hb in GM. Sex-related differences in muscle deoxygenation response may be heterogeneous among leg muscles in elderly subjects.

Keywords Aging • Cycling exercise • Muscle oxygen dynamics • Near infrared spectroscopy • Regional difference

S. Takagi (✉)
Faculty of Sport Sciences, Waseda University, 2-579-15, Mikajima, Tokorozawa, Saitama 359-1192, Japan

Department of Sports Medicine for Health Promotion, Tokyo Medical University, Shinjuku, Tokyo, Japan
e-mail: stakagi@aoni.waseda.jp

R. Kime • T. Osada • N. Murase • T. Katsumura
Department of Sports Medicine for Health Promotion, Tokyo Medical University, Shinjuku, Tokyo, Japan

M. Niwayama
Department of Electrical and Electronic Engineering, Shizuoka University, Hamamatsu, Shizuoka, Japan

S. Sakamoto
Faculty of Sport Sciences, Waseda University, 2-579-15, Mikajima, Tokorozawa, Saitama 359-1192, Japan

© Springer Science+Business Media, New York 2016
C.E. Elwell et al. (eds.), *Oxygen Transport to Tissue XXXVII*, Advances in Experimental Medicine and Biology 876, DOI 10.1007/978-1-4939-3023-4_7

1 Introduction

Near infrared spectroscopy (NIRS) has been widely used in measuring muscle oxy-genation during dynamic exercise. A few previous studies evaluated sex-related differences in muscle deoxygenation response using NIRS during cycling exercise in young subjects [1–3]. However, sex-related differences in muscle deoxygenation responses have not been established in elderly subjects. It has been reported that the muscle deoxygenation responses are heterogeneous among leg muscles [4]. Ideally, peak aerobic capacity should be matched to evaluate the sex-related differences because peak aerobic capacity potentially affects the hemodynamic and vascular responses, regardless of age and sex [5]. Therefore, the aim of this study was to compare the muscle deoxygenation responses in thigh and lower leg muscles during cycling exercise in peak aerobic capacity-matched elderly men and women.

2 Methods

2.1 Subjects

Untrained elderly men (n = 8; age 65 ± 2 years; height 169.4 ± 6.1 cm; weight 67.3 ± 8.4 kg; body mass index (BMI) 23.5 ± 2.7 kg/m^2, mean $\pm$ SD) and women (n = 8; age 66 ± 3 years; height 154.0 ± 4.4 cm; weight 53.8 ± 8.1 kg; BMI 23.3 ± 2.8 kg/m^2) participated in the study, which was approved by the Tokyo Medical University Local Research Ethics Committee, Japan. Peak VO$_2$ was matched between groups (describe below). A statin was taken by one subject in the men's group and one subject in the women's group. All volunteers were informed of the purpose and nature of the study and written informed consent was obtained.

2.2 Experimental Design

The subjects performed 10 or 15 W/min ramp bicycle exercise, after a 3-min warm up at 10 W, until exhaustion (Strength Ergo 8, Fukuda-Denshi, Tokyo, Japan). Pulmonary O$_2$ uptake (VO$_2$) was measured continuously during the experiments to determine peak VO$_2$ by using an automated gas analysis system (AE300S, Minato Medical Science, Osaka, Japan).

Muscle O$_2$ saturation (SmO$_2$) and relative changes from rest in oxygenated hemoglobin concentration (Δoxy-Hb), deoxygenated hemoglobin concentration (Δdeoxy-Hb), and total hemoglobin concentration (Δtotal-Hb) were monitored at vastus lateralis (VL) and gastrocnemius medialis (GM) in the left leg by multichannel near infrared spatial resolved spectroscopy (NIR_{SRS}). The NIR_{SRS} data were defined as the SmO_2 averaged over the last 10 s at rest, 20, 40, 60, 80, and 100 % of peak VO_2. In this study, peak workload was similar between groups (describe below), and therefore, we analyzed the muscle deoxygenation response as a function of percent of peak VO_2.

We used a two-wavelength (770 and 830 nm) light-emitting diode NIR_{SRS} (Astem Co., Japan). The probe consisted of one light source and two photodiode detectors, and the optode distances were 20 and 30 mm, respectively. The data sampling rate was 1 Hz. Even though fat layer thickness affects NIRS data because of light scattering, Niwayama et al. has recently reported that the effects of fat layer thickness can be corrected in relative changes in Hb and absolute value of SmO_2 by normalized measurement sensitivity [6]. In this study, we measured fat layer thickness at each measurement site in the muscles to correct these effects using an ultrasound device (LogiQ3, GE-Yokokawa Medical Systems, Japan). Even though an upper limit of fat layer thickness was designated as 10 mm to correct for the light-scattering effects in this study, fat layer thickness was within ~10 mm at each measurement site in all subjects.

2.3 Statistics

All data are given as means ± standard deviation (SD). To compare changes in NIRS variables during exercise between groups, a two-way repeated-measures analysis of variance was used with age and exercise intensity as factors. Where appropriate, the Bonferroni post hoc test was performed to determine specific significant differences. Differences in physical variables, peak VO_2, and peak workload were compared between groups using unpaired t-tests. For all statistical analyses, significance was accepted at $p < 0.05$.

3 Results

In VL, significantly lower SmO_2 was observed in men than women ($p < 0.05$), even though there was no significant sex × exercise intensity interaction for change in SmO_2 ($p = 0.96$). Moreover, there was a significant sex × exercise intensity interaction for change in Δdeoxy-Hb ($p < 0.05$) or Δtotal-Hb ($p < 0.05$), and a higher Δdeoxy-Hb or Δtotal-Hb was observed in men during high intensity exercise than women. In GM, no significant sex × exercise intensity interaction was observed in SmO_2 ($p = 0.64$), Δdeoxy-Hb ($p = 0.51$) or Δtotal-Hb ($p = 0.64$). In addition, no significant difference was observed in SmO_2 ($p = 0.45$), or in Δdeoxy-Hb ($p = 0.94$) or Δtotal-Hb ($p = 0.69$), in GM. No significant interaction or difference was observed in Δoxy-Hb at any measurement sites (Fig. 1).

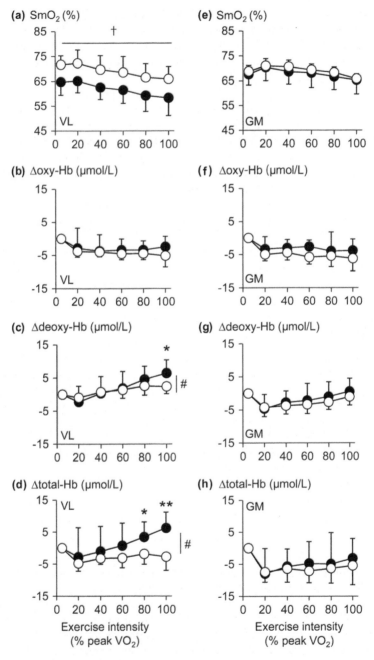

Fig. 1 Change in muscle O_2 saturation (SmO_2: **a**, **e**), oxygenated hemoglobin (oxy-Hb: **b**, **f**), deoxygenated hemoglobin (deoxy-Hb: **c**, **g**), and total hemoglobin (total-Hb: **d**, **h**) responses in VL (**a**, **b**, **c**, **d**) and GM (**e**, **f**, **g**, **h**) muscles during ramp cycling exercise. The *closed circles* show NIRS data in elderly men and the *open circles* show NIRS data in elderly women. There was a significant difference between groups (*p < 0.05, **p < 0.01). There was a significant sex × exercise intensity interaction (#p < 0.05). There was a main effect of sex (†p < 0.05)

Peak VO_2 per body weight was matched between groups (men: 21.3 ± 3.0 ml/kg/min, women: 22.2 ± 4.0 ml/kg/min, p = 0.63). As a result, there was no significant difference between groups in absolute peak VO_2 (1435 ± 261 vs. 1226 ± 245 ml/min, p = 0.16) or peak workload (men: 110 ± 18 W (ranged 75–127 W), women: 106 ± 19 W (ranged 72–124 W), p = 0.69). Even though BMI was matched between groups (p = 0.89), fat layer thickness was significantly higher in women than in men (VL: 4.05 ± 0.92 vs. 6.78 ± 2.37 mm, p < 0.05; GM: 3.51 ± 0.82 vs. 5.35 ± 1.34 mm, p < 0.01).

4 Discussion

In the present study, both Δdeoxy-Hb and Δtotal-Hb response in VL was more blunted in woman than men, and SmO_2 was higher throughout exercise in woman than in men. In addition, aerobic capacity and peak workload were matched between groups, and therefore, the effects of peak aerobic capacity and workloads on muscle O_2 dynamics may be negligible. A possible explanation of higher SmO_2 in women may be both higher arterial O_2 saturation and higher venous O_2 saturation than men [7]. Unfortunately, we did not measure these variables directly, and detailed mechanisms for the difference in SmO_2 are unclear. Even though the number of subjects is low, there was no significant relationship between fat layer thickness in the VL and corrected SmO_2 at resting (r = 0.29, p = 0.27) in all subjects (n = 16). Thus, the corrected SmO_2 is not critically affected by the fat layer thickness, and we believe that the correction algorithm is suitable for elderly people in this study. Additionally, in this study, Δdeoxy-Hb and Δtotal-Hb were lower in women than men, especially during high intensity exercise. Reduced estrogen level impairs leg blood flow and vasodilation response in elderly women [8, 9], and the age-related reduction in leg blood flow and vasodilation response tends to be larger in women than men [5]. In fact, mitochondrial content was also found to be reduced in ovariectomized rats [10]. Hence, the percentage of sex hormone may partly explain the difference in Δdeoxy-Hb and Δtotal-Hb. From our findings, we presume that elderly women have lower blood flow, lower vasodilation response, or lower mitochondrial content than peak aerobic capacity- and workloads-matched elderly men. In addition, lactate concentration after exercise may be still higher in elderly men than women, even though the difference in lactate concentrations is reduced as age advances [11]. Therefore, increased Δdeoxy-Hb via lactic acidosis (Bohr effects) may also be more blunted in women than men. However, there is a possibility that lower absolute VO_2 in women may be related to the lower Δdeoxy-Hb response in VL of women, even though the difference in absolute VO_2 between groups did not reach significance. Again, because of the low number of subjects, this area warrants further investigation.

There was some inconsistency of sex-related difference in muscle deoxygenation responses in VL among several previous studies. The disparities may be partly explained by the methods of normalized NIRS data. Peltonen et al. [2] reported that young women displayed lower SmO_2, Δdeoxy-Hb, and Δtotal-Hb response in VL than young men, as a function of % of peak VO_2. However, in their study, the effect of fat layer thickness on NIRS data was not corrected. In contrast, in the other previous study using a cuff ischemia method, there was no sex-related difference in muscle deoxygenation response in young subjects as a function of percent of peak VO_2 [1, 3]. In the present study of elderly subjects, there was also a lower deoxygenation response in women than men, using the method of the fat layer thickness correction to normalize NIRS data. Although it is difficult to compare the muscle deoxygenation response directly with previous results, to our knowledge, this is the first study to compare deoxygenation response between elderly subjects.

In GM, muscle deoxygenation response was not different between the groups, in contrast to VL. This means that sex-related difference in muscle deoxygenation response may be regional. In line with our findings, a previous study indicated that mitochondrial content and respiration were similar between middle-aged men and women in GM [12], while Green et al. reported that 16.4–18.9 % difference in the activities of Krebs cycle and glycolytic enzyme between untrained young men and women in VL [13]. Moreover, Takagi et al. reported that muscle deoxygenation response was similar between young men and elderly men in GM [4]. The balance between circulation and metabolism in GM muscle may not be affected by aging in both men and women. In addition, an alternative explanation for similar muscle deoxygenation in GM may be that GM muscle does not mainly contribute during cycling exercise, in contrast to VL muscle.

In conclusion, sex-related differences in muscle deoxygenation response were observed between elderly men and women, even though peak aerobic capacity was matched. Moreover, the sex-related differences in muscle deoxygenation responses may be heterogeneous among leg muscles.

Acknowledgments The authors are grateful for revision of this manuscript by Andrea Hope. This study was supported in part by Grant-in-Aid for scientific research from Japan Society for the Promotion of Science (246298) to S.T.

References

1. Murias JM, Keir DA, Spencer MD et al (2013) Sex-related differences in muscle deoxygenation during ramp incremental exercise. Respir Physiol Neurobiol 189:530–536
2. Peltonen JE, Hägglund H, Koskela-Koivisto T et al (2013) Alveolar gas exchange, oxygen delivery and tissue deoxygenation in men and women during incremental exercise. Respir Physiol Neurobiol 188:102–112
3. Bhambhani Y, Maikala R, Buckley S (1998) Muscle oxygenation during incremental arm and leg exercise in men and women. Eur J Appl Physiol Occup Physiol 78:422–431
4. Takagi S, Kime R, Murase N et al (2013) Aging affects spatial distribution of leg muscle oxygen saturation during ramp cycling exercise. Adv Exp Med Biol 789:157–162
5. Martin WH 3rd, Ogawa T, Kohrt WM et al (1991) Effects of aging, gender, and physical training on peripheral vascular function. Circulation 84:654–664

6. Niwayama M, Suzuki H, Yamashita T et al (2012) Error factors in oxygenation measurement using continuous wave and spatially resolved near-infrared spectroscopy. J Jpn Coll Angiol 52:211–215
7. Reybrouck T, Fagard R (1999) Gender differences in the oxygen transport system during maximal exercise in hypertensive subjects. Chest 115:788–792
8. Hickner RC, Kemeny G, McIver K et al (2003) Lower skeletal muscle nutritive blood flow in older women is related to eNOS protein content. J Gerontol A Biol Sci Med Sci 58:20–25
9. Moreau KL, Donato AJ, Tanaka H et al (2003) Basal leg blood flow in healthy women is related to age and hormone replacement therapy status. J Physiol 547(Pt 1):309–316
10. Cavalcanti-de-Albuquerque JP, Salvador ID, Martins EG et al (2014) Role of estrogen on skeletal muscle mitochondrial function in ovariectomized rats: a time course study in different fiber types. J Appl Physiol 116:779–789
11. Benelli P, Ditroilo M, Forte R et al (2007) Assessment of post-competition peak blood lactate in male and female master swimmers aged 40–79 years and its relationship with swimming performance. Eur J Appl Physiol 99:685–693
12. Thompson JR, Swanson SA, Casale GP et al (2013) Gastrocnemius mitochondrial respiration: are there any differences between men and women? J Surg Res 185:206–211
13. Green HJ, Fraser IG, Ranney DA (1984) Male and female differences in enzyme activities of energy metabolism in vastus lateralis muscle. J Neurol Sci 65:323–331

Chapter 8
Effects of Low Volume Aerobic Training on Muscle Desaturation During Exercise in Elderly Subjects

Shun Takagi, Ryotaro Kime, Norio Murase, Masatsugu Niwayama, Takuya Osada, and Toshihito Katsumura

Abstract Aging enhances muscle desaturation responses due to reduced O_2 supply. Even though aerobic training enhances muscle desaturation responses in young subjects, it is unclear whether the same is true in elderly subjects. Ten elderly women (age: 62 ± 4 years) participated in 12-weeks of cycling exercise training. Training consisted of 30 min cycling exercise at the lactate threshold. The subjects exercised 15 ± 6 sessions during training. Before and after endurance training, the subjects performed ramp cycling exercise. Muscle O_2 saturation (SmO_2) was measured at the vastus lateralis by near infrared spectroscopy during the exercise. There were no significant differences in SmO_2 between before and after training. Nevertheless, changes in peak pulmonary O_2 uptake were significantly negatively related to changes in SmO_2 ($r = -0.67$, $p < 0.05$) after training. Muscle desaturation was not enhanced by low volume aerobic training in this study, possibly because the training volume was too low. However, our findings suggest that aerobic training may potentially enhance muscle desaturation at peak exercise in elderly subjects.

Keywords Aging • Cycling training • Muscle oxygen saturation • Near infrared spectroscopy • Peak aerobic capacity

S. Takagi (✉)
Faculty of Sport Sciences, Waseda University, 2-579-15, Mikajima, Tokorozawa, Saitama 359-1192, Japan

Department of Sports Medicine for Health Promotion, Tokyo Medical University, Shinjuku, Tokyo, Japan
e-mail: stakagi@aoni.waseda.jp

R. Kime • N. Murase • T. Osada • T. Katsumura
Department of Sports Medicine for Health Promotion, Tokyo Medical University, Shinjuku, Tokyo, Japan

M. Niwayama
Department of Electrical and Electronic Engineering, Shizuoka University, Hamamatsu, Shizuoka, Japan

© Springer Science+Business Media, New York 2016
C.E. Elwell et al. (eds.), *Oxygen Transport to Tissue XXXVII*, Advances in Experimental Medicine and Biology 876, DOI 10.1007/978-1-4939-3023-4_8

1 Introduction

Muscle desaturation responses during whole-body exercise can be measured by near infrared spatial resolved spectroscopy (NIR_{SRS}), which has been widely utilized in many previous studies [1, 2]. However, few previous studies have evaluated the effects of exercise training on muscle desaturation responses using NIR_{SRS} during cycling exercise. A previous study found that aerobic exercise training enhances muscle desaturation during incremental cycling exercise in young subjects [1]. In another study, muscle desaturation responses were found to be enhanced in elderly subjects because muscle blood flow (i.e. O_2 supply to exercising muscle) was reduced due to aging [2]. In view of that, it is unclear whether aerobic exercise training enhances muscle desaturation responses in elderly subjects. The aim of this study was to examine the effects of aerobic exercise training on muscle O_2 dynamics in elderly subjects.

2 Methods

2.1 Subjects

Untrained elderly women (n = 10; age: 65 ± 2 years; height: 158.6 ± 8.0 cm; weight: 62.1 ± 11.9 kg, mean $\pm$ SD) participated in the study. This study protocol was approved by the institutional ethics committee, and was conducted in accordance with the Declaration of Helsinki. Two subjects were taking a statin, and one subject was taking an angiotensin II receptor antagonist and a calcium channel blocker. All subjects were informed of the purpose and nature of the study and written informed consent was obtained.

2.2 Experimental Design

The subjects performed 12-weeks of cycling exercise training for 30 min at the individual's estimated lactate threshold (LT). Estimated LT was determined as previous studies had reported [3, 4]. Training frequency was set at two exercise sessions/week for 12 weeks.

Before and after exercise training, the subjects performed 10 or 15 W/min ramp cycling exercise until exhaustion (Strength Ergo 8, Fukuda-Denshi, Japan). Pulmonary O_2 uptake (VO_2) was monitored continuously during the experiments to determine peak VO_2 by using an automated gas analysis system (AE300S, Minato Medical Science, Japan).

Muscle O_2 saturation (SmO_2) and relative changes from rest in oxygenated hemoglobin concentration (ΔOxy-Hb), deoxygenated hemoglobin concentration

(ΔDeoxy-Hb), and total hemoglobin concentration (ΔTotal-Hb) were measured at vastus lateralis (VL) in the left leg by NIR_{SRS} (Astem Co., Japan). The probe consisted of one light source and two photodiode detectors, and the distances between light source and detector were 20 and 30 mm, respectively. The data sampling rate was 1 Hz. The obtained signals were defined as the values averaged over the last 10 s. Changes in SmO_2 were calculated as SmO_2 at peak exercise before training subtracted from SmO_2 at peak exercise after training. Although fat layer thickness affects NIR_{SRS} data because of light scattering, Niwayama et al. have recently reported that the effects of fat layer thickness can be corrected in relative changes in Hb and SmO_2 [5]. The corrected relative changes in Hb were obtained by dividing the measured values by the normalized optical path length for muscle (S_{muscle}; when the fat layer thickness is zero), and the value of S_{muscle} can be calculated by only fat layer thickness. For the calculations of SmO_2, the measurements using NIR_{SRS} can be corrected by using the appropriate curve plotting the spatial slope of light intensity and absorption coefficient of the muscle for fat layer thickness. In this study, we measured fat layer thickness at each measurement site in VL muscles with an ultrasound device (LogiQ3, GE-Yokokawa Medical Systems, Japan). Then, we calculated the muscle O_2 dynamics with correction for light scattering effects. The specifications of correction for the influence of fat layer thickness have been fully described [5]. Even though an upper limit of fat layer thickness was designated as 10 mm to correct for the effects in this study, fat layer thickness was within ~10 mm at each measurement site in all subjects.

2.3   Statistics

All data are given as means ± standard deviation (SD). To compare changes in NIRS variables during exercise between groups, a 2-way repeated-measures analysis of variance was used with training and power output as factors. Where appropriate, the Bonferroni post hoc test was conducted. Because one subject could not exercise at more than 59 W before training, repeated measures between groups were limited to rest, 20, 30, 40, and 50 W compared as a function of power output. Differences in NIR_{SRS} and cardiorespiratory variables at peak exercise were compared between groups using paired t tests. Pearson's correlation coefficient was employed to determine the relationship between variables. For all statistical analyses, significance was accepted at $p < 0.05$.

3   Results

Even though training frequency was set at two exercise sessions/week for 12 weeks, unfortunately, the subjects exercised 15 ± 6 sessions during 12 weeks training as their schedules permitted. Estimated LT was significantly increased after training (before: 12.8 ± 3.1 ml/kg/min, after: 14.0 ± 2.1 ml/kg/min, $p < 0.05$), while peak VO_2 was not significantly increased (before: 20.1 ± 6.0 ml/kg/min, after: 21.1 ± 4.1 ml/kg/min, $p = 0.28$). Similarly, workload at estimated LT was significantly improved after training than before (before: 54 ± 15 W, after: 62 ± 16 W, $p < 0.05$), even though peak workload was not significantly increased (before: 98 ± 30 W, after: 101 ± 30 W, $p = 0.44$). Fat layer thickness was not significantly altered after training than before (7.41 ± 2.67 vs. 7.14 ± 2.67 mm, $p = 0.29$).

During submaximal exercise, there were no significant training × power output interactions for SmO_2 ($p = 0.82$), ΔOxy-Hb ($p = 0.46$), ΔDeoxy-Hb ($p = 0.23$), or ΔTotal-Hb ($p = 0.29$) between before and after training in all subjects. Moreover, no significant main effect for training was observed in SmO_2 ($p = 0.75$), ΔOxy-Hb ($p = 0.61$), ΔDeoxy-Hb ($p = 0.48$), or ΔTotal-Hb ($p = 0.50$). Also at peak exercise, no significant difference was found in SmO_2 ($p = 0.90$), ΔOxy-Hb ($p = 0.26$), ΔDeoxy-Hb ($p = 0.20$), or ΔTotal-Hb ($p = 0.17$) (Fig. 8.1).

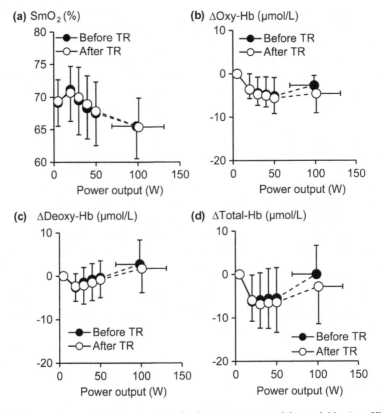

Fig. 8.1 Change in muscle O_2 saturation (SmO_2: **a**), oxygenated hemoglobin (oxy-Hb: **b**), deoxygenated hemoglobin (deoxy-Hb: **c**), and total hemoglobin (total-Hb: **d**) responses in vastus lateralis muscles during ramp cycling exercise before (*closed circles*) and after (*open circles*) exercise training

Fig. 8.2 Relationship between change in muscle O_2 saturation at peak exercise (values at after training minus values at before training) and improvement of peak VO_2 after aerobic exercise training

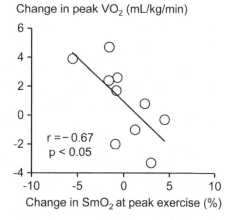

Improvement of peak VO_2 by exercise training was significantly positively related to the number of sessions of training ($r = 0.66$, $p < 0.05$) and negatively related to changes in SmO_2 ($r = -0.67$, $p < 0.05$) (Fig. 8.2). In addition, changes in ΔOxy-Hb ($r = -0.76$, $p < 0.05$) and $\Delta Total$-Hb ($r = -0.66$, $p < 0.05$) were significantly negatively associated with improvement of peak VO_2. Similarly, changes in ΔOxy-Hb ($r = -0.84$, $p < 0.01$) and $\Delta Total$-Hb ($r = -0.73$, $p < 0.05$) were significantly negatively associated with the number of sessions of training. However, $\Delta Deoxy$-Hb was not significantly related to peak VO_2 ($r = 0.05$, $p = 0.88$) or training frequency ($r = 0.09$, $p = 0.80$).

4 Discussion

In the present study, muscle desaturation during submaximal and peak exercise was not significantly changed by aerobic training. One possible interpretation may be that training volume was too low when compared to previous studies [1, 6]. Some previous studies demonstrated that peak VO_2 was significantly related to muscle desaturation responses in cross-sectional observation [7, 8]. In fact, in this study, peak VO_2 was not significantly increased by exercise training, and the number of training sessions was positively related to changes in peak VO_2 after training. These results suggest that no significant difference in muscle desaturation were observed in this study due to low volume exercise training. Another possibility to explain the lack of significant difference in muscle desaturation after aerobic training may be the subjects' characteristics. Previous studies reported that aerobic training enhanced muscle desaturation in healthy young subjects [1] and heart disease patients [6]. However, to our knowledge, there have been no published reports on effects of aerobic training on muscle deoxygenation responses in elderly subjects. This area warrants further investigation.

We also observed a significant negative relationship between improvement of peak VO_2 and changes in SmO_2 at peak exercise after exercise training, even though muscle desaturation was not largely enhanced after training in all elderly subjects. In addition, the number of training sessions was significantly positively related to improvement of peak VO_2 and negatively related to changes in SmO_2 at peak exercise. These findings lead us to speculate that aerobic training may potentially enhance muscle desaturation responses at peak exercise in elderly subjects.

Remarkably, there were significant relationships between enhancement of peak VO_2 and decreases in ΔOxy-Hb or ΔTotal-Hb after exercise training, while changes in ΔDeoxy-Hb were not significantly associated with improvement of peak VO_2. ΔOxy-Hb is an indicator of the balance between O_2 supply and O_2 utilization, and ΔTotal-Hb is indicator of blood volume. Additionally, muscle O_2 supply is also affected by mechanical stress [9]. In fact, in this study, increased peak workload by training was also significantly negatively related to changes in ΔOxy-Hb ($r = -0.63$, $p < 0.05$) or ΔTotal-Hb ($r = -0.64$, $p < 0.05$) by training. These data suggest that the change in muscle desaturation at peak exercise after low volume exercise training may have been mainly due to reduced O_2 supply, secondary to increased mechanical stress, such as intramuscular pressure.

In summary, muscle desaturation was not enhanced by low volume aerobic training in this study, possibly because the training volume was too low. However, there were significant relationships between change in SmO_2, improvement of peak VO_2 and the number of sessions. These results suggest that aerobic training may potentially enhance muscle desaturation responses at peak exercise in elderly subjects. However, in this study, the change in muscle desaturation at peak exercise may have been mainly due to reduced O_2 supply, secondary to increased mechanical stress, such as intramuscular pressure.

Acknowledgments The authors are grateful for revision of this manuscript by Andrea Hope. This study was supported in part by Grant-in-Aid for scientific research from Japan Society for the Promotion of Science (246298) to S.T. and Waseda University Grant for Special Research Projects (2014S-148) to S.T.

References

1. Kime R, Niwayama M, Fujioka M et al (2010) Unchanged muscle deoxygenation heterogeneity during bicycle exercise after 6 weeks of endurance training. Adv Exp Med Biol 662:353–358
2. Takagi S, Kime R, Murase N et al (2013) Aging affects spatial distribution of leg muscle oxygen saturation during ramp cycling exercise. Adv Exp Med Biol 789:157–162
3. Beaver WL, Wasserman K, Whipp BJ (1986) A new method for detecting anaerobic threshold by gas exchange. J Appl Physiol 60(6):2020–2027
4. Wasserman K, Whipp BJ, Koyl SN (1973) Anaerobic threshold and respiratory gas exchange during exercise. J Appl Physiol 35(2):236–243

5. Niwayama M, Suzuki H, Yamashita T et al (2012) Error factors in oxygenation measurement using continuous wave and spatially resolved near-infrared spectroscopy. J Jpn Coll Angiol 52:211–215
6. Mezzani A, Grassi B, Jones AM et al (2013) Speeding of pulmonary VO2 on-kinetics by light-to-moderate-intensity aerobic exercise training in chronic heart failure: clinical and pathophysiological correlates. Int J Cardiol 167(5):2189–2195
7. Kime R, Osada T, Shiroishi K et al (2006) Muscle oxygenation heterogeneity in a single muscle at rest and during bicycle exercise. Jpn J Phys Fit Sports Med 55(Suppl):S19–S22
8. Takagi S, Murase N, Kime R et al (2014) Skeletal muscle deoxygenation abnormalities in early post myocardial infarction. Med Sci Sports Exerc 46(11):2062–2069 doi:10.1249/MSS.0000000000000334
9. Saltin B, Radegran G, Koskolou MD et al (1998) Skeletal muscle blood flow in humans and its regulation during exercise. Acta Physiol Scand 162:421–436

Chapter 9
Differences in Contraction-Induced Hemodynamics and Surface EMG in Duchenne Muscular Dystrophy

Eva Van Ginderdeuren, Alexander Caicedo, Joachim Taelmans, Nathalie Goemans, Marlen van den Hauwe, Gunnar Naulaers, Sabine Van Huffel, and Gunnar Buyse

Abstract Duchenne muscular dystrophy (DMD) is the most common and devastating type of muscular dystrophy worldwide. In this study we have investigated the potential of the combined use of non-invasive near-infrared spectroscopy (NIRS) and surface electromyography (sEMG) to assess contraction-induced changes in oxygenation and myoelectrical activity, respectively in the biceps brachii of eight DMD patients aged 9–12 years and 11 age-matched healthy controls. Muscle tissue oxygenation index (TOI), oxyhemoglobin (HbO_2), and sEMG signals were continuously measured during a sustained submaximal contraction of 60 % maximal voluntary isometric contraction, and post-exercise recovery period. Compared to controls, DMD subjects showed significantly smaller changes in TOI during the contraction. In addition, during the reoxygenation phase some dynamic parameters extracted from the HbO_2 measurements were significantly different between the two groups, some of which were correlated with functional performances on a 6-min walking test. In conclusion, non-invasive continuous monitoring of skeletal muscle oxygenation by NIRS is feasible in young children, and significant differences in contraction-induced deoxygenation and reoxygenation patterns were observed between healthy controls and DMD children.

Keywords NIRS • sEMG • Duchenne muscular dystrophy • Oxygenation

E. Van Ginderdeuren • N. Goemans • M. van den Hauwe • G. Buyse
Department of Child Neurology, University Hospitals Leuven, Leuven, Belgium

A. Caicedo (✉) • J. Taelmans • S. Van Huffel
Department of Electrical Engineering (ESAT), STADIUS Center for Dynamical Systems, Signal Processing, and Data Analytics, KU Leuven, Leuven, Belgium

iMinds Medical IT, Leuven, Belgium
e-mail: acaicedo@esat.kuleuven.be

G. Naulaers
Department of Neonatology, University Hospitals Leuven, Leuven, Belgium

© Springer Science+Business Media, New York 2016
C.E. Elwell et al. (eds.), *Oxygen Transport to Tissue XXXVII*, Advances in Experimental Medicine and Biology 876, DOI 10.1007/978-1-4939-3023-4_9

1 Introduction

Duchenne muscular dystrophy (DMD) is an X-linked recessive muscle disease affecting 1 in 3500 newborn boys worldwide. Besides progressive weakness of the skeletal muscles leading to loss of ambulation by the age of 12 years, DMD patients develop cardiac and respiratory complications that lead to early morbidity and mortality [1]. DMD is caused by mutations in the DMD gene which encodes for dystrophin, a subsarcolemmal protein critical for muscle membrane integrity. One pathway to explain the pathogenesis of DMD involves contraction-induced damage to muscle fibers due to structural instability between the actin cytoskeleton and the sarcolemma. An alternative pathway involves enhanced sympathetic vasoconstriction by a downregulation of the neuronal nitric oxide synthase (NOS) in DMD, resulting in functional muscle ischemia determined by a mismatch between the metabolic demand and blood flow supply in exercising muscles [2].

Further advancement in both pathogenic understanding as in non-invasive disease-monitoring capabilities using a combination of near-infrared-spectroscopy (NIRS) and surface-EMG (sEMG) may contribute to the development and assessment of novel therapeutic approaches for DMD. So far, NIRS has been used in DMD subjects to assess, during contraction, significant changes in tissue oxygenation induced by defective reflex sympathetic activation in exercising muscles [2]. Changes in the reoxygenation phase due to contraction-induced alterations in oxidative metabolism have not yet been investigated. Combined with sEMG to mark the start and the end points for the contractions, we have therefore investigated in young children the potential of NIRS to assess differences in the oxygenation before, during and after the contraction of the biceps brachii of DMD patients versus age-matched healthy controls.

2 Methods

Data Concomitant measurements of sEMG and NIRS were obtained from eight DMD patients and 11 age-matched healthy controls, with age ranging from 9 to 12 years. The measurements follow the protocol depicted in Fig. 9.1. In short, a baseline measurement of 2 min was taken, followed by two consecutive maximum voluntary contractions (MVC), a recovery phase of 2 min, and a flexion of the elbow to 90°. Then the subjects were asked to sustain a contraction of 60 % MVC during 1 min, followed by a recovering period of 10 min. After the recordings finish the skinfold was measured twice on the central part of the biceps with a Harpenden Skinfold Caliper (British Indicator Ltd., Burgess Hill, West Sussex, UK). The mean value of the 2 measurements was used to correct for attenuations in the NIRS signals due to the fat layer [3]. Cutaneous sEMG electrodes were placed on the medial side of the right biceps brachii, according to the recommendations of SENIAM (Surface ElectroMyoGraphy for the Non-Invasive Assessment of

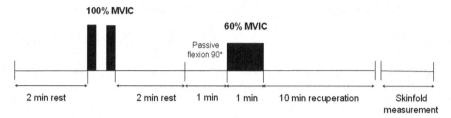

Fig. 9.1 Overview of the measurement Protocol: 2 min resting phase in supine position with arms horizontally; twice a maximal contraction while maintaining the hand in supination; 2 min resting phase; 1 min resting phase with the arm in a passive flexion of 90°; submaximal contraction during 1 min at 60 % of his maximal capacity, measured by a myometer; 10 min recuperation phase

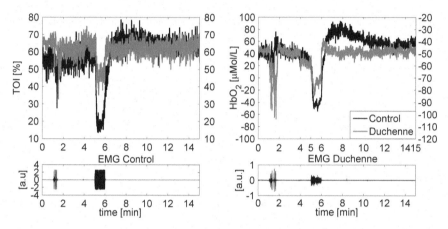

Fig. 9.2 Representative recordings from a control and a DMD subject. Hemodynamic parameters, TOI and HbO_2, are displayed in the *upper plots*, while the *lower plots* show the sEMG signals in arbitrary units (a.u.)

Muscles) [4]. The hemodynamic variables were measured using a NIR device (NIRO 300, Hammamatsu Photonics, Tokyo, Japan) with a NIRS probe (source-detector distance: 4 cm) that was placed on the lateral side of the right biceps brachii symmetrically in the direction of the muscle fibers at 1 cm of the midline. Figure 9.2 shows the recorded signals from a representative control subject and a DMD patient. Additionally, in the DMD group, a functional assessment was done at the time of measurement by one experienced evaluator using the 6-min walking test (6MWT), a measure used as an important clinical tool for progression of the disease [5].

Signal Processing From the sEMG signals we computed the entropy and the mean power frequency during the 1-min contraction period [6, 7]. We used these features in order to differentiate between control and DMD subjects. Additionally, we used the sEMG in order to localize the on-set and off-set of the contractions. Using this localization, we extracted dynamic parameters from the HbO_2 and TOI measurements. Figure 9.3 shows the features extracted from the HbO_2 measurements. We computed the change in HbO_2 caused by the contraction, we named this feature

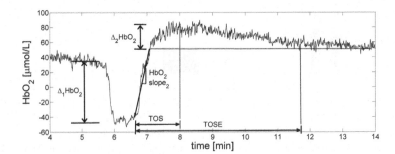

Fig. 9.3 Features extracted from the HbO_2 measurements, $\Delta_1 HbO_2$ represents the change in HbO_2 caused by the contraction, $\Delta_2 HbO_2$ is the difference between the maximum value reached by the HbO_2 after the contraction ended; TOS is the elapsed time between the release of the contraction and the maximum value measured for HbO_2 after contraction; and TOSE is the elapsed time between the release of the contraction and the stabilization of HbO_2

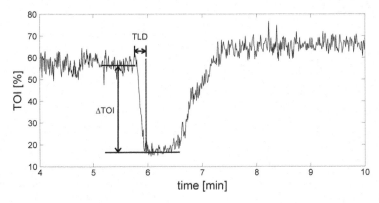

Fig. 9.4 Features extracted from the TOI measurements, ΔTOI is the change in TOI casued by the contraction, TLD is the elapsed time between the on-set of the contraction and when the TOI reaches a stable level during the contraction

$\Delta_1 HbO_2$; the difference between the maximum value reached by the HbO_2 after the contraction and the stabilization value after the contraction was called $\Delta_2 HbO_2$; the elapsed time between the release of the contraction and the maximum value measured for HbO_2 after contraction was called Time to OverShot (*TOS*); finally, the elapsed time between the release of the contraction and the stabilization of HbO_2 was called Time when OverShoot Ends (*TOSE*). Additionally, Fig. 9.4 displays the features extracted from the TOI measurements. The first feature was computed as the change between baseline levels and the stable TOI value reached during contraction, this value was named *ΔTOI*. Also the elapsed time between the on-set of the contraction and when the TOI reaches a stable level during the contraction was called Time of Linear Decay (*TLD*). Data were analyzed in Matlab 7.11.0 (Rs2010b) and SPSS IBM Statistics. Comparisons were made by using a Mann–Whitney U test in order to investigate whether there were differences in the features extracted between Control and DMD subjects. Correlations between parameters were analyzed by using a Pearson correlation coefficient. A p-value <0.05 was considered significant.

3 Results

The results for the sEMG and NIRS derived features are shown in Table 9.1. Before the contraction, there was no significant difference in TOI baseline level between both groups. During the contractions all individuals deoxygenated with ΔTOI values ranging from 2.50 to 41.67 %. DMD subjects presented a smaller ΔTOI than controls (ΔTOI p = 0.032) but without significant difference in time to reach the stabilization phase (TLD).

After the contraction, reoxygenation was observed in both groups. In all individuals except two, an overshoot of the HbO_2 signal was observed. It was observed that DMD subjects took significantly shorter times to reach the overshoot and return to baseline levels of oxygenation, reflected by a significant difference in TOS (p = 0.026) and TOSE (p = 0.003).

The EMG showed no significant difference in MPF between both groups. In addition, the entropy was not statistically significant between both groups, but entropy was positively correlated with ΔTOI (r = 0.550, p = 0.007), TOS (r = 0.455, p = 0.029) and TOSE (0.512, p = 0.013).

Additionally in the DMD group it was observed that the 6MWT, an interesting measure for disease progression, was positively correlated with the following parameters: ΔTOI (r = 0.832, p = 0.040), TOS (r = 0.884 with p = 0.019) and the TOSE (r = 0.878, p = 0.021). Figure 9.5 presents a scatter plot between the TOS and the 6MWT, with its regression line and 95 % confidence intervals.

Table 9.1 Results from the scores computed for the DMD and control subjects

Signal	Features	Control (n = 11)	DMD (n = 8)	p-value
EMG	MPF (initial) (Hz)	57.32 (46.86;62.63)	56.46 (45.45;68.31)	0.804
	MPF (final) (Hz)	46.02 (31.27;59.09)	42.21 (36.2;59.8)	0.563
	MPF (slope) (Hz)	−0.11 (−0.32;0.07)	−0.18 (−0.31;−0.02)	0.508
	Entropy	3.94 (3.35;4.87)	3.67 (2.63;4.01)	0.137
TOI	Baseline (%)	68,73 (57.24;77.89)	63.20 (61.09;71.75)	0.265
	ΔTOI (%)	34.69 (6.09;41.67)	11.62 (2.50;27.17)	**0.032**
	TLD (s)	13.20 (7.95;27.40)	11.25 (8.70;18.70)	0.772
HbO_2	$\Delta_1 HbO_2$ (μmol/L)	−45.61 (−120.2;−13.9)	−17.36 (−57.90;−3.99)	**0.021**
	$\Delta_2 HbO_2$ (μmol/L)	44.45 (7.78;175.30)	25.35 (3.58;119.20)	0.117
	TOS (s)	64.30 (12.70;170.00)	19.20 (10.80;65.60)	**0.026**
	TOSE (s)	285.30 (73.60;440.30)	68.15 (42.10;212.60)	**0.003**

Values are given as mean (min;max). The last column represents the p-value obtained when comparing the features from DMD and Control groups, bold p-values represent significant differences

Fig. 9.5 *Scatter plot* representing the relation between TOS and the 6MWT

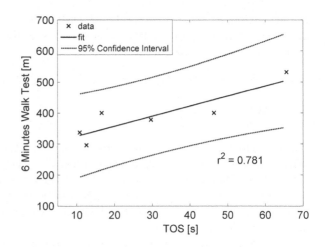

4   Conclusions

Non-invasive continuous monitoring of skeletal muscleoxygenation by NIRS, combined with EMG, can be used to study the response of a sustained contraction in controls and DMD subjects. We found evidence indicating that combined EMG and NIRS reveal information about the assessment of muscle state and disease progression in DMD subjects. Specifically, significant differences in contraction-induced deoxygenation and reoxygenation patterns were observed between healthy controls and muscle-diseased DMD children. Interestingly enough, it was observed that ΔTOI, TOS and TOSE correlated with the DMD subject's functional performances on a 6-min walking test (6MWT), making the combination of these two techniques a potential promising quantitative measure for disease-monitoring in the clinic as in future therapeutic. However, due to the small size of the population included in this study, the results presented in this paper should be validated in a larger population. A more detailed discussion about the results can be found in [8, 9].

Acknowledgments Research supported by a postdoctoral mandate of the Research Foundation Flanders (FWO); Research Council KUL: GOA MaNet, CoE PFV/10/002 (OPTEC). Flemish Government: FWO: travel grant. Belgian Federal Science Policy Office: IUAP P719/(DYSCO), 'Dynamical systems, control and optimization', 2012–2017. Belgian Federal Science Policy Office: IUAP P7/19 DYSCO. EU HIP Trial FP7-HEALTH/2007–2013 (n° 260777).

References

1. Emery AEH (2003) Clinical features. In: Emery AEH (ed) Duchenne muscular dystrophy. Oxford University Press, Oxford, pp 6–45

2. Sander M, Chavoshan B, Harris SA et al (2000) Functional muscle ischemia in neuronal nitric oxide synthase-deficient skeletal muscle of children with Duchenne muscular dystrophy. Proc Natl Acad Sci U S A 97:13818–13823
3. Vans Beekvelt MCP, Borghuis MS, Van Engelen BGM, Wevers RA, Colier WNJM (2001) Adipose tissue thickness affects in vivo quantitative near-IR spectroscopy in human skeletal muscle. Clin Sci 101:21–28
4. Freriks B, Hermens HJ (1999) European recommendations for surface ElectroMyoGraphy, results of the SENIAM project. Roessingh Research and Development, Enschede
5. McDonald CM, Henricson EK, Han JJ et al (2010) The 6-minute walk test as a new outcome measure in Duchenne muscular dystrophy. Muscle Nerve 41:500–510
6. Felici F, Quaresima V, Fattorini L, Sbriccoli P, Filligoi GC, Ferrari M (2007) Biceps brachii myoelectric and oxygenation changes during static and sinusoidal isometric exercises. J Electromyogr Kinesiol 92:e1–e11
7. Ihara S (1993) Information theory for continuous systems. World Scientific, Singapore/River Edge
8. Caicedo A (2013) Signal processing for monitoring cerebral hemodynamics in neonates. Ph.D. thesis, Faculty of Engineering, KU Leuven, Leuven, Belgium, 252 p
9. Van Ginderdeuren E, Caicedo A, Taelmans J, Goemans N, van den Hauwe M, Naulaers G, Van Huffel S, Buyse G (2013) Combined NIRS and surface EMG for Clinically assessing muscle oxygenation and myoelectrical activity in duchenne muscular dystrophy. Internal report

Chapter 10
Changes in Cortical Oxyhaemoglobin Signal During Low-Intensity Cycle Ergometer Activity: A Near-Infrared Spectroscopy Study

Atsuhiro Tsubaki, Haruna Takai, Sho Kojima, Shota Miyaguchi, Kazuhiro Sugawara, Daisuke Sato, Hiroyuki Tamaki, and Hideaki Onishi

Abstract Near-infrared spectroscopy (NIRS) is a widely used non-invasive method for measuring human brain activation based on the cerebral hemodynamic response during gross motor tasks. However, systemic changes can influence measured NIRS signals. We aimed to determine and compare time-dependent changes in NIRS signal, skin blood flow (SBF), and mean arterial pressure (MAP) during low-intensity, constant, dynamic exercise. Nine healthy volunteers (22.1 ± 1.7 years, 3 women) participated in this study. After a 4-min pre-exercise rest and a 4-min warm-up, they exercised on a bicycle ergometer at workloads corresponding to 30 % VO_2 peak for 20 min. An 8-min rest period followed the exercise. Cortical oxyhaemoglobin signals (O_2Hb) were recorded while subjects performed the exercise, using an NIRS system. Changes in SBF and MAP were also measured during exercise. O_2Hb increased to 0.019 mM cm over 6 min of exercise, decreased slightly from 13 min towards the end of the exercise. SBF continued to increase over 16 min of the exercise period and thereafter decreased till the end of measurement. MAP fluctuated from -1.0 to 7.1 mmHg during the exercise. Pearson's correlation coefficients between SBF and O_2Hb, and MAP and O_2Hb differed in each time phase, from -0.365 to 0.713. During low-intensity, constant, dynamic exercise, the profile of changes in measurements of O_2Hb, SBF, and MAP differed. These results suggested that it is necessary to confirm the relationship between O_2Hb and systemic factors during motor tasks in order to detect cortical activation during gross motor tasks.

Keywords Cortical oxyhaemoglobin • Skin blood flow • Mean arterial pressure • Low-intensity exercise • Near-infrared spectroscopy

A. Tsubaki (✉) • H. Takai • S. Kojima • S. Miyaguchi • K. Sugawara • D. Sato • H. Tamaki • H. Onishi
Institute for Human Movement and Medical Sciences, Niigata University of Health and Welfare, 1398 Shimami-cho, Kita-ku, Niigata-shi, Niigata 950-3198, Japan
e-mail: tsubaki@nuhw.ac.jp

© Springer Science+Business Media, New York 2016
C.E. Elwell et al. (eds.), *Oxygen Transport to Tissue XXXVII*, Advances in Experimental Medicine and Biology 876, DOI 10.1007/978-1-4939-3023-4_10

1 Introduction

Functional magnetic resonance imaging [1] and positron emission tomography [2, 3] are used to examine haemodynamic changes related to cortical neural activation during bicycle movement. However, these devices have some measurement constraints. Near-infrared spectroscopy (NIRS) is a good indicator to monitor real-time haemodynamic changes during gross motor tasks, and is widely used.

In NIRS, near-infrared beams are transmitted through the scalp and skull and O_2Hb signals are detected. These signals might indicate task-related cardiovascular responses occurring in the perfusion of extracranial layers. However, blood pressure fluctuations exert confounding effects on brain NIRS [4, 5], and some studies suggest that skin blood flow (SBF) or skin blood volume influences NIRS measurement [6, 7]. Physiological signals arising from cardiac and blood pressure modulations may interfere with the measurement of the haemodynamic response to brain activation.

In particular, blood pressure and scalp blood flow can increase during gross motor tasks. The relationship between NIRS signals and physiological signals must be clarified so that cortical activation can be detected on the basis of changes in O_2Hb concentrations during gross motor tasks.

This study aimed to determine and compare time-dependent changes in O_2Hb, SBF, and mean arterial pressure (MAP) during low-intensity, constant, dynamic exercise.

2 Methods

Nine healthy volunteers ([mean $\pm$ standard deviation] age 22.1 ± 1.7 years; height 167.2 ± 8.9 cm; weight 58.3 ± 7.2 kg; 3 women) participated in this study. All subjects were free from any known neurological, major medical, or cardiovascular diseases and were not taking any medications. Each subject received verbal and written explanations of this study. This study was approved by the Ethics Committee of Niigata University of Health and Welfare (17368-121108) and conformed to the standards set by the Declaration of Helsinki.

To detect exercise workload individually, peak oxygen consumption (VO_2peak) was determined using an incremental protocol on a cycle ergometer (Aerobike 75XLII; Combi, Japan) before the main experiments. Exhaustion was defined based on a previous study [8].

In the main exercise experiment, subjects performed constant exercise on a cycle ergometer. After a 4-min rest and a 4-min warm-up, they exercised at workloads corresponding to 30 % VO_2 peak for 20 min. An 8-min rest followed the exercise. During this experiment, NIRS signals, SBF, and MAP were measured.

A multichannel NIRS imaging system (OMM-3000; Shimadzu Co., Kyoto, Japan) with three wavelengths (780, 805, and 830 nm) was used to detect changes

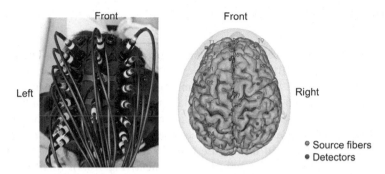

Fig. 10.1 NIRS optode placement and locations of the 12 light-source fibers and 12 detectors

in O_2Hb at a sampling rate of 190 ms. NIRS-determined O_2Hb is a sensitive indicator of changes in cerebral blood flow [9] and is the most sensitive indicator of sensory- and motor-related changes in regional cerebral blood flow [10, 11].

NIRS optodes, consisting of 12 light-source fibres and 12 detectors providing 34-channel simultaneous recording, were set in a 3×8 multichannel probe holder (Fig. 10.1). A 30-mm interoptode distance was used to measure cortical tissue oxygenation. We used a double-density probe holder [12, 13], consisting of two sets, one of which was shifted to half the optode distance from the origin. The Cz position of the international 10–20 system was used to ensure consistent optode placement among all subjects. The NIRS array map covered the right and left central and parietal areas of the scalp to measure cortical tissue oxygenation in motor-related areas.

Beat-to-beat MAP was recorded by volume clamping the finger pulse with a finger photoplethysmograph (Finometer; Finapres Medical Systems, Amsterdam, The Netherlands) on the left middle finger. Changes in SBF were measured at the forehead using a laser Doppler blood flow meter (Omegaflow FLO-CI; Omegawave Inc., Osaka, Japan). Analogue data were converted to digital data using an A/D converter (PowerLab; AD Instruments, Australia) at a 1000-Hz sampling rate.

To observe the effect of systemic changes on O_2Hb, the mean of all 34-channel O_2Hb values was calculated for each subject. O_2Hb concentration, SBF, and MAP were expressed as changes from the rest phase mean, and were calculated every 10 s. These values were compared by performing one-way analysis of variance (ANOVA) with time as an independent variable using the Statistical Package for Social Sciences (SPSS) ver. 21 (IBM Japan, Tokyo, Japan). The relationship between O_2Hb and SBF, and between O_2Hb and MAP were assessed using Pearson's correlation coefficients, with significance set at $p < 0.05$ during the pre-exercise rest period, warm-up period, main exercise period, and post-exercise rest period. The main exercise period and post-exercise rest period were divided into 4-min segments for analysis.

3 Results

One-way ANOVA showed a significant main effect of time on O_2Hb level ($F = 1.69$, $p < 0.05$), SBF ($F = 2.47$, $p < 0.05$), and MAP ($F = 1.27$, $p < 0.05$). O_2Hb increased to 0.019 mM cm over 6 min of exercise, decreased slightly from 13 min towards the end of the exercise, and continued to decrease during the first 20 s of the post-exercise rest phase. O_2Hb increased again, to 0.008 mM cm, over 2 min of the post-exercise rest and decreased again towards the end of measurement (Fig. 10.2). SBF continued to increase over 16 min of the exercise period, and thereafter decreased until the end of measurement (Fig. 10.3). MAP fluctuated between -1.0 and 7.1 mmHg during exercise (Fig. 10.4).

The correlation coefficients between SBF and O_2Hb, and MAP and O_2Hb differed in each time phase (Table 10.1). During the 30 % VO_2 peak cycling

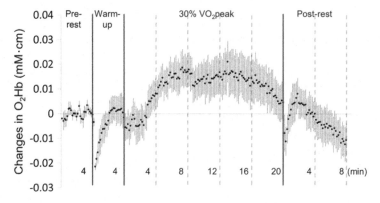

Fig. 10.2 Temporal changes in the averaged oxyhaemoglobin (O_2Hb) level. Values are presented as mean ± standard error of the mean (SEM)

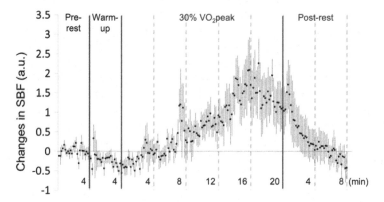

Fig. 10.3 Temporal changes in the averaged skin blood flow (SBF). Values are presented as mean ± standard error of the mean (SEM)

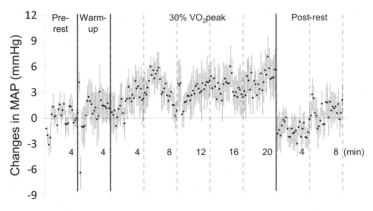

Fig. 10.4 Temporal changes in the averaged mean arterial pressure (MAP). Values are presented as mean ± standard error of the mean (SEM)

Table 10.1 Pearson's correlation coefficients between O_2Hb and SBF and between O_2Hb and MAP

	Pre-rest	Warm-up	30 % VO₂peak					Post-rest	
			(a)	(b)	(c)	(d)	(e)	(a)	(b)
SBF	0.250	−0.365	0.713*	0.671*	0.054	−0.365	0.589*	−0.439*	0.795*
MAP	0.760*	0.685*	0.652*	−0.296	0.760*	0.143	−0.437	0.022	0.008

O_2Hb oxyheamoglobin, *SBF* skin blood flow, *MAP* mean arterial pressure; (a), from start to 4 min; (b), after (a) to 8 min; (c), after (b) to 12 min; (d), after (c) to 16 min; (e), after (d) to 20 min
*$p < 0.05$

exercise, positive and moderate to strong correlations between O_2Hb and SBF were observed in the first, second, and last 4-min segments. These were also observed between O_2Hb and MAP in the first and third 4-min segments.

4   Discussion

This study is the first to report the relationship between the O_2Hb signal, measured by NIRS, and systemic factors during low-intensity, constant, dynamic exercise. The main findings were as follows: (1) the profile of changes in O_2Hb, SBF, and MAP differed, (2) the strength of the relationships between O_2Hb and SBF, and O_2Hb and MAP differed in each time phase.

In the present study, it was observed that O_2Hb changes were not constant despite subjects performing constant load cycle ergometer exercise. This result suggests that cerebral activation in the motor related area changes during the 20-min exercise. Changes in O_2Hb reflect changes in cortical neural activation [11, 14, 15] and depend on exercise intensity [16]. Habituation decreases activity in the ipsilateral motor cortex during a repetitive handgrip exercise [17]. The habituation of this motor task might cause the O_2Hb decline.

The correlation coefficients between O_2Hb and SBF were positive in the first, second, and last 4-min segments of the cycling exercise. O_2Hb concentration is correlated closely ($R^2 = 0.94$) with the integrated Doppler SBF signal in the frontal cortex [6]. Hirasawa et al. [18] showed that changes in O_2Hb and SBF are positively correlated during superficial temporal artery compression. O_2Hb changes in this study might be affected by SBF in the exercise phase.

Positive correlations were also observed between O_2Hb and MAP in the first and third 4-min segments of the cycling exercise. It was reported that blood pressure changes fluctuate brain NIRS signals in the visual cortex during visual stimulation [4]. Gross motor tasks produce systemic circulatory changes, including blood pressure changes. It is suggested that this blood pressure change leads to O_2Hb changes in this study.

During the 20-min of 30 % VO_2 peak cycling exercise, there were weak correlations between O_2Hb and SBF, and O_2Hb and MAP in the fourth 4-min segment. Therefore, we could measure the cerebral activation of the motor-related area with less SBF and/or MAP effects in this segment.

There were some limitations in this study. First, we measured the SBF on the forehead, not over the motor-related area, in order to prevent interference from near-infrared and laser light emitted from the laser Doppler flow meter. Second, we did not measure systemic blood flow changes during low-intensity, constant, dynamic exercise. Finally, although we could clarify the relationships between variables, the experimental design did not enable an assessment of the causality between variables.

In conclusion, O_2Hb changes were not constant during low-intensity, constant, dynamic exercise, and the correlation coefficients between both SBF and MAP, and O_2Hb differed in each time phase. It is necessary to confirm the relationship between O_2Hb and systemic factors during motor tasks in order to detect cortical activation during gross motor tasks, and the findings of the present study will assist with further analysis.

Acknowledgments This study was supported by a Grant-in-Aid for Young Scientists (B) from the Japan Society for the Promotion of Science and a Grant-in-Aid for Exploratory Research from the Niigata University of Health and Welfare.

References

1. Mehta JP, Verber MD, Wieser JA et al (2009) A novel technique for examining human brain activity associated with pedaling using fMRI. J Neurosci Methods 179:230–239
2. Christensen LO, Johannsen P, Sinkjaer T et al (2000) Cerebral activation during bicycle movements in man. Exp Brain Res 135:66–72

3. Hiura M, Nariai T, Ishii K et al (2014) Changes in cerebral blood flow during steady-state cycling exercise: a study using oxygen-15-labeled water with PET. J Cereb Blood Flow Metab 34:389–396
4. Minati L, Kress IU, Visani E et al (2011) Intra- and extra-cranial effects of transient blood pressure changes on brain near-infrared spectroscopy (NIRS) measurements. J Neurosci Methods 197:283–288
5. Tsubaki A, Kojima S, Furusawa AA et al (2013) Effect of valsalva maneuver-induced hemodynamic changes on brain near-infrared spectroscopy measurements. Adv Exp Med Biol 789:97–103
6. Takahashi T, Takikawa Y, Kawagoe R et al (2011) Influence of skin blood flow on near-infrared spectroscopy signals measured on the forehead during a verbal fluency task. Neuroimage 57:991–1002
7. Kirilina E, Jelzow A, Heine A et al (2012) The physiological origin of task-evoked systemic artefacts in functional near infrared spectroscopy. Neuroimage 61:70–81
8. Rupp T, Perrey S (2008) Prefrontal cortex oxygenation and neuromuscular responses to exhaustive exercise. Eur J Appl Physiol 102:153–163
9. Hoshi Y, Kobayashi N, Tamura M (2001) Interpretation of near-infrared spectroscopy signals: a study with a newly developed perfused rat brain model. J Appl Physiol 90:1657–1662
10. Miyai I, Suzuki M, Hatakenaka M et al (2006) Effect of body weight support on cortical activation during gait in patients with stroke. Exp Brain Res 169:85–91
11. Niederhauser BD, Rosenbaum BP, Gore JC et al (2008) A functional near-infrared spectroscopy study to detect activation of somatosensory cortex by peripheral nerve stimulation. Neurocrit Care 9:31–36
12. Kawaguchi H, Koyama T, Okada E (2007) Effect of probe arrangement on reproducibility of images by near-infrared topography evaluated by a virtual head phantom. Appl Opt 46:1658–1668
13. Ishikawa A, Udagawa H, Masuda Y et al (2011) Development of double density whole brain fNIRS with EEG system for brain machine interface. Conf Proc IEEE Eng Med Biol Soc 2011:6118–6122
14. Obrig H, Wolf T, Doge C et al (1996) Cerebral oxygenation changes during motor and somatosensory stimulation in humans, as measured by near-infrared spectroscopy. Adv Exp Med Biol 388:219–224
15. Miyai I, Tanabe HC, Sase I et al (2001) Cortical mapping of gait in humans: a near-infrared spectroscopic topography study. Neuroimage 14:1186–1192
16. Shibuya K, Kuboyama N, Tanaka J (2014) Changes in ipsilateral motor cortex activity during a unilateral isometric finger task are dependent on the muscle contraction force. Physiol Meas 35:417–428
17. Shibuya K (2011) The activity of the primary motor cortex ipsilateral to the exercising hand decreases during repetitive handgrip exercise. Physiol Meas 32:1929–1939
18. Hirasawa A, Yanagisawa S, Tanaka N et al (2015) Influence of skin blood flow and source-detector distance on near-infrared spectroscopy-determined cerebral oxygenation in humans. Clin Physiol Funct Imaging 35:237–244

Chapter 11
Validation of a New Semi-Automated Technique to Evaluate Muscle Capillarization

Sam B. Ballak, Moi H. Yap, Peter J. Harding, and Hans Degens

Abstract The method of capillary domains has often been used to study capillarization of skeletal and heart muscle. However, the conventional data processing method using a digitizing tablet is an arduous and time-consuming task. Here we compare a new semi-automated capillary domain data collection and analysis in muscle tissue with the standard capillary domain method. The capillary density (1481 ± 59 vs. 1447 ± 54 caps mm^{-2}; R^2:0.99; $P < 0.01$) and heterogeneity of capillary spacing (0.085 ± 0.002 vs. 0.085 ± 0.002; R^2:0.95; $P < 0.01$) were similar in both methods. The fiber cross-sectional area correlated well between the methods (R^2:0.84; $P < 0.01$) and did not differ significantly (~8 % larger in the old than new method at $P = 0.08$). The latter was likely due to differences in outlining the contours between the two methods. In conclusion, the semi-automated method gives quantitatively and qualitatively similar data as the conventional method and saves a considerable amount of time.

Keywords Validation • Capillarization • Muscle • Capillary density • Fiber cross-sectional area

1 Introduction

An adequate blood supply to the muscle is not only important for delivery of oxygen to the working muscle, but also for the removal of metabolites and heat. This exchange between blood and muscle fibers takes place in the capillaries and an

S.B. Ballak (✉)
School of Healthcare Science, Cognitive Motor Function Research Group, Manchester Metropolitan University, Manchester, UK

Faculty of Human Movement Sciences, Laboratory for Myology, MOVE Research Institute Amsterdam, VU University Amsterdam, Amsterdam, The Netherlands
e-mail: s.b.ballak@vu.nl

M.H. Yap • P.J. Harding • H. Degens
School of Healthcare Science, Cognitive Motor Function Research Group, Manchester Metropolitan University, Manchester, UK

© Springer Science+Business Media, New York 2016
C.E. Elwell et al. (eds.), *Oxygen Transport to Tissue XXXVII*, Advances in Experimental Medicine and Biology 876, DOI 10.1007/978-1-4939-3023-4_11

adequate muscle capillarization is thus crucial for muscle function. The capillary supply to a fiber is determined by the fiber its size, type, mitochondrial content and metabolic activity of surrounding fibers [1–3]. During hypoxia [4] and hypertrophy [2] capillary proliferation ensures adequate muscle oxygenation.

The method of capillary domains has been used to study the capillarization in skeletal [2, 3] and heart muscle [5]. The strengths of the method are that it not only provides measures of overall capillary supply, such as the capillary density (CD in caps mm^{-2}) and capillary to fiber ratio, but also the capillary supply to individual fibers. It is also unique in that it gives an indication of the heterogeneity of capillary spacing, which can have a significant impact on muscle oxygenation [6–9], and is an accurate method to estimate the oxygen supply areas of individual capillaries and is an indirect indicator of tissue oxygenation [10]. The data obtained with the method of capillary domain can be fed into models of tissue oxygenation [4, 6, 8]. The drawback of the method is, however, that data collection is a manual and time-consuming process. First, pictures have to be printed, fibers and capillaries manually traced on paper and then traced again on a digitizer. The coordinates of the capillaries and fiber outlines are then processed and analyzed with AnaTis (BaLoH Software, www.baloh.nl) [11].

Automation would significantly reduce the data processing time, potentially reduce human errors and improve the accuracy of the data processing. A semi-automated method would also provide the possibility to expand the analysis by introducing new parameters. Therefore, the aim of this study was to compare the conventional method with a new semi-automated Matlab® based software.

2 Methods

2.1 Immunohistochemistry

The left *m. plantaris* of ten 9-month-old C57BL/6j mice were excised, frozen in liquid nitrogen at optimal length and stored at $-80\,°C$ for further analysis. Sections of 10 μm were cut in a cryostat at $-20\,°C$ and stained with biotinylated lectin (*Griffonia simplicifolia*) to identify capillaries as described previously [3].

2.2 Analysis of Capillarization

The capillarization was analyzed first as described previously [2, 3]. In short, the coordinates of the capillaries and the outlines of the muscle fibers were manually delineated with a digitizing tablet (Summagraphics MM1201) and the data fed into AnaTis (BaLoH Software, www.baloh.nl) to calculate capillary domains. Domains are areas surrounding a capillary delineated from surrounding capillaries by

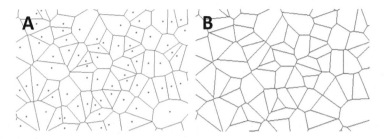

Fig 11.1 Calculated capillary domains using Voronoi tessellations in the conventional (**a**) and new (**b**) method

equidistant boundaries [5]. It also gives an index of the heterogeneity of capillary spacing as the logarithmic standard deviation of the domain radii ($Log_R SD$) and overall indices of capillarization, such as capillary density (CD; cap mm^{-2}). In addition, the program calculates the fiber cross-sectional areas (FCSA) and provides indices of the capillary supply to individual fibers: the local capillary to fiber ratio (LCFR). It is calculated as the sum of the fractions of the capillary domains overlapping a given fiber. The capillary fiber density (CFD; cap mm^{-2} of a given fiber) is the LCFR divided by the FCSA of that fiber. We developed a semi-automated program for capillary domain analysis, as originally described by Hoofd et al. [5] based on Voronoi tessellations. Two annotation tools were developed using Matlab$^{®}$ libraries. The first annotation tool was used to select the capillary and border coordinates. The second annotation tool delineated the fiber outlines. Subsequently, a data analysis function was implemented to calculate the capillary domain (Fig. 11.1). Capillary domains or fiber outlines crossing the border were considered border domains or fibers. The domain size, CD, FCSA, LCFR and CFD were calculated with custom Matlab$^{®}$ functions. All statistics and metrics were compiled from these data.

2.3   Statistical Analysis

The data of the two methods were compared with a paired Student's *t* test and correlations (R^2). Data are presented as mean $\pm$ standard error of the mean (SEM). A value of $P < 0.05$ was considered significant.

3   Results

Figure 11.2a shows a comparison of the CD obtained with the conventional and new method. The CD in the conventional and the new method (1481 ± 59 vs. 1447 ± 54 cap mm^{-2}) are highly correlated (R^2:0.99, $P < 0.01$). The same

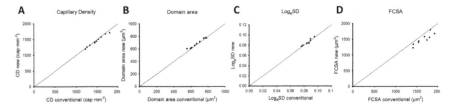

Fig 11.2 The correlation between the old and new method for CD (cap mm^{-2}) (R^2 = 0.99; $P < 0.01$) (**a**), domain area (μm^2) (R^2 = 0.97; $P < 0.01$) (**b**), Log$_R$SD (R^2 = 0.95; $P < 0.01$) (**c**) and FCSA (μm^2) (R^2 = 0.87; $P < 0.01$) (**d**) compared between the conventional (x-axis) and new method (y-axis). The line represents the line of identity

Table 11.1 Correlations and P-values of indices of capillary supply to individual fibers (*LCFR* local capillary to fiber ratio, *CFD* capillary fiber density) and % connective tissue (%CT) between the two methods in *m. plantaris* of 9-month-old male C57Bl/6j mice

	Conventional method	New method	R^2	P-value
%CT (%)	11.5 ± 0.6	16.7 ± 1.4	0.43	0.01
LCFR	2.503 ± 0.076	1.998 ± 0.070	0.80	0.00
CFD (mm^{-2})	1564 ± 56	1252 ± 46	0.86	0.00

applied to the capillary domain areas (R^2:0.97, $P < 0.01$; Fig. 11.2b). Figure 11.2c shows the heterogeneity of capillary spacing (Log$_R$SD) for the conventional and the new method (0.085 ± 0.002 vs. 0.085 ± 0.002). The Log$_R$SD for both methods are strongly correlated (R^2:0.95, $P < 0.01$). The correlation for the FCSA between the two methods is strong (R^2:0.87, $P < 0.01$; Fig. 11.2d) but the new method gives consistently, but not significantly lower FCSAs than the old method (1693 ± 66 vs. 1531 ± 55 μm^2; $P = 0.08$).

Table 11.1 shows that the percentage of connective tissue (%CT) did correlate between the two methods (R^2:0.43, $P = 0.01$). Although the LCFR and the CFD correlated strongly (R^2:0.80 and 0.86; $P < 0.01$), the new method gave consistently lower values for both LCFR and CFD than the old method ($P < 0.01$ and $P = 0.01$).

4 Discussion

The main finding of this study is that a newly developed Matlab® based semi-automatic method of capillary domains provides a quantitatively and qualitatively similar outcome for parameters of overall capillary supply as the traditional manual method. Although the FCSA is underestimated in the new method, affecting quantitatively also indices of the capillary supply to individual fibers, this is a systematic underestimate. Therefore, the qualitative outcome is similar for the two methods and the newly developed method can be readily used for comparative studies on changes in muscle or cardiac capillary supply.

The CD and $Log_R SD$ obtained by the two methods were virtually identical. This indicates that the coordinates recorded for each capillary were comparable between the methods. It also indicates that the calculation of the capillary domains is comparable between the two methods, since for the calculation of the $Log_R SD$ the surface areas of the individual domains and the radius of circles with corresponding surface areas have to be calculated.

There was, however, a small difference in the way the fiber outlines were traced; the FCSA was systematically underestimated in the new method. In the conventional method, accidental fiber overlap could be accounted for, while in the Matlab® based software fibers overlapping each other are joined and assessed as one large fiber. Even with good tracing skills, a researcher may avoid tracking the borders too close to each other to prevent this fusion of two separate fibers; in other words, the experimenter may intentionally draw the fibers slightly too small, resulting in a smaller fiber area. On the other hand, the conventional method could lead to a small overestimation of the size of the fibers since overlapping areas will be counted twice. Together, this stresses the importance of good tracing skills in both the new and the old method.

The difference in fiber size, caused by the difference in tracing the outlines of the fibers, also has an impact on the %CT, which was significantly higher in the new method. The differences in fiber size also work through in the LCFR and CFD, which were lower in the new than in the old method, even though the correlation between the two methods was strong. Thus, while there is a systematic qualitative difference between the two methods, the new method is readily applicable in comparative studies.

We are currently working on the tracing algorithm to adjust the systematic underestimation of the FCSA in the Matlab® based software. The main aim of this study was to evaluate the validity of this technique before looking into that issue.

It is clear that the method of capillary domains has limitations, as it does not consider the three-dimensional structure of the capillary network, differences in flow between capillaries or decrements in oxygen content of the blood from the arteriolar to the venous side of the capillary. Other investigators have tried to address this issue [12, 13], but this was outside the scope of the present study. Nevertheless, the method of capillary domains has successfully been used to calculate the domain size in venous and arterial capillaries [14] and using assumptions of the decrement in capillary oxygen tension from the arteriolar to venular side of the capillary, estimates can be made of the oxygenation in successive planes of tissue [15]. In addition, in a mathematical model also the impact of flow heterogeneity on tissue oxygenation can be incorporated [16] illustrating the usefulness of the collected data in the two dimensional plane to get an estimate of tissue oxygenation under different conditions.

The new Matlab® based program saves on average about 1.5 h per section/photo. This will increase even more when additional analyses are automated further. In addition, the Matlab® based program makes analysis more flexible and allows the implementation of new variables in the future. This new time-sparing method is the first step in fully automating the process and assessment of capillary domains. Future research should focus on fully automating the sampling of capillary and fiber outline coordinates.

5 Conclusion

In conclusion, the new semi-automated Matlab® based method is highly comparable to the standard method of capillary domain analysis when considering indices of overall muscle capillarization. However, due to differences in the way of tracing the fiber outlines there are small difference in FCSA, which also affect indices of capillary supply to individual fibers. Together, this new method appears to be a valid way to qualitatively and quantitatively analyze the capillarization in cardiac and skeletal muscle.

Acknowledgments This research was funded by the European Commission through MOVE-AGE, an Erasmus Mundus Joint Doctorate program (2011–2015). The Matlab codes are available from the authors on request.

References

1. Ahmed SK, Egginton S, Jakeman PM et al (1997) Is human skeletal muscle capillary supply modelled according to fibre size or fibre type? Exp Physiol 82(1):231–234
2. Degens H, Turek Z, Hoofd LJ et al (1992) The relationship between capillarisation and fibre types during compensatory hypertrophy of the plantaris muscle in the rat. J Anat 180 (Pt 3):455–463
3. Wust RC, Gibbings SL, Degens H (2009) Fiber capillary supply related to fiber size and oxidative capacity in human and rat skeletal muscle. Adv Exp Med Biol 645:75–80
4. Wust RC, Jaspers RT, van Heijst AF et al (2009) Region-specific adaptations in determinants of rat skeletal muscle oxygenation to chronic hypoxia. Am J Physiol Heart Circ Physiol 297 (1):H364–H374
5. Hoofd L, Turek Z, Kubat K et al (1985) Variability of intercapillary distance estimated on histological sections of rat heart. Adv Exp Med Biol 191:239–247
6. Degens H, Ringnalda BE, Hoofd LJ (1994) Capillarisation, fibre types and myoglobin content of the dog gracilis muscle. Adv Exp Med Biol 361:533–539
7. Degens H, Deveci D, Botto-van Bemden A et al (2006) Maintenance of heterogeneity of capillary spacing is essential for adequate oxygenation in the soleus muscle of the growing rat. Microcirculation 13(6):467–476
8. Al-Shammari AA, Gaffney EA, Egginton S (2012) Re-evaluating the use of Voronoi Tessellations in the assessment of oxygen supply from capillaries in muscle. Bull Math Biol 74 (9):2204–2231
9. Liu G, Mac Gabhann F, Popel AS (2012) Effects of fiber type and size on the heterogeneity of oxygen distribution in exercising skeletal muscle. PLoS One 7(9):e44375
10. Al-Shammari AA, Gaffney EA, Egginton S (2014) Modelling capillary oxygen supply capacity in mixed muscles: capillary domains revisited. J Theor Biol 356c:47–61

11. Hoofd L, Degens H (2013) Statistical treatment of oxygenation-related data in muscle tissue. Adv Exp Med Biol 789:137–142
12. Weerappuli DP, Popel AS (1989) A model of oxygen exchange between an arteriole or venule and the surrounding tissue. J Biomech Eng 111(1):24–31
13. Beard DA, Bassingthwaighte JB (2000) Advection and diffusion of substances in biological tissues with complex vascular networks. Ann Biomed Eng 28(3):253–268
14. Suzuki J, Gao M, Batra S et al (1997) Effects of treadmill training on the arteriolar and venular portions of capillary in soleus muscle of young and middle-aged rats. Acta Physiol Scand 159 (2):113–121
15. Hoofd L (1995) Calculation of oxygen pressures in tissue with anisotropic capillary orientation. II. Coupling of two-dimensional planes. Math Biosci 129(1):25–39
16. Hoofd L, Degens H (2009) The influence of flow redistribution on working rat muscle oxygenation. Adv Exp Med Biol 645:55–60

Chapter 12
Reduction in Cerebral Oxygenation After Prolonged Exercise in Hypoxia is Related to Changes in Blood Pressure

Masahiro Horiuchi, Shohei Dobashi, Masataka Kiuchi, Junko Endo, Katsuhiro Koyama, and Andrew W. Subudhi

Abstract We investigated the relation between blood pressure and cerebral oxygenation (COX) immediately after exercise in ten healthy males. Subjects completed an exercise and recovery protocol while breathing either 21 % (normoxia) or 14.1 % (hypoxia) O_2 in a randomized order. Each exercise session included four sets of cycling (30 min/set, 15 min rest) at 50 % of altitude-adjusted peak oxygen uptake, followed by 60 min of recovery. After exercise, mean arterial pressure (MAP; 87 ± 1 vs. 84 ± 1 mmHg, average values across the recovery period) and COX (68 ± 1 % vs. 58 ± 1 %) were lower in hypoxia compared to normoxia ($P < 0.001$). Changes in MAP and COX were correlated during the recovery period in hypoxia ($r = 0.568$, $P < 0.001$) but not during normoxia ($r = 0.028$, not significant). These results demonstrate that reductions in blood pressure following exercise in hypoxia are (1) more pronounced than in normoxia, and (2) associated with reductions in COX. Together, these results suggest an impairment in cerebral autoregulation as COX followed changes in MAP more passively in hypoxia than in normoxia. These findings could help explain the increased risk for postexercise syncope at high altitude.

Keywords Postexercise syncope • High altitude • Cerebral autoregulation • Vasodilation • Near infrared spectroscopy

1 Introduction

Postexercise hypotension (PEH) refers to a drop in arterial blood pressure that occurs immediately after exercise and may persist for up to 2 h in otherwise healthy individuals [1]. With effective cerebral autoregulation (CA), relatively constant

M. Horiuchi (✉) • S. Dobashi • M. Kiuchi • J. Endo • K. Koyama
Division of Human Environmental Science, Mt. Fuji Research Institute, Fujiyoshida, Japan
e-mail: mhoriuchi@mfri.pref.yamanashi.jp

A.W. Subudhi
Department of Biology, University of Colorado Colorado Springs, Colorado Springs, CO, USA

© Springer Science+Business Media, New York 2016
C.E. Elwell et al. (eds.), *Oxygen Transport to Tissue XXXVII*, Advances in Experimental Medicine and Biology 876, DOI 10.1007/978-1-4939-3023-4_12

cerebral blood flow and oxygenation (COX) may be maintained after exercise [2, 3] and prevent syncope secondary to PEH. However, few studies have examined the relation between arterial pressure and COX following exercise.

While two studies have demonstrated that CA is maintained following 40 min of moderate exercise [3] and marathon running [2], a recent study showed that CA was impaired after maximal exercise [4]. If CA is impaired after exercise, PEH could reduce cerebral blood flow and COX and increase the risk of syncope. This effect could be compounded in those exercising at high altitude, as hypoxia itself is known to impair CA [5]. This would be of particular concern to mountaineers for whom falls are life threatening.

To the best of our knowledge, no studies have examined the relationship between PEH and COX in hypoxia. We hypothesized that PEH would lead to greater reduction in COX following exercise in hypoxia relative to normoxia. To test this hypothesis, we used near infrared spectroscopy (NIRS) to monitor changes in regional COX following 120 min of exercise at the intensity of 50 % peak oxygen uptake (VO_2) performed at altitudes of 300 m (*breathing room air in the lab*) and 3200 m (*simulated using 14.1 % O_2*).

2   Methods

Ten healthy male subjects (mean ± standard error of the mean [SEM]; age 23 ± 1 years, height 174 ± 2 cm, weight 73 ± 5 kg) participated in this study. All procedures were approved by the ethical committee of Mt. Fuji Research Institute and were performed in accordance with the guidelines of the Declaration of Helsinki.

Each subject was studied on four separate occasions, at least 3 days apart. On days 1 and 2, subjects performed incremental exercise tests (30 W/min, 60 rpm) on a cycle ergometer (Ergomedic 828 E; Monark, Vansbro, Sweden) to determine peak VO_2 in normoxia or hypoxia. On days 3 and 4, subjects performed four 30-min sets of cycling (15 min rest between sets) at 50 % of altitude-adjusted peak VO_2, followed by 60 min of seated recovery. Throughout all protocols, subjects breathed either normoxic (room air) or hypoxic gas (14.1 % O_2) through a mask. An air compressor (0.75LP-750; Hitachi, Tokyo, Japan) and hypoxic generator (YHS-B05S; YKS, Nara, Japan) were used to deliver the respective gas mixture in a randomized balanced order.

Cerebral oxygenation (COX) was measured using NIRS (BOM-L1TRW; Omega Wave, Tokyo, Japan). The NIRS instrument used three laser-diodes (780, 810, and 830 nm) and calculated relative tissue levels of oxygenated hemoglobin (HbO_2) and deoxygenated hemoglobin (HHb) according to the modified Beer–Lambert law. Total Hb was calculated as the sum of HbO_2 and HHb, and the index of COX was expressed as (HbO_2/total Hb) × 100 (i.e., as a percentage). NIRS optodes were placed on the left forehead and shielded from ambient light with a black cloth. The light source was paired with two detectors, set 2 and 4 cm apart. The hemoglobin concentrations received by detector 1 were subtracted from those received by detector 2 to minimize the influence of skin blood flow (SkBF) [6]. In addition, forehead skin blood flow was monitored using laser flowmetry (ATBF-LC1; Unique Medical, Tokyo, Japan), ~2 cm medial from the NIRS probes. NIRS and SkBF signals were recorded at 1 Hz (es8; TEAC, Tokyo, Japan). Arterial blood pressure was measured at rest and every 10 min during recovery, using an oscillometric blood pressure monitoring device (HEM7420; Omron, Tokyo, Japan). Additionally, heart rate (HR) (RS800CX; POLAR Japan, Tokyo, Japan), arterial saturation (SpO_2) (PULSOX300; Konica Minolta, Tokyo, Japan), and partial pressure of end-tidal CO_2 ($P_{ET}CO_2$) (AE-300; Minato Medical Science, Osaka, Japan) were measured continuously through each protocol.

Mean arterial pressure (MAP) was calculated as [(2 × diastolic pressure) + systolic pressure]/3. SkBF were expressed in terms of the relative change from the resting baseline period prior to breathing normoxic or hypoxic gas. Cutaneous vascular conductance (CVC) was calculated by dividing SkBF by MAP.

Data are presented as mean ± SEM. Two-way repeated measures analysis of variance (ANOVA) (Sigma Stat ver 3.5; Hulinks, Chicago, IL, USA) were conducted to compare physiological responses between conditions (normoxia and hypoxia) and across time (at baseline and during recovery), with Bonferroni post hoc tests. Pearson's correlation coefficient was used to estimate the relation between changes in MAP and COX relative to their respective baseline values in normoxia and hypoxia. A P value of 0.05 was considered statistically significant.

3   Results

Figure 12.1 shows the baseline and postexercise MAP and COX responses in normoxia and hypoxia. MAP was similar at baseline but was lower at 10, 20, and 60 min post exercise in hypoxia compared to normoxia (87 ± 1 vs. 84 ± 1 mmHg, across the recovery period, $P < 0.001$). COX was similar at baseline but lower throughout the recovery period in hypoxia compared to normoxia (68 ± 1 % vs. 58 ± 1 %, $P < 0.001$).

The relations between changes in MAP and COX from baseline values through the recovery periods are shown in Figure 12.2. Changes in MAP and COX were correlated during the recovery period in hypoxia ($r = 0.568$, $P < 0.001$) but not in normoxia ($r = 0.028$, $P = 0.829$).

Table 12.1 shows baseline and postexercise SpO_2, HR, $P_{ET}CO_2$, and CVC responses in normoxia and hypoxia. All values were similar at baseline. Average SpO_2 across the recovery period was lower (95.7 ± 0.2 % vs. 83.6 ± 0.8 %, $P < 0.001$) and HR was higher (96 ± 2 vs. 106 ± 2 beats per min, $P < 0.001$) in hypoxia compared to normoxia. In contrast, average $P_{ET}CO_2$ in hypoxia tended to be lower (36.5 ± 0.4 % vs. 35.5 ± 0.5 %, $P = 0.068$) compared to normoxia. CVC in recovery was similar (7.4 ± 0.7 vs. 7.5 ± 0.5 mmHg, $P = 0.897$) between conditions.

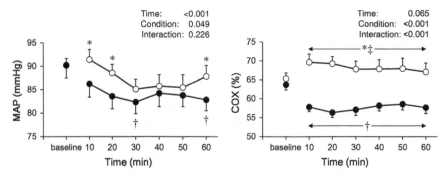

Fig. 12.1 Time course changes in mean arterial pressure (MAP; *left*) and cerebral oxygenation (COX; *right*) under normoxia (*white circles*) and hypoxia (*black circles*). Values are mean ± standard error of the mean. *$P < 0.05$ between normoxia and hypoxia, †$P < 0.05$ vs. baseline under hypoxia, ‡$P < 0.05$ vs. baseline under normoxia

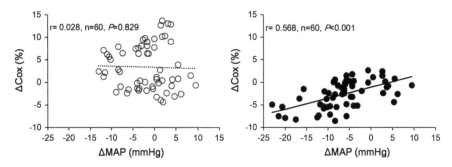

Fig. 12.2 The relation between ΔMAP and ΔCOX from baseline in normoxia (*left*) and hypoxia (*right*). Plotted data represent all subjects (n = 10) at each time point (10, 20, 30, 40, 50, and 60 min postexercise)

Table 12.1 Changes in cardiorespiratory variables and hemodynamics under normoxic and hypoxic conditions

		Preexercise	Postexercise		
		Baseline	10 min	30 min	60 min
SpO_2 (%)	N	97 ± 0.3	95 ± 0.5	96 ± 0.3	96 ± 0.3
	H	97 ± 0.3	83 ± 1.2*†	84 ± 1.5*†	85 ± 1.5*†
HR (bpm)	N	71 ± 2	105 ± 3	96 ± 4	90 ± 4
	H	73 ± 2	113 ± 3*†	103 ± 4*†	101 ± 3*†
$P_{ET}CO_2$ (mmHg)	N	39.3 ± 0.3	37.1 ± 0.9	36.8 ± 0.7†	36.2 ± 0.8†
	H	38.8 ± 0.5	35.0 ± 0.9*†	35.4 ± 0.7†	36.4 ± 0.9†
CVC (mV/mmHg)	N	7.1 ± 1.3	9.0 ± 1.5†	7.0 ± 1.0	6.2 ± 0.9
	H	7.8 ± 0.7	8.8 ± 1.0	7.3 ± 0.8	6.4 ± 0.7

Values are mean ± standard error of the mean
N normoxia, *H* hypoxia, *SpO_2* arterial saturation, *$P_{ET}CO_2$* partial pressure of end tidal carbon dioxide, *CVC* cutaneous vascular conductance
*$P < 0.05$ between normoxia and hypoxia, †$P < 0.05$ vs. baseline in each condition

4 Discussion

This study provides several important findings related to PEH. First, MAP and COX were significantly lower after exercise in hypoxia than in normoxia. Second, reductions in MAP were associated with reductions in COX in hypoxia only. The observation that changes in COX followed changes in MAP more passively in hypoxia is indicative of impaired autoregulation [7] and may help explain the increased risk of postexercise syncope at high altitude [8].

To the best of our knowledge, this is the first study demonstrating evidence of PEH during the first 20 min of recovery in hypoxia but not in normoxia. PEH is thought to be caused by a decrease in total vascular resistance and/or stroke volume (SV) [1]. While we did not observe a change CVC (the reciprocal of vascular resistance), the increased HR in hypoxia may have represented a compensatory response to offset reductions in SV and MAP associated with exercise in hypoxia [9]. Alternatively, previous studies have demonstrated hypoxia increases vasodilation in active skeletal muscle via the nitric oxide and prostaglandin pathways [10] and can increase splanchnic circulation [11]. These effects could have exacerbated the reduction in MAP. Additional studies are needed to ascertain the precise mechanisms explaining the larger degree of PEH in hypoxia relative to normoxia.

The correlations between ΔMAP and ΔCOX indicate that COX passively followed changes in MAP during the recovery period in hypoxia. The relation between MAP and COX has been used as an index of CA [7], and the tight correlation between ΔMAP and ΔCOX in hypoxia suggests impairment in CA [7]. Together our results suggest that the reduction in MAP along with impaired CA may lead to a passive reduction in COX, which, if severe enough, could result in syncope [12]. Additionally, our data shows that COX following exercise is increased in normoxia, but decreased in hypoxia. While a recent study demonstrated that cerebral blood flow (CBF) assessed by magnetic resonance imaging was unchanged or increased after 20 min aerobic exercise [13], this is the first study to show an a reduction in COX during recovery in hypoxia. This finding may help explain the increased risk of PEH at altitude. From a practical perspective, an increased risk of post-exercise syncope in hypoxia is of particular concern for those participating in high-risk activities, such as mountaineering, where falls may be life threatening. Accordingly, our findings suggest that identifying strategies that improve CA in hypoxia could reduce the risk of syncope at high altitude.

There are several methodological limitations that should be considered. First, we did not directly assess CA by measuring CBF responses. Previous studies that have evaluated the relation between MAP and CBF after exercise have reported conflicting results. Therefore, future studies with continuous monitoring of MAP, CBF, and COX are necessary to better understand the proposed connections with PEH and syncope.

Second, recent studies have demonstrated that COX assessed by NIRS may be affected by SkBF [14]. However, because we observed similar SkBF responses between conditions and the NIRS instrument corrected for the influence of cutaneous blood volume, we do not believe SkBF responses affected our conclusions.

In summary, recovery from exercise in hypoxia is associated with larger reductions in MAP and COX relative to normoxia. These findings suggest an impairment in CA, as COX followed changes in MAP more passively in hypoxia than in normoxia, and may help explain the increased risk for postexercise syncope at high altitudes.

Acknowledgments The authors thank all participants for their time and effort. This study was supported by the Japan Society for the Promotion of the Science (No. 26440268 to M.H. and No. 25350810 to K.K).

References

1. Halliwill JR (2001) Mechanisms and clinical implications of post-exercise hypotension in humans. Exerc Sports Sci Rev 29(2):65–70
2. Murrell C, Cotter JD, George K et al (2009) Influence of age on syncope following prolonged exercise: differential responses but similar orthostatic intolerance. J Physiol 587 (Pt 24):5959–5969
3. Willie CK, Ainslie PN, Taylor CE et al (2013) Maintained cerebrovascular function during post-exercise hypotension. Eur J Appl Physiol 113(6):1597–1604
4. Bailey DM, Evans KA, McEneny J et al (2011) Exercise-induced oxidative-nitrosative stress is associated with impaired dynamic cerebral autoregulation and blood–brain barrier leakage. Exp Physiol 96(11):1196–1207
5. Subudhi AW, Panerai RB, Roach RC (2010) Effects of hypobaric hypoxia on cerebral autoregulation. Stroke 41(4):641–646
6. Ando S, Hatamoto Y, Sudo M et al (2013) The effects of exercise under hypoxia on cognitive function. PLoS One 8(5):e63630
7. Steiner LA, Pfister D, Strebel SP et al (2009) Near-infrared spectroscopy can monitor dynamic cerebral autoregulation in adults. Neurocrit Care 10(1):122–128
8. Van Lieshout JJ, Wieling W, Karemaker JM et al (2003) Syncope, cerebral perfusion, and oxygenation. J Appl Physiol 94(3):833–848
9. Calbet JA, Boushel R, Radegran G et al (2003) Determinants of maximal oxygen uptake in severe acute hypoxia. Am J Physiol Regul Integr Comp Physiol 284(2):R291–R303
10. Crecelius AR, Kirby BS, Voyles WF et al (2011) Augmented skeletal muscle hyperaemia during hypoxic exercise in humans is blunted by combined inhibition of nitric oxide and vasodilating prostaglandins. J Physiol 589(Pt 14):3671–3683
11. Westendorp RG, Blauw GJ, Frolich M et al (1997) Hypoxic syncope. Aviat Space Environ Med 68(5):410–414
12. Szufladowicz E, Maniewski R, Kozluk E et al (2004) Near-infrared spectroscopy in evaluation of cerebral oxygenation during vasovagal syncope. Physiol Meas 25(4):823–836
13. MacIntosh BJ, Crane DE, Sage MD et al (2014) Impact of a single bout of aerobic exercise on regional brain perfusion and activation responses in healthy young adults. PLoS One 9(1): e85163
14. Miyazawa T, Horiuchi M, Komine H et al (2013) Skin blood flow influences cerebral oxygenation measured by near-infrared spectroscopy during dynamic exercise. Eur J Appl Physiol 113(11):2841–2848

Part II
Mathematical Models

Chapter 13
The Pathway for Oxygen: Tutorial Modelling on Oxygen Transport from Air to Mitochondrion

The Pathway for Oxygen

James B. Bassingthwaighte, Gary M. Raymond, Ranjan K. Dash, Daniel A. Beard, and Margaret Nolan

Abstract The 'Pathway for Oxygen' is captured in a set of models describing quantitative relationships between fluxes and driving forces for the flux of oxygen from the external air source to the mitochondrial sink at cytochrome oxidase. The intervening processes involve convection, membrane permeation, diffusion of free and heme-bound O_2 and enzymatic reactions. While this system's basic elements are simple: ventilation, alveolar gas exchange with blood, circulation of the blood, perfusion of an organ, uptake by tissue, and consumption by chemical reaction, integration of these pieces quickly becomes complex. This complexity led us to construct a tutorial on the ideas and principles; these first PathwayO$_2$ models are simple but quantitative and cover: (1) a 'one-alveolus lung' with airway resistance, lung volume compliance, (2) bidirectional transport of solute gasses like O_2 and CO_2, (3) gas exchange between alveolar air and lung capillary blood, (4) gas solubility in blood, and circulation of blood through the capillary syncytium and back to the lung, and (5) blood-tissue gas exchange in capillaries. These open-source models are at Physiome.org and provide background for the many respiratory models there.

Keywords Mechanics of ventilation • Oxygen transport in blood • Blood-tissue oxygen exchange • Oxidative phosphorylation

J.B. Bassingthwaighte (✉) • G.M. Raymond
Department of Bioengineering, University of Washington, Seattle, WA, USA
e-mail: Jbb2@uw.edu

R.K. Dash
Department of Physiology, Medical College of Wisconsin, Milwaukee, WI, USA

D.A. Beard
Department of Molecular and Integrative Physiology, University of Michigan, Ann Arbor, MI, USA

M. Nolan
Department of Bioengineering, University of Pennsylvania, Philadelphia, PA, USA

© Springer Science+Business Media, New York 2016
C.E. Elwell et al. (eds.), *Oxygen Transport to Tissue XXXVII*, Advances in
Experimental Medicine and Biology 876, DOI 10.1007/978-1-4939-3023-4_13

103

1   Introduction

Physiological models tend toward complexity. Carlson et al. [1] modeled ventilatory and alveolar-capillary exchanges showing that transport of O_2 and CO_2 to
tissue was influenced not only by respiration rate, composition of inspired gas,
tissue pH and CO_2 production, but also by the 1.5 times higher velocity of RBC than
plasma [2], which increases alveolar-arterial (A-a) differences in P_{O2}. This minor
effect is one of many that influence O_2 delivery and complicate attempts to
quantitate physiology. A more important, particularly useful development was an
efficient method for calculating the hemoglobin binding of oxygen and carbon
dioxide using invertible Hill-type equations [3, 4] accounting for intracapillary
gradients as the RBC progressed along the capillary-tissue exchange region. That
made it practical to combine these events with convective transport, axial diffusion
in the capillary, and with exchange and metabolism in the surrounding tissue
region [5].

Such models exemplify some of the complexity of modeling ventilatory, circulatory and metabolic gas exchange, but the price of physiological accuracy was the
difficulty in learning how to use the models. So, in the interests of assisting people
through the learning process, we are developing sets of relatively simple models
that extend from a one-alveolus mechanical 'lung' step-by-step to account for the
physiological behavior and lay a framework for the pathophysiology of disease and
the pharmacology of therapies. These models are a part of our lab's contribution to
the Physiome Projects, a world-wide grass roots consortium of efforts, including the
European Union's Virtual Human Project and NIGMS's Virtual Physiological Rat
program, to define integrative physiology quantitatively. The logic is that quantitative models are explicit hypotheses, but are inherently wrong in the sense of being
incomplete, inexact, or truly erroneous. The precise nature of quantitative model
hypotheses encourages their disproof, and so leads to the advancement of the
science. Models are merely transient stepping-stones.

Pursuant to this cause, the models that we provide are public, open source, freely
downloadable and reproducible. The language we use to define the models is
human readable, an XML variant called MML (Mathematical Modeling Language)
and the programs run under a freely downloadable simulation analysis system, JSim
[6], that uses a declarative language (rather than a procedural one) so the code is
easily readable and convertible to other languages. This system is designed to serve
an investigator through the steps of a project, from hypothesis and experiment
design to experimental data analysis, sensitivity analysis, verification testing, optimization for data fitting and validation testing, parameter confidence evaluation by
covariance and Monte Carlo analysis, and uncertainty quantification. The 'Project
File' nature of JSim project allows all the data, the model and the setups for the
analysis to be retained for personal retention and for public dissemination in what
we call "The Reproducible Exchangeable Package (REP)", which is simply the
operational project file, '*model*.proj', for each of the models. The primitive examples in this first section are lacking a central aspect of modeling, namely the data

and the relationships between the model and the data, but do portray fundamental principles underlying the real physiology.

2   The One-Alveolus Lung: Ventilation and Alveolar-Blood Exchange

This first set of models, PathwayO$_2$.1, provides a generic overview of the processes from inhaled air to consumption in the tissue, a grossly oversimplified view composed of the main elements of the processes. Five models of gradually increasing complexity illustrate elementary lung mechanics, flow of air, inhalation of a gas dissolved in air, transport into the blood, and distribution and reaction in the body. The models (labeled Physiome #0xxx) can be run over the web or downloaded at physiome.org. The simulation analysis system, JSim, is free also. The models are developed to answer questions that are posed as a part of the tutorial. Here is the code for model 1. The first and simplest in the series:

```
import nsrunit; unit conversion on;
math OneAlvLung.Assist{
   realDomain t sec; t.min = 0; t.max = 9; t.delta = 0.1; // independent
   variable, time
real Com  = 50  ml/mmHg,       // Compliance of the lung, linear
  Res    = 0.01 mmHg*sec/ml,  // Resistance of airway, a constant
  Patmo = 0    mmHg,           // Reference atmospheric pressure
                              external to body and ventilator
  VFRC = 3000 ml,             // Volume at rest, Functional
                              Residual Capacity, open glottis
extern real Pvent(t) mmHg,    // Driving Pressure from Ventilator. Set
                              generator at Run Time
  Fair(t)  ml/sec,            // Flow of air at mouth
  Pmouth(t) mmHg,             // Pressure at the mouth
  Plung(t) mmHg,              // Pressure in the lung
  Vlung(t) ml;                // Volume of air in lung, total
when(t = t.min) Vlung = VFRC;  // Initial Conditions
// Equations, Algebraic and ODEs
  Pmouth = Pref + Pvent;       // Pressure at mouth: Pvent set to 2 sec at
                              10 mmHg, zero for 4 sec
  Fair   = (Pmouth - Plung) / Res;  // Ohm's Law: current = driving force
                                    / resistance
  Vlung:t = Flow;              // Assumes incompressible air
  Plung  = Patmos + (Vlung-VFRC)/Com; // Linear Pressure/Volume
                                    relation around VFRC
} // program end
```

The human-readable syntax above declares variables, parameters, governing equations, and physical units for ventilation: pressure, flow (dV_{lung}/dt), and volume,

$$\frac{dV_{lung}}{dt} = \frac{P_{mouth} - P_{lung}}{Res}; \quad P_{mouth}(t) = P_{ref} + P_{drive}(t);$$

$$P_{lung}(t) = P_{ref} + \frac{V_{lung} - VFRC}{Com}.$$

Model 1. OneAlvLung.Assist (Physiome #0001) The code above is for a one-alveolus lung, a compliant stirred-tank, driven to expand by a positive pressure ventilator. The model lung is purely elastic (stretches instantly like a spring), it returns to its rest volume, functional reserve capacity, when the pressure falls to zero. The airway resistance makes volume changes slower than pressure changes. The tidal volume is proportional to the pressure because the elasticity is constant. (Real lungs are stiffer at higher volumes, i.e. the pressure-volume curve is concave upward.)

Question: With a linear compliance of 50 ml/mmHg, what is the tidal volume with a ventilator pressure excursion of 10 mmHg? (Fig. 13.1)

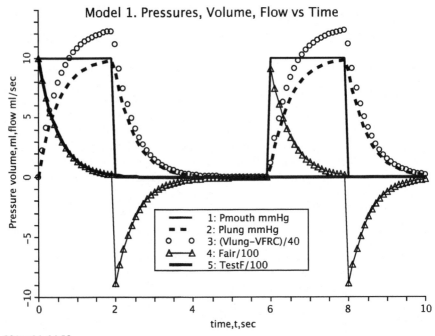

29Aug14, 14:53

Fig. 13.1 Model 1 variables versus time. The forcing ventilator pressure Pmouth is the solid square wave repeated each 6 s. The numerical solutions: *Flow triangles*; Volume of lung: *circles*; Pressure in lung, Plung, dashes. Test is *black line* exp(−t/(Res*com)), a verification test exactly fitting Flow(t) over the first two seconds. Model can be run at www.physiome.org Model #1

Model 2. OneAlvLung.Chest (Physiome #0002) A one-alveolus lung is driven periodically by chest expansion, not the ventilator. Expansion of the chest reduces the pressure in the intrapleural space sucking in air from the mouth. An intrapleural pressure drop of 10 mmHg gives the results as in Model 1.OneAlvLung.Assist.

Question: In this model code one can also use the positive pressure ventilator to assist. Describe the airflow and volume changes if the period of the mechanical ventilator were out of synchrony with the patient's chest-driven breathing.

Model 3. OneAlvLung.IronLung (Physiome #0195) The Iron Lung was used for ventilating polio patients with paralyzed chest muscles. It is a rigid iron tank surrounding the chest and body but leaving the head outside. Exhausting air from the tank every few seconds produced a negative pressure in the tank and expanded the chest passively, so creating a cycling negative intrapleural pressure.

Questions: Is the tidal volume with a tank pressure, Ptank, excursion of -10 mmHg the same as with an external positive pressure ventilator with a ventilator pressure of $+10$ mmHg? What happens when the patient breaths in synchrony with the exhaust phase of the iron lung? Given an initial volume of 20 litres in the tank surrounding the patient's body, and a total chest compliance of 50 ml/mmHg, how much air must be exhausted from the tank to give a normal 500 ml tidal volume on each breath? (Boyle's Law is $PV = nRT$, where $n = $ no. of moles of gas)

Model 4. OneAlvLung.GasExch (Physiome #0003) Inhalation carries a solute gas such as oxygen in the external air, and exhalation carries some of the alveolar contents back to the external air where it is lost. The gases, air and oxygen, have equal velocities. In the model code a reversible switch is set so the gaseous solute flows in the same direction as the ventilatory airflow. If there is no solute gas in the 'lung' initially, then it takes some time for the alveolar concentration to reach an oscillatory steady state; it takes less time at higher cycle rates and higher 'tidal volume' (the size of a breath). Passive oxygen uptake into the body with conductance 'PSO2lung' reduces the alveolar concentration, and would decrease the alveolar volume if the oxygen were not replaced by another gas.

Questions: Why are the switches needed in the computation? What happens if the 'valve' were leaky? How would you change the equations to do this? If inspiration were prolonged with a constant intrapleural pressure, oxygen is taken up into the body, but the alveolar volume is constant, and its pressure is the same as outside the body. What gases replace oxygen? Where do they come from?

Model 5. OneAlvLung.ExchBody.proj (Physiome #0206) Oxygen in the alveolar air exchanges with the pulmonary capillary blood VblLung, and is distributed by cardiac output, Fblood, to the body, here represented by a blood pool, Vbl ml, exchanging (PSO2tiss, ml/min) with a volume of body tissue, Vtiss ml, with consumption Gtiss, ml/min. Gtiss represents the concentration-dependent utilization of mitochondrial O_2 at cytochrome oxidase with an apparent Michaelis Km of about 0.5 mmHg [11]. Ventilation leads to a pseudo-steady-state in which the alveolar concentration PO2lung is less than in the outside air, PO2atmos, because of the consumption and there is no equilibration.

Questions: What parameter can be set so there is no gas uptake and alveolar gases equilibrate with outside air? With PSO2lung > 0 but no blood flowing, what regions equilibrate? With flow greater than 0, and PSO2tiss > 0, but with Gtiss $= 0$, the concentrations in all of the regions equilibrate, PO2atmos $=$ PO2lung $=$ PO2pulcap $=$ PO2tisscap $=$ PO2tiss. Calculate the unidirectional flux of gas in nanomoles/min across the alveolar barrier when Gtiss $= 100$ ml/min? What are the flux-limiting factors? With the default parameters provided, what is the ratio of airway impedance to membrane impedance, working this out from either a mathematical point of view or by changing the model parameters to minimize one impedance or the other?

3   Next Levels of Models

These five elementary models, although so over-simplified that they are almost useless for describing real data, do cover the types of processes involved in delivering an external gas to a site of metabolism: convection in air, permeation to enter blood, convection in the circulation, permeation to enter cells, and reaction. Where they fail is in the details or nuances: the driving forces have the wrong shape, the anatomy is not sufficiently complex, the structures and the processes are heterogeneous, and the domains for exchange are spatially distributed rather than instantaneously mixed lumped compartments. But they are coded in unit-balanced equations (the first step in code verification) and illustrate logically sequenced structuring of the system. They represent particular aspects of functional systems and provide a basis for considering gases inhaled each day, O2 and CO2, NH3, ozone, CO, toluene, etc.

In subsequent models being made available at Physiome.org we provide more detailed and practical descriptions of oxygen transport [3–5, 7–12], substrate transport and metabolism [13, 14]. Next comes mitochondrial consumption of O_2 to form CO_2 with stoichiometry dependent upon the RQ (respiratory quotient). The next set, $PathwayO_2.2$ has a multi-segment airway, humidifies the air, has red blood cells with haemoglobin to bind O_2 and CO_2, bicarbonate to buffer CO_2, O_2 gradients and pH shifts within capillaries, and diffusional resistance to exchange. $PathwayO_2.3mech$ adds dichotomous branching and axially distributed properties of airways, intra-pleural space (even pneumothorax) to provide more realistic mechanics (FEV, normal compliances and volume fractions), and allow pendelluft and other asymmetric situations. $PathwayO_2.3metab$ adds the interactions of O_2 and CO_2, pH, 2,3-DPG and temperature on binding to Hb and Mb, and a simplified oxidative phosphorylation with dependence on P_{O2} and pH. Level 3 teaching models are closer to the advanced research models [15], and reflect the ideas expressed long ago by incorporating spatially distributed exchange at the capillary level [1, 2, 16] and the dispersive nature of convective processes [17] . The design with modular components fosters the use of computationally fast components and allows substituting detailed complex components for the simple ones illustrating the principles. At higher level the models diverge, focusing on specific topics.

4   Conclusions: Teaching, Training, and Researching with Models

For physiologically realistic systems models to be comprehended and used effectively it is critical to convey not only code and instructions in usage but an understanding of the fundamentals. Sequences of models of increasing complexity are useful as open-source tutorials using reproducible component modules. They serve well in introductory classes in quantitative approaches to biology, biochemistry, physiology and pharmacology, and also in designing experiments and testing hypotheses in research.

Acknowledgment   This research has been supported by NIH/BE-01973 and BE-8407, an HL 073598, NSF 0506477, and the VPR project GM094503. B. Jardine installed JSim and the models, which can be downloaded from www.physiome.org. The inspiration for the title comes from the pioneering works of Professor Ewald Weibel, University of Bern [18].

References

1. Carlson BE, Anderson JC, Raymond GR, Dash RK, Bassingthwaighte JB (2008) Modeling oxygen and carbon dioxide transport and exchange using a closed loop circulatory system. In: Kang KA, Harrison DK, Bruley DF (eds) Oxygen transport to tissue XXIX. Springer, New York, pp 353–360
2. Goresky CA (1963) A linear method for determining liver sinusoidal and extravascular volumes. Am J Physiol 204(4):626–640
3. Dash RK, Li Z, Bassingthwaighte JB (2006) Simultaneous blood-tissue exchange of oxygen, carbon dioxide, bicarbonate, and hydrogen ion. Ann Biomed Eng 34:1129–1148
4. Dash RK, Bassingthwaighte JB (2010) Erratum to: blood HbO_2 and $HbCO_2$ dissociation curves at varied O_2, CO_2, pH, 2,3-DPG and temperature levels. Ann Biomed Eng 38:1683–1701
5. Bassingthwaighte JB, Chan JIS, Wang CY (1992) Computationally efficient algorithms for convection-permeation-diffusion models for blood-tissue exchange. Ann Biomed Eng 20(6):687–725
6. Butterworth E, Raymond GM, Jardine B, Neal ML, Bassingthwaighte JB (2013) JSim, an open-source modeling system for data analysis. F1000Res 2:288, 19pp
7. Deussen A, Bassingthwaighte JB (1996) Modeling ^{15}O-oxygen tracer data for estimating oxygen consumption. Am J Physiol Heart Circ Physiol 270:H1115–H1130
8. Li Z, Yipintsoi T, Bassingthwaighte JB (1997) Nonlinear model for capillary-tissue oxygen transport and metabolism. Ann Biomed Eng 25:604–619
9. Beyer RP, Bassingthwaighte JB, Deussen AJ (2002) A computational model of oxygen transport from red blood cells to mitochondria. Comput Methods Programs Biomed 67:39–54
10. Bassingthwaighte JB, Li Z (1999) Heterogeneities in myocardial flow and metabolism: exacerbation with abnormal excitation. Am J Cardiol 83:7H–12H

11. Beard DA (2005) A biophysical model of the mitochondrial respiratory system and oxidative phosphorylation. PLoS Comput Biol 1:252–264
12. Neal ML, Bassingthwaighte JB (2007) Subject-specific model estimation of cardiac output and blood volume during hemorrhage. Cardiovasc Eng 7:97–120
13. Kuikka J, Levin M, Bassingthwaighte JB (1986) Multiple tracer dilution estimates of D- and 2-deoxy-D-glucose uptake by the heart. Am J Physiol Heart Circ Physiol 250:H29–H42
14. Caldwell JH, Martin GV, Raymond GM, Bassingthwaighte JB (1994) Regional myocardial flow and capillary permeability-surface area products are nearly proportional. Am J Physiol Heart Circ Physiol 267:H654–H666
15. Kerckhoffs RCP, Neal ML, Gu Q, Bassingthwaighte JB, Omens JH, McCulloch AD (2007) Coupling of a 3D finite element model of cardiac ventricular mechanics to lumped systems models of the systemic and pulmonary circulation. Ann Biomed Eng 35:1–18
16. Reneau DD, Bruley DF, Knisely MH (1967) A mathematical simulation of oxygen release, diffusion, and consumption in the capillaries and tissue of the human brain. In: Chemical engineering in medicine and biology. Plenum Press, New York, pp 135–241
17. Bassingthwaighte JB (1966) Plasma indicator dispersion in arteries of the human leg. Circ Res 19:332–346
18. Weibel ER (1984) The pathway for oxygen. Structure and function in the mammalian respiratory system. Harvard University Press, Cambridge

Chapter 14
Simulation of Preterm Neonatal Brain Metabolism During Functional Neuronal Activation Using a Computational Model

T. Hapuarachchi, F. Scholkmann, M. Caldwell, C. Hagmann, S. Kleiser, A.J. Metz, M. Pastewski, M. Wolf, and I. Tachtsidis

Abstract We present a computational model of metabolism in the preterm neonatal brain. The model has the capacity to mimic haemodynamic and metabolic changes during functional activation and simulate functional near-infrared spectroscopy (fNIRS) data. As an initial test of the model's efficacy, we simulate data obtained from published studies investigating functional activity in preterm neonates. In addition we simulated recently collected data from preterm neonates during visual activation. The model is well able to predict the haemodynamic and metabolic changes from these observations. In particular, we found that changes in cerebral blood flow and blood pressure may account for the observed variability of the magnitude and sign of stimulus-evoked haemodynamic changes reported in preterm infants.

Keywords Mathematical model • fNIRS • Haemodynamics • Autoregulation • Stimulus – evoked functional response

This chapter was originally published under a CC BY-NC 4.0 license, but has now been made available under a CC BY 4.0 license. An erratum to this chapter can be found at DOI 10.1007/978-1-4939-3023-4_66.

T. Hapuarachchi (✉)
CoMPLEX, University College London, London, UK

Department of Medical Physics and Bioengineering, University College London, London, UK
e-mail: t.hapuarachchi@ucl.ac.uk

F. Scholkmann • S. Kleiser • A.J. Metz • M. Pastewski • M. Wolf
Division of Neonatology, Biomedical Optics Research Laboratory, University Hospital Zurich, Zurich, Switzerland

M. Caldwell • I. Tachtsidis
Department of Medical Physics and Bioengineering, University College London, London, UK

C. Hagmann
Clinic of Neonatology, University Hospital Zurich, Zurich, Switzerland

© The Author(s) 2016
C.E. Elwell et al. (eds.), *Oxygen Transport to Tissue XXXVII*, Advances in Experimental Medicine and Biology 876, DOI 10.1007/978-1-4939-3023-4_14

111

1   Introduction

Our research focuses on the development of a family of computational models of cerebral metabolism, primarily to investigate the effects of stimuli and physiological insults, and to inform the clinical treatment of brain injury. This work has so far centred on human adult [1] and piglet cerebral activity [2]. We have recently extended our focus to the preterm neonatal brain.

A number of studies investigating functional activity in neonates using functional near infrared spectroscopy (fNIRS) have observed different haemodynamic responses. Inconsistent results have been reported in literature regarding the characteristics of stimulus-evoked changes (i.e. magnitude and sign) in oxyhaemoglobin (HbO_2) and deoxyhaemoglobin (HHb). In particular, some studies report a decrease in HHb (an adult-like response) while others report the opposite. In order to research the mechanisms of these responses we have adapted an existing model of adult cerebral metabolism (BrainSignals) [1] to the preterm neonatal brain. In this paper, we (1) present a model of metabolism and haemodynamics in the preterm brain, (2) use the model to simulate observations of two published preterm functional response studies and (3) use the model to predict recently collected data from a stimulus-evoked haemodynamic response study in preterm neonates.

2   Modelling Functional Activation in the Developing Preterm Brain

The original BrainSignals model simulates blood circulation and energy metabolism. It uses a combination of differential equations and algebraic relations to mimic biochemical reactions and processes in a brain cell and the immediate vasculature. The model predicts in particular responses to changes in arterial blood pressure, oxygenation, carbon dioxide levels and functional activation. Figure 14.1 illustrates a simple schematic of the model. This model was adapted to the human neonate by altering a number of physiological parameters known to be significantly different in the young (a method similar to that employed in developing the piglet model [2]). These parameters are listed in Table 14.1. In particular, the reduction of normal arterial blood pressure (BP) was seen to have a significant effect on the behaviour of the model. Figure 14.2a shows the autoregulation curve of the adult model and the neonatal model for preterm neonates, comparable to approximations found in literature [11]. In order to simulate functional activation, the model uses a

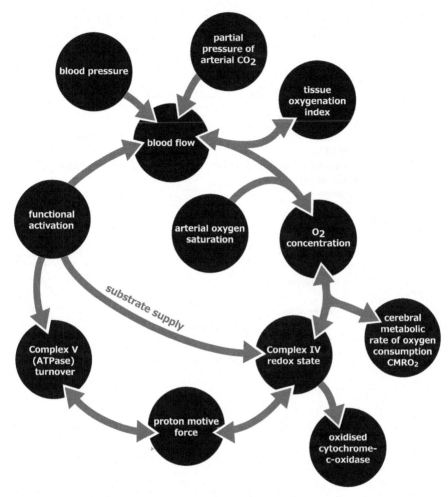

Fig. 14.1 A simple schematic of the model. Model inputs are blood pressure, arterial oxygenation saturation, partial pressure of arterial carbon dioxide and functional activation

dimensionless parameter u which represents demand. A change in u produces a response in vascular smooth muscle and affects ATP production by influencing the driving force for complex V in mitochondria.

Table 14.1 BrainSignals parameters modified to represent the preterm neonatal brain

Parameter	Description	Units	BrainSignals	Preterm neonate	Source
CBF_n	Normal cerebral blood flow (CBF)	ml 100 g^{-1} min^{-1}	49	19.8	[3]
$[CCO]_{tis}$	Total concentration of cytochrome-c-oxidase (CCO) in tissue	μM	5.5	2.2	[4, 5]
$Cu_{A,frac,n}$	Normal fraction of oxidised CCO		0.8	0.67	[5]
$CMRO_{2,n}$	Normal cerebral metabolic rate of oxygen consumption ($CMRO_2$)	μmol 100 g^{-1} min^{-1}	155	40.865	[6]
P_a and $P_{a,n}$	Mean arterial blood pressure	mmHg	100	30	[7]
[Hbtot] and $[Hbtot]_n$	Total concentration of haemoglobin in blood	mM	9.1	9.75	[8] [7]
$V_{blood,n}$	Normal brain blood fraction		0.04	0.0233	[9]
P_{ic} and $P_{ic,n}$	Intracranial pressure	mmHg	9.5	5.1	[10]

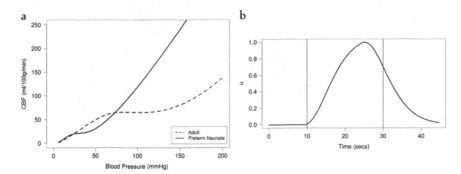

Fig. 14.2 (a) Autoregulation curve—cerebral blood flow (CBF) against blood pressure—for the adult BrainSignals model and the preterm neonate model. (b) Demand as a model input, using a haemodynamic response function, to simulate functional activation

3   Model Simulations

Kozberg et al. [12] conducted a study in postnatal rats age-equivalent to human newborns. Although our model is focused on the human neonate, we were interested in replicating the qualitative response observed in the animals as they were subjected to a somatosensory stimulus. The rats which exhibited a rise in BP in response to the stimulus also showed an increase in HbO$_2$ and total haemoglobin

(HbT) and a decrease in HHb, where the increase in HbO_2 was greater than the decrease in HHb (functional hyperemia). However, some rats showed a slight decrease or no change in BP. In these animals, the opposite results were observed—a decrease in HbO_2 and HbT and a rise in HHb. These conflicting results were attributed to a lack of functional hyperemia and an overarching effect of arterial vasoconstriction in the latter group. In order to model these results, we increased the model's demand parameter u to simulate functional activation with the shape of a steep rising haemodynamic response function (HRF). The demand was calculated as $u = 1.0 + \alpha$ HRF where α is a real number (Fig. 14.2b). We simulated BP also using the HRF ($P_a = P_{a,n} - \beta$ HRF). The values for α and β were optimised to achieve the best fit of our model simulations to the observed results. Arterial oxygen saturation (SaO_2) and partial pressure of carbon dioxide ($PaCO_2$) were assumed to remain constant. The model is able to predict changes in haemodynamics (ΔHbO_2, ΔHHb, ΔHbT) which we compare against experimental data. The model is also capable of simulating cerebral blood flow (CBF) and the cerebral metabolic rate of oxygen consumption ($CMRO_2$). For the first group of animals who showed an increase in blood pressure, we used $\alpha = 2$ and $\beta = 1.5$. Although we attempted to include vasoconstriction as observed in the study, we found that preventing the dilation of the vessels reversed our results. For the second group we reduced the arterial radius by 1 % ($r = r_n (1 - 0.01$ HRF$)$) and used $\beta = 0.18$. We did not need to increase the demand u, suggesting that an increase in oxygen consumption was not required to produce this response. Figures 14.3 and 14.4 illustrate these results, which are very closely comparable to the original data obtained by Kozberg et al. [12].

The second study we investigated was conducted in preterm human neonates by Roche-Labarbe et al. [13]. They observed a decrease in HHb and an increase in HbO_2, CBF, $CMRO_2$ and cerebral blood volume (CBV) in response to a somatosensory stimulus. We were able to replicate these results relatively well by simply increasing the model's demand ($\alpha = 0.5$) (Fig. 14.5). However, our predicted CBF and $CMRO_2$ was higher and HHb slightly lower than that observed.

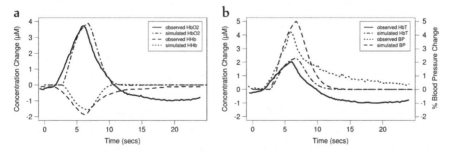

Fig. 14.3 Model simulated and observed haemodynamic response of the Kozberg et al. study [12], investigating functional response in rats, with an increase in demand and blood pressure (BP). (**a**) Changes in deoxy- and oxy- haemoglobin (HHb, HbO2) concentrations. (**b**) Changes in BP and total haemoglobin (HbT)

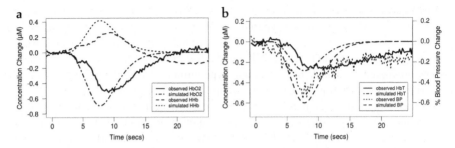

Fig. 14.4 Model simulated and observed haemodynamic response of the Kozberg et al. study [12], investigating functional response in rats, with a slight decrease in arterial radius and blood pressure (BP). (**a**) Changes in deoxy- and oxy- haemoglobin (HHb, HbO2) concentrations. (**b**) Changes in BP and total haemoglobin (HbT)

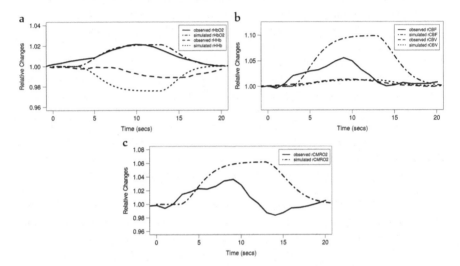

Fig. 14.5 Model simulated and observed haemodynamic response of the Roche-Labarbe et al. study [13], investigating functional response in human preterm neonates, with an increase in demand. (**a**) Relative changes in oxy- and deoxy- haemoglobin (rHbO2, rHHb), (**b**) cerebral blood volume (rCBV) and cerebral blood flow (rCBF) and (**c**) cerebral metabolic rate of oxygen consumption (rCMRO2)

We also simulated data from a functional near infrared spectroscopy (fNIRS) study conducted at University Hospital Zurich (USZ) which investigates the functional response in the preterm brain evoked by a visual stimulus. A blinking pocket LCD display was used as the stimulus and changes in tissue oxygenation and haemodynamics were measured over the prefrontal cortex using a novel spatially-resolved NIRS device (OxyPrem). Measurements of changes in HbO_2, HHb and HbT were averaged over repeated stimuli. Characteristics of two preterm subjects are detailed in Table 14.2.

Table 14.2 Physiological characteristics of the two preterm infant subjects

	Gestational age (weeks)	Actual age (weeks)	Weight (g)	Haematocrit (%)	Haemoglobin (g/dL)	FiO2 (%)	Baseline SpO2 (%)	Baseline heart rate (BPM)
Neonate 1	33.3	33.4	2380	52.5	17.08	21.0	95	148
Neonate 2	25.9	39.0	3460	30.0	9.70	28.0	92–95	165

We assumed that the SaO_2, $PaCO_2$ and BP remain constant in the first instance. In Neonate 1, the measurements showed an increase in ΔHbO_2 and ΔHbT during the stimulus, and a decrease in ΔHHb (Fig. 14.6). We were able to simulate this response in our model by a simple increase in demand ($\alpha = 0.7$). Modelled CBF and $CMRO_2$ showed a corresponding rise during the stimulus.

In Neonate 2, a similar increase in $CMRO_2$, ΔHbO_2 and ΔHbT was observed during the stimulus. However, an increase in ΔHHb was also observed (Fig. 14.7). We were able to simulate this response in ΔHHb by an increase in demand ($\alpha = 0.7$)

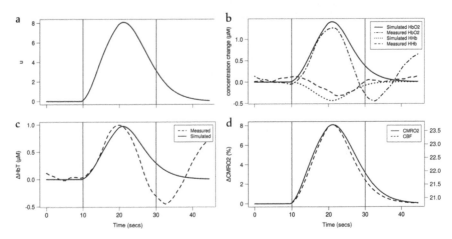

Fig. 14.6 Model simulation of Neonate 1 of the USZ study. Simulated haemodynamic response with (**a**) an increase in demand *u*. *Vertical lines* indicate stimulus period. Measured and simulated changes in (**b**) oxy- and deoxy- haemoglobin ($\Delta HbO2$, ΔHHb) and (**c**) total haemoglobin (ΔHbT). (**d**) Simulated cerebral metabolic rate of oxygen consumption (CMRO2) cerebral blood flow (CBF)

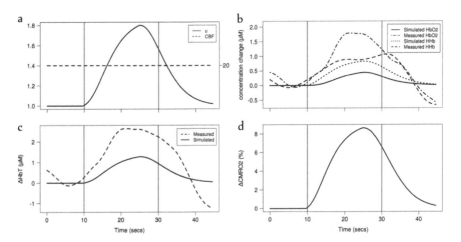

Fig. 14.7 Model simulation of Neonate 2 of the USZ study. Simulated haemodynamic response with (**a**) an increase in demand u and CBF maintained constant. *Vertical lines* indicate stimulus period. Measured and simulated changes in (**b**) oxy- and deoxy- haemoglobin ($\Delta HbO2$, ΔHHb) and (**c**) total haemoglobin (ΔHbT). (**d**) Simulated cerebral metabolic rate of oxygen consumption (CMRO2)

while maintaining CBF constant at its normal value throughout the stimulus. We were able to better predict ΔHbO_2 and ΔHbT by adding a decrease in blood pressure during the stimulus ($P_a = P_{a,n} - \beta\ HRF$ with $\beta = 7$) to match the magnitude of ΔHbO_2 and ΔHbT. However this decrease was too large to be physiologically likely within this timeframe (-7 mmHg).

4 Discussion

The autoregulatory capacity of the preterm neonatal brain remains unclear. However, as our model simulation (Fig. 14.2a) suggests, preterm neonates may be able to maintain constant blood flow only within a very narrow range of blood pressure values. Studies have shown that the response of HHb to a functional stimulus is sometimes 'inverted' in preterm neonates as compared to adults. Our efforts to simulate the haemodynamic responses observed in studies performed by Kozberg et al. [12] and Roche-Labarbe et al. [13] show that the preterm model is capable of simulating the varied functional responses observed. We observed that the model predicted an HHb decrease in response to the stimulus unless we imposed vasoconstriction (as observed by Kozberg et al. [12]). Decreasing the radius of the blood vessel resulted in the 'inverted' response. The model was also able to simulate fNIRS data from the USZ study relatively well. In Neonate 1 we observed a similar response of hyperemia as observed in the Roche-Labarbe et al. study. In Neonate 2, the observed rise in ΔHHb was simulated here by a constant CBF (Fig. 14.7). However, the magnitude of ΔHbO_2 and ΔHbT response was not sufficiently simulated. Neonates 1 and 2 are markedly different in both gestational and actual age (Table 14.2). The former, being older, is more likely to have a developed autoregulatory capacity although we note that both subjects showed an increased HbT response. Their differences in haematocrit and haemoglobin are also notable. Indeed it has been suggested previously that HbT may have an effect on the haemodynamic response in newborns [14]. However, by changing baseline HbT, we did not observe an effect on the magnitude or shape of the model's simulations.

In adapting the model to the human neonate we did not alter the normal radius of the blood vessel and other similar parameters, such as the thickness and muscular tension of the vessel wall. Our current work suggests that these baseline values do not have a significant effect on the results. However, these changes will be made in a forthcoming version of the model.

We aim to further investigate the functional response in neonates using our model. Our initial results suggest that the interaction of many variables affect this response including CBF, BP and the varied stages of development. This makes it very difficult to define a 'normal' functional response for all neonates.

Acknowledgments This work was supported by a UCL-UZH neuroscience collaborative grant. TH is supported by the doctoral training centre CoMPLEX at UCL and IT by the Wellcome Trust (088429/Z/09/Z).

References

1. Banaji M, Mallet A, Elwell CE et al (2008) A model of brain circulation and metabolism: NIRS signal changes during physiological challenges. PLoS Comput Biol 4(11):e1000212
2. Moroz T, Banaji M, Robertson NJ et al (2012) Computational modelling of the piglet brain to simulate near-infrared spectroscopy and magnetic resonance spectroscopy data collected during oxygen deprivation. J R Soc Interface 9(72):1499–1509
3. Greisen G (1986) Cerebral blood flow in preterm infants during the first week of life. Acta Paediatr Scand 75(1):43–51
4. Cooper CE, Springett R (1997) Measurement of cytochrome oxidase and mitochondrial energetics by near-infrared spectroscopy. Phil Trans R Soc Lond B 352:669–676
5. Springett R, Newman J, Cope M et al (2000) Oxygen dependency and precision of cytochrome oxidase signal from full spectral NIRS of the piglet brain. Am J Physiol Heart Circ Physiol 279:H2202–H2209
6. Elwell CE, Henty JR, Leung TS et al (2005) Measurement of CMRO2 in neonates undergoing intensive care using near infrared spectroscopy. Adv Exp Med Biol 566:263–268
7. Polin RA, Fox WW, Abman SH (2011) Fetal and neonatal physiology, 4th edn. Saunders, Philadelphia
8. Jopling J, Henry E, Wiedmeier SE et al (2009) Reference ranges for hematocrit and blood hemoglobin concentration during the neonatal period: data from a multihospital health care system. Pediatrics 123(2):e333–e337
9. Wyatt JS, Cope M, Delpy DT et al (1990) Quantitation of cerebral blood volume in human infants by near-infrared spectroscopy. J Appl Physiol (1985) 68(3):1086–1091
10. Easa D, Tran A, Bingham W (1983) Noninvasive intracranial pressure measurement in the newborn: an alternate method. Am J Dis Child 137(4):332–335
11. Volpe JJ (2008) Neurology of the newborn, 5th edn. Elsevier, Philadelphia
12. Kozberg MG, Chen BR, Deleo SE et al (2013) Resolving the transition from negative to positive blood oxygen level-dependent responses in the developing brain. Proc Natl Acad Sci U S A 110(11):4380–4385
13. Roche-Labarbe N, Fenoglio A, Radhakrishnan H et al (2014) Somatosensory evoked changes in cerebral oxygen consumption measured non-invasively in premature neonates. Neuroimage 85(Pt 1):279–286
14. Zimmermann BB, Roche-Labarbe N, Surova A et al (2012) The confounding effect of systemic physiology on the hemodynamic response in newborns. Adv Exp Med Biol 737:103–109

Chapter 15
Hemoglobin Effects on Nitric Oxide Mediated Hypoxic Vasodilation

Zimei Rong and Chris E. Cooper

Abstract The brain responds to hypoxia with an increase in cerebral blood flow (CBF). However, such an increase is generally believed to start only after the oxygen tension decreases to a certain threshold level. Although many mechanisms (different vasodilator and different generation and metabolism mechanisms of the vasodilator) have been proposed at the molecular level, none of them has gained universal acceptance. Nitric oxide (NO) has been proposed to play a central role in the regulation of oxygen supply since it is a vasodilator whose production and metabolism are both oxygen dependent. We have used a computational model that simulates blood flow and oxygen metabolism in the brain (BRAINSIGNALS) to test mechanism by which NO may elucidate hypoxic vasodilation. The first model proposed that NO was produced by the enzyme nitric oxide synthase (NOS) and metabolized by the mitochondrial enzyme cytochrome c oxidase (CCO). NO production declined with decreasing oxygen concentration given that oxygen is a substrate for nitric oxide synthase (NOS). However, this was balanced by NO metabolism by CCO, which also declined with decreasing oxygen concentration. However, the NOS effect was dominant; the resulting model profiles of hypoxic vasodilation only approximated the experimental curves when an unfeasibly low K_m for oxygen for NOS was input into the model. We therefore modified the model such that NO generation was via the nitrite reductase activity of deoxyhemoglobin instead of NOS, whilst keeping the metabolism of NO by CCO the same. NO production increased with decreasing oxygen concentration, leading to an improved reproduction of the experimental CBF versus PaO_2 curve. However, the threshold phenomenon was not perfectly reproduced. In this present work, we incorporated a wider variety of oxygen dependent and independent NO production and removal mechanisms. We found that the addition of NO removal via oxidation to nitrate

Z. Rong (✉)
Centre for English Language Education, University of Nottingham Ningbo China, Ningbo 315100, China

School of Biological Sciences, University of Essex, Colchester CO4 3SQ, UK
e-mail: zimeirong@hotmail.com

C.E. Cooper
School of Biological Sciences, University of Essex, Colchester CO4 3SQ, UK

© Springer Science+Business Media, New York 2016
C.E. Elwell et al. (eds.), *Oxygen Transport to Tissue XXXVII*, Advances in Experimental Medicine and Biology 876, DOI 10.1007/978-1-4939-3023-4_15

121

mediated by oxyhemoglobin resulted in the optimum fit of the threshold phenomenon by the model. Our revised model suggests, but does not prove, that changes in NO concentration can be the primary cause of the relationship between pO_2 and cerebral blood flow.

Keywords Hemoglobin • Nitric oxide • Hypoxic vasodilation • Nitrite • Nitrite reductase activity

1 Introduction

Hypoxia is a pathological situation in which the whole body or a region of the body is deprived of an adequate oxygen supply. Mostly, the body or part of the body responds to hypoxia in an auto regulation way with vasodilation. The brain is the organ most sensitive to the oxygen tension and consequently there have been many studies reporting the relationship between cerebral blood flow (CBF) and arterial oxygen partial pressure (e.g. [1]).

Many hypotheses at a molecular level (different vasodilators, such as NO, ATP; different production and removal mechanisms of vasodilator) have been suggested to explain the mechanism of hypoxic vasolidation. Whilst strong evidence suggests that nitric oxide (NO) plays a significant role in this response, the formation and removal of this gas are both pO_2 dependent, making predictions as to its level and influence during hypoxia problematic. To quantitatively test mechanistic hypotheses we therefore incorporated NO vasodilation into BRAINSIGNALS [7], a simple model of brainblood flow and metabolism, designed to be informed by non-invasive real time clinical measurements. In theory, NO can be generated and removed in various ways. Some NO generation mechanisms, such as enzymatic generation of NO from nitric oxide synthase (NOS), are unlikely to lead to vasodilation because oxygen is a substrate for NO [2]; therefore NO generation is expected to decrease as pO_2 falls. Some NO removal mechanisms also increase when pO_2 falls and similarly cannot lead to hypoxic vasodilation. However, NO production from nitrite by deoxyhemoglobin [3] and the conversion of NO to nitrite by the oxidised copper centre (CuB) of mitochondrial cytochrome c oxidase [4], both fit the theoretical profile of NO production and consumption mechanisms that could drive hypoxic vasodilation. We previously incorporated NO generation by deoxyhemoglobin [5] and NO removal by oxidised CCO [6] into our BRAINSIGNALS [7] model to reproduce the experimental CBF versus pO_2 curves digitised from literature sources. However, we were still unable to quantitatively reproduce the threshold phenomenon seen in the experimental CBF versus PaO_2 curve.

In this paper we incorporate an additional NO removal mechanism into the BRAINSIGNALS model, the conversion of NO to nitrate by red cell oxyhemoglobin to test how the simulated CBF versus PaO_2 curve improves the fit to the experimental data.

2   Mathematical Model

Although CO_2 is an important vasodilator and hypercapnia, and therefore a major part of BRAINSIGNALS model [7], we limited our current work to pure isocapnic pO_2 effects. Before using a mathematical model to reproduce hypoxic vasolidation we therefore required an experimental isocapnic CBF versus PaO_2 curve. We digitized the experimental data from Brown reporting hypoxic vasolidation in humans [1]. Then we analyzed various NO generation and removal mechanisms and incorporated them into the BRAINSIGNALS model to test their feasibility. Previously we analyzed the CBF versus PaO_2 relationship with NO generated from NOS and found that vasodilation only occurs at extremely low pO_2 with a low K_m [2]. Low K_m indicates constant NO generation. Although the simulation itself did not work well, methodologically the work shows that we can analyze the hypoxic vasodilation in two steps. We can analyze a NO generation mechanism first. If it causes vasodilation, we incorporate that mechanism into the BRAINSIGNALS model to simulate the CBF versus PaO_2 curve and then compare them with the experimental data.

We analyzed NO generation mechanisms, which can be classed into three categories according to whether and how its rate depends on the oxygen tension. (1) NO is generated independently of oxygen concentration, which can be grouped into the constant NO generation group and does not lead to vasodilation. (2) NO is generated where O_2 is a substrate. This mechanism does not lead to vasodilation because NO production rate decreases with decreasing oxygen concentration. (3) NO is generated by reduction of nitrite by deoxyhemoglobin, which leads to vasodilation.

We analyzed NO removal mechanisms. NO is a substrate and is expected to increase significantly after the oxygen concentration decreases down to the threshold as required for the mechanism. The concentration of the other factor causing NO metabolism must decrease when oxygen tension decreases. Previously we tested NO removal by oxidized CCO and were unable to reproduce the threshold phenomenon properly. Here we added another NO removal path by oxyhemoglobin to improve fit of the simulated CBF versus pO_2 curve to the experimental one.

This work is based on the reported BRAINSIGNALS model [7] and recently the NO induced vasodilation [2, 5], which was incorporated into the model. We also follow previous notations [2, 5, 7]. In order to describe NO induced hypoxic vasodilation mathematically we used the Monod, Wyman, Changeux (MWC) model of allosteric transition in hemoglobin. Three parameters in the MWC model [8, 9] are: a dimensionless oxygen concentration $\alpha = pO_2/K_R$, a conformation equilibrium constant $L = T_0/R_0$, and a dissociation constant ratio $c = K_R/K_T$, where K_R and K_T are defined as the microscopic dissociation constants of oxygen from the Relaxed (R) and Tense (T) state quaternary structures of hemoglobin, respectively. Before we express NO generation and metabolism in the model we used the MWC model notation to express hemoglobin oxygen binding.

$$Y = \frac{\alpha(1+\alpha)^3 + Lc\alpha(1+c\alpha)^3}{(1+\alpha)^4 + L(1+c\alpha)^4} = 1 - D_R - D_T \tag{15.1}$$

where D_R and D_T are the fraction of R and T state hemoglobin in the deoxy state.
   Now we can define the deoxyhemoglobin reduction of nitrite to NO as

$$\frac{d[NO]}{dt} = [Hb][NO_2^-](k_R D_R + k_T D_T) = p(k_R D_R + k_T D_T) \tag{15.2}$$

where k_R and k_T are the micro reaction rate constants of the unliganded R-state and T-state heme sites with nitrite, respectively. $p = [Hb][NO_2^-]$ is a parameter defined for modeling. NO is metabolized by oxidized CCO and oxyhemoglobin as

$$\frac{d[NO]}{dt} = -[Cytoa_{3o}][NO]k_{NOreac} \tag{15.3}$$

where k_{NOreac} is the rate constant of NO metabolism by the oxidized enzyme

$$\frac{d[NO]}{dt} = -[Hb][NO](1 - D_R + D_T)k_O = -q[NO](1 - D_R + D_T) \tag{15.4}$$

where $q = [Hb]k_o$ is a parameter defined for modeling.
   The stimuli η is a function of the radius r of a blood vessel and also a function of the delayed concentration of NO, v_{NO}.

$$\tanh\left(\frac{\eta}{2}\right) = \frac{\frac{\left((P_a+P_v)/2-P_{ic}\right)r}{\sqrt{r^2+2r_oh_o+h_o^2}-r} - \sigma_{eo}\left(\exp\left(\frac{K_\sigma(r-r_o)}{r_o}\right)-1\right) + \sigma_{coll}}{\frac{T_{maxo}}{\sqrt{r^2+2r_oh_o+h_o^2}-r}\exp\left(-\left|\frac{r-r_m}{r_t-r_m}\right|^{n_m}\right)} - 1 \tag{15.5}$$

where P_a, P_v and P_{ic} are arterial, venous and extravascular blood pressure respectively. r_0 is the radius in the elastic tension relationship. h_0 is the vascular wall thickness when radius is r_0. σ_e is the elastic stress in vessel wall. σ_{eo} is the parameter in relationship determining σ_e. σ_{coll} is the parameter in the pressure-elastic tension relationship. K_σ is the parameter controlling sensitivity of σ_e to radius r. T_{max0} is a constant. r_m, r_t and n_m are parameters determining the shape of the curve, the muscular force developed in the vessel wall T_m versus the vessel radius r.

$$\eta = R_{NO}\left(\frac{v_{NO}}{v_{NOn}} - 1\right)$$

(15.6)

where the parameter R_{NO} represents the sensitivity to change of NO concentration.

Equations (15.5) and (15.6) therefore define how the blood vessel radius r is linked to the delayed NO concentration, v_{NO}.

3   Results and Discussions

In the first simulation on NO induced hypoxic vasodilation we worked on the human system because of the importance of the work and the experimental data available. In the second simulation we used a dog system because of the quality of the reported experimental data, CBF versus PaO_2 curve. In this paper, we carefully consider three sets of experimental parameters: (1) an observed CBF versus pO_2 curve including the absolute value of normal CBF; (2) The MWC parameters (L, c and K_R) and the Hill parameters (n and p50); (3) the intrinsic rate constants for NO production by R-state and T-state hemoglobin (k_R and k_T). By carefully searching reported data from literature we decide to work on a human model. The experimental CBF versus pO_2 curve was reported by Brown [1] with a normal value of 45 ml/min/100g and was digitized as a reference for our simulation with the x-axis changed from the oxygen content to the oxygen partial pressure. Hemoglobin oxygen affinity was characterized by the Hill parameters, $n = 2.84$ and p50 $= 28$ mmHg [9] and the MWC model parameters $L = 433,894$, $c = 0.00805203$ and $K_R = 0.994333$ mmHg [10]. We were able to obtain the nitrite reductase activity $k_R = 18$ $M^{-1}s^{-1}$ and $k_T = 0.33$ $M^{-1}s^{-1}$ by fitting the nitrite reductase activity of deoxyhemoglobin versus oxygen saturation to experimental data [10]. In our previous work we showed that $k_R = 18$ $M^{-1}s^{-1}$ and $k_T = 6.6$ $M^{-1}s^{-1}$ provided an improved fit [5] and therefore these parameters will be used here.

The deoxyhemoglobin nitrite reductase activity to produce NO is shown in Eq. (15.1) and the oxyhemoglobin and oxidized cytochrome c oxidase metabolism of NO is shown in Eqs. (15.2) and (15.3), respectively. We incorporated the NO production and removal mechanisms into the BRAINSIGNALS model as shown in Fig. 15.1a. An experimental CBF ~ PaO_2 in humans (diamond) was constructed from Brown [1]. We used the BRAINSIGNALS model to simulate hypoxic vasolation as shown in Fig. 15.1b with the parameters in Table 15.1. The solid line in Fig. 15.1b is the simulated CBF ~ PaO_2 curve with NO removed by both oxyhemoglobin and oxidized CCO and the dotted line incorporates NO removal by the oxidized CCO only. We can see that there is now an improvement on the fit, in particular, the threshold phenomenon.

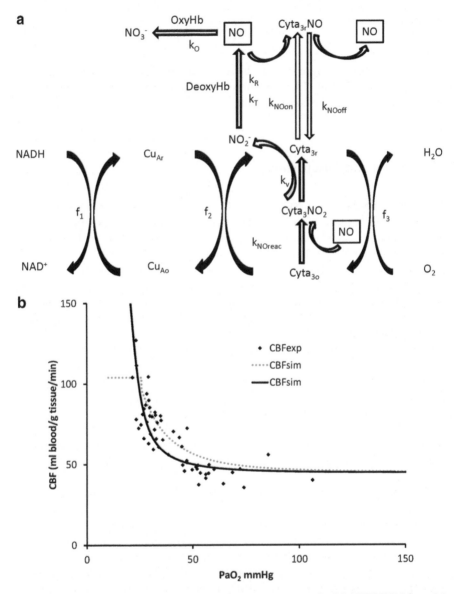

Fig. 15.1 (**a**) Schematic diagram of the changes (*open arrow*) to the BRAINSIGNALS model (*solid arrow*). The chemical reaction rates for NO generation by deoxyhemoglobin are $k_R = 18$ $M^{-1}s^{-1}$ and $k_T = 6.6$ $M^{-1}s^{-1}$ [5]. The chemical reaction rates for competitive inhibition of cytochrome oxidase by NO are $k_{NOon} = 0.04$ $nM^{-1}s^{-1}$ and $k_{NOoff} = 0.16$ s^{-1} [6]. The chemical reaction rates for uncompetitive inhibition are $k_{NOreac} = 2.0 \times 10^{-4}$ $nM^{-1}s^{-1}$ and $k_v = 0.75$ s^{-1} [6]. The chemical reaction rate for NO metabolism by intraerythrocytic oxyhemoglobin is $k_O = 110$ $M^{-1}s^{-1}$, which was obtained from a fit of the simulated CBF versus PaO$_2$ to the experimental data. (**b**) Experimental CBF versus PaO$_2$ (*diamonds*) in humans was reported by Brown [1]. The solid line is the simulated CBF versus PaO$_2$ with NO generated by deoxyhemoglobin and removed by oxyhemoglobin and oxidized CCO (parameters in Table 15.1). The dotted line is the simulated CBF versus PaO$_2$ with NO generated by deoxyhemoglobin and removed by oxidized CCO (parameters in Table 15.1) [5]

Table 15.1 Parameters for reproduction of cerebral blood flow versus oxygen partial pressure curve with the BRAINSIGNALS model

k_R ($M^{-1}s^{-1}$)	k_T ($M^{-1}s^{-1}$)	n	p50 (mmHg)	L	c	K_R (mmHg)
18	6.6	2.84	27.37	433,894	0.00805203	0.994333
p (M^2)	q (s^{-1})	R				
40	0	0.5	Previous			
10	1	2.5	Current			

4 Conclusions

In this paper, we analyzed the NO induced hypoxic vasodilation in two steps. First of all, we semi-quantitatively analyzed various NO generation and metabolisms versus oxygen tension to determine which mechanism could lead to vasodilation. Then, we incorporated the three potential hypoxic vasodilation mechanisms into the BRAINSIGNALS model in an attempt to reproduce the experimental CBF versus pO_2 curve. Our resulting model is consistent with, but does not prove, that NO generated by deoxyhemoglobin and removed by oxyhemoglobin and oxidized mitochondrial cytochrome c oxidase is the cause of hypoxic vasodilation.

Acknowledgment This work is financially supported by the Leverhulme Trust.

References

1. Brown M, Wade J, Marshall J (1985) Fundamental importance of arterial oxygen content in the regulation of cerebral blood flow in man. Brain 108:81–93
2. Rong Z, Banaji M, Moroz T, Cooper CE (2013) Can mitochondrial cytochrome oxidase mediate hypoxic vasodilation via nitric oxide metabolism? Adv Exp Med Biol 765:231–238
3. Gladwin MT, Kim-Shapiro DB (2008) The functional nitrite reductase activity of the heme-globins. Blood 112:2636–2647
4. Palacios-Callender M, Hollis V, Mitchison M et al (2007) Cytochrome c oxidase regulates endogenous nitric oxide availability in respiring cells: a possible explanation for hypoxic vasodilation. Proc Natl Acad Sci U S A 104:18508–18513
5. Rong Z, Cooper CE (2013) Modeling hemoglobin nitrite reductgase activity as a mechanism of hypoxic vasodilation? Adv Exp Med Biol 789:361–368
6. Antunes F, Boveris A, Cadenas E (2007) On the biological role of the reaction of NO with oxidized cytochrome c oxidase. Antioxid Redox Signal 9:1569–1579
7. Banaji M, Mallet A, Elwell CE et al (2008) A model of brain circulation and metabolism: NIRS signal changes during physiological challenges. PLoS Comput Biol 4(11):e1000212. doi:10.1371/journal.pcbi.1000212
8. Monod J, Wyman J, Changeux J (1965) On the nature of allosteric transitions: a plausible model. J Mol Biol 12:88–118
9. Patel RP, Hogg N, Kim-Shapiro DB (2011) The potential role of the red blood cell in nitrite-dependent regulation of blood flow. Cardiovasc Res 89:507–515
10. Rong Z, Wilson MT, Cooper CE (2013) A model for the nitric oxide producing nitrite reductase activity of hemoglobin as a function of oxygen saturation. Nitric Oxide 33:74–80

Chapter 16
A Discussion on the Regulation of Blood Flow and Pressure

Christopher B. Wolff, David J. Collier, Mussadiq Shah, Manish Saxena, Timothy J. Brier, Vikas Kapil, David Green, and Melvin Lobo

Abstract This paper discusses two kinds of regulation essential to the circulatory system: namely the regulation of blood flow and that of (systemic) arterial blood pressure. It is pointed out that blood flow requirements sub-serve the nutritional needs of the tissues, adequately catered for by keeping blood flow sufficient for the individual oxygen needs. Individual tissue oxygen requirements vary between tissue types, while highly specific for a given individual tissue. Hence, blood flows are distributed between multiple tissues, each with a specific optimum relationship between the rate of oxygen delivery (DO_2) and oxygen consumption (VO_2). Previous work has illustrated that the individual tissue blood flows are adjusted proportionately, where there are variations in metabolic rate and where arterial oxygen content (CaO_2) varies. While arterial blood pressure is essential for the provision of a sufficient pressure gradient to drive blood flow, it is applicable throughout the arterial system at any one time. Furthermore, It is regulated independently of the input resistance to individual tissues (local arterioles), since they are regulated locally, that being the means by which the highly specific adequate local requirement for DO_2 is ensured. Since total blood flow is the summation of all the individually regulated tissue blood flows cardiac inflow (venous return) amounts to total tissue blood flow and as *the heart puts out what it receives* cardiac output is therefore determined at the tissues. Hence, regulation of arterial blood pressure is independent of the distributed independent regulation of individual tissues. It is proposed here that mechanical features of arterial blood pressure

C.B. Wolff (✉) • D.J. Collier • M. Shah • M. Saxena • T.J. Brier
Barts and the London School of Medicine and Dentistry, William Harvey Research Institute, Centre for Clinical Pharmacology, William Harvey Heart Centre, Queen Mary University of London, London, UK
e-mail: chriswolff@doctors.org.uk

V. Kapil • M. Lobo
Barts and the London School of Medicine and Dentistry, William Harvey Research Institute, Centre for Clinical Pharmacology, William Harvey Heart Centre, Queen Mary University of London, London, UK

Barts BP Clinic, St Bartholomew's Hospital, Barts Health NHS Trust, London, UK

D. Green
Anaesthetics Department, King's College School of Medicine and Dentistry, London, UK

© Springer Science+Business Media, New York 2016
C.E. Elwell et al. (eds.), *Oxygen Transport to Tissue XXXVII*, Advances in Experimental Medicine and Biology 876, DOI 10.1007/978-1-4939-3023-4_16

129

regulation will depend rather on the balance between blood volume and venous wall tension, determinants of venous pressure. The potential for this explanation is treated in some detail.

Keywords Blood flow • Arterial blood pressure • Tissue response • Regulation • Discussion

1 Introduction

1.1 Blood Flow Regulation

Blood flow regulation is subordinate to the oxygen requirements of individual tissues. For the majority of individual tissues the rate of oxygen delivery (DO_2) is normally sustained at a specific, individual, ratio relationship to oxygen consumption (VO_2) [1–4]. For skeletal muscle the ratio is close to 1.5, for brain 3 and heart 1.6. These ratios are sustained over a moderate physiological range of metabolic rate changes by proportional changes in blood flow. Where arterial oxygen content (CaO_2) is reduced, blood flow increases such that the DO_2 to VO_2 ratio is sustained. For example, for moderate levels of activity, normal skeletal muscle oxygen delivery (1.5 times VO_2) is sustained in the face of reduced oxygen content (CaO_2, either anaemia, hypoxia or both combined), down to levels only 60 % of normal CaO_2 [5]. Compensation for low CaO_2 is also sustained for the cardiac blood supply down to 60 % of normal CaO_2 [3] whereas, for the brain, the lower limit for CaO_2 is around 90 % of normal [6].

The regulation of blood flow, sustaining appropriate rates of oxygen delivery, resides at the tissues. Blood flow regulation at the tissues has been shown to be independent of the nervous system [2]. Guyton's group showed that there is the same 'correlation between local blood flow and arterial oxygen saturation in animals under total spinal anesthesia as seen in normal animals' [7, 8]. An example of the determination of cardiac output and oxygen delivery by the metabolic requirements of tissues is found in the study of exercise by Donald and Shepherd [9]; illustrated in Fig. 16.1.

Here the tissue demand from exercise is met, whether or not cardiac innervation is present. The difference is that there is a much larger stroke volume for the denervated heart, since the heart rate fails to increase normally.

Since there is independent control of blood flow at the tissues, the idea that sympathetic vaso-constrictor nerves are the principal regulator of blood flow by acting on the arterioles or on myocardial contractility [10, 11] requires revision. There is a related problem in view of known arteriolar constriction which occurs with stimulation of the sympathetic nerve supply to arterioles. The main problem, however, is the need for hypotheses addressing means by which blood pressure regulation occurs independently of the regulation of blood flow.

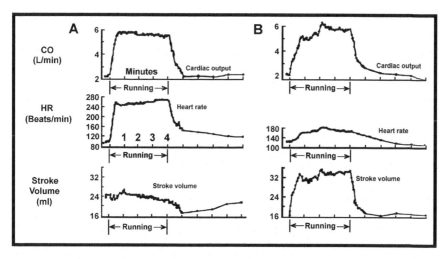

Fig. 16.1 Results of exercise (running) in normal dogs are shown in (**A**), where the cardiac output increase is mainly due to increased heart rate. This contrasts with the cardiac denervated dogs (**B**) where most of the increased cardiac output results from increased stroke volume. After Donald and Shepherd [3, 9], with permission. In both cases cardiac output increased to meet the metabolic rate changes

1.2   The Sympathetic Nervous System and Arterial Blood Pressure Regulation

A wealth of evidence attests to the role of the autonomic nervous system, especially the sympathetic division, in the regulation of arterial blood pressure [11]. There is the well-known short-term regulation, via afferent information from the arterial baro-receptors, with brain stem processing and both parasympathetic and sympathetic efferent output. Where arterial blood pressure rises, or the arterial baro-receptor mechanism is stimulated by some other means, the inhibitory response results in reflex lowering of arterial pressure. There is also the longer term renal mediation, partly via sympathetic efferent innervation, with additional humoral involvement. The renal involvement affects blood volume, which in turn affects arterial blood pressure [12]. There are also effects mediated by volume receptors on the venous side of the circulation. There is a mass of detail available in the literature concerning, in particular, central connections involving major sites, such as the nucleus of the tractus solitarius (NTS), the rostral ventro-lateral medulla (RVLM) and the hypothalamus [11].

It is known that there is sympathetic innervation of the arterioles, causing the theoretical difficulty mentioned above, since the resistance offered by arterioles is adjusted precisely such that DO_2 bears the correct relationship to VO_2. This problem will be considered later. A second known innervation is of the peripheral veins, where sympathetic stimulation raises venous pressure [11, 13]. Furthermore, "small veins and venules are more sensitive to sympathetic activation than arterioles" [14, 15]. The role of changes in venous pressure will be discussed in relation to means by which arterial blood pressure may be regulated independently of blood flow changes.

2   Analyses

2.1   *Arterial and Venous Blood Volumes*

Altered Venous Capacity

The dominant role of peripheral venous pressure in determining arterial BP is discussed in Guyton et al. [2]. In relation to the interdependence of arteries and veins it is worth considering the relative volumes of the whole arterial tree and that of both the venous and arterial systems. Figure 16.2 illustrates the approximately five- to six-fold difference in venous and arterial volumes.

Assuming the venous system volume is five times the arterial volume, knowing the venous and arterial percentages, we can calculate the effect of small changes in venous volume. Starting with the idea of a total blood volume of 5 l, the arterial volume, at 12 %, is 600 ml, whereas the venous volume, at 60 %, will be 3000 ml. If we reduce the venous volume by 150 ml, and this is transferred into the arteries, it will have increased arterial volume by 25 % (to 750 ml) and thus arterial blood pressure might be expected to increase by more than 25 % since arterial compliance decreases as its volume increases. The initial action, reducing venous volume, would require reduction of venous wall compliance (or increase in wall tension or 'tone'). Hottenstein and Kreulen [15] emphasize the particular role of the sympathetic activation here. Although this is a very simple model it shows how we would expect a small change in venous volume to generate a large change in arterial blood pressure.

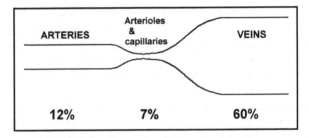

Fig. 16.2 The relative volumes of arteries and veins, considered as percentages of total blood volume, constitute a total of around 72 % [10]. Arteries constitute 12 % and veins 60 % of total vascular capacity (Elsewhere venous volume is said to constitute 70 % [14]

Altered Blood Volume

If the venous wall initial 'tone' remains unaltered but blood volume increases, for whatever reason, this will stretch the vein walls and the pressure on the venous side of the circulation will increase. This will also cause a re-distribution, with a sharing of the extra volume between arteries and veins, according to their relative compliance. While mean arterial compliance is much lower than venous any extra fluid will raise arterial pressure.

2.2   Arteriolar Interaction Between Tissue Control and Sympathetic Innervation

It is known that there is sympathetic innervation of both arteriolar and peripheral venous wall smooth muscle and, as pointed out above, this raises the question as to how the tissues sustain precisely appropriate blood flows in the face of alterations in sympathetic drive to arterioles. It is possible that, under particular circumstances, there is local modulation of arteriolar and venous tone, such that the tissue concerned still achieves the appropriate input resistance. Overall sympathetically mediated increases in arteriolar tone could be compatible with facilitating appropriate blood flow in individual tissues, under conditions of altered systemic arterial blood pressure. In other words the sympathetic efferent input/output reflex may well function specifically at locations where the situation changes, such as in the leg vasculature with postural change (such as standing up from the sitting position). Postural change alters the head of pressure [16]. A second example is the classic effect of the Valsalva manoeuvre in normal subjects. Here, blood flow is strongly opposed by the raised intra-thoracic pressure since arterial blood pressure falls to low levels. Recovery of blood flow is assumed to result from the restoration of systemic arterial blood pressure towards the end of the manoeuvre. This is attributed to the well documented sympathetically driven increase in tone in splanchnic and renal veins, with restoration of arterial blood pressure. Recent unpublished work shows that blood flow is also restored as arterial blood pressure recovers in the late stage of the Valsalva manoeuvre.

Hence we have the possibility that changes in arteriolar resistance mediated by sympathetic innervation can be over-ridden by the tissue concerned and that situations can occur where tissue blood flow control is facilitated. During exercise venous tone is presumably greater than at rest, avoiding dilatation from the increased blood flow. This area remains open to further discussion and investigation.

One of the situations where abnormal blood volume re-distribution occurs is with the induction of anaesthesia. There is known reduction of sympathetic outflow [17–19] and cardiac output and arterial blood pressure fall together. This is a situation where the fall in arterial blood pressure is entirely due to relaxation of the venous system walls, since there has been no change in blood volume. The relaxed venous system having a larger capacitance (with increased compliance/reduced tone) results in a redistribution of the blood volume. The small loss of arterial volume will have a disproportionate effect in lowering the arterial blood pressure. It also means that the tissue requirement becomes deficient with artificial reduction of cardiac output (slower arrival of tissue output at the heart, and less blood in the arterial system to sustain perfusion). Abnormal blood volume distribution is also recognised in the setting of arteriovenous (AV) fistulae or large AV malformations and is associated with reduction in arterial blood pressure. Here the primary mechanism is the reduction in total systemic vascular resistance resulting in a modicum of arterial blood loss to the veins increasing venous volume, rather than any change in venous wall tone. Here cardiac output is increased. This finding has been exploited to therapeutic benefit in a novel technology to treat hypertension: the ROX arteriovenous coupler [20].

3   Discussion

High venous pressures [21] occur in hypertension [12]. Blood volume and peripheral venous pressure are critical determinants of arterial blood pressure regulation ([2]; and many others). Aspects of regulation of arterial blood pressure have been measured in great detail over many years. On the other hand, similarly, a great deal of work has been undertaken on blood flow, including the well-known, auto-regulation with little change in organ blood flow in the face of widely varying arterial blood pressure.

There has been little data available to relate blood flow and arterial blood pressure regulation to each other until recent work supporting and augmenting that of Guyton et al. [2] on the tendency for oxygen supply and metabolic rate to be related. Once it was clear that the rate of oxygen delivery for most tissues was held at a precise ratio to consumption it was apparent that the rate of oxygen extraction (VO_2/DO_2) was usually constant. This is simply the inverse of DO_2/VO_2, found to be constant for most tissues. That meant that one could assess this property indirectly from knowledge of oxygen extraction. For example cerebral oxygen extraction can be calculated from near infra-red spectroscopic data and arterial oxygen saturation [6].

The crucial point favoring a new way of examining arterial blood pressure regulation, the loss of availability of blood flow change, limits the problem. This paper has not shown any definitive model of arterial blood pressure regulation, merely made suggestions concerning some of the aspects of mechanical responses which seem to be needed to provide it. It is hoped that a solution to the problem of the regulation of arterial blood pressure will emerge from integration of presently available data and any further measurements which emerge as necessary for an adequate description.

References

1. Starling EH (1923) The wisdom of the body: the Harveian oration. Br Med J ii:685–690
2. Guyton AC, Jones CE, Coleman TG (1973) Circulatory physiology: cardiac output and its regulation. W. B. Saunders, Philadelphia
3. Wolff CB (2007) Normal cardiac output, oxygen delivery and oxygen extraction. Adv Exp Med Biol 599:169–182
4. Wolff CB (2013) Oxygen delivery: the principal role of the circulation. Adv Exp Med Biol 789:37–42. doi:10.1007/978-1-4614-7411-1_6
5. Wolff CB (2003) Cardiac output, oxygen consumption and muscle oxygen delivery in submaximal exercise: normal and low O_2 states. Adv Exp Med Biol 510:279–284
6. Wolff CB, Richardson N, Kemp O, Kuttler A, McMorrow R, Hart N, Imray CHE (2007) Near infra-red spectroscopy and arterial oxygen extraction at altitude. Adv Exp Med Biol 599:183–189
7. Ross JM, Linhart JW, Guyton AC (1962) Autoregulation of blood flow by oxygen lack. Am J Physiol 202:21–24
8. Ross JM, Fairchild HM, Weldy J et al (1962) Autoregulation of blood flow by oxygen lack. Am J Physiol 202(1):21–24
9. Donald DE, Shepherd JT (1964) Initial cardiovascular adjustment to exercise in dogs with chronic cardiac denervation. Am J Physiol 207(6):1325–1329
10. Mohrman DE, Heller LJ (1986) Cardiovascular physiology, 2nd edn. McGraw-Hill, New York
11. Guyenet PG (2006) The sympathetic control of blood pressure. Nat Rev Neurosci 7:335–346
12. Yamamoto J, Trippodo NC, Ishise S, Frolich ED (1980) Total vascular pressure-volume relationship in the conscious rat. Am J Physiol 238:H823–H828
13. Fink GD (2009) Sympathetic activity, vascular capacitance, and long-term regulation of arterial pressure. Hypertension 53:307–312
14. King AJ, Fink GD (2006) Chronic low-dose angiotensin ii infusion increases venomotor tone by neurogenic mechanisms. Hypertension 48:927–933. doi:10.1161/01.HYP.0000243799.84573.f8
15. Hottenstein OD, Kreuken DL (1987) Comparison of the frequency dependence of venous and arterial responses to sympathetic nerve stimulation in guinea pigs. J Physiol 384:153–167
16. Levick JR (1991) An introduction to cardiovascular physiology. Butterworth, New York/London
17. Ebert TJ, Muzi M, Berens R et al (1992) Sympathetic responses to induction of anesthesia in humans with propofol or etomidate. Anesthesiology 76(5):725–733
18. Elliott J (1997) Alpha-adrenoceptors in equine digital veins: evidence for the presence of both alpha$_1$ and alpha$_2$-receptors mediating vasoconstriction. J Vet Pharmacol Ther 20:308–317
19. Goodchild CS, Serrao JM (1989) Cardiovascular effects of propofol in the anaesthetized dog. Br J Anaesth 63:87–92
20. Lobo MD, Sobotka PA, Stanton A et al (2015) Central arteriovenous anastomosis for the treatment of patients with uncontrolled hypertension (the ROX CONTROL HTN study): a randomised controlled trial. Lancet 385(9978):1634–1641
21. Sharpey-Schaffer EP (1963) Venous tone: effects of reflex changes, humoral agents and exercise. Br Med J 19(2):145–148

Part III
Critical Care Adult

Chapter 17
Near Infrared Light Scattering Changes Following Acute Brain Injury

David Highton, Ilias Tachtsidis, Alison Tucker, Clare Elwell, and Martin Smith

Abstract Acute brain injury (ABI) is associated with changes in near infrared light absorption reflecting haemodynamic and metabolic status via changes in cerebral oxygenation (haemoglobin oxygenation and cytochrome-c-oxidase oxidation). Light scattering has not been comprehensively investigated following ABI and may be an important confounding factor in the assessment of chromophore concentration changes, and/or a novel non-invasive optical marker of brain tissue morphology, cytostructure, hence metabolic status. The aim of this study is to characterize light scattering following adult ABI. Time resolved spectroscopy was performed as a component of multimodal neuromonitoring in critically ill brain injured patients. The scattering coefficient (μ'_s), absorption coefficient and cerebral haemoglobin oxygen saturation (SO_2) were derived by fitting the time resolved data. Cerebral infarction was subsequently defined on routine clinical imaging. In total, 21 patients with ABI were studied. Ten patients suffered a unilateral frontal infarction, and mean μ'_s was lower over infarcted compared to non-infarcted cortex (injured 6.9/cm, non-injured 8.2/cm p = 0.002). SO_2 did not differ significantly between the two sides (injured 69.3 %, non-injured 69.0 % p = 0.7). Cerebral infarction is associated with changes in μ'_s which might be a novel marker of cerebral injury and will interfere

---

This chapter was originally published under a CC BY-NC 4.0 license, but has now been made available under a CC BY 4.0 license. An erratum to this chapter can be found at DOI 10.1007/978-1-4939-3023-4_66.

D. Highton (✉)
Neurocritical Care, University College Hospitals, Queen Square, London, UK
e-mail: d.highton@ucl.ac.uk

I. Tachtsidis • A. Tucker
Medical Physics and Bioengineering, University College London, Malet Place, London, UK

C. Elwell
Department of Medical Physics and Biomedical Engineering, University College London, London, UK

M. Smith
Neurocritical Care, University College Hospitals, Queen Square, London, UK

Medical Physics and Bioengineering, University College London, Malet Place, London, UK

NIHR University College London Hospitals Biomedical Research Centre, London, UK

© The Author(s) 2016
C.E. Elwell et al. (eds.), *Oxygen Transport to Tissue XXXVII*, Advances in Experimental Medicine and Biology 876, DOI 10.1007/978-1-4939-3023-4_17

with quantification of haemoglobin/cytochrome c oxidase concentration. Although further work combining optical and physiological analysis is required to elucidate the significance of these results, μ'_s may be uniquely placed as a non-invasive biomarker of cerebral energy failure as well as gross tissue changes.

Keywords Near infrared spectroscopy • Brain injury • Scattering • Cerebral ischaemia

1   Introduction

The optical characteristics of cerebral tissues following acute brain injury (ABI) have been of considerable clinical interest, because they can be exploited to interrogate cerebral oxygenation in-vivo, non-invasively. An optical window in the near infrared spectrum (700–900 nm) facilitates measurement of the dominant absorbing chromophores in this region—oxy/deoxy-haemoglobin and cytochrome c oxidase, by their relative specific absorption spectra—thus inferring information about the haemodynamic and metabolic status of the brain [1]. However light transport through complex biological media is highly dependent on light scattering as well as absorption, which has not been comprehensively investigated following ABI, and may be an important confounding factor in the assessment of chromophore concentration changes [2], and/or a novel non-invasive optical marker of brain tissue morphology, cytostructure, hence metabolic status [3]. The aim of this study is to characterize light scattering in brain tissue following adult ABI.

2   Methods

A total of 21 critically ill, ventilated brain injured patients were recruited following ethical approval and representative consent. Time resolved spectroscopy (TRS-20, Hamamatsu Photonics KK) was performed as a component of multimodal neuromonitoring in critically ill brain injured patients. Serial recordings were taken whilst in critical care. The reduced scattering coefficient (μ'_s), absorption coefficient, cerebral haemoglobin oxygen saturation (SO$_2$) and total haemoglobin concentration ([HbT]) were derived by fitting the time resolved data (diffusion equation for light transport in a semi-infinite homogeneous medium, fitting the entire temporal point spread function, as standard within the Hamamatsu software). Three bilateral time-resolved recordings were made over frontal cortex at 4 cm source detector separation with a 5 s acquisition time. The mean of these three recordings were used for comparison. Cerebral infarction was subsequently defined on routine clinical imaging and paired comparison of μ'_s was performed in patients with unilateral infarction, using the paired t-test. μ'_s

The wavelength dependence of μ'_s was analysed using least squares regression from the group data, across the three wavelengths measured (761, 801 and 834 nm), the 95 % confidence intervals and F-statistic is reported. To facilitate comparison with the literature three previously described models were fitted: $\mu'_s(\lambda) = a\lambda + b$,

$\mu_s(\lambda) = b\lambda^{-a}$ normalized to the mean scattering value at 801 nm as in Matcher et al. [4], and $\mu'_s(\lambda) = b(\lambda/500nm)^{-a}$ from Jacques et al. [5].

3   Results

Patients' characteristics are summarised in Table 17.1 and the μ'_s and SO_2 data are summarised in Fig. 17.1. Ten patients suffered a unilateral frontal infarction, and mean μ'_s was lower over infarcted compared to non-infarcted cortex (Table 17.2). SO_2 (injured 69.3 %, non-injured 69.0 % p = 0.7) and [HbT] (injured 64.6 μmol^{-1}, non-injured 51.9 μmol^{-1} p = 0.09) did not differ significantly. The time course of μ'_s data is shown in Fig. 17.1 suggesting a trend of increasing μ'_s with respect to time in the infarct group:- however the patient group is heterogeneous, as is the onset of infarction.

The wavelength dependence of μ'_s and fitted parameters are shown in Table 17.3. Using the linear model there was no significant difference in wavelength dependence defined by parameter a (injured -9.3×10^{-3}, non-injured -9.2×10^{-3} p = 0.88)

4   Conclusions

Cerebral infarction is associated with significant reduction in μ'_s below previously reported values for normal adults [4]. Unilateral infarction was not associated with a similar difference in SO_2 or [HbT] indicating either less discriminatory ability of these complex physiological variables or error in the assumptions underlying their derivation. The wavelength dependence of μ'_s was approximately linear over the narrow band of wavelengths studied and revealed similar parameter estimates to other reports of cerebral tissues in vivo [5] and in vitro [6]. This did not vary between injured and non-injured cortex. Clearly these findings have potentially important implications for analysis using differential spectroscopy over injured brain as variation in μ'_s violates assumptions required to calculate concentration changes.

The physiological origin of μ'_s reduction is also of considerable interest. Studies on exposed or in vitro cerebral cortex identify subcellular structures as the dominant scatterers in the near infrared spectrum with the mitochondria as a key contributor—given its typical size of 500–1000 nm. Thus scattering has been used experimentally to define mitochondrial volume and density—hence metabolic compromise [3]. Translation of this paradigm into the clinic might hold

Table 17.1 Patient characteristics. Reported as median [quartile] or number (percentage)		
	Age (years)	58 [47, 70]
	Diagnosis	
	Subarachnoid haemorrhage (%)	11 (52)
	Intracerebral haemorrhage (%)	8 (38)
	Ischaemic stroke (%)	1 (5)
	Traumatic brain injury (%)	1 (5)
	In-hospital mortality (%)	4 (19)

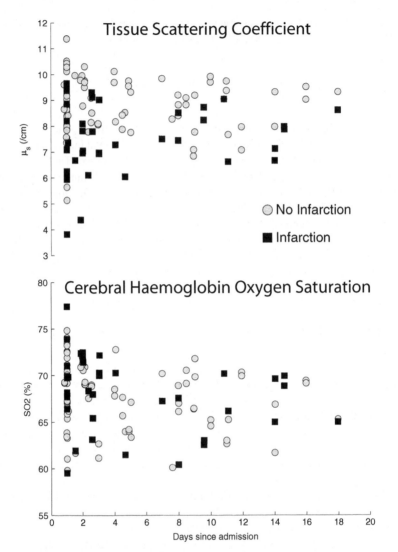

Fig. 17.1 Mean μ'_s (834 nm) and SO_2 for each day post intensive care admission in cerebral hemispheres demonstrating infarction versus no infarction. The lower scattering in infarcted cerebral hemisphere is evident, whereas there is no appreciable pattern with SO_2

Table 17.2 μ'_s measured over infarcted and non-infarcted cortex

	μ'_s761 nm (cm^{-1})	μ'_s801 nm (cm^{-1})	μ'_s834 nm (cm^{-1})
Infarct	7.43 (1.74)	7.18 (1.64)	6.87 (1.57)
No infarct	8.84 (1.51)	8.52 (1.45)	8.20 (1.43)
p-value	0.003	0.003	0.002

Brackets denote standard deviation

Table 17.3 Wavelength dependence of μ'_s fitted to three established models

	$\mu'_s(\lambda) = a\lambda + b$	$\mu'_s(\lambda) = b\lambda^{-a}$	$\mu'_s(\lambda) = b(\lambda/500\text{nm})^{-a}$
a (cm^{-1} nm^{-1})	-0.0085 (-9.0×10^{-3} to -8.1×10^{-3})	0.77 (0.73–0.81)	0.77 (0.31–1.22)
b (cm^{-1} nm^{-1})	15.6 (15.3–16.0)	1485 (1134–1942)	12.4 (10.1–12.5)
F-statistic	1442	1424	11.14
p-value	<0.0001	<0.0001	0.001

Brackets denote the 95 % confidence limits of the model parameters. All models fit the data significantly

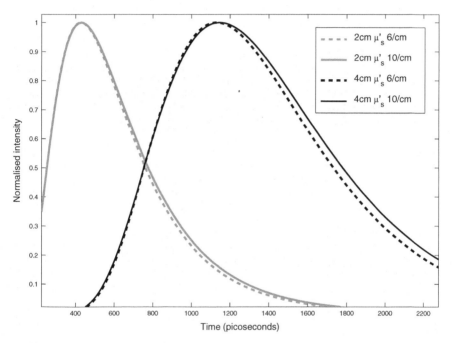

Fig. 17.2 Two-layer model simulation of the effect of cerebral μ'_s on TPSF data at source detector separations of 2 and 4 cm. It can be seen that the TPSF has much greater sensitivity to μ'_s changes in the deep layer at 4 cm source detector separation than 2 cm. A two-layer semi-infinite slab geometry was used in this simulation, layer 1 (1 cm depth, μ_a 0.1/cm and μ'_s 10/cm), layer 2 (5 cm depth, μ_a 0.1/cm and μ'_s 6/cm or 10/cm) using a model from Liemert et al. [8]

considerable promise, but it is important to consider the differences in measurement technique and underlying assumptions fitting the time resolved data. Specifically fitting the time resolved temporal point spread function (TPSF) assuming a homogenous semi-infinite medium may be an oversimplification of cerebral structure following brain injury and especially after surgical intervention. A specific issue which has been discussed is the sensitivity of the TPSF to μ'_s changes in the brain, rather than overlying extracerebral tissues. Simulated photon diffusion considering a two-layer structure and 2 cm source detector separation predicts little sensitivity in the TPSF to μ'_s below 1 cm depth [7] implying that much of our findings could be superficial tissue changes. However repeating simulations in a simple two-layer slab geometry shows greater sensitivity of the TPSF to changes in scattering to the 4 cm source detector separation we have used (Fig. 17.2).

Further work is required to identify the physiological significance of our findings. Individualized optical modeling based on anatomical data from radiological imaging is feasible, and may define the optical situation with further clarity. Combining optical and physiological modeling may hold the key to elucidate individualized physiological explanations for our multimodal optical and physiological data. Thus optical scattering might have considerable potential as a novel independent optical signal of cerebral ischaemia/infarction following ABI, by characterising the resultant changes in mitochondrial morphology and cellular structure in vivo—an approach which appears more discriminatory than traditional haemoglobin based measures.

Acknowledgments This work was supported by the Wellcome Trust undertaken at University College London Hospitals and partially funded by the Department of Health's National Institute for Health Research Centres funding scheme via the UCLH/UCL Biomedical Research Centre. The authors are indebted to the medical and nursing staff of the Neurocritical Care Unit at the National Hospital for Neurology & Neurosurgery, and to the study patients and their families. We would like to thank Hamamatsu Photonics for the loan of their spectrometer.

References

1. Ghosh A, Elwell CE, Smith M (2012) Review article: cerebral near-infrared spectroscopy in adults: a work in progress. Anesth Analg 115:1373–1383
2. Metz AJ, Biallas M, Jenny C et al (2013) The effect of basic assumptions on the tissue oxygen saturation value of near infrared spectroscopy. Adv Exp Med Biol 765:169–175
3. Johnson LJ, Chung W, Hanley DF et al (2002) Optical scatter imaging detects mitochondrial swelling in living tissue slices. Neuroimage 17:1649–1657
4. Matcher SJ, Cope M, Delpy DT (1997) In vivo measurements of the wavelength dependence of tissue-scattering coefficients between 760 and 900 nm measured with time-resolved spectroscopy. Appl Opt 36:386–396
5. Torricelli A, Pifferi A, Taroni P et al (2001) In vivo optical characterization of human tissues from 610 to 1010 nm by time-resolved reflectance spectroscopy. Phys Med Biol 46:2227–2237
6. Jacques SJ (2013) Optical properties of biological tissues: a review. Phys Med Biol 58:R37–R61
7. Selb J, Ogden TM, Dubb J, Fang Q et al (2014) Comparison of a layered slab and an atlas head model for Monte Carlo fitting of time-domain near-infrared spectroscopy data of the adult head. J Biomed Opt 19:16010–016010
8. Liemert A, Kienle A (2010) Light diffusion in N-layered turbid media: frequency and time domains. J Biomed Opt 15:025002

Chapter 18
Comparison of Cerebral Oxygen Saturation and Cerebral Perfusion Computed Tomography in Cerebral Blood Flow in Patients with Brain Injury

Alexey O. Trofimov, George Kalentiev, Oleg Voennov, and Vera Grigoryeva

Abstract The purpose of this study was to determine the relationship between cerebral tissue oxygen saturation and cerebral blood volume in patients with traumatic brain injury. Perfusion computed tomography of the brain was performed in 25 patients with traumatic brain injury together with simultaneous $SctO_2$ level measurement using cerebral near-infrared oxymetry. The mean age of the injured persons was 34.5 ± 15.6 years (range 15–65); 14 men, 11 women. The Injury Severity Score (ISS) values were 44.4 ± 9.7 (range 25–81). The Glasgow Coma Score (GCS) mean value before the study was 10.6 ± 2.1 (range 5–13). $SctO_2$ ranged from 51 to 89 %, mean 62 ± 8.2 %. Cerebral blood volume (CBV) values were 2.1 ± 0.67 ml/100 g (min 1.1; max 4.3 ml/100 g). Cerebral blood flow (CBF) was 31.99 ± 13.6 ml/100 g $\times$ min. Mean transit time (MTT) values were 5.7 ± 4.5 s (min 2.8; max 34.3 s). The time to peak (TTP) was 22.2 ± 3.1 s. A statistically significant correlation was found between $SctO_2$ level and cerebral blood volume (CBV) level ($R = 0.9$; $p < 0.000001$). No other significant correlations were found between brain tissue oxygenation and other parameters of brain perfusion.

Keywords Head injury • NIRS • Perfusion CT

1 Introduction

The reduction of mortality from severe intracranial hemorrhages achieved in the last decade is associated by many authors with the capability of the timely detection and correction of cerebral ischemia.

A.O. Trofimov (✉) • G. Kalentiev • O. Voennov • V. Grigoryeva
Department of Neurosurgery, Nizhniy Novgorod State Medical Academy, 10/1, Minin str., Nizhniy Novgorod 603950, Russia
e-mail: xtro7@mail.ru

© Springer Science+Business Media, New York 2016
C.E. Elwell et al. (eds.), *Oxygen Transport to Tissue XXXVII*, Advances in Experimental Medicine and Biology 876, DOI 10.1007/978-1-4939-3023-4_18

According to Taussky [1] the most informative methods, which enable to reveal cerebral ischemia, are perfusion computed tomography (CT) and cerebral near-infrared spectroscopy (NIRS).

The principle of the perfusion CT consists in the quantitative measurement of changes in the X-ray density of tissue during the passage of an intravenously administered contrast agent [2]. Its theoretical backgrounds were elaborated by Axel as early as 1979, but the practical implementation only became possible with the introduction of multi-slice spiral scanners with more than eight detectors [3].

The implementation of multi-slice CT-scanners has enabled to assess the cerebral blood flow both in a particular site and in the brain as a whole [4, 5]. It should be noted that, although the NIRS is not capable of the direct assessment of the cerebral blood flow, it shows the hemoglobin oxygen saturation in all parts of the vascular bed in a certain part of the brain. NIRS enables to make quantitative assessment of the reduction of oxyhemoglobin in the capillary and intravenous bed. These events have been considered, and the relationship between cerebral tissue oxygen saturation ($SctO_2$) and the state of the cerebral microcirculation seems to be the most probable. However, there are few studies on this issue and they relate only to acute cerebrovascular accidents [1].

The aim of the investigation was to compare the cerebral tissue saturation and perfusion data and to analyse the relationship between the state of the cerebral blood flow and cerebral oxygenation in patients with head injury.

2   Materials and Methods

We studied the results of cerebral perfusion CT and cerebral tissue oxygen saturation measurements using NIRS simultaneously performed for 25 patients with head injury who were treated at the Nizhny Novgorod Regional Trauma Center in 2011–2012. The mean age of the patients with head injury was 34.5 ± 15.6 (range 15–65 years); there were 11 women and 14 men. The wakefulness level according to the Glasgow Coma Score (GCS) averaged 10.6 ± 2.1 (range 5–13) before the study. The severity of their state according to the Injury Severity Score (ISS) was 44.4 ± 9.7 (min score 25, max 81). Cerebral oximetry was performed using the Fore-Sight 2030 MS (CAS Medical Systems Inc., Branford, CT, USA) and included the detection of the $SctO_2$ level in the region of the frontal lobe pole by standard device according to the manufacturer's recommendations. Figure 18.1 presents a photograph of the set-up.

A total of 25 patients underwent perfusion CT by 64-slice tomograph Toshiba Aquilion TSX-101A (Toshiba Medical Systems, Europe BV, the Netherlands). The perfusion examination report included an initial contrast-free CT of the brain. The so-called 'area of interest' was established based on detected areas of focal traumatic injuries.

Also performed were four extended scanning of such areas of interest, 32 mm in thickness, within 55 s with a contrast agent administered (the Brain Perfusion

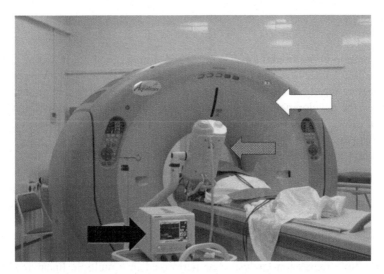

Fig. 18.1 Photograph of the system. The *white arrow* indicates the computer tomograph, the *black arrow* indicates the cerebral oximeter, and the *other arrow* indicates the syringe-injector

mode). The scanning parameters were: 120 kVp, 70 mA, 70 mAs, 1000 ms. The contrast agent (Ultravist 370, Shering AG, Germany) was administered with an automatic syringe-injector (Stellant, One Medrad, Indianola, PA, USA) into a peripheral vein through a standard catheter (20G) at a rate of 4–5 ml/s in a dose of 30–50 ml per single examination.

In order to verify the causes of the impaired cerebral perfusion, the CT angiography of cerebral and great cervical vessels was performed for all the patients after the perfusion examination.

After scanning the data were transferred to a work station Vitrea 2 (Vital Imaging, Inc., ver. 4.1.8.0). Artery and vein marks were automatically recorded followed by the manual control of indices in the Time-Concentration Diagram.

Then, color-coded perfusion charts were automatically constructed, which described values of the regional blood flow velocity (rCBF), regional circulating blood volume (rCBV), mean contrast transit time (MTT), and time to peak contrast concentration (TTP). Those parameters were qualitatively and quantitatively evaluated in all cerebral lobes: frontal, temporal, parietal, occipital lobes and basal ganglia areas. Particular attention was paid to perfusion parameters in the cortical areas of the frontal lobes at a depth up to 25 mm immediately adjacent to optosensor locations. Statistical analysis was performed using methods of nonparametric statistics, and regression analysis. The program Statistica 7.0 (StatSoft Inc., USA, 2004) was used for analyses. A level of ≤ 0.01 was considered to be significant.

3   Results

The saturation in the frontal lobe tissue ranged from 51 to 89 %; it was 62 ± 8.2 % over the left frontal lobe and 61 ± 5.2 % over the right frontal lobe. Values below the ischemic threshold ($SctO_2$ less than 60 %) were detected in five patients. This indicated the development of the ischemia in the area of perfusion by the anterior cerebral artery. The total analysis of PCT data revealed the following values of perfusion parameters.

Cerebral blood flow values ranged from 11.6 to 80.8 ml/100 g × min (average values ranged from 31.99 ± 13.6 ml/100 g × min). Average values of the regional blood volume were 2.1 ± 0.67 ml/100 g (min 1.1 max 4.3 ml/100 g). The mean transit time averaged 5.7 ± 4.5 s (min 2.8, max 34.3 s). The time to peak concentration ranged from 16.9 to 33 s and was 22.3 ± 3.1 s.

Analysis of the perfusion charts revealed that in 16 patients there were zones with cerebral ischemia characteristics.

A multiple regression analysis was performed in a multiple linear regression block. Cerebral perfusion values and the brain saturation level in the frontal lobes were pairwise compared. The $SctO_2$ level was selected as a response indicator. As predictors the perfusion indices rCBF, rCBV, TTP, MTT were selected.

In the case of no focal injuries in the frontal lobes, the cerebral tissue saturation levels ($SctO_2$) were in proportional dependence of the regional volume of blood (rCBV) circulating in the cortical areas of the frontal lobes (Spearman's $R = 0.9$; $p < 0.000001$).

There was no significant correlation between the $SctO_2$ level and other indices of cerebral perfusion in the same zones: rCBF ($R = 0.24$; $p = 0.03$), TTP ($R = 0.1$; $p = 0.11$), MTT ($R = -0.04$; $p = 0.69$).

4   Discussion

In our study the level of saturation of blood in cerebral tissue in patients with a head injury correlated with the regional volume of blood circulating in the frontal lobes (Spearman's $R = 0.9$; $p < 0.000001$).

However, we determined no significant correlation between the $SctO_2$ level and rCBF ($R = 0.24$; $p = 0.03$), TTP ($R = 0.1$; $p = 0.11$), MTT ($R = -0.04$; $p = 0.69$).

Our conclusions are correlated with the results of the cerebral oximetry and positron emission tomography in healthy volunteers, which have shown that the regional blood volume (CBV) correlates with the cerebral tissue oxygen saturation level.

However, having investigated the relationship between the cerebral oxygenation level and CT parameters of the brain perfusion in patients with cerebral accidents, Taussky [1] found a credible relationship between $SctO_2$ and rCBF ($p < 0.0001$) but found no linear correlation with other perfusion parameters.

It should be noted that the volume rate of flow and regional blood volume should theoretically be a direct relationship. It follows from the equation for calculating perfusion parameters (CBF = CBV/MTT) [5] but, in practice, this relationship is ambiguous.

In our opinion, such inconsistencies may be overcome, when considering that the regional volumetric cerebral blood flow (CBF) in contrast to the regional blood volume (CBV) also depends on the state of the arterial bed [4] and, therefore, may vary considerably with the development of cerebral vasospasm, which is very typical for non-traumatic intracranial hemorrhage.

5   Conclusions

The cerebral tissue saturation in patients with a head injury correlates with indices of regional blood volume in the cerebral tissue (CBV) ($R = 0.9$; $p < 0.000001$).

No significant correlation was detected between the level of the cerebral tissue oxygen saturation and other indices of cerebral perfusion in patients with a head injury. NIRS may serve as a screening method for an indirect noninvasive assessment of the regional blood volume level in patients with a head injury.

References

1. Taussky P (2012) Validation of frontal near-infrared spectroscopy as noninvasive bedside monitoring for regional cerebral blood. Neurosurg Focus 32:1–6
2. Hoeffner E, Case I, Jain R (2004) Cerebral perfusion CT: technique and clinical applications. Radiology 231:632–644
3. Wintermark M, Albers G, Alexandrov A (2008) Acute stroke imaging research roadmap. Am J Neuroradiol 29:23–30
4. Constantoyannis C, Sakellaropoulos G (2007) Transcranial cerebral oximetry and transcranial doppler sonography in patients with ruptured cerebral aneurysms and delayed cerebral vasospasm. Med Sci Monit 13(10):35–40
5. Potapov A, Kornienko V, Zakharova N (2012) Peculiarities of regional cerebral blood flow, indices of intracranial and cerebral perfusion pressure in case of severe brain injury. Radio Diagn Ther 3:79–92

Chapter 19
Detection of ROSC in Patients with Cardiac Arrest During Chest Compression Using NIRS: A Pilot Study

Tsukasa Yagi, Ken Nagao, Tsuyoshi Kawamorita, Taketomo Soga, Mitsuru Ishii, Nobutaka Chiba, Kazuhiro Watanabe, Shigemasa Tani, Atsuo Yoshino, Atsushi Hirayama, and Kaoru Sakatani

Abstract Return of spontaneous circulation (ROSC) during chest compression is generally detected by arterial pulse palpation and end-tidal CO_2 monitoring; however, it is necessary to stop chest compression during pulse palpation, and to perform endotracheal intubation for monitoring end-tidal CO_2. In the present study, we evaluated whether near-infrared spectroscopy (NIRS) allows the detection of ROSC during chest compression without interruption. We monitored cerebral blood oxygenation in 19 patients with cardiac arrest using NIRS (NIRO-200NX, Hamamatsu Photonics, Japan). On arrival at the emergency room, the attending

T. Yagi (✉) • T. Soga • K. Watanabe
Department of Cardiology, Nihon University Hospital, 1-6 Kanda Surugadai, Chiyoda-ku, Tokyo 101-8309, Japan

Department of Emergency and Critical Care Medicine, Nihon University Hospital, Tokyo, Japan
e-mail: ygt0108@gmail.com

K. Nagao • S. Tani
Department of Cardiology, Nihon University Hospital, 1-6 Kanda Surugadai, Chiyoda-ku, Tokyo 101-8309, Japan

T. Kawamorita • M. Ishii • N. Chiba
Department of Emergency and Critical Care Medicine, Nihon University Hospital, Tokyo, Japan

A. Yoshino
Division of Neurosurgery, Department of Neurological Surgery, Nihon University School of Medicine, Tokyo, Japan

A. Hirayama
Division of Cardiology, Department of Medicine, Nihon University School of Medicine, Tokyo, Japan

K. Sakatani
Division of Neurosurgery, Department of Neurological Surgery, Nihon University School of Medicine, Tokyo, Japan

NEWCAT Institute, Nihon University College of Engineering, Koriyama, Japan

© Springer Science+Business Media, New York 2016
C.E. Elwell et al. (eds.), *Oxygen Transport to Tissue XXXVII*, Advances in Experimental Medicine and Biology 876, DOI 10.1007/978-1-4939-3023-4_19

151

physicians immediately assessed whether a patient was eligible for this study after conventional advanced life support (ALS) and employed NIRS to measure cerebral blood oxygenation (CBO) in the bilateral frontal lobe in patients. We found cerebral blood flow waveforms in synchrony with chest compressions in all patients. In addition, we observed abrupt increases of oxy-hemoglobin concentration and tissue oxygen index (TOI), which were associated with ROSC detected by pulse palpation. The present findings indicate that NIRS can be used to assess the quality of chest compression in patients with cardiac arrest as demonstrated by the detection of synchronous waveforms during cardiopulmonary resuscitation (CPR). NIRS appears to be applicable for detection of ROSC without interruption of chest compression and without endotracheal intubation.

Keywords Cardiac arrest • Cardiopulmonary resuscitation • Near-infrared spectroscopy • Quality of cardiopulmonary resuscitation • Return of spontaneous circulation

1   Introduction

Cardiac arrest is a major public health problem worldwide. Despite decades of efforts to promote cardiopulmonary resuscitation (CPR) science and education, the survival rate for out-of-hospital cardiac arrest remains low [1, 2]. In Japan, the JCS-ReSS study showed that a favorable neurological outcome at 30 days was extremely rare in patients with out-of-hospital cardiac arrest who arrived at the emergency hospital in cardiac arrest [3–5]. The 2010 CPR Guidelines indicated that rescuers should attempt to minimize the frequency and duration of interruptions in compressions to maximize the number of compressions delivered per minutes [1, 2]. Although return of spontaneous circulation (ROSC) during chest compression is generally detected by arterial pulse palpation, it is necessary to stop chest compression during pulse palpation. In addition, healthcare providers may take too long to check for pulse, and have difficulty determining if a pulse is present or absent. Furthermore, the 2010 CPR Guidelines indicated that it is reasonable to consider using quantitative waveform capnography in intubated patients to monitor CPR quality, optimize chest compressions, and detect ROSC during chest compression or when rhythm check reveals an organized rhythm [1, 2]. If the pressure of end-tidal CO_2 ($PETCO_2$) abruptly increases to a normal value (35–40 mmHg), it is reasonable to consider that this is an indicator of ROSC [1, 2]. However, it is necessary to perform endotracheal intubation for monitoring $PETCO_2$. In the present study, we evaluated whether near-infrared spectroscopy (NIRS) allows the detection of ROSC during chest compression without interruption. NIRS, an optical technique, is potentially an attractive tool for this purpose because it allows noninvasive, continuous measurements of cerebral blood oxygenation (CBO) changes with high time resolution and is easy to use [6, 7].

2   Methods

2.1   Patients

Between November 2009 and March 2014, we employed NIRS (NIRO-200NX, Hamamatsu Photonics, Japan) to measure CBO in the bilateral frontal lobe in patients transported to the emergency room (ER) after out-of-hospital cardiac arrest. The patients were included in a prospective observational study. They were enrolled in this study when they met the following criteria: presumed cardiac etiology of cardiac arrest according to the Utstein style guidelines [8]; persistent cardiac arrest on arrival at the ER; and successful ROSC after arrival at the ER with conventional advanced life support (ALS) and/or extracorporeal CPR (ECPR) using an emergency cardiopulmonary bypass (CPB) [9–11]. Exclusion criteria were a tympanic-membrane temperature below 30 °C on arrival at the ER; non-cardiac etiology of cardiac arrest; or pregnancy. Patients were also excluded if their families refused to give informed consent for participation in this study.

2.2   Procedures

On arrival at the ER, the attending physicians assessed as soon as possible whether a patient was eligible for this study after conventional ALS, and employed NIRS to measure CBO in the bilateral frontal lobe in patients. CPB was initiated when defibrillation by bystander and/or emergency medical personnel using an automated external defibrillator had been unsuccessful, and ROSC could not be achieved within 10 min of arrival [10, 11].

2.3   Statistical Analysis

All analyses were performed using the SPSS software package (version 20.0 J SPSS, Chicago, IL, USA). Continuous variables are expressed as mean ± SD.

3   Results

During the study period, 19 patients met the above criteria and NIRS was employed to measure their CBO. Among them, 2 patients achieved ROSC after arrival at the ER with conventional ALS, and 17 patients were performed ECPR using emergency CPB. Characteristics of these patients are presented in Table 19.1. The mean

Table 19.1 Baseline characteristics of the study population

Characteristics	Patients (n = 19)
Age (years)	60.9 ± 14.6
Male sex (no. (%))	18 (94.7)
Prehospital treatment (no. (%))	
Defibrillations	16 (84.2)
Administration of intravenous epinephrine	10 (52.6)
Initial cardiac rhythm	
VF/pulseless VT	16 (84.2)
PEA	2 (10.5)
Asystole	1 (5.3)
Time interval (min)	
From collapse to implementation of CPB (n = 17)	51.8 ± 15.6
From arrival at the ER to implementation of CPB (n = 17)	17.6 ± 7.3
From collapse to ROSC (n = 2)	62.5 ± 32.5
Cause of cardiac arrest	
Acute coronary syndrome	12 (63.1)
Cardiomyopathy	2 (10.5)
Acute aortic dissection	2 (10.5)
Others	3 (15.8)

VF/pulseless VT ventricular fibrillation/pulseless ventricular tachycardia, *PEA* pulseless electrical activity, *CPB* cardiopulmonary bypass, *ER* emergency room, *ROSC* return of spontaneous circulation

age was 60.9 ± 14.6 years. The proportion of male patients was 94.7 %. The proportion of patients due to acute coronary syndrome (ACS) was 63.1 %.

Figure 19.1 show a typical case where ROSC was achieved after arrival at the ER with conventional ALS, and NIRS was employed to measure CBO in the bilateral frontal lobe. The cerebral blood flow waveform was in synchrony with chest compressions, and abrupt increases of oxy-hemoglobin concentration and tissue oxygenation index (TOI) were seen in association with ROSC detected by pulse palpation. The cerebral blood flow waveform was in synchrony with chest compressions in all patients, and both patients with successful ROSC showed abrupt increases of NIRS-detected oxy-hemoglobin concentration and TOI.

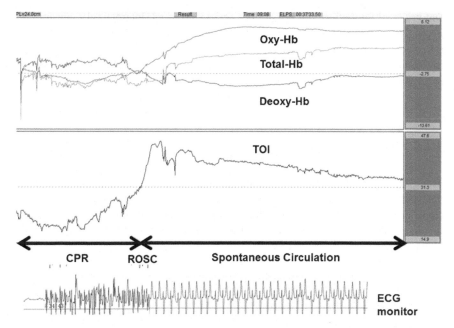

Fig. 19.1 NIRS-detected cerebral blood flow waveform is in synchrony with chest compressions in a patient who achieved return of spontaneous circulation (ROSC). Abrupt increases of oxy-hemoglobin concentration and tissue oxygenation index (TOI) were associated with ROSC detected by pulse palpation

4 Discussion

Our observation of synchronous NIRS-detected waveforms during CPR, in accordance with another recent study [12], suggests that NIRS would be suitable to reliably assess the quality of the chest compression in patients with cardiac arrest. This is important, because NIRS represents easy-to-use technology for noninvasive, continuous measurements of CBO changes with high time resolution, and it should be highly advantageous for real-time CPR prompting and feedback, and for the detection of ROSC without interruption of chest compression and without endotracheal intubation. The 2010 CPR Guidelines indicated that the quality of unprompted CPR is often poor, and methods should be developed to improve the quality of CPR delivered to victims of cardiac arrest [1, 2]. In particular, the guidelines indicated that real-time CPR prompting and feedback technology, such as visual and auditory prompting devices, can improve the quality of CPR [1, 2].

In the present study, we detect ROSC, which occurred in 2 of 19 patients, by using NIRS without interruption of chest compression, although endotracheal intubation was performed in all patients. It has been reported that NIRS can be used to measure CBO in cardiac arrest patients with/without endotracheal intubation [11–13], so NIRS should be suitable for detection of ROSC without endotracheal intubation. Further studies seem warranted. Interestingly, our preliminary study indicated that increase of TOI during implementation of CPB might reflect an improvement in cerebral blood flow [11]. TOI increased following implementation of CPB in 17 patients who received ECPR in addition to emergency CPB.

4.1   Study Limitations

There are several limitations to our study. First, it was not a multicenter study for resuscitation after out-of-hospital cardiac arrest. Second, our findings should be considered preliminary because of the small sample size: there were only 19 patients in the present study, and only 2 achieved ROSC after arrival at the ER with conventional ALS.

5   Conclusions

Our findings support the idea that NIRS would be suitable for allow the detection of ROSC without interruption of chest compression and without endotracheal intubation.

References

1. International Liaison Committee on Resuscitation (2010) 2010 International consensus on cardiopulmonary resuscitation and emergency cardiovascular care science with treatment recommendations. Circulation 122:S250–S605
2. American Heart Association (2010) 2010 American Heart Association guidelines for cardiopulmonary resuscitation and emergency cardiovascular care. Circulation 122:S639–S946
3. Kitamura T, Iwami T, Kawamura T, Implementation Working Group for All-Japan Utstein Registry of the Fire and Disaster Management Agency et al (2010) Nationwide public access defibrillation in Japan. N Engl J Med 362:994–1004
4. Iwami T, Kitamura T, Kawamura T et al (2012) Chest compression-only cardiopulmonary resuscitation for out-of-hospital cardiac arrest with public-access defibrillation: a nationwide cohort study. Circulation 126:2844–2851
5. Japanese Circulation Society Resuscitation Science Study Group (2013) Chest-compression-only bystander cardiopulmonary resuscitation in the 30:2 compression-to-ventilation ratio era. Nationwide observational study. Circ J 77:2742–2750
6. Samra SK, Dorjie P, Zelenock GB et al (1996) Cerebral oximetry in patients undergoing carotid endarterectomy. Stroke 27:49–55
7. Nakamura S, Kano T, Sakatani K et al (2009) Optical topography can predict occurrence of watershed infarction during carotid endarterectomy: technical case report. Surg Neurol 71:540–542

8. Cummins RO, Chamberlain DA, Abramson NS et al (1991) Recommended guidelines for uniform reporting of data from out-of-hospital cardiac arrest: the Utstein Style – a statement for health professionals from a task force of the American Heart Association, the European Resuscitation Council, the Heart and Stroke Foundation of Canada, and the Australian Resuscitation Council. Circulation 84:960–975
9. Nagao K, Hayashi N, Kanmatsuse K et al (2000) Cardiopulmonary cerebral resuscitation using emergency cardiopulmonary bypass, coronary reperfusion therapy and mild hypothermia in patients with cardiac arrest outside the hospital. J Am Coll Cardiol 36:776–783
10. Nagao K, Kikushima K, Watanabe K et al (2010) Early induction of hypothermia during cardiac arrest improves neurological outcomes in patients with out-of-hospital cardiac arrest who undergo emergency cardiopulmonary bypass and percutaneous coronary intervention. Circ J 74:77–85
11. Yagi T, Nagao K, Sakatani K et al (2013) Changes of cerebral oxygen metabolism and hemodynamics during ECPR with hypothermia measured by near-infrared spectroscopy: a pilot study. Adv Exp Med Biol 789:121–128
12. Koyama Y, Wada T, Lohman BD et al (2013) A new method to detect cerebral blood flow waveform in synchrony with chest compression by near-infrared spectroscopy during CPR. Am J Emerg Med 31:1504–1508
13. Ito N, Nishiyama K, Callaway CW et al (2014) Noninvasive regional cerebral oxygen saturation for neurological prognostication of patients with out-of-hospital cardiac arrest: a prospective multicenter observational study. Resuscitation 85:778–784

Chapter 20
Myocardial Microcirculation and Mitochondrial Energetics in the Isolated Rat Heart

J.F. Ashruf and C. Ince

Abstract Normal functioning of myocardium requires adequate oxygenation, which in turn is dependent on an adequate microcirculation. NADH-fluorimetry enables a direct evaluation of the adequacy of tissue oxygenation while the measurement of quenching of Pd-porphyrine (P_pIX) phosphorescence enables quantitative measurement of microvascular pO_2. Combination of these two techniques provides information about the relation between microvascular oxygen content and parenchymal oxygen availability in Langendorff hearts. In normal myocardium there is heterogeneity at the microcirculatory level resulting in the existence of microcirculatory weak units, originating at the capillary level, which reoxygenate the slowest upon reoxygenation after an episode of ischemia. Sepsis and myocardial hypertrophia alter the pattern of oxygen transport whereby the microcirculation is disturbed at the arteriolar/arterial level. NADH fluorimetry also reveals a disturbance of mitochondrial oxygen availability in sepsis. Furthermore it is shown that these techniques can also be applied to various organs and tissues in vivo.

Keywords Microcirculation • Mitochondrial redox state • NADH fluorescence • Sepsis • Hypertrophia

1 Introduction

Oxygen is transported to the cells via the microcirculation, which is composed of terminal arterioles, capillaries and venules. Cellular ischemia or hypoxia results in an increase of the mitochondrial $NADH/NAD^+$ ratio (redox state), which can be measured using NADH-fluorimetry [1]. Using this technique, it was found that

J.F. Ashruf (✉)
Department of Surgery, OZG Hospital, Groningen, The Netherlands
e-mail: jesseashruf@gmail.com

C. Ince (✉)
Department of Translational Physiology, Academic Medical Center, University of Amsterdam, Amsterdam, The Netherlands
e-mail: c.ince@amc.uva.nl

© Springer Science+Business Media, New York 2016   159
C.E. Elwell et al. (eds.), *Oxygen Transport to Tissue XXXVII*, Advances in Experimental Medicine and Biology 876, DOI 10.1007/978-1-4939-3023-4_20

there are areas in the healthy myocardium that are more vulnerable to dysoxia; these areas are reproducible in time and place and originate at the capillary level and are therefore closely associated with the architecture of the microcirculation. These weak units show the largest relative reduction in flow (independent of absolute flow levels) during compromising conditions, with dysoxia initially developing at the venous end of the capillary [2, 3]. Myocardial hypertrophia, work and sepsis have a profound effect upon the distribution of oxygen in the microcirculation as is shown by measuring the mitochondrial redox state under these conditions.

Another factor influencing NADH fluorescence intensity is photo-bleaching. Earlier studies showed a gradual decline in fluorescence intensity over time; it was suggested that this was caused by a steady decline of endogenous substrate eventually making the heart totally dependent on substrate in the perfusate. We showed however that this decline is not actually caused by an altered ratio, but by photo-bleaching caused by continuous illumination of the organ surface with UV-light. Applying discontinuous illumination nearly completely eliminated this effect [4, 5].

In myocardial hypertrophia a dysregulation of oxygen distribution can be found at the arteriolar level. Alleviation of ischemia at the arteriolar level can be induced by vasodilation but probably also by preventing acidification of the interstitium by fatty-acid oxidation (thereby preventing lactic acid formation). Scavenging of oxygen free radicals by SOD also improves oxygenation of critical areas in hypertrophic myocardium.

Increase in work results in a homogeneous decrease in $NADH/NAD^+$ ratio when precautions are taken to avoid development of ischemia/dysoxia. This is of importance because the development of ischemia can cause one to assume that increase in work actually increases the $NADH/NAD^+$ ratio.

The $NADH/NAD^+$ ratio can also be measured in vivo in heart and gut of rats and it is shown that changes in the ratio are associated with changes in oxygen content of parenchyma and circulation using Pd-phorphyrine phosphorescence measurements.

Sepsis has a complex effect on the redox state: on the one hand it is shown that sepsis induces ischemia through microcirculatory changes; on the other hand there is a strong suggestion that sepsis also changes mitochondrial functioning resulting in a decreased rate of oxygen consumption.

2   Distribution of Oxygen to Myocardium in Normal and Hypertrophic Rat Hearts

Measurement of the intracellular autofluorescence of reduced pyridine nucleotide on myocardial tissue surfaces (NADH; 360-nm excitation, 460-nm emission), pioneered by Barlow et al. [1], allows identification of areas of cellular hypoxia, because reduction of NADH is dependent on an adequate supply of oxygen. Using

NADH fluorescence photography of the surface of Tyrode-perfused rat hearts, Barlow et al. [1] found that lowering coronary perfusion results in patchy NADH fluorescence. Steenbergen et al. [6] found that this heterogeneity in the hypoxic state occurred under ischemic as well as during high-flow hypoxic conditions.

Can et al. found that this heterogeneity originates at the capillary level and that the ischemic patches occurring during reperfusion after hypoxia represent microcirculatory units furthest away from the oxygen supply. These microcirculatory units also were the first areas to become ischemic during increased cardiac work in Langendorff hearts; it was also shown that these areas were identical to those occurring during reperfusion and during embolization with microspheres roughly having the size of capillaries [2, 7, 8].

Hypertrophia is known to be associated with a decreased muscle mass to vascularization ratio and thus a change of myocardial perfusion when compared to normal hearts [9–11]. Hulsmann et al. [12] found that hypertrophic Langendorff rat hearts displayed ischemic areas during normoxic perfusion whereas normal heart under the same circumstances did not. The relative hypoperfusion in these areas was shown to be caused by acidosis and oxygen free radicals (OFR), since measures taken to prevent acidosis (by stimulating fatty acid oxidation) and OFR production (by adding super oxide dismutase to the perfusate) dramatically improved local hypoxia. In a further study [7] it was shown that perfusion is altered at the arteriolar and arterial level in hypertrophy.

An important observation in experiments with Langendorff hearts using NADH fluorescence imaging is that the Langendorff heart is borderline hypoxic, and in case of higher workloads, even partly hypoxic. This can seriously affect the interpretation of the results and can even render the experiment useless [2, 5, 7, 8, 12].

The $NADH/NAD^+$ ratio is an effective parameter for the evaluation of tissue hypoxia. It was also known that this ratio can be influenced by other factors besides the lack of oxygen. For instance, earlier studies showed that increased cardiac work was associated with an increased NADH/NAD+ ratio: it was suggested that this increased ratio actually drives the electron chain and thus ATP production. However, in our studies [2, 4, 5] we found that this increase was caused by development of ischemia in borderline normoxic Langendorff heart which became hypoxic during increased workload. Instead we found that during increased workload, when securing adequate oxygen supply to the myocardium, the NADH/NAD+ ratio actually decreased, making it very unlikely that during increased workload NADH drives the electron chain of oxidative phosphorylation. Several studies, for example the study by Territo et al. [13], have shown that the changing extramitochondrial $[Ca^{2+}]$ during changes in workload probably is the most important signal to cause an increased oxidative phosphorylation.

Another factor influencing this ratio is the type of substrate used in the perfusate: pyruvate increases the ratio more than glucose [4]. In our study of the effect of cardiac work on the NADH/NAD+ ratio, we observed that during perfusion with glucose as a substrate, the work-related decrease of the ratio was much slower than during perfusion with pyruvate as a substrate. An explanation for this might be that pyruvate enters the Krebs cycle directly, thus reducing NAD^+ relatively quickly, whereas glucose first enters the glycolysis, which is a slower process when compared to the Krebs cycle and speeds up more slowly in case of increased workload. However, more investigations are needed to explain this observation.

3   Distribution of Oxygen to Myocardium in Septic Rat Hearts

Avontuur et al. [14] showed that hearts of septic rats during Langendorff perfusion develop regional ischemia when coronary flow is reduced with N^{ω}-nitro-L-arginine (NNLA) or Methylene blue (MB). NNLA and MB both inhibit the effects of nitric oxide (NO). This finding suggests that endotoxemia promotes myocardial ischemia in vulnerable areas of the heart after inhibition of the NO pathway or direct vasoconstriction. A further study by Avontuur et al. showed that sepsis induces massive coronary vasodilation due to increased myocardial NO synthesis, resulting in autoregulatory dysfunction.

Indeed, in a recent study by us (unpublished data, to be submitted) Langendorff perfused hearts from septic rats develop hypoxic areas which appear larger than the heterogeneous areas during reperfusion and embolization with microspheres of 5.9 μm diameter [2, 8]. This agrees with the finding that autoregulation is disturbed [14]; autoregulation is situated at the arterial and arteriolar level and we found that the hypoxic areas were larger than those elicited by embolization of capillaries and are comparable in size to those found in hypoperfused areas in hypertrophy and during embolization of arterioles and arteries [7, 8]. The result of these effects of sepsis on the microcirculation can be describes as shunting of the microcirculation [15], which means that certain portions of the myocardial capillary network remain hypoperfused and other portions receive a higher than needed (for adequate oxygenation) capillary flow. We also found that mitochondrial functioning in septic hearts appeared to be disturbed, as the development of hypoxia during interrupted perfusion was slower than in normal control hearts.

4   In Vivo Evaluation of Hypoxia and Ischemia

NADH videofluorimetry has been shown to be an effective method to assess tissue hypoxia in numerous studies ex vivo. Clinical applicability however requires in vivo application of this measurement and therefore we performed several studies to assess the possibility of in vivo application of NADH videofluorimetry on various organs (heart, intestines, skeletal muscle) [16–21]. It was shown, for instance, that with this technique it was possible to detect local hypoxia in vivo in myocardium induced by selective ligation of coronary arteries in rat heart

[18]. However, in vivo NADH fluorescence is disturbed by movement, hemodynamic and oximetric effects. A method was developed to compensate for these factors by means of utilizing the NADH fluorescence/UV reflectance ratio, making it possible to monitor the mitochondrial redox state of intact blood-perfused myocardium [20].

Furthermore, measurement of quenching of Pd-porphyrine (P_pIX) phosphorescence enabled us to evaluate microvascular pO_2. Combined with NADH fluorescence measurements this enabled us to correlate the mitochondrial energy state to the microvascular pO_2 [17–19].

van der Laan et al. [21] explored the possibility to evaluate ischemia and reperfusion injury in skeletal muscle (in rat) and found a clear correlation between tissue hypoxia and NADH fluorescence intensity.

5 Recent Developments

A new technique for evaluation of the microcirculation is the sidestream dark-field (SDF) imaging, mainly applied to observe the sublingual microcirculation. It has provided great insight into the importance of this physiological compartment in (perioperative) medicine and could prove to become a useful tool in treatment of shock, based on evaluation of the microcirculation [22].

A major disadvantage of the use of NADH video fluorimetry for obtaining information about mitochondrial bioenergetics and oxygenation is that it is a non-quantitative technique relying on relative changes in fluorescence signals. Recently we introduced a new method for a quantitative measurement of mitochondrial pO_2 (mit pO_2) values. The method is based on the oxygen dependent decay of delayed fluorescence of endogenously present mitochondrial protoporphyrin IX (P_pIX) [23]. Proof of concept measurements were initially performed by Mik et al. in single cells and later validated in vivo in rat liver and recently also in the heart [24, 25]. Deconvolution of the decay curves allowed mit pO_2 histograms to be generated and showed a heterogeneous distribution of mit pO_2 values. Ischemia-reperfusion injury was shown to induce mit PO_2 values with hypoxic as well as hyperoxic values [24].

One of the remarkable measurements of the mit pO_2 values was the finding of mit pO_2 values much higher (20–30 mmHg) than previously thought (less than 5 mmHg) and secondly the finding of markedly heterogeneous mit pO_2 values, similar to the distributions of microvascular pO_2 measurements using P_pIX quenching of phosphorescence we had measured in heart [23]. Indeed (early) observations of the heterogeneous state of myocardial and mitochondrial energy states underscores the importance of heterogeneity of oxygen delivery and utilization [3]. It is expected that the combination of NADH fluorimetry and mit pO_2 measured by P_pIX phosphorescence will provide new information concerning the transport and consumption of oxygen by the heart.

References

1. Barlow CH, Chance B (1976) Ischemic areas in perfused rat hearts: measurement by NADH fluorescence photography. Science 193:909–910
2. Ince C, Ashruf JF, Avontuur JAM et al (1993) Heterogeneity of the hypoxic state in rat heart is determined at capillary level. Am J Physiol 264 (Heart Circ Physiol 33):H294–H301
3. Zuurbier CJ, van Iterson M, Ince C (1999) Functional heterogeneity of oxygen supply-consumption ratio in the heart. Cardiovasc Res 44(3):488–497
4. Ashruf JF, Coremans JM, Bruining HA et al (1995) Increase of cardiac work is associated with decrease of mitochondrial NADH. Am J Physiol 269(3 Pt 2):H856–H862
5. Ashruf JF, Coremans JM, Bruining HA et al (1996) Mitochondrial NADH in the Langendorff rat heart decreases in response to increases in work: increase of cardiac work is associated with decrease of mitochondrial NADH. Adv Exp Med Biol 388:275–282
6. Steenbergen CG, Deleeuw C, Barlow B et al (1977) Heterogeneity of the hypoxic state in perfused rat heart. Circ Res 4:606–615
7. Ashruf JF, Ince C, Bruining HA et al (1994) Ischemic areas in hypertrophic Langendorff rat hearts visualized by NADH videofluorimetry. Adv Exp Med Biol 345:259–262
8. Ashruf JF, Ince C, Bruining HA (1999) Regional ischemia in hypertrophic Langendorff-perfused rat hearts. Am J Physiol 277 (Heart Circ Physiol 46):H1532–H1539
9. Anderson PG, Bishop SP, Digerness SB (1987) Transmural progression of morphologic changes during ischemic contracture and reperfusion in the normal and hypertrophied rat heart. Am J Pathol 129(1):152–167
10. Dellsperger KC, Marcus LM (1990) Effects of left ventricular hypertrophy on the coronary circulation. Am J Cardiol 65:1504–1510
11. Einzig SJ, Leonard J, Tripp MR et al (1981) Changes in regional myocardial blood flow and variable development of hypertrophy after aortic banding in puppies. Cardiovasc Res 15:711–723
12. Hulsmann WC, Ashruf JF, Bruining HA et al (1993) Imminent ischemia in normal and hypertrophic Langendorff rat hearts: effects of fatty acids and superoxide dismutase monitored by NADH surface fluorescence. Biochim Biophys Acta 1181:273–278
13. Territo PR, French SA, Dunleavy MC et al (2001) Calcium activation of heart mitochondrial oxidative phosphorylation: rapid kinetics of mVO2, NADH, AND light scattering. J Biol Chem 276(4):2586–2599
14. Avontuur JA, Bruining HA, Ince C (1995) Inhibition of nitric oxide synthesis causes myocardial ischemia in endotoxemic rats. Circ Res 76(3):418–425
15. Ince C, Sinaasappel M (1999) Microcirculatory oxygenation and shunting in sepsis and shock. Crit Care Med 27(7):1369–1377
16. Ince C, Coremans JM, Bruining HA (1992) In vivo NADH fluorescence. Adv Exp Med Biol 317:277–296
17. Bruining HA, Pierik GJ, Ince C et al (1992) Optical spectroscopic imaging for non-invasive evaluation of tissue oxygenation. Chirurgie 118(5):317–322, discussion 323
18. Ince C, Ashruf JF, Sanderse EA et al (1992) In vivo NADH and Pd-porphyrin video fluori-/phosphorimetry. Adv Exp Med Biol 317:267–275
19. Ince C, van der Sluijs JP, Sinaasappel M et al (1994) Intestinal ischemia during hypoxia and experimental sepsis as observed by NADH videofluorimetry and quenching of Pd-porphine phosphorescence. Adv Exp Med Biol 36:105–110
20. Coremans JM, Ince C, Bruining HA (1997) (Semi-)quantitative analysis of reduced nicotinamide adenine dinucleotide fluorescence images of blood-perfused rat heart. Biophys J 72 (4):1849–1860
21. van der Laan L, Coremans A, Ince C et al (1998) NADH videofluorimetry to monitor the energy state of skeletal muscle in vivo. J Surg Res 74(2):155–160
22. Ashruf JF, Ince C, Bruining HA (2013) New insights into the pathophysiology of cardiogenic shock: the role of the microcirculation. Curr Opin Crit Care 19(5):381–386

23. Mik EG, Stap J, Sinaasappel M et al (2006) Mitochondrial PO2 measured by delayed fluorescence of endogenous protoporphyrin IX. Nat Methods 3(11):939–945
24. Mik EG, Johannes T, Zuurbier C et al (2008) In vivo mitochondrial oxygen tension measured by a delayed fluorescence lifetime technique. Biophys J 95(8):3977–3990
25. Mik EG, Ince C, Eerbeek O et al (2009) Mitochondrial oxygen tension within the heart. J Mol Cell Cardiol 46(6):943–951

Part IV
Cancer Metabolism

Chapter 21
Tissue Discs: A 3D Model for Assessing Modulation of Tissue Oxygenation

M.J. Gandolfo, A.H. Kyle, and A.I. Minchinton

Abstract The presence of hypoxia in solid tumours is correlated with poor treatment outcome. We have developed a 3-D tissue engineered construct to quantitatively monitor oxygen penetration through tumour tissue using the exogenous 2-nitroimidazole bioreductive probe pimonidazole and phosphorescence quenching technologies. Using this in vitro model we were able to examine the influence of the biguanides metformin and phenformin, antimycin A and KCN, on the distribution and kinetics of oxygen delivery as prototypes of modulators of oxygen metabolism.

Keywords Hypoxia • Oxygen sensing • Tumour • Metabolism • Radiosensitivity

1 Introduction

The presence of molecular oxygen increases the radiosensitivity of mammalian cells three-fold [1, 2]. This 'oxygen effect' has significant implications for radiation therapy since it is well established that many solid cancers harbour hypoxic cells. Measured tissue oxygenation is an independent prognostic indicator of poor outcome in cancers of the head and neck [3], cervix [4], prostate [5] and sarcoma [6].

Oxygen tension decreases with distance from functioning blood vessels rendering cells increasingly radioresistant. Numerous experimental strategies have been examined to increase the oxygen delivery to cells distal to blood vessels such as high-oxygen content breathing [7], hyperbaric oxygen breathing [8, 9], administration of synthetic oxygen carriers [10, 11] and administration of oxygen mimetic compounds such as the radiosensitizer misonidazole [12]. A difficulty in many of theses strategies is quantitatively assessing the effect of these treatments on oxygen delivery.

We have developed a tissue engineered 3D model that can be used to model oxygen gradients and tissue metabolism encountered in solid tumours [13, 14]. The oxygenation, proliferation and necrosis in these cultures have previously been described [13, 14]. We have now developed a real time oxygen sensing system to

M.J. Gandolfo • A.H. Kyle • A.I. Minchinton (✉)
Department of Integrative Oncology, BC Cancer Research Centre, Vancouver, BC, Canada
e-mail: aim@bccrc.ca

© Springer Science+Business Media, New York 2016
C.E. Elwell et al. (eds.), *Oxygen Transport to Tissue XXXVII*, Advances in
Experimental Medicine and Biology 876, DOI 10.1007/978-1-4939-3023-4_21

quantitatively assess the effects of drugs on oxygen diffusion. To exemplify this model we examined the effects on oxygen tension of the anti-diabetic drugs metformin and phenformin, and antimycin A and KCN, all inhibitors of mitochondrial respiration. Our measurements were validated using the exogenous marker of hypoxia pimonidazole.

2 Methods

2.1 3-Dimensional Tissue Disc Cultures

Tissue culture inserts (Millipore Millicell^{CM} 12 mm diameter, pore size 0.4 µm; Millipore, Nepean, Ontario, Canada) were coated with 150 µl of collagen (rat tail type I; Sigma Chemical), dissolved in 60 % ethanol to 0.75 mg/ml and allowed to dry overnight. HCT116 cells, 7.5×10^5 in 0.5 ml growth media, were then added to the inserts and incubated for 12 h to allow the cells to attach to the coated membrane. The cultures were then incubated for 7 days in custom-built growth chamber to form tissue discs ~ 200 µm in thickness. Each growth vessel contained a frame that held the inserts completely immersed in 120 ml of stirred media (700 rpm, 25-mm stir bar) under continual gassing (5 % CO_2, balance air) at 37 °C. Drugs were added to the growth vessels at different concentrations for 2 h. After 2 h, bromodeoxyuridine (BrdUrd) and pimonidazole were added at 100 and 50 µM, respectively, for an additional 2 h. Tissue discs were then removed from the growing vessels and frozen.

2.2 Immunohistochemistry

Tissue disc cryosections (10 µm thick), were air dried for 12 h and then fixed in formalin for 15 min at room temperature and DNA was denatured to allow BrdUrd detection after autoclave heat treatment 125 °C for 5 min in 10 mM sodium citrate pH 6. Slides were immediately transferred to a PBS/0.05 % Tween-20 bath to permeabilize. Subsequent steps were each followed by a 5-min wash in PBS/0.05 % Tween-20. BrdUrd incorporated into DNA was detected using a 1:500 dilution of monoclonal rat anti-BrdUrd (clone BU1/75; abcam) and pimonidazole was detected using monoclonal mouse anti-PIMO FITC tagged (Hypoxyprobe) followed by 1:500 dilution of goat anti rat Alexa 546 antibody (Invitrogen) and 10 µg/ml Hoechst 33342 to label nuclei. Slides were then imaged.

2.3 Image Acquisition and Analysis

The imaging system consists of a robotic fluorescence microscope (Zeiss Axioimager Z1, Oberkochen, Germany), a cooled, monochrome CCD camera (Retiga 4000R, QImaging, Vancouver, BC, Canada), a motorized slide loader and x-y stage (Ludl Electronic Products, Hawthorne, NY, USA) and customized ImageJ software (public domain program developed at the U.S. National Institutes of Health, available at http://rsb.info.nih.gov/ij/) running on a Macintosh computer (Apple Inc., Cupertino, CA, USA). The system allowed tiling of adjacent microscope fields of view. Using this system, images of entire tissue disc cryosections were captured at a resolution of 1.5 µm/pixel using a 10× objective. For Hoechst 33342 we used excitation 380 nm and emission 440 nm, pimonidazole was excitation 485 nm and emission 521 nm, and BrdUrd was excitation 560 nm and emission 607 nm. Using ImageJ and user supplied algorithms, images from pimonidazole and BrdUrd staining from each tissue disc section were overlaid and staining artifacts manually removed. Pimonidazole and BrdUrd positive staining were identified by selecting pixels that were 5 standard deviations above tissue background levels. The data were tabulated so as to determine the fraction of positive pixels of the total number pixels found relative to the top of the tissue disc.

2.4 Oxygen Sensing

To study the change in oxygenation caused by the test drugs on the tissue discs we used a fluorescence-based NeoFox fiber optic oxygen sensing system (Ocean Optics, FL, USA). For these experiments each 3D tissue disc was placed against a glass window whose surface was covered by an oxygen sensing patch (RedEye, Ocean Optics), Fig. 21.2a. A fiber optic probe was then used to stimulate fluorophores in the O_2 patch to determine pO_2 status at the adjacent surface of the tissue disc. In the configuration depicted in Fig. 21.2a the patch is separated from the oxygen source, the stirred reservoir, by the tissue and hence the amount of oxygen reaching the patch will vary with the rate of oxygen consumption within the tissue. A second version of the system was developed using a multiplexed array of pO_2 sensors so that up to 6 3D tissue discs could be monitored. For this configuration Tris (4,7-diphenyl-1,10-phenanthroline)ruthenium(II) dichloride doped silicone [15] was used to coat windows instead of the RedEye patches, and a stepper motor drive was used to move the fiber optic probe between windows so as to monitor pO_2 at each 3D discs at a rate of once per minute.

3   Results

3.1   Mitochondrial Inhibitors Decrease Oxygen Utilization by Tumour Cells Resulting in Reduction of the Hypoxic Fraction in Tissue Discs

Pimonidazole staining confirmed that both metformin and phenformin decrease the hypoxic fraction by reducing cellular oxygen consumption in a dose-dependent manner (Fig. 21.1a, b). Phenformin is seen to reduce hypoxia staining by pimonidazole with increasing effect over the dose range of 10–500 µM compared to metformin which requires millimolar concentrations. Interestingly, S-phase cells status (indicated by active incorporation of BrdUrd staining) shows that DNA replication occurs even at the highest doses of electron transfer inhibitors, where significant inhibition of mitochondrial respiration occurs. The results of analysis of the dose dependency of the two different mitochondrial respiration inhibitors show

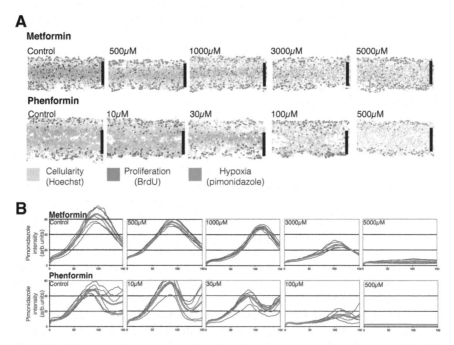

Fig. 21.1 (**a**) Pattern of the hypoxia marker pimonidazole and the proliferation marker BrdU in tissue cryosections following increasing exposure of the mitochondrial inhibitors metformin and phenformin. The HCT116 discs were grown under physiological oxygen (5 % O_2). *Scale bar* indicates 150 µm. (**b**) Analysis of dose response for reduction of hypoxia following the mitochondrial inhibitors metformin and phenformin. Each *blue line* represents one sample and the *red line* represents the average on the group. In both cases drug exposure was 4 h with hypoxia labeling by pimonidazole and BrdU labeling proliferation for the final 2 h

that approximately 50 % inhibition occurs at 100 μM for phenformin, whereas 3 mM metformin is required to see a similar effect (Fig. 21.1b).

3.2 Kinetics of Reoxygenation of the Mitochondrial Inhibitors by Oxygen Sensing

Direct pO$_2$ sensing confirmed that both antidiabetic agents, antimycin A and KCN reduce oxygen consumption by tumour cells and thereby increase oxygen penetration throughout the tissue disc in a dose-dependent manner (Fig. 21.2b). Phenformin was found to have an effect on oxygen utilization within minutes of drug exposure compared to metformin, which took hours to be effective. This is consistent with slow cellular uptake of metformin, which is not cell permeable and requires active uptake via OCT1 transporter [16]. The effect of antimycin A and KCN are shown in Fig. 21.2c and d and show more rapid inhibition of oxygen

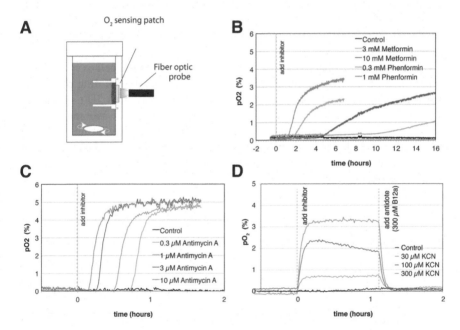

Fig. 21.2 Kinetics of re-oxyenation of tissue following inhibition of mitochondrial respiration. (**a**) Diagram of the photoluminescence based oxygen sensor system used to measure oxygenation within tissue in real time. (**b**) Time course of reoxygenation within 3D HCT116 tissue discs for metformin and phenformin. (**c**) Time course of reoxygenation within 3D HCT116 tissue discs for antimyacin A. (**d**) Time course of reoxygenation within 3D HCT116 tissue discs for potassium cyanide. The discs were maintained in media under 5 % oxygen with continuous stirring throughout the experiment. Note that the distance through which the oxygen has to diffuse in the oxygen sensor experiments doubles, since one of the sides is against the oxygen sensor patch thus preventing media from diffusing through

consumption than the antidiabetics. The inhibition of oxygen consumption by KCN could be abrogated by the addition of the antidote Vitamin B12a.

4 Conclusions

In this study we looked at the impact of a panel of mitochondrial inhibitors on tissue oxygenation using an in vitro 3D tissue model and two oxygen responsive methodologies. Pimonidazole was employed to provide spatial information of the profile of tissue hypoxia that formed within the discs using a fixed 2-h labelling window while transient temporal information was obtained using a fiber-optic based phosphorescence quenching technique that lacked spatial information. By combining the methods it was then possible to follow the modulation in the kinetics of oxygen consumption within tissue caused by metformin and phenformin, and antimycin A and KCN. All decrease oxygen utilization in a concentration dependent manner resulting in reduction of the hypoxic fraction consistent with their ability to inhibit mitochondrial respiration. This model can therefore be useful in modelling how mitochondrial respiratory inhibitors could be applied to radiosensitize tumours in a clinical setting.

Our findings show that tissue discs allow the study of tumour cell oxygenation in a controlled 3D microenvironment, avoiding the complexities of drug pharmacokinetics and metabolism encountered using in vivo models. This 3D tissue model incorporates the gradients in oxygenation, nutrient supply and waste removal that would be encountered in a solid tumour. In addition, data obtained using this model indicate that modulation of mitochondrial efficiency is feasible and can lead to an increase in tissue oxygenation without severe changes to tissue architecture or ability to maintain tumour cell proliferation.

References

1. Gray LH, Conger AD, Ebert M, Hornsey S, Scott OC (1953) Concentration of oxygen dissolved in tissues at the time of irradiation as a factor in radiotherapy. Br J Radiol 26:638–648
2. Crabtree H, Cramer W (1933) Action of radium on cancer cells. Proc R Soc B 113:238–250
3. Nordsmark M, Overgaard M, Overgaard J (1996) Pretreatment oxygenation predicts radiation response in advanced squamous cell carcinoma of the head and neck. Radiother Oncol 41:31–39
4. Hockel M, Schlenger K, Aral B et al (1996) Association between tumor hypoxia and malignant progression in advanced cancer of the uterine cervix. Cancer Res 56:4509–4515
5. Movsas B, Chapman JD, Horwitz EM et al (1999) Hypoxic regions exist in human prostate carcinoma. Urology 53:11–18
6. Brizel DM, Scully SP, Harrelson JM et al (1996) Tumor oxygenation predicts for the likelihood of distant metastases in human soft tissue sarcoma. Cancer Res 56:941–943

7. Hil SA, Collingridge DR, Vojnovic B, Chaplin DJ (1998) Tumour radiosensitization by high-oxygen-content gases: influence of the carbon dioxide content of the inspired gas on PO2, microcirculatory function and radiosensitivity. Int J Radiat Oncol Biol Phys 40:943–951
8. Bennett M, Feldmeier J, Smee R, Milross C (2008) Hyperbaric oxygenation for tumour sensitization to radiotherapy: a systematic review of randomized controlled trials. Cancer Treat Rev 34:577–591
9. Coles C, Williams M, Burnet N (1999) Hyperbaric oxygen therapy. Combination with radiotherapy in cancer is of proved benefit but rarely used. BMJ 318:1076–1077
10. Horinouchi H, Yamamoto H, Komatsu T, Huang Y, Tsuchida E, Kobayashi K (2007) Enhanced radiation response of a solid tumor with the artificial oxygen carrier 'albumin-heme'. Cancer Sci 99:1274–1278
11. Tsuchida E, Sou K, Nakagawa A et al (2009) Artificial oxygen carriers, hemoglobin vesicles and albumin–hemes, based on bioconjugate chemistry. Bioconjug Chem 20:1419–1440
12. Michaels H, Peterson E, Ling C et al (1981) Combined effects of misonidazole radiosensitization and hypoxic cell toxicity in mammalian cells. Radiat Res 88:354–368
13. Kyle AH, Baker JHE, Minchinton AI (2012) Targeting quiescent tumour cells via oxygen and IGF-I supplementation. Cancer Res 72:801–809
14. Kyle AH, Huxham LA, Chiam AS, Sim DH, Minchinton AI (2004) Direct assessment of drug penetration into tissue using a novel application of three dimensional cell culture. Cancer Res 64:6304–6309
15. Klimant I, Wolfbeis O (1995) Oxygen-sensitive luminescent materials based on silicone-soluble Ruthenium Diimine complexes. Anal Chem 67:3160–3166
16. Sogame Y, Kitamura A, Yabuki M et al (2009) A comparison of uptake of metformin and phenformin mediated by hOCT1 in human hepatocytes. Biopharm Drug Dispos 30:476–484

Chapter 22
Hypoxia-Driven Adenosine Accumulation: A Crucial Microenvironmental Factor Promoting Tumor Progression

Peter Vaupel and Arnulf Mayer

Abstract Systematic studies on the oxygenation status of solid tumors have shown that the development of hypoxic/anoxic tissue subvolumes is a pathophysiological trait in a wide range of human malignancies. As a result of this characteristic property, adenosine (ADO) accumulation (range: 50–100 μM) occurs caused by intra- and extracellular generation of ADO. Extracellular nucleotide catabolism by hypoxia-/HIF-1α-sensitive, membrane-associated ecto- 5′- nucleotidases most probably is the major source of ADO in the halo of cancer cells upon specific genetic alterations taking place during tumor growth. Extracellular ADO can act through autocrine and paracrine pathways following receptor-binding and involving different intracellular signalling cascades. Hypoxia-driven receptor activation can lead to a broad spectrum of strong immune-suppressive properties facilitating tumor escape from immune control (e.g., inhibition of $CD4^+$, $CD8^+$, NK and dendritic cells, stimulation of Treg cells). In addition, tumor growth and progression is promoted by ADO-driven direct stimulation of tumor cell proliferation, migration, invasion, metastatic dissemination, and an increase in the production of molecules stimulating tumor angiogenesis. Hypoxia- driven ADO accumulation in the tumor microenvironment thus plays a critical role in tumor growth and progression at multiple pathophysiological levels.

Keywords Adenosine accumulation • Tumor hypoxia • Tumor growth • Tumor progression • Tumor immune escape

1 Introduction

Systematic studies on the oxygenation status of human tumors have shown that the existence of hypoxic/anoxic tissue subvolumes is a pathophysiological trait [1]. Hypoxic tissue areas show complex spatial and temporal heterogeneities

P. Vaupel (✉) • A. Mayer (✉)
Department of Radiooncology and Radiotherapy, University Medical Center,
Langenbeckstrasse 1, 55131 Mainz, Germany
e-mail: vaupel@uni-mainz.de; arnmayer@uni-mainz.de

© Springer Science+Business Media, New York 2016   177
C.E. Elwell et al. (eds.), *Oxygen Transport to Tissue XXXVII*, Advances in
Experimental Medicine and Biology 876, DOI 10.1007/978-1-4939-3023-4_22

[2, 3]. As a consequence of these properties, tumor cells switch to (mal-)adaptive processes and develop aggressive phenotypes and treatment resistance [2–6]. In a recent review article, a synopsis was presented of the various mechanisms by which tumor hypoxia *per se* and through indirect actions (e.g., increased HIF-1 or Sp1 expression) may contribute to a more aggressive phenotype, to an increased resistance to anticancer therapies and finally to poor patient outcome [5]. In this context, it has been outlined that a hypoxia-associated adenosine (ADO) accumulation may be one of the central drivers for an inhibition of antitumor immune response.

2   Hypoxia-Promoted Adenosine Accumulation in Tumors

Extracellular ADO- concentrations in *normal tissues* are found to be in the range of 10–100 nM [7–11]. Adenosine receptors involved at these low concentrations are the "high-affinity receptors" A1, A2A and A3. In severe hypoxia, ADO-levels may increase up to 5–10 μM, the lowest level that is necessary to activate the "low-affinity-receptor" A2B [12–14]. In contrast, ADO-concentrations in *experimental tumors* were in the μM-range. Using snap-frozen tissue samples of rat DS-sarcomas, Busse and Vaupel [15] found ADO-levels of 50–100 μM (see Fig. 22.1). In these experiments, [ADO] correlated with tumor size, and the extent of hypoxia, respectively [16]. In experiments using microdialysates of human HT-29 colon cancers in nude mice, Blay et al. [9] found ADO-concentrations of 9–13 μM, also indicating that tumors—in contrast to normal tissues—accumulate ADO in concentrations high enough to even stimulate the "low-affinity-receptor" A2B. In the latter experiments ADO-concentrations did not correlate with tumor size.

Hypoxia-promoted adenosine accumulation is the result of (a) *intracellular generation* from AMP by a cytosolic 5′-AMP nucleotidase and inhibition of adenosine kinase with subsequent ADO-release into the extracellular space through a nucleoside transporter (Fig. 22.2), and (b) *extracellular cleavage* by hypoxia-/HIF-1-sensitive, membrane-associated ecto-5′-nucleotidases (CD39 = EC 3.6.1.5 and CD73 = EC 3.1.3.5) upon channel-mediated transport of adenine nucleotides (ATP, ADP and AMP) or leakage due to nonspecific membrane damage from the intra- into the extracellular compartment of cancer and stromal cells [14, 17].

ADO-producing stromal cells with hypoxia-driven CD39/CD73 activities include Treg cells, T17-helper cells and myeloid-derived immune-suppressive cells (MDSCs). Extracellular nucleotide catabolism most probably is the major source of ADO in the halo of cancer cells due to specific genetic alterations taking place during tumor growth [12].

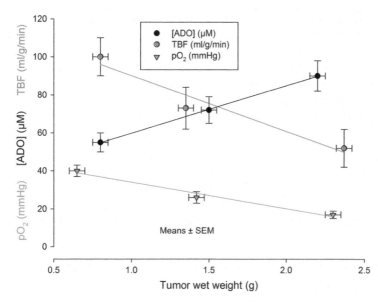

Fig. 22.1 Tissue adenosine levels ([ADO], *black dots*), oxygen partial pressures (pO$_2$, *green triangles*), and blood flow (TBF, *red dots*) as a function of tumor weight (rat DS- sarcomas)

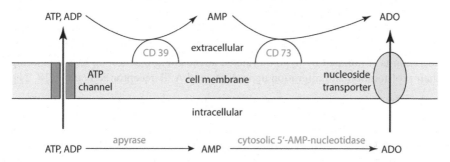

Fig. 22.2 Hypoxia-mediated adenosine (ADO) generation from adenine nucleotides ATP, ADP and AMP in the intra- and extracellular space of solid tumors. CD39, CD73: HIF-dependent ecto-5′-nucleotidases; apyrase: ATPase/ADPase

3 Hypoxia-Driven Adenosinergic Effects on Tumor Cells

Direct effects of ADO on cancer cells are manifold. Mechanisms derived from experimental data are listed in Table 22.1.

Table 22.1 Hypoxia-driven adenosinergic effects on cancer cells (selection). ADO is mainly acting at A3 receptors

Adenosinergic effects	ADO receptor involved	References
Stimulation of cell proliferation	A3, A2A, A1	[12, 14, 18, 19]
Activation of cell cycle	A1	
Increase in HIF-1alpha expression	A3	
Increase in production of pro-angiogenic factors	A3	
Stimulation of tumor cell migration	A3	
Facilitation of metastatic dissemination	A3	
Stimulation of glycolytic flux	A1	

Note: G-protein-coupled A1- and A3-receptors induce a G_i-mediated inhibition of adenylyl cyclase (cAMP-decrease) or G_o/G_q-mediated activation of phospholipase C (increase in InsP3 and DAG), while A2A- and A2B-receptors induce a G_s-mediated activation of adenylyl cyclase (cAMP-increase). All receptors couple to mitogen-activated protein kinase (MAPK) pathways [17, 20]

4 Hypoxia-Driven Adenosinergic Effects on Tumor Angiogenesis

To grow over a size of 1–2 mm in diameter, solid tumors need a blood supply from surrounding vessels, i.e., tumor angiogenesis. The formation of new blood vessels is mainly sustained by the production and secretion of pro-angiogenic factors originating from tumor and stromal cells. These factors include VEGF, bFGF, angiopoietin and IL-8 preferentially secreted by endothelial cells (ECs). This secretion is stimulated by hypoxia-related adenosinergic action on ECs promoting their proliferation and migration upon A2A- and A2B-receptor activation [18, 21].

5 Hypoxia-Driven Adenosinergic Effects on Immune Cells

A broad spectrum of immune-suppressive properties facilitates tumor escape from immune control. To a great extent, this tumor evasion is triggered by immune-suppressive measures of adenosine, mostly through A2A- and A2B-receptor activation following ADO production by cancer cells, T17-helper cells and MDSCs. ADO binds to immune cells and modifies their activity as listed in Table 22.2. In addition, ADO promotes chronic inflammation fostering tumor growth and increases the activity of immune-suppressive cytokines (e.g., TGFß, IL-10).

Table 22.2 Hypoxia-driven adenosinergic effects on immune cells (selection). ADO is mainly acting at A2- receptors

Target cells	Adenosinergic effects	ADO receptor involved	References
Neutrophils	Increase in production of metalloproteases promoting metastasis	A2B, A1	[12–14, 18–20, 22–24]
Macrophages	Activation of a pro-tumor phenotype (M2 macrophages)	A2A	
MDSCs	Expansion of functions promoting tumor tolerance (tumor evasion)	A2B	
NK cells	Decrease in anti-tumor activities	A2A	
CD4+, CD8+ cells	Decrease in anti-tumor activities	A2A	
Dendritic cells	Decrease in activity	A2B	
Treg cells	Facilitation of function, promotion of tumor tolerance	A2A	

6 Other Biologically Relevant Catabolites of Adenine Nucleotide Degradation

As mentioned above, the extracellular ADO accumulation results from the hypoxia-driven adenine nucleotide degradation. During this process, which ends in the formation of *uric acid*, a potent antioxidant in the extracellular space, *reactiveoxygenspecies* (ROS) and *protons* (H^+) are also generated at several stages during this degradation [15]. These products also exert major impacts on tumor growth and therapeutic efficacy.

7 Perspectives

In the light of the data presented and considering recently published articles [19, 25–28], the following therapeutic strategies may help to mitigate (or even eradicate) the tumor-promoting activities of ADO: (a) respiratory hyperoxia as an adjuvant during anti-tumor immunotherapy may convert an immune-suppressive into an immune-permissive tumor microenvironment, (b) blockade of "high-affinity" A2A-receptors with A2A-R antagonists, specific monoclonal antibodies or A2A-siRNA [20, 26, 27, 29–31], (c) use of extracellular ADO-degrading drugs (stabilized adenosine deaminase) [26], and (d) inhibition of ADO-generating "tandem" ecto-enzymes CD39 and CD73 using pharmacologic, small molecule inhibitors, specific monoclonal antibodies or siRNA [20, 32].

Therapeutic interventions described in (b)–(d) may improve the efficacy of anti-cancer immunotherapies by interrupting ADO-dependent immune-evasion [25, 30–32]. In addition, they can result in distinct inhibition of tumor growth and metastasis. They also can reduce microvessel density in tumors and can enhance the efficacy of chemotherapeutic drugs (e.g., doxorubicin, methotrexate, oxaliplatin) [20]. Inhibition of tumor growth and of metastasis can also be achieved using A3-receptor antagonists [20]. Taken together, preclinical in vivo studies have indicated that targeting the adenosinergic pathways has a considerable therapeutic potential when used as monotherapy and in combination with conventional cancer therapies or with immunotherapy. The challenge concerns the ability to selectively inhibit those targets of the adenosinergic pathways that promote tumor progression [19].

References

1. Vaupel P, Höckel M, Mayer A (2007) Detection and characterization of tumor hypoxia using pO_2 histography. Antioxid Redox Signal 9(8):1221–1235
2. Vaupel P, Mayer A, Höckel M (2004) Tumor hypoxia and malignant progression. Methods Enzymol 381:335–354
3. Vaupel P, Mayer A (2007) Hypoxia in cancer: significance and impact on clinical outcome. Cancer Metastasis Rev 26(2):225–239
4. Höckel M et al (1996) Association between tumor hypoxia and malignant progression in advanced cancer of the uterine cervix. Cancer Res 56(19):4509–4515
5. Mayer A, Vaupel P (2013) Hypoxia, lactate accumulation, and acidosis: siblings or accomplices driving tumor progression and resistance to therapy? Adv Exp Med Biol 789:203–209
6. Semenza GL (2012) Hypoxia-inducible factors: mediators of cancer progression and targets for cancer therapy. Trends Pharmacol Sci 33(4):207–214
7. Hauber W (2002) Adenosin: ein Purinnukleosid mit neuromodulatorischen Wirkungen. Neuroforum 8:228–234
8. Schulte G (2004) Adenosin, Adenosinrezeptoren und Adenosinrezeptor-aktivierte Signalwege. Biospektrum 2(40):159–161
9. Blay J, White TD, Hoskin DW (1997) The extracellular fluid of solid carcinomas contains immunosuppressive concentrations of adenosine. Cancer Res 57(13):2602–2605
10. Lasley RD et al (1998) Comparison of interstitial fluid and coronary venous adenosine levels in in vivo porcine myocardium. J Mol Cell Cardiol 30(6):1137–1147
11. MacLean DA, Sinoway LI, Leuenberger U (1998) Systemic hypoxia elevates skeletal muscle interstitial adenosine levels in humans. Circulation 98(19):1990–1992
12. Spychala J (2000) Tumor-promoting functions of adenosine. Pharmacol Ther 87 (2–3):161–173
13. Ghiringhelli F et al (2012) Production of adenosine by ectonucleotidases: a key factor in tumor immunoescape. J Biomed Biotechnol. doi:10.1155/2012/473712
14. Di Virgilio F (2012) Purines, purinergic receptors, and cancer. Cancer Res 72(21):5441–5447
15. Busse M, Vaupel P (1996) Accumulation of purine catabolites in solid tumors exposed to therapeutic hyperthermia. Experientia 52(5):469–473
16. Vaupel P (1994) Blood flow, oxygenation, tissue pH distribution and bioenergetic status of tumors, vol 23, Lecture. Ernst Schering Research Foundation, Berlin
17. Schulte G, Fredholm BB (2003) Signalling from adenosine receptors to mitogen-activated protein kinases. Cell Signal 15(9):813–827
18. Gessi S et al (2011) Adenosine receptors and cancer. Biochim Biophys Acta 1808 (5):1400–1412
19. Muller-Haegele S, Muller L, Whiteside TL (2014) Immunoregulatory activity of adenosine and its role in human cancer progression. Expert Rev Clin Immunol 10(7):897–914

20. Antonioli L et al (2013) Immunity, inflammation and cancer: a leading role for adenosine. Nat Rev Cancer 13(12):842–857
21. Allard B, Turcotte M, Stagg J (2012) CD73-generated adenosine: orchestrating the tumor-stroma interplay to promote cancer growth. J Biomed Biotechnol. doi:10.1155/2012/485156
22. Lee C-T, Mace T, Repasky EA (2010) Hypoxia-driven immunosuppression: a new reason to use thermal therapy in the treatment of cancer? Int J Hyperthermia 26(3):232–246
23. Sitkovsky MV et al (2008) Hypoxia-adenosinergic immunosuppression: tumor protection by T regulatory cells and cancerous tissue hypoxia. Clin Cancer Res 14(19):5947–5952
24. Kumar V (2013) Adenosine as an endogenous immunoregulator in cancer pathogenesis: where to go? Purinergic Signal 9(2):145–165
25. Young A et al (2014) Targeting cancer-derived adenosine: new therapeutic approaches. Cancer Discov 4(8):879–888
26. Sikovsky MV et al (2014) Hostile, hypoxia-A2-adenosinergic tumor biology as the next barrier to overcome for tumor immunologists. Cancer Immunol Res 2(7):598–605
27. Ohta A, Sitkovsky M (2014) Extracellular adenosine-mediated modulation of regulatory T cells. Front Immunol. doi:10.3389/fimmu.2014.00304
28. Antonioli L et al (2014) Adenosine pathway and cancer: where to go from here? Expert Opin Ther Targets 18(9):973–977
29. Iannone R et al (2013) Blockade of A2b adenosine receptor reduces tumor growth and immune suppression mediated by myeloid-derived suppressor cells in a mouse model of melanoma. Neoplasia 15(12):1400–1409
30. Sitkovsky M et al (2008) Adenosine A2A receptor antagonists: blockade of adenosinergic effects and T regulatory cells. Br J Pharmacol 153(S1):S457–S464
31. Merighi S et al (2003) A glance at adenosine receptors: novel target for antitumor therapy. Pharmacol Ther 100(1):31–48
32. Häusler SF et al (2014) Anti-CD39 and anti-CD73 antibodies A1 and 7G2 improve targeted therapy in ovarian cancer by blocking adenosine-dependent immune evasion. Am J Transl Res 6(2):129–139

Chapter 23
Approaching Oxygen-Guided Intensity-Modulated Radiation Therapy

Boris Epel, Gage Redler, Charles Pelizzari, Victor M. Tormyshev, and Howard J. Halpern

Abstract The outcome of cancer radiation treatment is strongly correlated with tumor oxygenation. The aim of this study is to use oxygen tension distributions in tumors obtained using Electron Paramagnetic Resonance (EPR) imaging to devise better tumor radiation treatment. The proposed radiation plan is delivered in two steps. In the first step, a uniform 50 % tumor control dose (TCD_{50}) is delivered to the whole tumor. For the second step an additional dose boost is delivered to radioresistant, hypoxic tumor regions. FSa fibrosarcomas grown in the gastrocnemius of the legs of C3H mice were used. Oxygen tension images were obtained using a 250 MHz pulse imager and injectable partially deuterated trityl OX63 (OX71) spin probe. Radiation was delivered with a novel animal intensity modulated radiation therapy (IMRT) XRAD225Cx microCT/radiation therapy delivery system. In a simplified scheme for boost dose delivery, the boost area is approximated by a sphere, whose radius and position are determined using an EPR O_2 image. The sphere that irradiates the largest fraction of hypoxic voxels in the tumor was chosen using an algorithm based on Receiver Operator Characteristic (ROC) analysis. We used the fraction of irradiated hypoxic volume as the true positive determinant and the fraction of irradiated normoxic volume as the false positive determinant in the terms of that analysis. The most efficient treatment is the one that demonstrates the shortest distance from the ROC curve to the upper left corner of the ROC plot. The boost dose corresponds to the difference between TCD_{90} and TCD_{50} values. For the control experiment an identical radiation dose to the normoxic tumor area is delivered.

Keywords Radiation therapy • Oxygen guided therapy • Oxygen imaging • EPRimaging

B. Epel • G. Redler • C. Pelizzari • H.J. Halpern (✉)
Center for EPR Imaging In Vivo Physiology, University of Chicago, Chicago, IL, USA

Department of Radiation and Cellular Oncology, University of Chicago, Chicago, IL, USA
e-mail: h-halpern@uchicago.edu

V.M. Tormyshev
Center for EPR Imaging In Vivo Physiology, University of Chicago, Chicago, IL, USA

Novosibirsk Institute of Organic Chemistry, Novosibirsk, Russia

© Springer Science+Business Media, New York 2016
C.E. Elwell et al. (eds.), *Oxygen Transport to Tissue XXXVII*, Advances in
Experimental Medicine and Biology 876, DOI 10.1007/978-1-4939-3023-4_23

1 Introduction

Common radiation delivery protocols used in cancer treatment deliver homogeneous radiation dose to a tumor [1]. This ensures destruction of cancerous cells
but does not take into account different radioresistance in different portions of a
tumor. It is known that tumors treated to a uniform 50 % tumor control dose
(TCD_{50}) exhibit different control probability depending on their oxygenation
(Fig. 23.1) [2]. Hypoxia desensitizes tumors to radiation and mandates higher
treatment doses. Knowledge of the spatial distributions of radioresistant tumor
portions in combination with Intensity-Modulated Radiation Therapy (IMRT)
may be used for targeted destruction of radiation-resistant areas (and sparing
healthy tissues, dose painting) [3]. We expect partial oxygen pressure (pO_2) in
tumor portions to be an excellent targeting parameter. This article describes the
design of the experiment for the validation of oxygen-guided IMRT on mice.

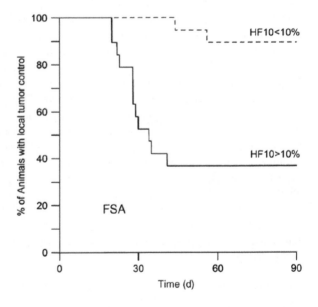

Fig. 23.1 Kaplan-Meier
plot showing the percentage
of animals with local tumor
control as a function of time
after treatment with a single
TCD50 dose (33.8 Gy) of
X-rays [2]. Hypoxic
fraction of voxels below
10 torr 10 % (HF10) was
used for analysis. FSa
fibrosarcoma tumor mode
was used

2 Methods

2.1 Experiment

Figure 23.2 presents the flow chart of the experiment. FSa fibrosarcomas grown in the gastrocnemius of the legs of C3H mice were used. 10-min pO_2 images were obtained using a 250 MHz pulse EPR imager (Fig. 23.3a) [4] and injectable partially deuterated trityl OX63 (OX71) spin probe synthesized by the Novosibirsk Institute of Organic Chemistry. A spin–lattice relaxation based oxygenimaging protocol was used [5]. For tumor definition, an anatomic MRI image was taken prior to oxygen image. EPR and MRI images were registered with the help of fiducials embedded into a vinyl polysiloxane dental mold [6]. The ArbuzGUI MATLAB toolbox developed by the Center for EPR Imaging in vivo Physiology at the University of Chicago was used for image registration. For administration of radiation treatment an XRAD225Cx micro-CT/therapy delivery system (Figs. 23.3b and 23.4a) was used. For the first radiation treatment step, a uniform irradiation of

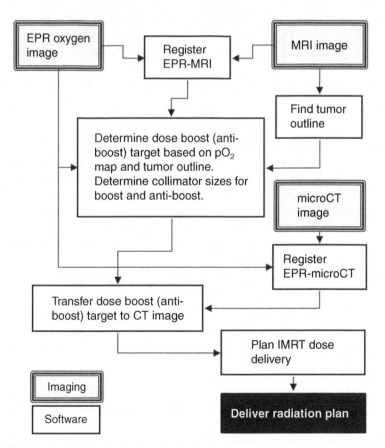

Fig. 23.2 Flow chart of the treatment protocol

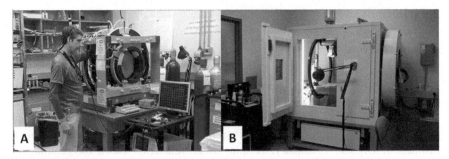

Fig. 23.3 (**a**) 250 MHz pulse Electron Paramagnetic Resonance oxygen imager. (**b**) Precision X-ray XRAD225Cx image-guided biologic irradiator/CT imager

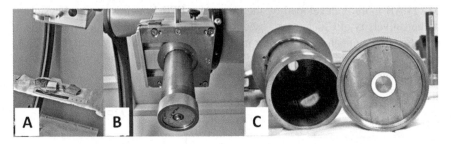

Fig. 23.4 Irradiation setup. (**a**) Animal placed on the animal support. (**b**) Irradiation of head with the collimator installed. (**c**) Collimator set-up for anti-boost treatment protocol. The diaphragm consists of two concentric lead elements to deliver shell-shaped radiation dose

the whole tumor to a 50 % control dose (TCD_{50}) dose of 33.8 Gy was used. This dose was delivered using two opposed beams that cover the whole tumor in anterior-posterior, posterior-anterior alignments. Then a boost dose corresponding to the difference between TCD_{90} and TCD_{50} was delivered to hypoxic areas (see Sects. 2.2 and 2.3). The boost dose was delivered by a 360° arc beam. To target the IMRT boost, a CT image was taken. Using similar fiducial technology the CT was registered to EPR image. Then the coordinates of the boost dose were transferred from the EPR to the CT image.

Animal experiments followed USPHS policy, and were approved by the Institutional Animal Care and Use Committee.

2.2 Boost Dose Definition Algorithm

For boost dose delivery we devised a simplified scheme, where the boost volume is approximated by a sphere. The XRAD225Cx has a single beam treatment capability with 1 mm resolution. The delivery of arbitrary radiation pattern with this resolution is a very lengthy procedure. Mouse tumors models that we treat have a diameter of about 8 mm and exhibit a very heterogeneous oxygen environment (Fig. 23.5a). In most cases the core of the tumor is very hypoxic and can be approximated by a sphere (Fig. 23.5b).

From possible positions of the radiation boost sphere, we choose the one that irradiates the largest fraction of hypoxic voxels in the tumor pO_2 image, while minimizing exposure of well-oxygenated voxels. Voxels with the oxygen tension below 10 Torr were considered hypoxic, voxels above that threshold—well oxygenated. The sphere radius was chosen using an algorithm based on Receiver Operator Characteristic (ROC) analysis [7]. We used the fraction of irradiated hypoxic volume as the true positive fraction and the fraction of irradiated normoxic volume as the false positive fraction in the terms of that analysis (Fig. 23.6). The most efficient treatment is the one that demonstrates the shortest distance from the ROC curve to the upper left corner of the plot (Fig. 23.6). For the control experiment an identical radiation dose to the normoxic tumor area is delivered.

The control scheme for validation of the proposed method efficiency includes radiation delivery to tumor or normal tissue of equal volume, where the hypoxic areas of the tumor are avoided by delivering a shell-shaped radiation dose to well-oxygenated areas equal to the boost dose. The collimator used for anti-boost delivery is presented in Fig. 23.4b, c. It consists of two coaxial elements: one with a circular opening and circular radiation blocker (Fig. 23.4c). The cross-sections of radiation beam for boost and anti-boost therapies are shown in the Fig. 23.5b.

Fig. 23.5 EPR oxygen image of the leg tumor. (**a**) Oxygen map with tumor contour (*white contour*) transferred from the registered MRI image. (**b**) Boost (*red line*) and anti-boost (*shaded area*) as determined by the boost planning software

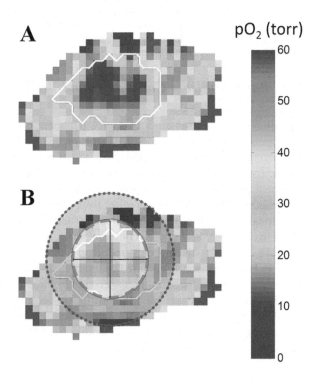

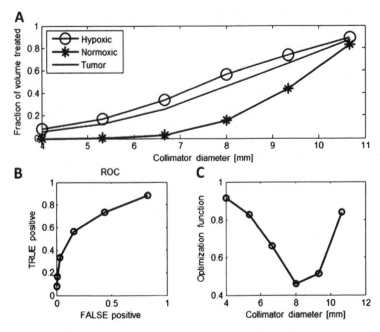

Fig. 23.6 Dose boost planning algorithm for both hypoxic region targeting and hypoxic region avoidance. (**A**) Plot of fractions of whole tumor voxels -, hypoxic voxels o and normoxic voxels * as a function of spherical boost diameter (**B**) Receiver operator characteristic (ROC) plotting the fraction of hypoxic voxels (true positive) vs fraction of well oxygenated voxels (false positive) for each boost diameter. (**C**) Minimizing optimization fraction (distance from the upper left corner of the ROC) at optimum of 8 mm diameter

2.3   Radiation Dose Calculation and Verification

Radiation planning used an open-source treatment planning system developed in MATLAB, which was provided by Precision X-ray (North Branford, CT, USA). The software allows identification of a target point and treatment volume on a mouse cone-beam CT (CBCT) scan taken with the animal immobilized in the treatment position after it has been transferred from the EPRI O_2 imager to the treatment machine. The user can select from a set of available beam sizes (normally cylindrical, though this is not a requirement) and choose a set of incident directions, which may be discrete angles or continuous arcs, all parallel to the transverse CBCT plane, i.e., normal to the inferior-superior axis of the animal. Given a user-specified dose per beam, previously acquired calibration dosimetry data are used to compute the integrated exposure in milliamp-seconds required for each discrete field or arc. A radiation protocol is saved on the instrument control PC which is read in by the XRad225Cx control software to automatically deliver the specified beams, sequencing from one beam or arc to the next and turning the X-ray beam on and off for the required time and beam current.

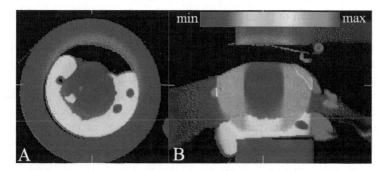

Fig. 23.7 Axial and sagittal views of the anti-boost dose intensity profile as obtained from Monte Carlo calculations. The dose is delivered using a 360° arc turn of the irradiation head located in the plane perpendicular to the plane of the *left panel* (**A**). (**B**) Shows sagittal dose distribution. The EGSnrc system from the National Research Council of Canada was used for calculations

Dose distributions are verified using Monte Carlo calculations performed offline. The calculations were too long to be carried during the oxygen targeting protocol. The EGSnrc system from the National Research Council Canada [8] is used for these calculations. A model of the XRad225Cx treatment head including the Comet MXR-225/22 X-ray tube, primary collimator, aluminum and copper filters, and treatment apertures (cones) for each available radiation field size have been built using the BEAMnrc user code. For each field size and beam quality, a phase space file is computed which summarizes the energy, spatial and orientation distribution of radiation crossing the aperture exit. Histories of 500 million to 1 billion incident electrons are used to generate each phase space file. These pre-computed phase space files are then used in the DOSXYZnrc user code [9] to simulate radiation transport and dose deposition through the case-specific animal geometry derived from the pretreatment CBCT [10] used for treatment planning. Dose is typically accumulated on a 3D, isotropic 0.2 mm grid for 100–500 million X-ray histories incident from the phase space plane. Results can be superimposed in 3D with the CBCT image, which is in the same coordinate system, as illustrated in Fig. 23.7. Any other image volume registered with the CBCT, in particular the EPR oxygen map, can also be superimposed with the Monte Carlo generated dose distribution for display and analysis.

3   Discussion

The delivery of the oxygen-guided radiotherapy to small rodents presents a substantial technological challenge. The typical size of tumors in our studies is about 8 mm in diameter. To achieve spatial localization of treatment not worse than 15 % of tumor diameter, all stages of the protocol have to exhibit better than 1 mm precision. The resolution of CT and MRI images are considerably higher than

1 mm. The resolution of the EPR image is about 1.2 mm. However, the fiducial image used for image registration is acquired in higher resolution (0.66 mm). According to our experience, the registration procedure for tumor bearing leg gives a precision comparable with the resolution of the lowest resolved image, about 0.6 mm. The leg tumor is not susceptible to breathing motion artifacts and its position is well preserved by the vinyl polysiloxane cast during the whole imaging and treatment protocol. Finally, the spatial resolution of the radiation delivery system, considering beam penumbra and the interaction of the radiation beam with tissues and other materials in the radiation window, is estimated to be ~ 1 mm. Thus the overall spatial resolution of the oxygen guided intensity-modulated radiation therapy is about 1 mm. This resolution appears to be consistent with our results demonstrating correlation between the outcome of radiation therapy and tissue oxygenation in tumors [2].

4 Conclusions

This work presents the first practical implementation of EPRoxygen image-guided radiation therapy. The experiments determining the outcome of the proposed scheme are currently underway.

Acknowledgments Supported by NIH grants P41 EB002034 and R01 CA98575. Monte Carlo computations are performed on a computer cluster administered by the Department of Radiology and partially supported by the University of Chicago Comprehensive Cancer Center (NCI P30 CA014599).

References

1. Khan FM (2010) The physics of radiation therapy, 4th edn. Lippincott Williams & Wilkins, Philadelphia
2. Elas M, Magwood JM, Butler B et al (2013) EPR oxygen images predict tumor control by a 50% tumor control radiation dose. Cancer Res 73:5328–5335
3. Ling CC, Humm J, Larson S et al (2000) Towards multidimensional radiotherapy (MD-CRT): biological imaging and biological conformality. Int J Radiat Oncol Biol Phys 47:551–560
4. Epel B, Sundramoorthy SV, Mailer C et al (2008) A versatile high speed 250-MHz pulse imager for biomedical applications. Concepts Magn Reson B 33B:163–176
5. Epel B, Bowman MK, Mailer C et al (2014) Absolute oxygen R_{1e} imaging in vivo with pulse electron paramagnetic resonance. Magn Reson Med 72:362–368
6. Haney CR, Fan X, Parasca AD et al (2008) Immobilization using dental material casts facilitates accurate serial and multimodality small animal imaging. Concepts Magn Reson B 33B:138–144
7. Metz CE (1978) Basic principles of ROC analysis. Semin Nucl Med 8:283–298

8. Kawrakow I (2000) Accurate condensed history Monte Carlo simulation of electron transport. I. EGSnrc, the new EGS4 version. Med Phys 27:485–498

9. Kawrakow I, Walters BRB (2006) Efficient photon beam dose calculations using DOSXYZnrc with BEAMnrc. Med Phys 33:3046–3056

10. Rogers DWO, Faddegon BA, Ding GX et al (1995) Beam - a Monte-Carlo code to simulate radiotherapy treatment units. Med Phys 22:503–524

Chapter 24
Hypoxia-Associated Marker CA IX Does Not Predict the Response of Locally Advanced Rectal Cancers to Neoadjuvant Chemoradiotherapy

Arnulf Mayer, Peter Vaupel, Katja Oberholzer, Maren Ebert, Marc Dimitrow, Andreas Kreft, Wolfgang Mueller-Klieser, and Heinz Schmidberger

Abstract Hypoxia-associated proteome changes have been shown to be associated with resistance to chemo- and radiotherapy. Our study evaluated the role of the hypoxia-inducible (HIF)-1 target gene carbonic anhydrase (CA) IX in the prediction of the response to neoadjuvant chemoradiotherapy in locally advanced rectal cancer (stages II and III). A total of 29 pretreatment biopsy specimens were stained for CA IX by immunohistochemistry, converted to digital images and evaluated in a quantitative fashion using image analysis software. Contrary to our expectations, a trend towards a correlation between better tumor regression (>50 %) and higher expression of CA IX ($p = 0.056$) was found. CA IX was also present more frequently in pathological tumor stage T1 (pT1) tumors ($p = 0.048$). Conversely, no association with lymph node metastasis was identified. In conclusion, as a single marker, CA IX expression is not able to identify a hypoxia-related treatment resistant phenotype in rectal cancer.

Keywords Tumor hypoxia • Rectal cancer • Carbonic anhydrase IX • Neoadjuvant chemoradiotherapy • Tumor regression grading

A. Mayer (✉) • P. Vaupel (✉) • M. Dimitrow • H. Schmidberger
Department of Radiooncology and Radiotherapy, University Medical Center, Langenbeckstrasse 1, 55131 Mainz, Germany
e-mail: arnmayer@uni-mainz.de; vaupel@uni-mainz.de

K. Oberholzer • M. Ebert
Department of Diagnostic and Interventional Radiology, University Medical Center, Mainz, Germany

A. Kreft
Institute of Pathology, University Medical Center, Mainz, Germany

W. Mueller-Klieser
Institute of Physiology and Pathophysiology, University Medical Center, Mainz, Germany

© Springer Science+Business Media, New York 2016
C.E. Elwell et al. (eds.), *Oxygen Transport to Tissue XXXVII*, Advances in Experimental Medicine and Biology 876, DOI 10.1007/978-1-4939-3023-4_24

1 Introduction

5-FU-based neoadjuvant chemoradiotherapy is the current standard of care for locally advanced rectal cancer (clinical stages II and III). In conjunction with surgery according to the principle of total mesorectal excision, this combined modality treatment achieves low local failure rates in the range of 6–8 %. However, the rate of distant metastases, which frequently are the limiting factor for overall survival, approaches 40 % [1–3]. Studies have shown that the degree of histopathological response to neoadjuvant chemoradiotherapy is inversely proportional to the probability of the occurrence of distant metastases [4]. Patients with poor tumor regression may require a more intensive systemic treatment. Conversely, patients who do achieve a complete pathological response have an excellent prognosis. It is currently being discussed whether subsequent tumor resection may be dispensable in these latter patients, provided that frequent clinical follow-up examinations are carried out [5, 6]. Identification of pathophysiological mechanisms involved in the resistance to neoadjuvant chemoradiation is of high clinical interest, since this may permit the development of predictive clinical tests for the stratification of patients. Hypoxia-dependent proteome alterations were repeatedly correlated with the development of resistance to radio- and chemotherapy in experimental and clinical studies [7, 8]. Markers of the hypoxic response, therefore, represent candidate predictive tests. In the present study, we investigated whether the hypoxia-associated endogenous marker carbonic anhydrase (CA) IX is useful to predict the response to neoadjuvant chemoradiotherapy.

2 Methods

The expression of CA IX was determined in pretreatment biopsies of 29 patients with rectal cancer (stages II and III) by immunohistochemistry (IHC). Patients received neoadjuvant radiotherapy to the pelvis up to a total dose of 50.4 Gy, five times weekly (1.8 Gy per fraction). 5-FU was administered as a continuous infusion over 5 days at a dose of 1000 mg per square meter of body surface area per day (with a maximum daily dose of 1800 mg) in the first and the fifth weeks of radiotherapy.

Regarding IHC technique, 3 μm-thick sections were prepared from paraffin blocks and dried overnight at 37 °C (Histology core facility, University Medical Center, Mainz). On the next day, specimens were dewaxed in two changes of fresh xylene and then rehydrated in a descending alcohol series. Retrieval of antigenic binding sites was performed by heating specimens in citrate buffer (pH 6.0) in a steamer (Braun FS 10, Braun, Kronberg, Germany) for 40 min. The primary anti-CA IX antibody (Cat.-No. 3829-1, Epitomics, Burlingame, CA, USA) was incubated overnight at 4 °C. The biotin-free, micropolymer-based Vector Immpress reagent (Vector Laboratories, Burlingame, CA) was used for the detection and

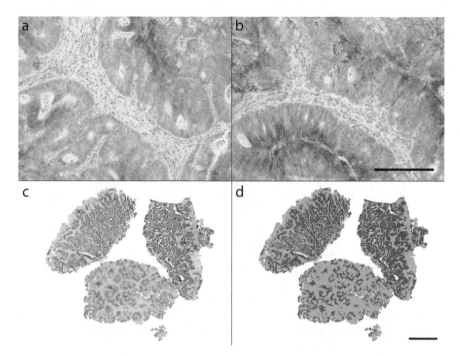

Fig. 24.1 (**a**, **b**) Membranous expression of CA IX (*brown*) is preferentially found at a distance from the tumor stroma which carries microvessels. (**c**) Image of a digital scan of a rectal cancer biopsy. (**d**) Quantification of antigen expression using Image Pro Premier; CA IX positive areas are shown in *blue*, negative areas (including tumor stroma) is shown in *green*. See text for details. Scale bars for panels *a* and *b*: 200 μm, *c* and *d* 2 mm

visualization of primary antibody binding. Negative control specimens were incubated in PBS without the primary antibody under the same conditions. Diaminobenzidine (DAB) was used as the peroxidase substrate. Slides were counterstained with Mayer's hematoxylin, dehydrated in an ascending alcohol series, and covered with a coverslip using Roti-Histokitt mounting medium (Carl Roth, Karlsruhe, Germany). Only immunostaining compatible with the known biological function and corresponding membranous localization of CA IX was considered as being evaluable as marker expression [see Fig. 24.1a and b; images captured using a microscope (AxioImager, Zeiss, Oberkochen, Germany) equipped with a digital camera (AxioCam HRc, Zeiss) connected to a Microsoft Windows-based PC running AxioVision (Zeiss)]. For the quantitative analysis, digital images of the specimens were acquired using a histology scanner (OpticLab H850, Plustek, Taipei, Taiwan; Fig. 24.1c) and quantitatively evaluated using Image Pro Premier 9.1 (Media Cybernetics, Rockville, MD; Fig. 24.1d). Proportions of antigen-positive and -negative areas were determined relative to the total tumor area. The data obtained were correlated with the degree of tumor regression (0–25, 26–50, 51–75, 76–100 %, Fig. 24.2), tumor stage and lymph node status.

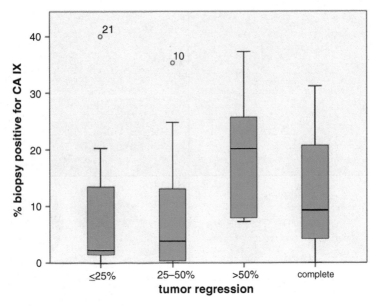

Fig. 24.2 *Box*-and-*whisker plots* showing the percentage of CA IX-positive tumor areas (of the total biopsy area) in four classes of histopathological tumor regression following neoadjuvant chemoradiotherapy and assessed in surgical specimens of total mesorectal excision surgery

3 Results

Expression of CA IX was higher in tumors exhibiting a tumor regression > 50 % or a complete tumor regression (Fig. 24.2) compared to tumors in the lower two classes of the regression grading. After dichotomization of tumor regression grades into two groups (0–50 vs. 51–100 %), this observation became a clear statistical trend towards a higher expression of CA IX in the group with a good response (51–100 % regression) to neoadjuvant chemoradiation (Mann–Whitney U-test, p = 0.056). In addition, there was a significantly higher expression of CA IX in pT1 tumors compared to higher pT stages (Kruskal Wallis test, p = 0.048). Conversely, no correlation could be identified between the expression of CA IX and lymph node status.

4 Discussion

Contrary to our initial expectations derived from established knowledge about hypoxia-associated therapeutic resistance, results of our analysis indicated a clear trend towards a positive correlation between the expression of CA IX and the response to neoadjuvant chemoradiotherapy. Our data also contradict the findings of a prior study by Guedj et al. [9], who investigated the expression of CA IX in 40 pretreatment biopsies. These authors could show a significantly lower expression of CA IX in patients who achieved a pathologic complete response or had only microscopic residual tumor foci after neoadjuvant 5-FU-based (long-course) chemoradiotherapy. There is no obvious explanation for the differences between these two studies, although the methodological details of CA IX quantification are not clearly stated in the study of Guedj et al. [9]. Since CA IX is regarded as the central element of a proton extrusion mechanism in cancer cells [10], it is worth mentioning that an enhanced uptake of 5-FU has been described under conditions of lowered extracellular pH [11], which may have contributed to higher chemosensitivity of CA IX expressing cells. Our study also showed the unexpected finding of a higher CA IX expression in pT1 tumors. Similar data have been reported by Rasheed et al. [12], who found a larger proportion of CA IX positive tumors in earlier tumor stages. These authors, however, also reported a significant correlation between CA IX expression and negative lymph nodes, while we could not show an association between these parameters. Visual assessment of the expression pattern of CA IX in rectal cancers also revealed hints at differential causative mechanisms. While a 'hypoxia-dependent' pattern with increasing intensity in diffusion-limited areas of the tissue was observed regularly, we also detected a uniform pattern which suggested no such dependency, even within the same tumor biopsy. However, additional markers, e.g., CD34 and other hypoxia-associated proteins need to be assessed to possibly identify different types of CA IX expression which may have different pathophysiological implications. This question should preferably be addressed by a multiparametric immunohistochemical/immunofluorescence assay. Based on our current data, however, CA IX alone seems inappropriate for the prediction of the response to neoadjuvant chemoradiotherapy.

References

1. Sauer R, Becker H, Hohenberger W et al (2004) Preoperative versus postoperative chemoradiotherapy for rectal cancer. N Engl J Med 351(17):1731–1740
2. Bosset J-F, Collette L, Calais G et al (2006) Chemotherapy with preoperative radiotherapy in rectal cancer. N Engl J Med 355(11):1114–1123
3. Gerard J-P, Conroy T, Bonnetain F et al (2006) Preoperative radiotherapy with or without concurrent fluorouracil and leucovorin in T3-4 rectal cancers: results of FFCD 9203. J Clin Oncol 24(28):4620–4625
4. Fokas E, Liersch T, Fietkau R et al (2014) Tumor regression grading after preoperative chemoradiotherapy for locally advanced rectal carcinoma revisited: updated results of the CAO/ARO/AIO-94 trial. J Clin Oncol 32(15):1554–1562
5. Habr-Gama A, Sabbaga J, Gama-Rodrigues J et al (2013) Watch and wait approach following extended neoadjuvant chemoradiation for distal rectal cancer: are we getting closer to anal cancer management? Dis Colon Rectum 56(10):1109–1117
6. Maas M, Beets-Tan RGH, Lambregts DMJ et al (2011) Wait-and-see policy for clinical complete responders after chemoradiation for rectal cancer. J Clin Oncol 29(35):4633–4640

7. Semenza GL (2012) Hypoxia-inducible factors: mediators of cancer progression and targets for cancer therapy. Trends Pharmacol Sci 33(4):207–214
8. Mayer A, Vaupel P (2011) Hypoxia markers and their clinical relevance. In: Osinsky S, Friess H, Vaupel P (eds) Tumor hypoxia in the clinical setting. Akademperiodyka, Kiev, pp 187–202
9. Guedj N, Bretagnol F, Rautou PE et al (2011) Predictors of tumor response after preoperative chemoradiotherapy for rectal adenocarcinomas. Hum Pathol 42(11):1702–1709
10. Potter CP, Harris AL (2003) Diagnostic, prognostic and therapeutic implications of carbonic anhydrases in cancer. Br J Cancer 89(1):2–7
11. Ojugo AS, McSheehy PM, Stubbs M et al (1998) Influence of pH on the uptake of 5-fluorouracil into isolated tumour cells. Br J Cancer 77(6):873–879
12. Rasheed S, Harris AL, Tekkis PP et al (2009) Assessment of microvessel density and carbonic anhydrase-9 (CA-9) expression in rectal cancer. Pathol Res Pract 205(1):1–9

Chapter 25
Impact of Oxygenation Status on ^{18}F-FDG Uptake in Solid Tumors

Marie-Aline Neveu, Vanesa Bol, Anne Bol, Vincent Grégoire, and Bernard Gallez

Abstract The influence of changes in tumor oxygenation (monitored by EPR oximetry) on the uptake of ^{18}F-FDG tracer was evaluated using micro-PET in two different human tumor models. The ^{18}F-FDG uptake was higher in hypoxic tumors compared to tumors that present a pO_2 value larger than 10 mmHg.

Keywords EPR • ^{18}F-FDG • PET • Carbogen • Tumor oxygenation

1 Introduction

High fluorodeoxyglucose (^{18}F-FDG) uptake may be a direct consequence of the upregulation of the glucose transporters (GLUTs) stimulated by hypoxia. However, high ^{18}F-FDG uptake can also arise even under non-hypoxic condition through a situation known as aerobic glycolysis or the Warburg effect. The issue of the relationship between glucose uptake, GLUTs expression and hypoxia within tumors has been debated in the literature [1–4]. To assess the influence of manipulation of tumor oxygenation on the uptake of ^{18}F-FDG, we evaluated the uptake of ^{18}F-FDG using micro-PET imaging under different breathing conditions in parallel with pO_2 measurements with EPR oximetry in two different human tumor models, MDA-MB-231 and SiHa models.

M.-A. Neveu • B. Gallez (✉)
Biomedical Magnetic Resonance Group, Louvain Drug Research Institute, Université Catholique de Louvain, Brussels, Belgium
e-mail: bernard.gallez@uclouvain.be

V. Bol • A. Bol • V. Grégoire
Radiation Oncology Department & Center for Molecular Imaging, Institute of Experimental and Clinical Research, Université Catholique de Louvain, Brussels, Belgium

© Springer Science+Business Media, New York 2016
C.E. Elwell et al. (eds.), *Oxygen Transport to Tissue XXXVII*, Advances in Experimental Medicine and Biology 876, DOI 10.1007/978-1-4939-3023-4_25

2 Material and Methods

A total of 10^7 MDA-MB-231 cells (ATCC, Manassas, USA) or 10^7 SiHa cells (ATCC, Manassas, USA), amplified in vitro, were collected by trypsinization, washed three times with Hanks balanced salt solution and resuspended in 200 μL of a 1:1 mixture of Matrigel (BD Biosciences) and Hanks balanced salt solution. The tumor cells were inoculated subcutaneously into the hind thigh of nude NMRI female mice (Janvier, Le Genest-Saint-Isle, France). The experiments were performed when tumors reached 7 mm in diameter.

Mice were scanned twice for the breathing challenge, air versus carbogen breathing, with 1 day between each condition. The details of the protocol are presented in Fig. 25.1. For each breathing condition, EPR measurements were performed before PET imaging and final anatomical images were acquired by CT scan. Mice were fasted overnight before measurements. Animals were anesthetized by inhalation of isoflurane (Forene, Abbot, England) mixed with either air (21 % oxygen) or carbogen (5 % CO_2 in oxygen), depending on the breathing condition tested, in a continuous flow (2 L/min). Fasted animals were warmed (approximately 35 °C) throughout the anesthesia period.

In vivo tumor pO_2 was monitored by EPR oximetry using charcoal as the oxygen-sensitive probe [5, 6]. EPR spectra were recorded using a 1.1 GHz EPR spectrometer (Magnettech, Berlin, Germany). A charcoal suspension (100 mg/mL) was injected intratumorally (60 μL) 24 h before experiments. For EPR reading, the tumor under study was placed in the center of the extended loop resonator. For air condition, basal measurements were performed. For carbogen condition, pO_2 measurements were started after a 10 min inhalation period. According to calibration curves [5], the EPR line width was converted to pO_2.

Whole-body PET imaging was performed on a dedicated small-animal PET scanner (Mosaic, Philips Medical Systems, Cleveland, USA) with a spatial resolution of 2.5 mm (FWHM). The PET scans were followed by whole-body acquisitions using a helical CT scanner (NanoSPECT/CT Small Animal Imager, Bioscan Inc., DC, USA). For each breathing condition, anesthetized mice were injected 120 μL intraperitoneally with 300–400 μCi of ^{18}F-FDG (Betaplus Pharma, Brussels, Belgium).

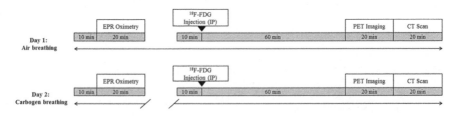

Fig. 25.1 Experimental Protocol

A 10 min transmission scan was first obtained in a single mode using a 370 MBq ^{137}Cs source for attenuation correction. A 10 min static PET acquisition was then performed after a 60 min resting period. After the correction with attenuation factors obtained from the transmission scan, images were reconstructed using a fully 3D iterative algorithm (3D-RAMLA) in a $128 \times 128 \times 120$ matrix, with a voxel of 1 mm^3. After PET acquisition, anesthetized animals were transferred on the same bed from the PET scanner to the CT scanner (X-ray tube voltage: 55 kVp; number of projections: 180; exposure time 1000 ms). The CT projections were reconstructed with a voxel size of $0.221 \times 0.221 \times 0.221$ mm^3.

Regions of Interest (ROIs) were delineated on fused PET/CT images using PMOD software (PMOD™, version 3.403, PMOD technologies Ltd, Zurich, Switzerland). 2D ROIs were established on consecutive transversal slices using a 50 % isocontour tool (ROI including the pixel values larger than 50 % of the maximum pixel) that semi-automatically defined a 3D Volume of Interest (VOI) around the tissue of interest. The global tracer uptake was assessed in tumors and expressed as standardized uptake values (SUV).

Paired t-tests were used to compare mean changes between groups (air vs. carbogen) for each tumor model and non-paired t-tests were used to compare mean changes between the two tumor models. Analysis was performed using the Graphpad software. Results were expressed as mean value of parameter $\pm$ SEM. For all tests, results with p-values < 0.05 (*), <0.01 (**), or <0.001 (***) were considered significant. The scatter plots of measured pO$_2$ versus SUV were traced using data from all tumors of both groups. The process of finding the best fit was done by using CurveExpert software (version 1.4).

3 Results

For each breathing condition, air or carbogen breathing, the oxygen status of MDA-MB-231 tumors (n = 16) and SiHa tumors (n = 11) was assessed by using EPR spectroscopy, followed by PET imaging for tracer uptake study.

A significant change in pO$_2$ was observed during the breathing challenge for each tumor model (Fig. 25.2). Basal pO$_2$ measured was 4 ± 1 mmHg for MDA-MB-231 tumors and 5 ± 1 mmHg for SiHa tumors. Under carbogen breathing, MDA-MB-231 and SiHa tumors reached pO$_2$ values around 10 ± 1 mmHg and 16 ± 1 mmHg, respectively.

Acute changes in global ^{18}F-FDG uptake linked to carbogen challenge were found in this study. In Fig. 25.3, we can observe a significant decrease in the uptake of ^{18}F-FDG under carbogen compared to air breathing in both tumor models. SUV measured on PET images were 0.675 ± 0.023 under air and 0.548 ± 0.017 under carbogen for MDA-MB-231 tumors, and 0.678 ± 0.023 under air and 0.553 ± 0.019 under carbogen for SiHa tumors. There were no differences in ^{18}F-FDG uptake between the two tumor models.

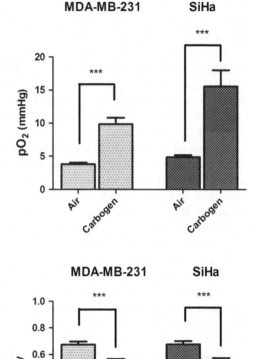

Fig. 25.2 The mean values $\pm$ SEM of pO_2 measured by EPR during the breathing challenge in MDA-MB-231 tumors (*light grey*) and SiHa tumors (*dark grey*)

Fig. 25.3 ^{18}F-FDG uptake under normoxia and hyperoxia in MDA-MB-231 tumors (*light grey*) and SiHa (*dark grey*). A lower uptake of ^{18}F-FDG was observed under hyperoxic conditions compared to normoxic conditions in both tumor models

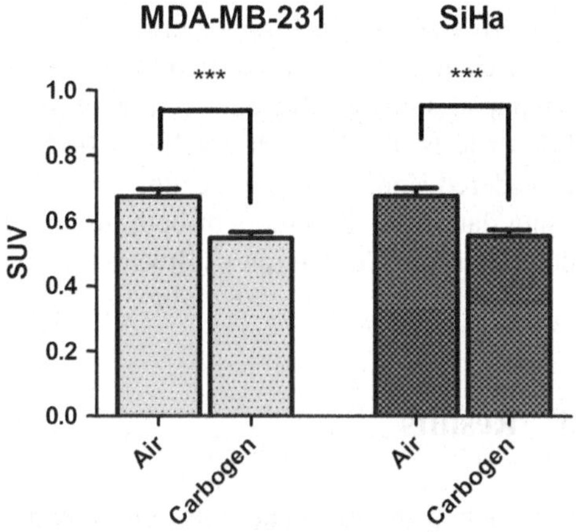

In Fig. 25.4, the relationship between global ^{18}F-FDG uptake and pO_2 measurements obtained from individual tumors (mice breathing air or carbogen) is presented as a non-linear fit (modified exponential, $r = 0.557$). The ^{18}F-FDG uptake was higher in hypoxic tumors compared to tumors with pO_2 larger than 10 mmHg.

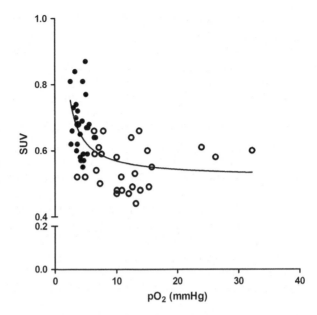

Fig. 25.4 Relationship between ^{18}F-FDG uptake and pO$_2$ values during air (*filled symbol*) and carbogen breathing (*open symbol*). The ^{18}F-FDG uptake was higher in hypoxic tumors compared to tumors with pO$_2$ larger than 10 mmHg. According to CurveExpert software, the best fit is $y = 0.52\,e^{0.93/x}$

4   Discussion

In this study, carbogen breathing was used as a modulator of hypoxia in the two tumor models and pO$_2$ variation has been assessed by EPR oximetry. To our knowledge, it is the first time that EPR, a highly sensitive method to measure pO$_2$ values in vivo [7], has been associated to global ^{18}F-FDG uptake measurement in vivo. Furthermore, the experimental design was built as a dynamic follow-up of the tumors during a breathing challenge. Here we assessed the correlation between global PET uptake and pO$_2$ values for each tumor during the breathing challenge.

We found that the uptake of ^{18}F-FDG was higher in tumors with a pO$_2$ value inferior to 10 mmHg. This observation is consistent with the upregulation of GLUT-1 that is associated with hypoxia. Nevertheless, we also found that the uptake of ^{18}F-FDG was lower after a short period of carbogen breathing. This observation emphasizes that the uptake is not only depending on the GLUT-1 expression, but depends on the rapid adaptation of the metabolism of the tumor cells when oxygen became available as well, phenomenon known as the Pasteur Effect (glycolysis inhibition in presence of oxygen) [8]. Furthermore, as mentioned by Thews et al., GLUT-1 expression is also controlled by other microenvironmental parameters not only oxygen dependent [4]. The change in ^{18}F-FDG uptake was true for both tumor models although their metabolic phenotype characterized in vitro indicate that the MDA-MB-231 tumor model is highly glycolytic [8] compared to the SiHa model that possess an oxidative phenotype [9].

Our results could also suggest that, beyond tumor delineation based on metabolism evaluation, ^{18}F-FDG uptake could also indirectly reflect the oxygenation status of tumors, a major factor involved in tumor progression and resistance to therapy. The fact that ^{18}F-FDG could indirectly reflect the level of hypoxia has already been discussed in several studies [4, 10–19]. Our results are consistent with the study of Christian et al. that showed that the ^{18}F-FDG uptake was higher under severe hypoxia compared to normoxic conditions [19]. However, this trend was found only when considering the whole tumor, as authors also found that the correlation at the microscopic level was poor when considering the co-localization of ^{18}F-FDG uptake and of the nitroimidazole ^{14}C-EF3 [19]. On the other hand, Thews et al. found that ^{18}F-FDG uptake did not vary with the tumor volume in DS-sarcoma, whereas the oxygenation status was determined to be impaired with increasing tumor volume in this tumor model [4]. However, no dynamic follow up of the tumors was assessed in this study in order to evaluate the impact hypoxia on ^{18}F-FDG uptake in the same animals over time. The fact that ^{18}F-FDG could also be used as a hypoxia biomarker could be attractive, but our results should not be over-interpreted. They just highlight the rapid plasticity of the tumor cells to adapt their metabolism to the oxygen environment.

In conclusion, our study showed that global ^{18}F-FDG uptake was higher in hypoxic tumors compared to tumors with pO_2 values above 10 mmHg, and that acute changes in ^{18}F-FDG uptake were observed during carbogen challenge, demonstrating a rapid metabolic adaptation in both tumor models investigated.

Acknowledgments This study was supported by grants from the Belgian National Fund for Scientific Research (FNRS).

References

1. Dierckx RA, Van de Wiele C (2008) FDG uptake, a surrogate of tumour hypoxia? Eur J Nucl Med Mol Imaging 35:1544–1549
2. Mayer A, Höckel M, Wree A et al (2005) Microregional expression of glucose transporter-1 and oxygenation status: lack of correlation in locally advanced cervical cancers. Clin Cancer Res 11:2768–2773
3. Mayer A, Höckel M, Vaupel P (2008) Endogenous hypoxia markers: case not proven! Adv Exp Med Biol 614:127–136
4. Thews O, Kelleher DK, Esser N et al (2003) Lack of association between tumor hypoxia, GLUT-1 expression and glucose uptake in experimental sarcomas. Adv Exp Med Biol 510: 57–61
5. Jordan BF, Baudelet C, Gallez B (1998) Carbon-centered radicals as oxygen sensors for in vivo electron paramagnetic resonance: screening for an optimal probe among commercially available charcoals. Magn Reson Mater Phys Biol Med 7:121–129
6. Gallez B, Jordan BF, Baudelet C et al (1999) Pharmacological modifications of the partial pressure of oxygen in murine tumors: evaluation using in vivo EPR. Magn Reson Med 42: 627–630
7. Gallez B, Baudelet C, Jordan BF (2004) Assessment of tumor oxygenation by electron paramagnetic resonance: principles and applications. NMR Biomed 17:240–262
8. Gatenby RA, Gillies RJ (2004) Why do cancers have high aerobic glycolysis? Nat Rev Cancer 4:891–899

9. Sonveaux P, Végran F, Schroeder T et al (2008) Targeting lactate-fueled respiration selectively kills hypoxic tumor cells in mice. J Clin Invest 118:3930–3942
10. Busk M, Horsman MR, Kristjansen PE et al (2008) Aerobic glycolysis in cancers: implications for the usability of oxygen-responsive genes and fluorodeoxyglucose-PET as markers of tissue hypoxia. Int J Cancer 122:2726–2734
11. Busk M, Horsman MR, Jakobsen S et al (2008) Cellular uptake of PET tracers of glucose metabolism and hypoxia and their linkage. Eur J Nucl Med Mol Imaging 35:2294–2303
12. Waki A, Fujibayashi Y, Yonekura Y et al (1997) Reassessment of FDG uptake in tumor cells: high FDG uptake as a reflection of oxygen-independent glycolysis dominant energy production. Nucl Med Biol 24:665–670
13. Clavo AC, Brown RS, Wahl RL (1995) Fluorodeoxyglucose uptake in human cancer cell lines is increased by hypoxia. J Nucl Med 36:1625–1632
14. Burgman P, Odonoghue JA, Humm JL et al (2001) Hypoxia-induced increase in FDG uptake in MCF7 cells. J Nucl Med 42:170–175
15. Chan LW, Hapdey S, English S et al (2006) The influence of tumor oxygenation on (18)F-FDG (fluorine-18 deoxyglucose) uptake: a mouse study using positron emission tomography (PET). Radiat Oncol 1:3
16. Gagel B, Piroth M, Pinkawa M et al (2007) pO2 polarography, contrast enhanced color duplex sonography (CDS), [18F] fluoromisonidazole and [18F] fluorodeoxyglucose positron emission tomography: validated methods for the evaluation of therapy-relevant tumor oxygenation or only bricks in the puzzle of tumor hypoxia? BMC Cancer 7:113
17. de Geus-Oei LF, Kaanders JH, Pop LA et al (2006) Effects of hyperoxygenation on FDG-uptake in head-and-neck cancer. Radiother Oncol 80:51–56
18. Li XF, Ma Y, Sun X et al (2010) High 18F-FDG uptake in microscopic peritoneal tumors requires physiologic hypoxia. J Nucl Med 51:632–638
19. Christian N, Deheneffe S, Bol A et al (2010) Is (18)F-FDG a surrogate tracer to measure tumor hypoxia? Comparison with the hypoxic tracer (14)C-EF3 in animal tumor models. Radiother Oncol 97:183–188

Chapter 26
Direct Evidence of the Link Between Energetic Metabolism and Proliferation Capacity of Cancer Cells In Vitro

Géraldine De Preter, Pierre Danhier, Paolo E. Porporato, Valéry L. Payen, Bénédicte F. Jordan, Pierre Sonveaux, and Bernard Gallez

Abstract The aim of the study was to assess the link between the metabolic profile and the proliferation capacity of a range of human and murine cancer cell lines. First, the combination of mitochondrial respiration and glycolytic efficiency measurements allowed the determination of different metabolic profiles among the cell lines, ranging from a mostly oxidative to a mostly glycolytic phenotype. Second, the study revealed that cell proliferation, evaluated by DNA synthesis measurements, was statistically correlated to glycolytic efficiency. This indicated that glycolysis is the key energetic pathway linked to cell proliferation rate. Third, to validate this hypothesis and exclude non-metabolic factors, mitochondria-depleted were compared to wild-type cancer cells, and the data showed that enhanced glycolysis observed in mitochondria-depleted cells is also associated with an increase in proliferation capacity.

Keywords Cancer metabolism • Proliferation • Electron paramagnetic resonance • Oxidative phosphorylation • Aerobic glycolysis

1 Introduction

Since Otto Warburg's proposal in the 1920s [1], it was commonly believed that glycolysis is the main or exclusive energy (ATP) provider in cancer cells, reflecting an irreversible damage to mitochondrial respiration [2, 3]. More recently with the advent of high resolution respirometry, studies rather revealed functional and active cancer mitochondria (as reviewed in [4]). Even if enhanced glycolysis is observed

G. De Preter • P. Danhier • B.F. Jordan • B. Gallez (✉)
Biomedical Magnetic Resonance Research Group, Université catholique de Louvain (UCL), Avenue Mounier 73 (B1.73.08), 1200 Brussels, Belgium
e-mail: bernard.gallez@uclouvain.be

P.E. Porporato • V.L. Payen • P. Sonveaux
Institut de Recherche Expérimentale et Clinique (IREC), Pole of Pharmacology, Université catholique de Louvain (UCL), Brussels, Belgium

© Springer Science+Business Media, New York 2016   209
C.E. Elwell et al. (eds.), *Oxygen Transport to Tissue XXXVII*, Advances in
Experimental Medicine and Biology 876, DOI 10.1007/978-1-4939-3023-4_26

in numerous cancer cell lines [5], Zu and Guppy have shown that oxidative phosphorylation (OxPhos) could also predominate for ATP supply in many tumor cell lines [6]. In the present study, biochemical methods were used to characterize tumor cells metabolism. The metabolic profiles of six different tumor cell lines were drawn thanks to the measurement of oxygen consumption (EPR oximetry) [7], glucose utilization and lactate production. Energetic pathway flux rate analyses were correlated to proliferation assays to potentially show a direct, experimental link between the bioenergetic status and the ability of cancer cells to grow in vitro.

2 Methods

SiHa human cervix squamous cell carcinoma (ATCC), MDA-MB-231 human breast cancer (ATCC), TLT (transplantable liver tumor) mouse hepatocarcinoma [8], FSa II mouse fibrosarcoma [9], KHT mouse sarcoma [10], NT2 mouse mammary tumor [11] and mitochondrial DNA-depleted (ρ0) SiHa tumor cells [12] were grown as described. All cultures were kept at 37 °C in 5 % CO_2 atmosphere. To avoid artifactual variations due to different proliferation rates or culture conditions, all experiments were carried out on cells at 80–90 % confluence. All metabolic and proliferation assays were achieved separately on early passaged cancer cells using a unique medium (DMEM without glutamine, containing 4.5 g/L glucose supplemented with 10 % heat inactivated FBS and 1 % penicillin-streptomycin).

The oxygen consumption rate (OCR) of intact whole cells was measured using a Bruker EMX EPR spectrometer operating at 9.5 GHz as previously described [13]. Briefly, adherent cells were trypsinized and resuspended in fresh medium (10^7 cells/mL). 100 μL of the cell suspension was mixed with 100 μL of 20 % dextran to avoid agglomeration and was sealed in a glass capillary tube in the presence of 0.2 mM of a nitroxide probe acting as an oxygen sensor (^{15}N 4-oxo-2,2,6,6-tetramethylpiperidine-d_{16}-^{15}N-1-oxyl, CDN isotopes, Pointe-Claire, Quebec, Canada). Cells were maintained at 37 °C during the acquisition of the spectra. EPR linewidth was measured every minute and reported on a calibration curve to obtain the oxygen concentration [14]. OCR was determined by the absolute value of the slope of the decrease in oxygen concentration in the closed capillary tube over time. Mitochondrial oxygen consumption was obtained by subtracting OCR after rotenone treatment (20 μM, 20 min) to total OCR of the cells. Rotenone, a mitochondrial complex I inhibitor, was purchased from Sigma (Diegem, Belgium) and was dissolved in DMSO.

Glucose consumption and lactate production were measured from supernatants of cultured cells. For this purpose, medium was removed and fresh medium was added. Supernatant was sampled directly and 24 h after the cells were returned in the incubator. Glucose and lactate concentrations were quantified on deproteinized samples using enzymatic assays on a CMA600 analyzer (CMA Microdialysis AB, Solna, Sweden). To efficiently detect differences in metabolites concentrations between SiHa WT and SiHa ρ0 supernatants, a low glucose (1 g/L) medium was used.

Cell proliferation was assayed with a 5-bromo-2′-deoxyuridine (BrdU)-ELISA based kit (Roche) following the provider's instructions. Cells were incubated in the presence of BrdU (nucleotide analog) during 4 or 6 h depending on the experiment. The amount of BrdU incorporated in the cells allowed the quantitation of the DNA synthesis in replicating cells.

All results are from at least three independent experiments. Results are expressed as means ± standard error of the mean (SEM). Correlation and t-test analysis were performed using the GraphPad Prism 5 software. $P < 0.05$ was considered to be statistically significant.

3   Results

The characterization of the energetic phenotype of the cancer cell lines was performed using the combination of their mitochondrial respiration and glycolytic activity. Mitochondrial oxygen consumption rate (mitoOCR) was obtained by measuring the rotenone-sensitive OCR of intact whole cells using in vitro EPR oximetry. For each cell line, results are expressed by the absolute value of the mean slope of the linear reduction in oxygen concentration over time. Different oxidative capacities were found even if O_2 availability was comparable for all cell lines. The glycolytic activity of the cells was evaluated by enzymatic dosages to measure glucose utilization and lactate secretion. Glycolytic efficiency (efficacy of conversion of glucose into lactate) was determined as the lactate produced to glucose consumed ratio. Highest efficiency corresponds to a ratio of 2, as one molecule of glucose can produce a maximum of two molecules of lactate during glycolysis. Again, different amounts of lactate produced from the utilization of one mole of glucose were observed among the cell lines. Interestingly, when the data of the two metabolic pathways analyses were plotted together, we observed that mitoOCR significantly correlated to glycolytic efficiency (correlation coefficient = -0.91, $P < 0.05$) (Fig. 26.1a). The strong negative correlation suggests a tight regulation of the metabolic pathway in all cell lines. Indeed, when mitochondrial respiration was reduced, glycolysis was systematically enhanced and inversely. Also, by combining these two parameters together, different metabolic profiles were drawn. MDA-MB-231, FSAII and KHT cancer cells essentially exhibited a glycolytic metabolism (Warburg phenotype) with low mitoOCR and high glycolytic efficiency. On the other hand, NT2, TLT and SiHa cancer cells showed an oxidative metabolism with high mitoOCR and low glycolytic efficiency.

To experimentally determine with our cancer cells the possible coupling between metabolism and proliferation, correlations between the metabolic parameters and DNA synthesis were performed. For this purpose, proliferation was assessed with a metabolism-independent method. Cells were incubated in the presence of a nucleotide analog (BrdU) and the amount of incorporated BrdU

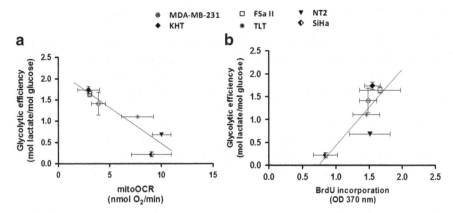

Fig. 26.1 (a) Correlation plots between glycolytic efficiency (expressed as the lactate secreted to glucose consumed ratio) and mitochondrial oxygen consumption rate (measured by EPR oximetry on intact whole cancer cells) obtained from six cancer cell lines (n = 3). (b) Correlation plots between glycolytic efficiency and proliferation (quantified by the amount of BrdU incorporated into DNA after a 4-h incubation) (n = 3)

was quantified to evaluate division rates. A significant positive correlation between glycolytic efficiency and proliferation was found (correlation coefficient = 0.82, $P < 0.05$) (Fig. 26.1b), but not between mitoOCR and proliferation (correlation coefficient = −0.58, $P > 0.05$). The trend towards a negative relationship between OxPhos and proliferation was likely a biological compensatory consequence of the enhanced/decreased glycolytic activity in the cells. However, we also highlighted that even in highly glycolytic and proliferative cells, oxidative respiration was still active, which means that, as already suggested [15], enhanced glycolysis cannot support tumor growth without adequate OxPhos. In the next experiment, to exclude non-metabolic factors, proliferation of SiHa wild-type (oxidative phenotype) and mitochondria-depleted SiHa ρ0 [12] were compared. It appeared that SiHa ρ0 exhibited lower mitoOCR due to their mitochondrial depletion (Fig. 26.2a) as well as enhanced glycolytic efficiency (Fig. 26.2b), which was also associated with increased proliferation (Fig. 26.2c).

The association of aerobic glycolysis and increased cancer cell proliferation has already been described: it mostly arises from the observation that accelerated glycolysis and metabolic plasticity provide anabolic intermediates for cell biomass expansion [16, 17], as also found in normal rapidly dividing cells such as activated immune cells [18]. In this context, the increased production of lactate from glucose molecules observed in our proliferative cancer cells could sound wasteful, as this conversion provides fewer carbons for biosynthetic pathways branching from glycolysis. However, in rapidly dividing cells, lactate generation has a predominant role as it allows the regeneration of NAD$^+$ via the enzyme lactate dehydrogenase, which is necessary for continued flux through glycolysis. As postulated by Lunt and Vander Heiden [15], the large amounts of lactate excreted by cancer cells may allow faster incorporation of glucose metabolites into biomass by efficiently

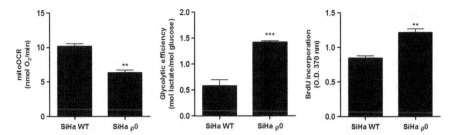

Fig. 26.2 Switching metabolism from an oxidative to a glycolytic phenotype increases proliferation rate in vitro. (**a**) Mitochondrial oxygen consumption rate measured on SiHa wild-type (n = 3) and mitochondria-depleted SiHa (ρ0, n = 3) using EPR oximetry and determined using rotenone 20 μM treatment (20 min). (**b**) Glycolytic efficiency expressed as the lactate secreted to glucose consumed ratio measured on SiHa wild-type (n = 3) and mitochondria-depleted SiHa (ρ0, n = 3). (**c**) Proliferation of SiHa wild-type (n = 3) and mitochondria-depleted SiHa (ρ0, n = 3) quantified by the amount of BrdU incorporated into DNA after a 6-h incubation

maintaining a NAD^+/NADH redox balance along with higher glycolytic flux. This hypothesis conciliates the observed systematic increase in glycolytic efficiency in the rapidly proliferating cancer cells used in this study.

To conclude, we experimentally and systematically correlated the metabolic signature of cancer cells to the proliferation rate and validated that glycolysis is probably the key energetic pathway linked to proliferation rate, where increased glycolysis is associated with increased proliferation capacity and inversely. Nevertheless, the debate on cause and effect mechanisms between higher proliferation rate and metabolic switch in tumors is still open and it would be interesting to investigate the causal connection between these parameters in further studies. Importantly, artificial culture conditions used in in vitro studies are different from the tumor microenvironment, which evolves temporally and spatially during carcinogenesis. For example, the vascular system and the resulting oxygen and nutrients availability are also major determinants of cell proliferation [19]. In vivo, systematic analysis of metabolism and proliferation using accurate methods and considering tumor features such as glucose deprivation, acidic pH and hypoxia would greatly help to better understand tumor progression.

Acknowledgments This work was supported by grants from the Télévie, the Belgian National Fund for Scientific Research (F.R.S.-FNRS), the Fonds Joseph Maisin, the *Fondation Belge contre le Cancer*, the Saint-Luc Foundation, the *Actions de Recherches Concertées-Communauté Française de Belgique* (ARC 04/09-317), and the European Research Council (FP7/2007-2013 ERC Independent Researcher Starting Grant No. 243188 TUMETABO). B.F.J. and P.S. are Research Associates and P.D. a Research Fellow of the F.R.S.-FNRS. G.D. is a Télévie Research Fellow.

References

1. Warburg O (1925) Uber den Stoffwechsel der Carcinomzelle. Klin Wochenschr 4:534–536
2. Penta JS, Johnson FM, Wachsman JT et al (2001) Mitochondrial DNA in human malignancy. Mutat Res 488:119–133
3. Xu RH, Pelicano H, Zhou Y et al (2005) Inhibition of glycolysis in cancer cells: a novel strategy to overcome drug resistance associated with mitochondrial respiratory defect and hypoxia. Cancer Res 65:613–621
4. Moreno-Sanchez R, Marin-Hernandez A, Saavedra E, Pardo JP et al (2014) Who controls the ATP supply in cancer cells? Biochemistry lessons to understand cancer energy metabolism. Int J Biochem Cell Biol 50:10–23
5. Moreno-Sanchez R, Rodriguez-Enriquez S, Marin-Hernandez A et al (2007) Energy metabolism in tumor cells. FEBS J 274:1393–1418
6. Zu XL, Guppy M (2004) Cancer metabolism: facts, fantasy, and fiction. Biochem Biophys Res Commun 313:459–465
7. Diepart C, Verrax J, Calderon PB et al (2010) Comparison of methods for measuring oxygen consumption in tumor cells in vitro. Anal Biochem 396:250–256
8. Taper HS, Woolley GW, Teller MN et al (1966) A new transplantable mouse liver tumor of spontaneous origin. Cancer Res 26:143–148
9. Volpe JP, Hunter N, Basic I et al (1985) Metastatic properties of murine sarcomas and carcinomas. I. Positive correlation with lung colonization and lack of correlation with s.c. tumor take. Clin Exp Metastasis 3:281–294
10. Rockwell S, Kallman RF (1972) Growth and cell population kinetics of single and multiple KHT sarcomas. Cell Tissue Kinet 5:449–457
11. Reilly RT, Gottlieb MB, Ercolini AM et al (2000) HER-2/neu is a tumor rejection target in tolerized HER-2/neu transgenic mice. Cancer Res 60:3569–3576
12. King MP, Attardi G (1989) Human cells lacking mtDNA: repopulation with exogenous mitochondria by complementation. Science 246:500–503
13. James PE, Jackson SK, Grinberg OY et al (1995) The effects of endotoxin on oxygen consumption of various cell types in vitro: an EPR oximetry study. Free Radic Biol Med 18:641–647
14. Jordan BF, Gregoire V, Demeure RJ et al (2002) Insulin increases the sensitivity of tumors to irradiation: involvement of an increase in tumor oxygenation mediated by a nitric oxide-dependent decrease of the tumor cells oxygen consumption. Cancer Res 62:3555–3561
15. Fogal V, Richardson AD, Karmali PP et al (2010) Mitochondrial p32 protein is a critical regulator of tumor metabolism via maintenance of oxidative phosphorylation. Mol Cell Biol 30:1303–1318
16. Lunt SY, Vander Heiden MG (2011) Aerobic glycolysis: meeting the metabolic requirements of cell proliferation. Annu Rev Cell Dev Biol 27:441–464
17. Hume DA, Weidemann MJ (1979) Role and regulation of glucose metabolism in proliferating cells. J Natl Cancer Inst 62:3–8
18. Hedeskov CJ (1968) Early effects of phytohaemagglutinin on glucose metabolism of normal human lymphocytes. Biochem J 110:373–380
19. Tannock IF (1968) The relation between cell proliferation and the vascular system in a transplanted mouse mammary tumour. Br J Cancer 22:258–273

Chapter 27
Acidosis Promotes Metastasis Formation by Enhancing Tumor Cell Motility

A. Riemann, B. Schneider, D. Gündel, C. Stock, M. Gekle, and O. Thews

Abstract The tumor microenvironment is characterized by hypoxia, acidosis as well as other metabolic and biochemical alterations. Its role in cancer progression is increasingly appreciated especially on invasive capacity and the formation of metastasis. The effect of acidosis on metastasis formation of two rat carcinoma cell lines was studied in the animal model. In order to analyze the pH dependency of different steps of metastasis formation, invasiveness, cell adhesion and migration of AT-1 prostate cancer cells as well as possible underlying cell signaling pathways were studied in vitro.

Acidosis significantly increased the formation of lung metastases of both tumor cell lines in vivo. In vitro, extracellular acidosis neither enhanced invasiveness nor affected cell adhesion to a plastic or to an endothelial layer. However, cellular motility was markedly elevated at pH 6.6 and this effect was sustained even when extracellular pH was switched back to pH 7.4. When analyzing the underlying mechanism, a prominent role of ROS in the induction of migration was observed. Signaling through the MAP kinases ERK1/2 and p38 as well as Src family kinases was not involved. Thus, cancer cells in an acidic microenvironment can acquire enhanced motility, which is sustained even if the tumor cells leave their acidic microenvironment e.g. by entering the blood stream. This increase depended on elevated ROS production and may contribute to the augmented formation of metastases of acidosis-primed tumor cells in vivo.

Keywords Acidosis • Metastasis formation • Migration • ROS • MAP kinases

A. Riemann (✉) • B. Schneider • D. Gündel • M. Gekle • O. Thews
Julius-Bernstein-Institute of Physiology, University of Halle, Magdeburger Straße 6, 06112 Halle (Saale), Germany
e-mail: anne.riemann@medizin.uni-halle.de

C. Stock
Institute of Physiology II, University of Münster, Münster, Germany

© Springer Science+Business Media, New York 2016
C.E. Elwell et al. (eds.), *Oxygen Transport to Tissue XXXVII*, Advances in Experimental Medicine and Biology 876, DOI 10.1007/978-1-4939-3023-4_27

215

1 Introduction

The tumor microenvironment differs profoundly from the environment found in normal tissue. Insufficient supply with oxygen and nutrients due to structural and functional aberration of the vasculature result in metabolic reprogramming. In conclusion, the elevated formation of lactic acid by glycolysis in combination with impaired removal of protons due to the poor perfusion leads to extracellular acidosis with pH values in the range of 6.5–6.8 [1, 2]. Previous studies showed that not only hypoxia, but also sole acidosis can aggravate metastasis, often by fostering degradation of extracellular matrix by proteases [3–6]. Metastases formation comprises multiple steps that might be susceptible to acidosis like tumor cell detachment and migration, local invasiveness, intravasation and circulation in the blood stream, extravasation and tumor growth at a distant site. The aim of the present study was to analyze the impact of acidosis on metastasis with special emphasis on invasive behavior, adhesion and tumor cell migration. Additionally, the involved signaling pathways were analyzed.

2 Methods

2.1 Cell Culture

The subline AT-1 of the rat R-3327 Dunning prostate carcinoma and the rat breast carcinoma cell line Walker-256 were grown in RPMI medium supplemented with 10% fetal calf serum (FCS) and additional 2 mM L-glutamine for Walker-256 cells. After 24 h of serum depletion, cells were incubated for 3–6 h in bicarbonate-HEPES- (pH 7.4) or MES- (pH 6.6) buffered Ringer solutions.

2.2 Animals and Metastasis Formation

AT-1 cells were tumorigenic in Copenhagen rats, while for Walker-256 cells Wistar rats were used. To determine tumor growth, 4×10^6 cells were injected into the dorsum of the hind foot and tumor volume was measured starting from day 3 after injection. Formation of lung metastases was analyzed by fluorescence imaging, after injection of 2×10^6 pTurboFP635-transfected cells into the tail vein. Since pTurboFP635 fluorescence was too weak for in vivo fluorescence imaging, the animals were sacrificed at the indicated time points and the lungs were explanted for fluorescence detection. The studies were approved by the regional ethics committee and conducted according to UKCCCR guidelines [7] and the German Law for Animal Protection.

2.3   Experimental Settings

Adhesion on plastics or endothelial layers (EA.hy926 cells) was assessed after 2 h by rinsing with PBS and subsequent counting of the adherent cells using the Casy cell counting system. Invasion was monitored in BioCoat Matrigel invasion chambers after 48 h with 10% FCS as chemoattractant. For both experimental settings, AT-1 cells were pre-incubated for 3 h at pH 7.4 or pH 6.6 and then transferred to media. Migratory speed was determined by time lapse microscopy taking pictures every 5 min over a time period of 100 min. To study whether effects on motility were sustained, cells were incubated for 3 h at pH 6.6 followed by pH 7.4 for the indicated time intervals and finally motility was assessed as indicated above. ROS formation was determined after incubation for 30 min with the fluorescent dye DCFDA-AM whose fluorescence (excitation 485 nm; emission 535 nm) was measured using a microplate reader.

3   Results

To study the effect of an acidic tumor environment on the formation of metastases, an in vivo tumor model expressing far-red fluorescent protein TurboFP635 for the investigation of metastasis was established. After stable transfection of AT-1 and Walker-256 cells, fluorescence intensity, stability of construct expression and reaction on acidosis were determined for the individual clones. Interestingly, cell growth of the selected pTurbo-Walker-256 clone in vitro was decelerated when compared to wild type cells, while AT-1 cells showed no differences. The same was found for tumor growth in vivo (Fig. 27.1a, b), however both fluorescently labeled cell lines retained their tumorigenic potential in rats. Exposing AT-1 or Walker-256 pTurbo-cells to an acidic extracellular pH for 6 h prior to tail vein injection resulted in a distinct increase in the formation of lung metastases (Fig. 27.1c, d). This increase for AT-1 tumors was not the result of tightened cell adhesion, nor to a plastic surface (Δ(pH 6.6-pH 7.4) $= +12 \pm 9\%$, n $= 6$) nor to an endothelial layer ($+4 \pm 2\%$, n $= 6$). The invasive behavior was also not significantly different ($+2 \pm 7\%$, n $= 14$). However, AT-1 cells showed increased motility when subjected to an acidic environment ($+32 \pm 4$ %, p $= 5 \times 10^{-9}$, n $= 37$). In order to analyze whether this acidosis-increased migratory speed may contribute to the formation of far-distance metastases which are established by the tumor cells after having entered temporarily the blood with a pH of 7.4, tumor cells were incubated at pH 7.4 or pH 6.6 for 3 h and subsequently kept at pH 7.4 for an additional 3 or 24 h. Pre-incubation of tumor cells at pH 6.6 led to an increased motility even when cells were subsequently kept at pH 7.4 for at least 3 h (Fig. 27.2). The increase disappeared after 24 h after acidic incubation. To elucidate which signaling pathways may be involved in acidosis-induced acceleration of tumor cell migration, MAP kinases ERK1/2 and p38 as well as Src family of protein tyrosine kinases

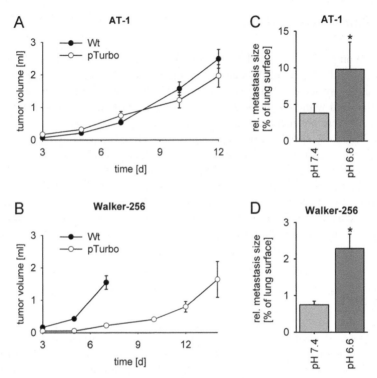

Fig. 27.1 Tumor growth of AT-1 and Walker-256 cells. (**a, b**) Tumor volume after subcutaneous implantation of wild type (Wt) and pTurboFP635-transfected (pTurbo) tumor cells; $n = 4$–11 for AT-1, $n = 3$–14 for Walker-256. (**c, d**) Formation of lung metastases after injection of pre-incubated (6 h pH 7.4 or pH 6.6) pTurboFP635-transfected cells in the tail vein expressed by the fraction of lung surface covered with tumor for AT-1 4 days after injection ($n = 4$–5) and for Walker-256 21 days after injection ($n = 3$); (*) $p < 0.05$

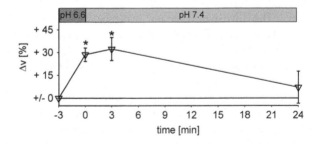

Fig. 27.2 Relative migratory speed of AT-1 cells with previous acidic incubation (-3 to 0 h). The motility was determined after subsequent change to media pH 7.4 (at 3 and 24 h) and values were normalized to respective controls (3 h pH 7.4 and subsequent change to media pH 7.4); $n = 4$–37; (*) $p < 0.05$

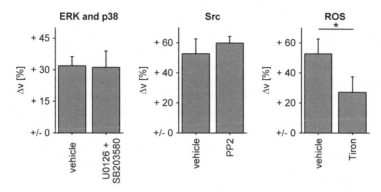

Fig. 27.3 Relative migratory speed of AT-1 cells after 3 h at pH 6.6 in the presence of ERK1/2 and p38 inhibitors U0126 and SB203580 (n = 5–15), Src family inhibitor PP2 (n = 5–6) and ROS scavenger Tiron (n = 5–6) (*) p < 0.05

were blocked with U0126, SB203580 or PP2/AG1878, respectively (Fig. 27.3) leading to the conclusion that these pathways were not involved. However, acidosis leads to an increase in ROS formation (+114 ± 20%, n = 12) and this was critical for enhanced motility since scavenging ROS with Tiron abrogated acidosis-mediated acceleration of tumor cell migration (Fig. 27.3).

4 Conclusion

Exposure of AT-1 and Walker-256 tumor cells to an acidic environment fostered metastases formation in vivo. Since AT-1 cells displayed no acidosis-induced alterations in invasiveness or adhesion in culture, this effect might be due to the elevated motility. The increase in motility was sustained (for at least 3 h) even when switching extracellular pH to physiological pH values. Therefore acidic pre-treatment might drive tumor cells to more aggressive invasive behavior. Similar results were depicted by Moellering et al. with melanoma cells, although acidic pre-treatment was prolonged to 24 h [5]. The elevated migration of AT-1 tumor cells depended on an increased ROS production based on intracellular acidification induced by the reduction of extracellular pH. These ROS originate presumably from mitochondrial complex I and not from altered activity of NO synthase or NADPH oxidases [8]. Induction of migration and metastasis by increased mito-chondrial ROS production was recently shown by Porporato et al. [9]. However, their findings were based on Src activation, which was not critical for migration in AT-1 tumor cells. Additionally, MAPK ERK1/2 and p38 were not involved in accelerated migration although both were activated in response to acidosis [10]. However, in these experiments increased ROS production was only measured directly after 3 h incubation at pH 6.6. In our present study migration remained accelerated for at least 3 h after returning to normal pH (7.4) (Fig. 27.2). It cannot be decided conclusively whether ROS production would also stay elevated during this period of time. This question has to be addressed in further studies. In conclusion, acidosis-induced increase in metastasis supports the assumption that microenvironmental parameters strongly affect tumor cell behavior. From these findings supportive treatment modalities targeting the extracellular acidosis [11, 12] might be suitable to improve the therapeutic outcome and by this the patient's prognosis.

Acknowledgments This study was supported by the Deutsche Krebshilfe (Grants 106774/ 106906), the BMBF (ProNet-T^3 Ta-04) and the Wilhelm-Roux program of the Medical School, Universität Halle-Wittenberg.

References

1. Vaupel P, Kallinowski F, Okunieff P (1989) Blood flow, oxygen and nutrient supply, and metabolic microenvironment of human tumors: a review. Cancer Res 49:6449–6465
2. Gerweck LE, Seetharaman K (1996) Cellular pH gradient in tumor versus normal tissue: potential exploitation for the treatment of cancer. Cancer Res 56:1194–1198
3. Gatenby RA, Gawlinski ET, Gmitro AF et al (2006) Acid-mediated tumor invasion: a multidisciplinary study. Cancer Res 66:5216–5223
4. Giusti I, D'Ascenzo S, Millimaggi D et al (2008) Cathepsin B mediates the pH-dependent proinvasive activity of tumor-shed microvesicles. Neoplasia 10:481–488
5. Moellering RE, Black KC, Krishnamurty C et al (2008) Acid treatment of melanoma cells selects for invasive phenotypes. Clin Exp Metastasis 25:411–425
6. Rofstad EK, Mathiesen B, Kindem K, Galappathi K (2006) Acidic extracellular pH promotes experimental metastasis of human melanoma cells in athymic nude mice. Cancer Res 66:6699–6707
7. Workman P, Aboagye EO, Balkwill F et al (2010) Guidelines for the welfare and use of animals in cancer research. Br J Cancer 102:1555–1577
8. Riemann A, Schneider B, Ihling A et al (2011) Acidic environment leads to ROS-induced MAPK signaling in cancer cells. PLoS One 6:e22445
9. Porporato PE, Payen VL, Pérez-Escuredo J et al (2014) A mitochondrial switch promotes tumor metastasis. Cell Rep 8:754–766
10. Sauvant C, Nowak M, Wirth C et al (2008) Acidosis induces multi-drug resistance in rat prostate cancer cells (AT1) in vitro and in vivo by increasing the activity of the p-glycoprotein via activation of p38. Int J Cancer 123:2532–2542
11. Estrella V, Chen T, Lloyd M et al (2013) Acidity generated by the tumor microenvironment drives local invasion. Cancer Res 73:1524–1535
12. Robey IF, Baggett BK, Kirkpatrick ND et al (2009) Bicarbonate increases tumor pH and inhibits spontaneous metastases. Cancer Res 69:2260–2268

Part V
Cellular Hypoxia

Chapter 28
Preoperative Stress Conditioning in Humans: Is Oxygen the Drug of Choice?

G.A. Perdrizet

Abstract Complications following invasive medical and surgical procedures are common and costly. No clinical protocols exist to actively condition patients prior to these high risk interventions. Effective preconditioning algorithms have been repeatedly demonstrated in animal models for more than a quarter century, where brief exposures to hyperthermia (heat shock), ischemia (ischemic preconditioning) or hypoxia have been employed. Heat shock pretreatment confers protection against experimental acute ischemia-reperfusion, endotoxin challenge and other stressors. The resulting state of protection is short lived (hours) and is associated with new gene expression, typical of a cell stress response (CSR). We aim to use the CSR to actively precondition patients before surgery, a process termed stress conditioning (SC). SC is a procedure in which tissues are briefly exposed to a conditioning stressor and recovered to permit the development of a transient state of resistance to ischemia-reperfusion injury. Successful SC of humans prior to surgery may reduce postoperative complications related to periods of hypotension, hypoxia, or ischemia. Stressors such as heat shock, acute ischemia, endotoxin, heavy metals or hypoxia can induce this protected state but are themselves harmful and of limited clinical utility. The identification of a stressor that could induce the CSR in a non-harmful manner seemed unlikely, until high dose oxygen was considered. Human microvascular endothelial cells (HMEC-1) exposed to high dose oxygen at 2.4 ATA × 60–90 min developed increased resistance to an oxidant challenge in vitro (peroxide). The molecular changes described here, together with our understanding of the CSR and SC phenomena, suggest high dose oxygen may be the drug of choice for clinical preconditioning protocols and should be systematically tested in clinical trials. Oxygen dosing includes the following ranges: room air exposure is 0.21 ATA, clinical oxygen therapy 0.3–1.0 ATA (normobaric hyperoxia) and hyperbaric oxygen is 1.5–3.0 ATA (ATA—atmosphere absolute).

G.A. Perdrizet (✉)
Department of Emergency Medicine, Center for Wound Healing and Hyperbaric Medicine, UCSD, San Diego, CA, USA

Department of Molecular and Cell Biology, University of Connecticut, Storrs, CT, USA
e-mail: gperdrizet@ucsd.edu

© Springer Science+Business Media, New York 2016
C.E. Elwell et al. (eds.), *Oxygen Transport to Tissue XXXVII*, Advances in Experimental Medicine and Biology 876, DOI 10.1007/978-1-4939-3023-4_28

Keywords Heat shock • Hyperbaric oxygen • Stress response • Ischemia •
Reperfusion • Adverse events • Molecular chaperones • Preconditioning

*"Cum melius et utilius sit in tempore occurrere quam
post causam vulneratam quaerrere remedium."*
(Prevention is better than cure.)
Bracton, circa 1240 [1]

1 A Clinical Need

While the incidence of anesthesia-related deaths has fallen significantly during the
past 50 years, perioperative morbidity and mortality rates have not [2]. Postoperative adverse events, such as myocardial infarction, stroke, acute spinal ischemia and acute kidney injury are common, costly and detract from clinical outcomes
[3]. In the USA, a point prevalence study of Medicare Beneficiaries performed at
hospital discharge (Oct. 2008) documented a 13.5 % incidence of adverse events
during hospitalization, with 28 % related to surgical procedures [4]. Similar rates
have been reported by the United Kingdom (11.7 % overall and 16 % surgical) and
Australia (16.6 %) [5, 6]. A recent report on adverse events related to spinal surgery
at a single institution over a 4 year period (2007–2010) found that procedure-related
complications accounted for 16 % ($8.38 million) of the total cost of care and added
4684 bed days to the total duration of hospital stay. Independent of the type of
surgical procedure, all surgical patients are exposed to some degree of acute
ischemia and reperfusion injury (IRI) and subsequent acute inflammation. The
severity of injury will vary depending upon extent of the ischemic event and
upon the patient's resistance to ischemia. Local and systemic acute inflammation
exacerbates this damage, often leading to microvascular damage, organ dysfunction
and death. There is no dearth of reports in the literature addressing this situation
[7]. The rising prevalence of elderly, obese, and chronically-ill individuals requiring invasive procedures is further adding to the challenges of providing safe and
successful surgical treatment. Efforts to minimize complication rates are being
made at many levels, however complications will occur. Clinical approaches rely
primarily on maximizing an individual's medical status prior to surgery. Classi-
cally, a passive approach has been taken towards an individual's stress tolerance
immediately prior to a surgical intervention. The concept of preconditioning the
patient before a surgical procedure to enable a safer post-operative recovery is
attracting growing interest [8, 9]. Prevention is better than cure.

2 Stress Conditioning- Defined and Reviewed [10, 11]

Stress conditioning (SC) is the purposeful exposure of a cell, tissue or organism to a conditioning stress and period of recovery prior to undergoing a more stressful, challenging event, such as a surgical intervention. The goal is to enhance an individual's resistance to injury and promote an uncomplicated postoperative recovery. Following SC, the resulting state of protection from acute IRI is called the protected phenotype (PP). The PP is transient, typically lasting hours, and is nonspecific, thus providing opportunity for cross protection. Initial studies were performed using whole-body hyperthermia as the conditioning stressor, due to the extensive knowledge base that exists regarding the hyperthermic treatment of cancer [12]. This state of protection was initially induced in a rodent model of transplant organ preservation; hyperthermia protected against acute IRI [13]. This technique was extended to a model of aortic surgery and provided protection of the rabbit spinal cord during acute IRI [14]. This approach has been applied to multiple surgically-relevant translational models and has been reviewed in detail elsewhere [10, 11, 15, 16].

These initial SC protocols were based on the kinetics of two well described phenomenon; thermotolerance (TT) and heat shock response (HSR) [17]. Therapeutic efficacy of hyperthermic treatments of cancer was limited by the development of resistance to thermal killing by malignant cells. Normal tissues also acquired the same thermotolerance [18, 19]. The HSR added mechanistic insight into the problem of thermotolerance in cancer therapy. The HSR originally describe the altered pattern of chromosomal puffing that was observed following the accidental exposure of fruit flies to whole body hyperthermia [20]. This phenomenon was viewed as a useful model to study gene expression, but was considered to be a laboratory oddity with little physiologic or medical significance [21, 22]. Geneticists had noted that flies exposed to a heat shock (HS, brief supraphysiologic heat exposure) would become resistant to lethal levels of hyperthermia. This remarkable state of protection was temporally associated with the chromosomal puffing pattern which led to the discovery of the heat shock genes and their proteins (HSP). Thermal stress was found to induce protein denaturation as the primary trigger of the HSR. The HSPs were subsequently renamed and classified as molecular chaperones. The molecular chaperones provide the fundamental basis for the cellular stress response (CSR) and the PP [23]. A logical application of the PP would be to precondition individuals prior to surgical interventions where acute IRI was anticipated. Based on the TT and HSR phenomena, we began testing methods that would enable SC of humans prior to surgical interventions.

The term "stress conditioning" indicates that there are two universal responses upon which all preconditioning phenomena are based: (1) the triphasic response to stress as described by Hans Selye and (2) the CSR [24, 25]. The protection observed in SC models is likely related to the second phase of the classic triphasic response to stress, extensively annotated by Dr. Selye [26]. All biologic systems respond to stress in a similar fashion. The CSR represents the cellular and molecular basis for the triphasic stress response [27]. The universal nature of these responses means all cellular life forms are capable of a CSR and thus of acquiring the PP. Nature has conserved a powerful, intrinsic cellular response that can change a potentially lethal intervention into a survivable one. SC protocols aim to control this response to improve clinical outcomes following planned invasive procedures. The challenge is to determine if the CSR/PP can be induced in a clinically relevant, non-injurious fashion. Exposing the elderly patient with chronic disease to a classical heat shock is not practical, nor favorable from a risk benefit perspective. Can the CSR/PP be induced in a non-harmful manner?

3 Preconditioning: Brief Look

The concept of preconditioning developed as a result of experimental studies on cardiac ischemia/reperfusion injury. Investigators encountered a state of paradoxical protection that followed brief exposures to hypoxia [28]. Thus hypoxic and ischemic preconditioning phenomena were born [29, 30]. Early and delayed time courses were described and multiple mechanisms identified. While promising, exposing patients to hypoxia is neither practical nor safe. Ischemic pre-conditioning is invasive and difficult to control, requiring an invasive or surgical procedure of its own [9]. Both methods do harm to achieve an overall net effect of improved post-ischemic function. Never the less, extensive work has continued to modify the application of ischemic preconditioning, in the form of remote and post conditioning algorithms [8, 9, 31]. While theoretically appealing, the clinical application of ischemic preconditioning is fraught with logistical difficulties and risk.

Preconditioning is not unique to low oxygen states. Endotoxin tolerance and cytokines are pre-conditioning agents when provided at the right dose and time [32–34]. Toxins, heavy metals, ethanol and even ionizing radiation have been used successfully, in experimental models, to induce the PP. Presently, the human risks for all of these approaches is unknown and clinical benefit unproven. The goal of developing a safe, practical method for SC that is minimally harmful seemed paradoxical, until oxygen was considered.

4 Oxygen as a Stress

Oxygen is toxic. Oxygen is a medical gas with a well characterized toxicity profile. Brief exposures to supra-physiological levels of oxygen are well tolerated, but would the CSR be triggered? The practice of hyperbaric medicine routinely exposes elderly, obese, diabetic patients to 2.4 ATA of oxygen for 90 min with a very low adverse event profile. In fact, observing an individual receiving hyperbaric oxygen therapy (HBOT), one wonders if the treatment is causing any effect, as most patients read, watch television or sleep during therapy. The fact that clinical doses of HBOT were found to induce the expression of heat shock genes was an unexpected, but welcomed, observation [35, 36]. Could HBOT induce the CSR/PP? It has been previously shown that normobaric hyperoxia, 50–100 % oxygen (0.5–1.0 ATA) can both damage the lung (pulmonary oxygen toxicity) and also protect the lung from subsequent oxygen toxicity [37–39].

Multiple cell culture and animal models support the ability of HBOT to induce the PP, see below. We have focused on protecting human microvascular endothelium to preserve the integrity of the microcirculation during acute IRI events. Human microvascular endothelial cells (HMEC-1) were exposed to high dose oxygen (2.4 ATA × 60 min) and developed increased resistance to an in vitro oxidant challenge (t-butyl-hydroperoxide, 0.05–0.24 mM × 4 h). Importantly 1.0 ATA oxygen (normobaric hyperoxia-100 % oxygen at ambient, atmospheric pressure) did not provide this protection. The 2.4 ATA, but not 1.0 ATA, oxygen exposure was also associated with increased cellular proliferation in vitro [40, 41]. A genome-wide microarray analysis was performed at 1 and 24 h after HBOT. Over 8000 genes demonstrated significantly altered levels of expression. Pathway analysis identified Nrf-2 as a primary responder and is consistent with the observed increased resistance to oxidant injury in vitro. Validation of several top performing genes was carried out using quantitative PCR (e.g., Nrf2, HMOX1, HSPA1A).

Initial work in an animal model of hepatic ischemia demonstrated that HBOT pretreatment could provide ischemic protection of the liver in rodents [42]. Since that time, HBOT has been used in many animal models to successfully precondition the brain, heart, kidney, spinal cord and skin flaps from acute IRI [43–52]. Several small clinical studies have been reported and support the safety of pre-operative administration of HBOT prior to open heart surgery and cardiopulmonary bypass [53–56].

The most serious therapeutic risk of HBOT is CNS oxygen toxicity. The risk for a generalized seizure is reported to be extremely low. The reported incidence varies over a range of 0.01–0.2 % incidence. Elderly, frail and infirm individuals are routinely exposed to a tenfold increase in oxygen concentration (above ambient air at sea level) without harm. The same individuals become threatened by a mere 15–20 % reduction in ambient oxygen concentration. Why should toxic, high-dose oxygen be so well tolerated by humans? Mammalian cells survived eons of selection pressure by exposure to relative hyperoxia and have evolved cellular defenses to control the harm from oxygen toxicity. Oxygen, at the correct dose, may be uniquely qualified to provide a non-harmful stimulus of the CSR and PP. Normobaric oxygen while less toxic and more convenient to use clinically does not produce consistent and reliable changes in gene expression nor induction of the PP at the brief exposure times reported here. Prolonged exposures (multiple hours to days) to normobaric hyperoxia (50–100 % at 1 atm or 0.5–1.0 ATA) can lead to some degree of pulmonary protection but risks the development of pulmonary oxygen toxicity. Given this narrow therapeutic window, normobaric hyperoxia is a less attractive method for the development of clinical stress conditioning protocols given its minimal benefit and significant risk profile compared to hyperbaric oxygen.

5 More Mechanisms

Changes in gene expression described above for the HMEC-1 cells have been reported in diverse models testing HBOT preconditioning. HBOT pretreatment protected rodent primary spinal neurons against in vitro oxidant stress (hydrogen peroxide) and was dependent upon elevated expression of HSP32 (aka hemeoxygenase-1) [57]. HBOT preconditioning provided protection of acute spinal cord ischemia in the rodent and was associated with elevated antioxidant enzymes (Mn-SOD and catalase) and inhibition of programmed cell death [44]. HBOT preconditioning induced increased levels of the Class III histone deacetylase, SirT1, and provided protection in a model of middle cerebral artery occlusion in the rodent. This protection was also associated with reduce programmed cell death in neuronal cells [58]. The same investigators have also shown that HBOT mediated neuroprotection is associated with increased formation of autophagosomes and enhancement of autophagy [59]. HBOT preconditioning provided protection in a mouse model of acute spinal cord ischemia and was associated with Nrf2 activation within astrocytes [60]. HBOT preconditioning reduced infarct volumes in a rodent model of focal cerebral ischemia and was associated with elevated levels of HIF-1alpha and erythropoietin [61, 62].

Although the dose of oxygen used in these studies is very high, exposure times are brief. The potential for HBOT to exacerbate IRI by enhancing ROS production, while logical, does not appear to occur when directly studied [63]. The molecular changes described here, together with our understanding of the CSR and SC, suggest that high dose oxygen, in the form of HBOT, may be the drug of choice for clinical preconditioning protocols and should be systematically tested.

6 L'envoi

Pretreatment of animals and humans with high dose oxygen, as HBOT at 2.4 ATA, is safe and confers significant functional protection in many models of acute IRI. Multiple phenomena from experimental biology such as TT, HSR, the triphasic stressresponse and the CSR all support the SC concept. Ischemic protection is associated with a myriad of cellular and molecular changes consistent with the development of a PP. Oxygen, as a drug, is safe and widely accessible. The present knowledge base appears adequate to support Phase I and II clinical trials to establish HBOT-based SC algorithms.

References

1. Bracton 1240 (1966) Oxford dictionary of English proverbs (De Legibus), 2nd edn. Clarendon Press, Glocestershire, UK, p 251
2. Bartels K, Karhausen J, Clambey ET et al (2013) Perioperative organ injury. Anesthesiology 119(6):1474–1489
3. Nearman H, Klick JC, Eisenberg P, Pesa N (2014) Perioperative complications of cardiac surgery and postoperative care. Crit Care Clin 30(3):527–555
4. Levinson DR (2010) Adverse events in hospitals: national incidence among medicare beneficiaries. OIG, Department of Health and Human Services, OEI 06-09-00090, http://oig.hhs.gov/oei/reports/oei-06-00090.pdf
5. Vincent C, Neale G, Woloshynowych M (2001) Adverse events in British hospitals: preliminary retrospective record review. BMJ 322:517–519
6. Wilson RM, Runciman WB, Gibberd RW et al (1995) The quality in Australian health care study. Med J Aust 163:458–471
7. Tarakji KG, Sabik JF 3rd, Bhudia SK et al (2011) Temporal onset, risk factors, and outcomes associated with stroke after coronary artery bypass grafting. JAMA 305(4):381–390
8. Ovize M, Thibault H, Przyklenk K (2013) Myocardial conditioning: opportunities for clinical translation. Circ Res 113(4):439–450
9. Dezfulian C, Garrett M, Gonzalez NR (2013) Clinical application of preconditioning and postconditioning to achieve neuroprotection. Transl Stroke Res 4(1):19–24
10. Perdrizet GA (1997) Heat shock response and organ preservation: models of stress conditioning. R.G. Landes/Chapman and Hall, Austin/New York
11. Perdrizet GA, Rewinski MJ, Noonan EJ, Hightower LE (2007) The biology of the heat shock response and stress conditioning. In: Calderwood S (ed) Protein reviews, cell stress proteins. Springer, New York, pp 7–56
12. Henle KJ, Dethlefsen LA (1978) Heat fractionation and thermotolerance: a review. Cancer Res 38:1843–1851
13. Perdrizet GA, Heffron TG, Buckingham FC et al (1989) Stress conditioning: a novel approach to organ preservation. Curr Surg 46:23–26
14. Perdrizet GA, Lena CJ, Shapiro D, Rewinski MJ (2002) Preoperative stress conditioning prevents paralysis following experimental aortic surgery. J Thorac Cardiovasc Surg 124:162–170
15. Perdrizet GA (1997) Heat shock response and organ preservation: models of stress conditioning, medical intelligence unit. RG Landes, Austin
16. Perdrizet GA (1995) Heat shock and tissue protection. New Horiz 3:312–320
17. Landry J, Bernier D, Chrétien P et al (1982) Synthesis and degradation of heat shock proteins during development and decay of thermotolerance. Cancer Res 42:2457–2461
18. Urano M (1986) Kinetics of thermotolerance in normal and tumor tissues: a review. Cancer Res 46:474–482
19. Crile G Jr (1963) The effects of heat and radiation on cancers implanted on the feet of mice. Cancer Res 23:372–380
20. Ritossa F (1964) Experimental activation of specific loci in polytene chromosomes of Drosophila. Exp Cell Res 35:601–607
21. Schlesinger MJ, Ashburner J, Tissières A (eds) (1982) Heat shock from bacteria to man. Cold Spring Harbor Laboratory Press, Cold Spring Harbor
22. Lindquist S (1986) The heat-shock response. Annu Rev Biochem 55:1151–1191
23. Voellmy R, Boellman F (2007) Chaperone regulation of heat shock protein response. In: Csermely P, Vigh L (eds) Molecular aspects of the stress response: chaperones, membranes and networks. Landes Bioscience/Springer Science + Business Media, Austin/New York, pp 89–99
24. Selye H (1936) A syndrome produced by diverse nocuous agents. Nature 138:32

25. Hightower LE, Guidon PT, Whelan SA et al (1985) Stress responses in avian and mammalian cells. In: Atkinson BC, Walden DB (eds) Changes in eukaryotic gene expression in response to environmental stress. Academic, Orlando, pp 197–210

26. Selye H (1976) Stress in health and disease. Butterworth, Boston

27. Perdrizet GA (1997) Hans Selye and beyond: responses to stress. Cell Stress Chaperones 2 (4):1–6

28. Neely JR, Grotyohann LW (1984) Role of glycolytic products in damage to ischemic myocardium. Circ Res 55:816–824

29. Kato H, Liu Y, Araki T, Kogure K (1991) Temporal profile of the effects of pretreatment with brief cerebral ischemia on the neuronal damage following secondary ischemic insult in the gerbil: cumulative damage and protective effects. Brain Res 553(2):238–242

30. Murry CE, Jennings RB, Reimer KA (1986) Preconditioning with ischemia: a delay of lethal cell injury in ischemic myocardium. Circulation 74(5):1124–1136

31. Zhao H (2013) Hurdles to clear before clinical translation of ischemic post conditioning against stroke. Transl Stroke Res 4(1):63–70

32. Frank L, Yam J, Roberts RJ (1978) The role of endotoxin in protection of adult rats from oxygen-induced lung toxicity. J Clin Invest 61:269–275

33. Brown JM, Grosso M, Terada LS et al (1989) Endotoxin pretreatment increases endogenous myocardial catalase activity and decreases ischemia-reperfusion injury of isolated rat hearts. Proc Natl Acad Sci U S A 86:2516–2520

34. White CW, Ghezzi P, McMahon S et al (1989) Cytokines increase rat lung antioxidant enzymes during exposure to hyperoxia. J Appl Physiol 66(2):1003–1007

35. Lin H, Chang CP, Lin HJ, Lin MT, Tsai CC (2012) Attenuating brain edema, hippocampal oxidative stress, and cognitive dysfunction in rats using hyperbaric oxygen preconditioning during simulated high-altitude exposure. J Trauma Acute Care Surg 72(5):1220–1227

36. Shinkai M, Shinohia N, Kanoh S et al (2004) Oxygen stress effects proliferation rates and heat shock proteins in lymphocytes. Aviat Space Environ Med 75:109–113

37. Frank L, Massaro D (1980) Oxygen toxicity. Am J Med 69(1):117–126

38. Clark JM, Lambertsen CJ, Gelfand R, Troxel AB (2006) Optimization of oxygen tolerance extension in rats by intermittent exposure. J Appl Physiol (1985) 100(3):869–879

39. Hendricks PL, Hall DA, Hunter WL Jr, Haley PJ (1977) Extension of pulmonary O2 tolerance in man at 2 ATA by intermittent O2 exposure. J Appl Physiol Respir Environ Exerc Physiol 42 (4):593–599

40. Godman CA et al (2010) Hyperbaric oxygen induces a cytoprotective and angiogenic response in human microvascular endothelial cells. Cell Stress Chaperones 15:431–442

41. Godman CA et al (2010) Hyperbaric oxygen treatment induces antioxidant gene expression. Ann N Y Acad Sci 1197:178–183

42. Chen MF, Chen HM, Ueng SW, Shyr MH (1998) Hyperbaric oxygen pretreatment attenuates hepatic reperfusion injury. Liver 18(2):110–116

43. Wada K, Ito M, Miyazawa T et al (1996) Repeated hyperbaric oxygen induces ischemic tolerance in gerbil hippocampus. Brain Res 740(1–2):15–20

44. Wang L, Li W, Kang Z et al (2009) Hyperbaric oxygen preconditioning attenuates early apoptosis after spinal cord ischemia in rats. J Neurotrauma 26(1):55–66

45. Sun L, Xie K, Zhang C et al (2014) Hyperbaric oxygen preconditioning attenuates postoperative cognitive impairment in aged rats. Neuroreport 25(9):718–724

46. Kim CH, Choi H, Chun YS et al (2001) Hyperbaric oxygenation pretreatment induces catalase and reduces infarct size in ischemic rat myocardium. Pflugers Arch 442(4):519–525

47. Kang N, Hai Y, Liang F et al (2014) Preconditioned hyperbaric oxygenation protects skin flap grafts in rats against ischemia/reperfusion injury. Mol Med Rep 9(6):2124–2130

48. Cabigas BP, Su J, Hutchins W et al (2006) Hyperoxic and hyperbaric-induced cardioprotection: role of nitric oxide synthase 3. Cardiovasc Res 72(1):143–151

49. Dong H, Xiong L, Zhu Z et al (2002) Preconditioning with hyperbaric oxygen and hyperoxia induces tolerance against spinal cord ischemia in rabbits. Anesthesiology 96(4):907–912

50. Nie H, Xiong L, Lao N et al (2006) Hyperbaric oxygen preconditioning induces tolerance against spinal cord ischemia by upregulation of antioxidant enzymes in rabbits. J Cereb Blood Flow Metab 26(5):666–674
51. He X, Xu X, Fan M et al (2011) Preconditioning with hyperbaric oxygen induces tolerance against renal ischemia-reperfusion injury via increased expression of heme oxygenase-1. J Surg Res 170(2):e271–e277
52. Yamashita S, Hirata T, Mizukami Y et al (2009) Repeated preconditioning with hyperbaric oxygen induces neuroprotection against forebrain ischemia via suppression of p38 mitogen activated protein kinase. Brain Res 1301:171–179
53. Li Y, Dong H, Chen M et al (2011) Preconditioning with repeated hyperbaric oxygen induces myocardial and cerebral protection in patients undergoing coronary artery bypass graft surgery: a prospective, randomized, controlled clinical trial. J Cardiothorac Vasc Anesth 25 (6):908–916
54. Yogaratnam JZ, Laden G, Guvendik L et al (2010) Hyperbaric oxygen preconditioning improves myocardial function, reduces length of intensive care stay, and limits complications post coronary artery bypass graft surgery. Cardiovasc Revasc Med 11(1):8–19
55. Jeysen ZY, Gerard L, Levant G et al (2011) Research report: the effects of hyperbaric oxygen preconditioning on myocardial biomarkers of cardioprotection in patients having coronary artery bypass graft surgery. Undersea Hyperb Med 38(3):175–185
56. Alex J, Laden G, Cale AR et al (2005) Pretreatment with hyperbaric oxygen and its effect on neuropsychometric dysfunction and systemic inflammatory response after cardiopulmonary bypass: a prospective randomized double-blind trial. J Thorac Cardiovasc Surg 130 (6):1623–1630
57. Huang G, Xu J, Xu L et al (2014) Hyperbaric oxygen preconditioning induces tolerance against oxidative injury and oxygen-glucose deprivation by up-regulating heat shock protein 32 in rat spinal neurons. PLoS One 9(1), e85967. doi:10.1371/journal.pone.0085967
58. Yan W, Fang Z, Yang Q et al (2013) SirT1 mediates hyperbaric oxygen preconditioning-induced ischemic tolerance in rat brain. J Cereb Blood Flow Metab 33(3):396–406
59. Yan W, Zhang H, Bai X et al (2011) Autophagy activation is involved in neuroprotection induced by hyperbaric oxygen preconditioning against focal cerebral ischemia in rats. Brain Res 1402:109–121
60. Xu J, Huang G, Zhang K et al (2014) Nrf2 activation in astrocytes contributes to spinal cord ischemic tolerance induced by hyperbaric oxygen preconditioning. J Neurotrauma 31:1343–1353
61. Peng Z, Ren P, Kang Z et al (2008) Up-regulated HIF-1alpha is involved in the hypoxic tolerance induced by hyperbaric oxygen preconditioning. Brain Res 1212:71–78
62. Gu GJ, Li YP, Peng ZY et al (2008) Mechanism of ischemic tolerance induced by hyperbaric oxygen preconditioning involves upregulation of hypoxia-inducible factor-1alpha and erythropoietin in rats. J Appl Physiol (1985) 104(4):1185–1191
63. Sun L, Wolferts G, Veltkamp R (2014) Oxygen therapy does not increase production and damage induced by reactive oxygen species in focal cerebral ischemia. Neurosci Lett 577:1–5

Chapter 29
In Vivo Imaging of Flavoprotein Fluorescence During Hypoxia Reveals the Importance of Direct Arterial Oxygen Supply to Cerebral Cortex Tissue

K.I. Chisholm, K.K. Ida, A.L. Davies, D.B. Papkovsky, M. Singer, A. Dyson, I. Tachtsidis, M.R. Duchen, and K.J. Smith

Abstract Live imaging of mitochondrial function is crucial to understand the important role played by these organelles in a wide range of diseases. The mitochondrial redox potential is a particularly informative measure of mitochondrial function, and can be monitored using the endogenous green fluorescence of oxidized mitochondrial flavoproteins. Here, we have observed flavoprotein fluorescence in the exposed murine cerebral cortex in vivo using confocal imaging; the mitochondrial origin of the signal was confirmed using agents known to manipulate mitochondrial redox potential. The effects of cerebral oxygenation on flavoprotein fluorescence were determined by manipulating the inspired oxygen concentration. We report that flavoprotein fluorescence is sensitive to reductions in cortical oxygenation, such that reductions in inspired oxygen resulted in loss of flavoprotein fluorescence with the exception of a preserved 'halo' of signal in periarterial regions. The findings are consistent with reports that arteries play an important role in supplying oxygen directly to tissue in the cerebral cortex, maintaining mitochondrial function.

Keywords Oxygen • Mitochondria • Brain • Vasculature • Confocal microscope

---

This chapter was originally published under a CC BY-NC 4.0 license, but has now been made available under a CC BY 4.0 license. An erratum to this chapter can be found at DOI 10.1007/978-1-4939-3023-4_66.

K.I. Chisholm (✉) • A.L. Davies • M. Singer • A. Dyson • I. Tachtsidis • M.R. Duchen • K.J. Smith
University College London, London, UK
e-mail: k.chisholm@ucl.ac.uk

K.K. Ida
University College London, London, UK

University of São Paulo, São Paulo, Brazil

D.B. Papkovsky
University College Cork, Cork, Ireland

© The Author(s) 2016
C.E. Elwell et al. (eds.), *Oxygen Transport to Tissue XXXVII*, Advances in Experimental Medicine and Biology 876, DOI 10.1007/978-1-4939-3023-4_29

1 Introduction

Mitochondrial pathology has been implicated in a wide range of diseases, including multiple sclerosis [1], Parkinson's disease [2], and sepsis [3], emphasizing the need for greater understanding of the role of mitochondrial function in vivo.

The electron transport chain (ETC) is one of the main regulators of mitochondrial function, and can be indirectly assessed using confocal microscopy and membrane potential-sensitive dyes, such as tetramethylrhodamine methyl ester (TMRM), or endogenous fluorescent indicators of ETC redox potential, including oxidized flavoproteins and reduced nicotinamide adenine dinucleotide (NAD(P)H). Reduced NAD(P)H and oxidized flavoproteins have fluorescent properties [4] that differ from those of their oxidized and reduced counterparts, respectively, permitting cellular redox potential to be mapped with the spatial and temporal resolution afforded by confocal microscopy. ETC efficiency depends on oxygen availability; therefore hypoxic conditions can lead to reduction of the ETC due to accumulation of electrons and this can be visualized using flavoprotein and/or NAD(P)H fluorescence [5].

Oxygen is normally supplied at a rate sufficient to maintain tissue levels above the critical value necessary for mitochondrial function [6], and this supply has historically been attributed to the capillary network [7]. However, more recent evidence suggests that substantial oxygen diffusion can also occur across arteries and arterioles [8–11].

Here we examined mitochondrial function in vivo, as assessed by endogenous flavoprotein fluorescence, in response to changes in the inspired oxygen fraction (FiO_2) to explore the role of arteries in the supply of cortical tissue oxygen.

2 Method

C57bl/6 mice (~20 g) were housed in a 12 h light/dark cycle with food and water *ad libitum*. All experiments were performed in accordance with the UK Home Office Animals (Scientific Procedures) Act (1986).

Mice were anaesthetised (~2 % isoflurane in room air, or 2 g/kg urethane and 20 mg/kg ketamine i.p.; the injectable anaesthetic was only used when imaging NAD(P)H together with flavoproteins), and placed on a homeothermic heating mat to maintain rectal temperature at 37 °C. An incision was made in the scalp, the skull cleaned of connective tissue, and affixed to a titanium bar for stability with dental cement (Contemporary Ortho-Jet Powder, USA) mixed with cyanoacrylate glue (Loctite, Henkel Ltd., UK). The cortex was exposed by partial craniotomy (~5 mm diameter) over the right hemisphere, and the dura was moistened and cleaned with saline. A circular glass coverslip (6 mm) was placed over a ring of petroleum jelly to prevent evaporation during imaging. In a subset of experiments, oxygen-sensitive microbeads impregnated with a phosphorescent dye, PtPFPP (ex: 543 nm; em:

650 nm, collected with 585 nm long pass filter) were spread on the dura (5 μl of 5 mg/ml aqueous suspension) prior to placement of the coverslip. Alternatively, TMRM (T-668, Molecular Probes, Invitrogen, UK; 1 μM incubated on the cortex for 30 min; ex 561 nm, em 584–656) was applied after removal of the dura. Following surgery, the animals were moved to a custom-made stage for confocal microscopy.

In experiments employing cyanide (NaCN) or carbonyl cyanide 4-(trifluoromethoxy)phenylhydrazone (FCCP), a coverslip was not used, and the dura was removed. A well was created around the exposed cortex using silicone (Body Double, Smooth-On Inc., USA) and filled with 40 μl saline to which 2 μl of NaCN or FCCP (working concentration of 5 mM and 10 μM, respectively) were added during time lapse imaging. Five of these images were averaged, seconds or minutes after application, depending on the stabilization of the image. FiO_2 was controlled by mixing oxygen and nitrogen as indicated (100, 21, 15, 21, 10, 21 and 5 % oxygen, each for 5 min).

The endogenous flavoprotein signal (ex: 488 nm, em: 505–570 nm) was imaged with a LSM 5 Pascal laser-scanning confocal microscope (Zeiss, Germany), using time series recordings with an in-plane resolution of 512 by 512 pixels and an optical slice thickness of 896 μm. Endogenous NAD(P)H (ex: 720, em: 430–480) was imaged using a Zeiss 510 NLO META equipped with a Coherent Chameleon Ti:sapphire laser.

Images were processed using Fiji/ImageJ Version 1.48v. Time lapse sequences were aligned using the 'Stackreg'-Plugin. Statistical significance was assessed using the IBM SPSS Statistics 22 package.

3   Results

Under normoxic conditions, endogenous green fluorescence was uniformly distributed across the surface of the cerebral cortex, with the superficial vasculature clearly defined in negative contrast. Arteries were distinguishable from veins based on their morphology, and their uniform outline, which was typically highlighted by brightly fluorescent walls.

The origin of endogenous green fluorescence was explored by administrating agents known to change the redox state of flavoproteins. Application of NaCN (reducing the ETC) significantly decreased fluorescence intensity (~35 %), whereas application of FCCP (oxidizing the ETC) significantly increased fluorescence intensity (~23 %; Fig. 29.1). These data are consistent with the assumption that green autofluorescence originates from oxidized mitochondrial flavoproteins.

Although increasing FiO_2 had no effect on the signal, reducing FiO_2 (to ≤ 10 %) resulted in a marked decrease in flavoprotein fluorescence. This decrease preferentially affected tissue distal to arteries, with a 'halo' of preserved fluorescence in tissue adjacent to arteries and arterioles (Fig. 29.2), and typically appeared at an FiO_2 of 5–10 %.

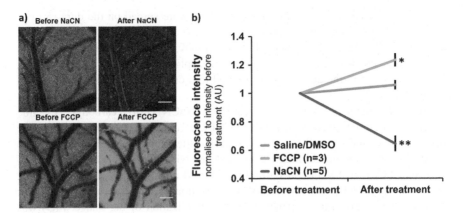

Fig. 29.1 (a) Fluorescence intensity in response to NaCN and FCCP. Scale bar = 100 μm. (b) Quantification of fluorescence intensity before and after application of saline/DMSO or NaCN/ FCCP to the cortex. Data are normalised to signal intensity before treatment and displayed as mean ± SEM. Statistical significance was assessed using a paired sample t-test (*$p \leq 0.05$, **$p \leq 0.01$)

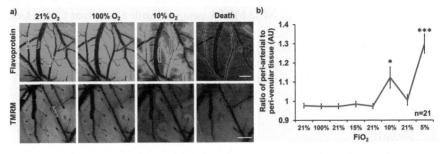

Fig. 29.2 (a) Flavoprotein (*green*) and TMRM (*red*) fluorescence in response to changes in FiO_2. Scale bar = 200 μm. (b) The ratio of periarterial to perivenular tissue flavoprotein fluorescence intensity (examples indicated in **a**), *red* = periarterial and *blue* = perivenular. Data are displayed as mean ± SEM. Statistical significance was assessed using a paired sample t-test (*$p \leq 0.05$, ***$p \leq 0.001$)

To explore whether changes in flavoprotein fluorescence were associated with changes in mitochondrial membrane potential, we examined the effects of hypoxia on TMRM fluorescence. The same arterial 'halos' were observed with TMRM at 5–10 % FiO_2 as were seen when imaging flavoproteins (Fig. 29.2a).

The changes in the distribution of the flavoprotein fluorescence at reduced FiO_2 varied inversely with NAD(P)H fluorescence (Fig. 29.3a). As expected, changing the FiO_2 also resulted in corresponding changes in cortical tissue oxygen concentration, as measured by oxygen-sensitive phosphorescent beads (Fig. 29.3b). When FiO_2 was increased, a larger change in emission intensity was observed in beads within the 'halo' of preserved flavoprotein fluorescence surrounding arteries than beads located distal to arteries (Fig. 29.3c). At an FiO_2 of 5 %, a greater response in

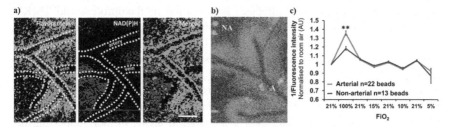

Fig. 29.3 (a) Flavoprotein (*green*) and NAD(P)H (*blue*) fluorescence in response to hypoxia. (b) Oxygen-sensitive phosphorescent beads (*red*) on the hypoxic cortex. A = Periarterial, NA = Nonarterial. (c) Bead emission intensity in response to changes in FiO_2. Data are displayed as mean ± SEM. Statistical significance was assessed using an independent sample t-test (**$p \leq 0.01$)

fluorescence was observed in beads in nonarterial regions, but this difference was not significant (Fig. 29.3c).

4 Discussion

Using endogenous flavoprotein fluorescence we have demonstrated that cortical mitochondrial function is selectively impaired at low FiO_2, with preservation of mitochondrial redox potential around arteries and arterioles, demonstrating the role of these structures in the direct supply of oxygen to cerebral cortex tissue.

The use of flavoprotein fluorescence as an indicator of mitochondrial redox potential is well established [4, 12–14] but its application in vivo has been limited. To our knowledge flavoprotein autofluorescence has not been used previously to assess cortical redox state in response to changes in FiO_2 at the high spatial and temporal resolutions of confocal microscopy.

To confirm the mitochondrial origin of the flavoprotein signal, we assessed signal changes in response to variations in mitochondrial redox state, induced by the well characterised agents NaCN and FCCP. The signal source was further validated by simultaneous measurement of flavoprotein and NAD(P)H fluorescence, revealing an inverse relationship and permitting mapping of the redox ratio of the cerebral cortex in vivo.

Under normoxic conditions, oxygen supply to the brain was sufficient to maintain functioning mitochondria throughout the cortex. Increasing FiO_2 accordingly had no effect on flavoprotein fluorescence, presumably because oxygen availability was not a rate limiting factor in the function of the ETC.

A slight decrease in FiO_2 to 15 % also had little influence on the mitochondrial redox potential. However, a further decrease of FiO_2 to $\leq$ 10 % induced a characteristic change, with preservation of oxidized flavoprotein in periarterial tissue, and reduction of flavoproteins in distal tissue and near veins. This pattern was also seen with TMRM, suggesting other measures of mitochondrial function such as membrane potential are also affected. Preservation of flavoprotein and TMRM fluorescence around arteries is not consistent with the historical assumption that oxygen exchange is limited to capillaries [7]. Rather, our data support recent evidence that arteries play a major role in supplying cortical oxygen directly to tissue [8–11].

As expected, decreasing FiO_2 also decreased cortical oxygenation as measured by oxygen-sensitive beads. Tissue oxygenation in periarterial regions increased to a greater extent during hyperoxia than in nonarterial areas, further supporting the suggestion that oxygen exchange occurs along arteries [8–11]. However, no measurable difference in periarterial and nonarterial tissue responsiveness was detected at ≤ 10 % FiO_2, despite the decrease in oxidized flavoprotein remote from arteries.

In conclusion, changes in mitochondrial redox potential, as demonstrated by a regionally selective decrease in flavoprotein fluorescence, are evident in the hypoxic cerebral cortex. The simultaneous preservation of oxidized flavoproteins in tissue surrounding arteries is consistent with the direct delivery of oxygen from arteries to adjacent tissue.

Acknowledgments This work was supported by the Wellcome Trust grants from the University College London Grand Challenges and Multiple Sclerosis Society of Great Britain and Northern Ireland.

References

1. Mahad D, Lassmann H, Turnbull D (2008) Review: mitochondria and disease progression in multiple sclerosis. Neuropathol Appl Neurobiol 34:577–589
2. Schapira AHV, Gu M, Taanman JW et al (1998) Mitochondria in the etiology and pathogenesis of Parkinson's disease. Ann Neurol 44:S89–S98
3. Fink MP (2002) Bench-to-bedside review: cytopathic hypoxia. Crit Care 6:491–499
4. Reinert KC, Dunbar RL, Gao WC et al (2004) Flavoprotein autofluorescence imaging of neuronal activation in the cerebellar cortex in vivo. J Neurophysiol 92:199–211
5. Kasischke KA, Lambert EM, Panepento B et al (2011) Two-photon NADH imaging exposes boundaries of oxygen diffusion in cortical vascular supply regions. J Cereb Blood Flow Metab 31:68–81
6. Erecinska M, Silver IA (2001) Tissue oxygen tension and brain sensitivity to hypoxia. Respir Physiol 128:263–276
7. Krogh A (1919) The number and distribution of capillaries in muscles with calculations of the oxygen pressure head necessary for supplying the tissue. J Physiol Lond 52:409–415

8. Ivanov KP, Derry AN, Vovenko EP et al (1982) Direct measurements of oxygen-tension at the surface of arterioles, capillaries and venules of the cerebral-cortex. Pflugers Arch Eur J Physiol 393:118–120

9. Ivanov KP, Sokolova IB, Vovenko EP (1999) Oxygen transport in the rat brain cortex at normobaric hyperoxia. Eur J Appl Physiol Occup Physiol 80:582–587

10. Vovenko E (1999) Distribution of oxygen tension on the surface of arterioles, capillaries and venules of brain cortex and in tissue in normoxia: an experimental study on rats. Pflugers Arch Eur J Physiol 437:617–623

11. Sakadzic S, Roussakis E, Yaseen MA et al (2010) Two-photon high-resolution measurement of partial pressure of oxygen in cerebral vasculature and tissue. Nat Methods 7:755–759

12. Chance B, Schoener B, Oshino R et al (1979) Oxidation-reduction ratio studies of mitochondria in freeze-trapped samples – NADH and flavoprotein fluorescence signals. J Biol Chem 254:4764–4771

13. Huang SH, Heikal AA, Webb WW (2002) Two-photon fluorescence spectroscopy and microscopy of NAD(P)H and flavoprotein. Biophys J 82:2811–2825

14. Scholz R, Thurman RG, Williams JR et al (1969) Flavin and pyridine nucleotide oxidation-reduction changes in perfused rat liver. I. Anoxia and subcellular localization of fluorescent flavoproteins. J Biol Chem 244:2317

Chapter 30
Cardiovascular Adaptation in Response to Chronic Hypoxia in Awake Rats

Saki Hamashima and Masahiro Shibata

Abstract To examine how cardiovascular adaptation to chronic hypoxia might evolve, the responses to blood pressure (Pt) and hematocrit (Ht) during long-term systemic exposure to hypoxia were observed in awake rats. Furthermore, the total peripheral vascular resistance (TPR) was estimated using direct measurements of systemic blood pressure (Ps) and blood flow (Qs) in carotid artery based on Darcy's law (TPR = Ps/Qs) to evaluate the remodeling procedure in the microcirculation. BP and Ht under normoxic conditions were kept almost constant, while hypoxic exposure immediately increased Ht to 58 % and, thereafter, it remained stable. The TPR values showed no significant differences between hypoxic and normoxic conditions. These results suggest that effects of high viscosity caused by increasing Ht on peripheral vascular resistance can be compensated by inducing microvascular remodeling with the arteriolar dilation and capillary angiogenesis.

Keywords Hypoxia • Total peripheral vascular resistance • Microvascular remodelling • Awake rats

1   Introduction

Whether the long-term exposure of hypoxia results in significant harmful effects to the cardiovascular system or induces protective properties that mediate vascular remodeling remains unclear [1]. In studies of the pathophysiological effects, hypoxic exposure has been used to develop an animal model for pulmonary hypertension [2], while hypoxic training is believed to be capable of potentiating greater performance improvements [3]. In addition, the morbidity rates of hypertension and coronary heart disease are reportedly lower in populations residing at high altitudes [4]. The purpose of the present study was to examine how cardiovascular adaptation to chronic hypoxia might develop. For this purpose, the blood pressure (Pt) and hematocrit (Ht) responses during long-term systemic exposure to hypoxia were observed in awake rats. Furthermore, the total peripheral vascular resistance (TPR)

S. Hamashima • M. Shibata (✉)
Department of Bioscience and Engineering, Shibaura Institute of Technology, Tokyo, Japan
e-mail: shibatam@sic.shibaura-it.ac.jp

© Springer Science+Business Media, New York 2016
C.E. Elwell et al. (eds.), *Oxygen Transport to Tissue XXXVII*, Advances in Experimental Medicine and Biology 876, DOI 10.1007/978-1-4939-3023-4_30

241

was estimated using direct measurements of systemic blood pressure (Ps) and blood flow (Qs) in carotid artery based on Darcy's law (TPR = Ps/Qs) to evaluate the remodeling procedure in the microcirculation.

2 Material and Methods

2.1 Experimental Protocols

All animal procedures were approved by the Shibaura Institute of Technology Animal Care and Use Committee. A total of 29 male Wister rats weighing 130–220 g (5–8 weeks old) were used in this study. Fourteen rats were used to measure the Pt and Ht responses to chronic exposure of hypoxia for 2 weeks. After Pt and Ht measurements during chronic exposure of hypoxia, these rats were provided to determine the TPR value. The TPR values under normoxic conditions were obtained by the remaining 15 rats. The chronic exposure to hypoxia of rats was carried out by housing the animals in a hypoxic chamber (ProOx110, USA) for 2 weeks, in which this chamber was kept at an oxygen concentration of 10 % by injecting nitrogen. The oxygen concentration of 10 % is equivalent to oxygen concentration at an altitude of 5500 m. The breeding room including the animal chamber and, the Pt and Ht measuring area were maintained a temperature of 20 °C.

2.2 Blood Pressure (Pt) and Hematocrit (Ht) Measurements

Pt and Ht were measured every 3 days during 2 weeks exposure to hypoxia. Pt was measured non-invasively using the tail-cuff oscillometric method (Softron, Japan) and the obtained Pt value was represented as the mean blood pressure. The Ht value was obtained with the conventional method by blood sampling from a tail vein. Only during a measuring period the rats were anesthetized by inhalation of ether and kept on the heated mat to prevent the reduction of body temperature.

2.3 Determination of Total Peripheral Resistance (TPR)

The changes in total peripheral vascular resistance (TPR) under hypoxic and normoxic conditions were determined to measure the Ps and Qs in carotid artery under anesthesia with urethane. First, Qs measurement was performed using a transit time blood flowmeter for small arteies (Transonic Systems, USA). The flow probe was attached to a carotid artery and obtained the flow signal continuously with mean value. On the other hand, to measure the Ps, the catheter was cannulated to carotid artery and connected to the pressure transducer (Ohmeda, USA). In this measurement, Ps value was also represented by the mean blood pressure. On the other hand, based on these measured Ps and Qs values, Darcy's law was applied to determine a total peripheral vascular resistance as TPR = Ps/Qs. All animals in which the Ps or Qs value fell to <60 mmHg or <3 mL/min during experiment were excluded.

2.4 Data Analysis

All data are reported as means ± SD. Data within each group were analyzed by analysis of variance for repeated measurements (ANOVA). Differences between groups were determined using a t-test with the Bonferroni correction. Differences with a p-value of <0.05 were considered statistically significant.

3 Results

3.1 Changes in Blood Pressure and Hematocrit to Hypoxia

Figure 30.1 shows the changes in mean Pt and Ht under hypoxic conditions in awake rats. Mean Pt and Ht under normoxic conditions before exposure of hypoxia remained almost constant (Pt: 78 ± 8.0 mmHg, Ht: 45 ± 3.5 %), while hypoxic exposure immediately increased Ht to 58 % and, thereafter, it remained stable. The mean Pt during hypoxic exposure showed no significant changes: from 81 ± 8.6 mmHg at the beginning to 84 ± 8.4 mmHg at the end.

3.2 Comparison of Peripheral Vascular Resistance (TPR)

To compare the peripheral vascular resistance under normoxic and hypoxic conditions, the blood flow (Qs) and mean blood pressure (Ps) in carotid artery were measured directly in anesthetized rats. The measured Qs and Ps values with carotid artery during 2 weeks hypoxic exposure are shown in Fig. 30.2. Although the blood flow after 2 weeks hypoxic exposure has a tendency to decrease, there were no significant differences compared with those of normoxic conditions (hypoxia: Qs and Ps: 5.2 ± 2.3 mL/min and 96 ± 22 mmHg, normoxia: 6.8 ± 1.9 mL/min and 89 ± 14 mmHg). The total peripheral resistance (TPR) values obtained by Qs and Ps are shown in Fig. 30.3. Again, there are no significant differences between TPR values under hypoxic and normoxic conditions (21 ± 10 and 15 ± 5 mmHg min mL^{-1}).

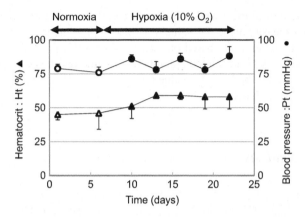

Fig. 30.1 Changes in mean blood pressure (Pt) measured by the tail-cuff method and hematocrit under normoxic and hypoxic conditions. All measurements were carried out under anesthesia by inhalation of ether

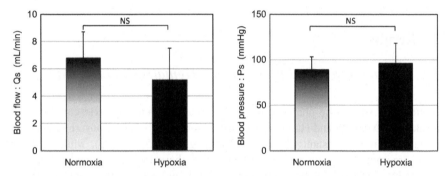

Fig. 30.2 The direct measured blood flow (Qs) and mean blood pressure (Ps) in carotid artery under normoxic conditions and after 2 weeks hypoxic exposure. There were no significant differences between normoxic and hypoxic values

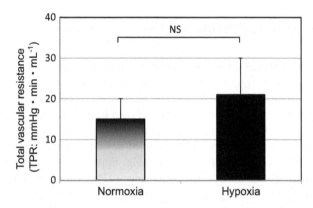

Fig. 30.3 The total peripheral resistance (TPR) values under normoxic conditions and after 2 weeks hypoxic exposure obtained by Qs and Ps. There are no significant differences between the values under hypoxic and normoxic conditions

4 Discussion

In the present study, we aimed to clarify how cardiovascular adaptation to chronic hypoxia would evolve. For this purpose, the responses to blood pressure and hematocrit during 2 weeks systemic exposure to hypoxia were observed in awake rats. The total peripheral vascular resistances under hypoxic and normoxic conditions were compared to evaluate the remodeling procedure in the microcirculation. The Ht values under hypoxic exposure immediately increased to 58 % and, thereafter, remained stable. These changes in Ht are well-known responses to hypoxia or hypobaric conditions [5, 6]. On the other hand, the blood pressure was kept almost constant during 2 weeks exposure to hypoxia in this study. This constancy of blood pressure implies successful adaptation to chronic hypoxia through protective mechanisms that mediate vascular remodeling. There are many reports that the hypoxia induced microvascular angiogenesis [7, 8] and vasodilation [9], in which one of the major factors is the high vascular wall shear stress induced by increased blood viscosity. Increased shear stress increases the endothelial nitric oxide production. The successful process of cardiovascular adaptation to chronic hypoxia considered from the present study is shown in Fig. 30.4.

Fig. 30.4 The successful process of cardiovascular adaptation to chronic hypoxia

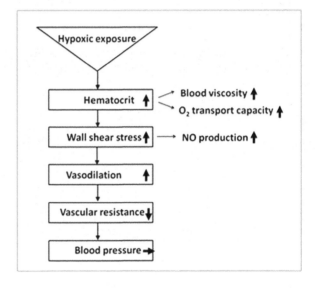

5 Conclusion

These results suggest that the effects of high viscosity caused by increasing Ht on peripheral vascular resistance can be compensated by inducing microvascular remodeling with arteriolar dilation and capillary angiogenesis.

References

1. Essop Faadiel M (2007) Cardiac metabolic adaptation in response to chronic hypoxia. J Physiol 584:715–726
2. Ostadal B, Ostadalova I, Dhalla NS (1999) Development of cardiac sensitivity to O_2 deficiency: comparative and ontogenetic aspect. Physiol Rev 79:635–659
3. Lara B, Salinero JJ, Del Coso J (2014) Altitude is positively correlated to race time during the marathon. High Alt Med Biol 15:64–69
4. Mortimer EA Jr, Monson RR, MacMahon B (1977) Reduction in mortality from coronary heart disease in men residing at high altitude. N Engl J Med 296:581–585
5. Núñez-Espinosa C, Douziech A, Ríos-Kristjánsson JG et al (2014) Effect of intermittent hypoxia and exercise on blood rheology and oxygen transport in trained rats. Respir Physiol Neurobiol 192:112–117
6. Takuwa H, Masanoto K, Yamazaki K et al (2013) Long-term adaptation of cerebral hemodynamics response to somatosensory stimulation during chronic hypoxia in awake mice. J Cerebr Blood Flow Metab 33:774–779
7. Harik SI, Hritz MA, LaManna JC (1995) Hypoxia-induced brain angiogenesis in the adult rat. J Physiol 485:525–530
8. Boero JA, Ascher J, Arregui A et al (1999) Increased brain capillaries in chronic hypoxia. J Appl Physiol 86:1211–1219
9. Shibata M, Ichioka S, Ando J et al (2005) Non-linear regulation of capillary perfusion in relation to ambient pO_2 changes in skeletal muscle. Eur J Appl Physiol 94:352–355

Chapter 31
Prototyping the Experimental Setup to Quantify the Tissue Oxygen Consumption Rate and Its Preliminary Test

N. Watanabe, M. Shibata, S. Sawada, and K. Mizukami

Abstract In order to establish a reliable and practical method to make a diagnosis on the viability of an amputated extremity, we propose a method to evaluate the oxygen consumption rate. To validate this concept, we prototyped an experimental system with which the oxygen transfer rate into tissue can be assessed by the rate of change of the decrease in dissolved oxygen (DO) concentration within the buffer fluid surrounding the target tissue. The purpose of this study is to examine the feasibility of our prototyped experimental system by comparison between fresh and non-fresh rat skeletal muscles. The results show that the fresher tissue transferred more oxygen to the tissue, which suggests that tissue oxygen consumption is highly related to tissue freshness and can indirectly assess the tissue viability.

Keywords Oxygendiffusion • Dissolved oxygen • Tissue freshness • Tissue viability • Temperature

1 Introduction

Evaluation of the viability of an amputated extremity is currently very difficult because there is no blood flow in such fragmented tissue, and no standard diagnostic procedure has been established so far. As an index for the viability of such tissue, a

N. Watanabe (✉) • M. Shibata
Department of Bioscience and Engineering, College of Systems Engineering and Science, Shibaura Institute of Technology, 307 Fukasaku, Minuma-Ku, Saitama 337-8570, Japan

Systems Engineering and Science, Graduate school of Engineering and Science, Shibaura Institute of Technology, Saitama, Japan
e-mail: nobuo@sic.shibaura-it.ac.jp

S. Sawada
Systems Engineering and Science, Graduate school of Engineering and Science, Shibaura Institute of Technology, Saitama, Japan

K. Mizukami
Department of Bioscience and Engineering, College of Systems Engineering and Science, Shibaura Institute of Technology, 307 Fukasaku, Minuma-Ku, Saitama 337-8570, Japan

© Springer Science+Business Media, New York 2016   247
C.E. Elwell et al. (eds.), *Oxygen Transport to Tissue XXXVII*, Advances in
Experimental Medicine and Biology 876, DOI 10.1007/978-1-4939-3023-4_31

possible candidate is the oxygen consumption capability. It is well known that oxygen transfer inside our body consists of convection and diffusion phenomena [1]. Therefore, without blood flow (convection), oxygen transfer occurs with only the diffusion mechanism, and the transfer region would be too limited to satisfy the needs within our body. On the other hand, it is possible to transfer oxygen only by diffusion into fragmented tissue (such as an amputated extremity) if the tissue surface area is large enough. Therefore, many methods for the evaluation of cell viability have been reported so far by optical means, such as the phosphorescence quenching technique [2, 3]. However, the preparation of individual cells takes time and the development of a more convenient and practical method, applicable as a clinical diagnostic procedure is desirable. It is proposed that it may be possible to employ a small volume (such as a fragment from the amputated extremity). For this purpose, we propose using such a fragment in an indirect assessment method where dissolved oxygen is measured within buffer fluid in a closed chamber. This should allow evaluation of the rate of oxygen consumption of the sampled tissue. The concept is explained in more detail below.

1.1 Concept to Assess the Viability of a Part of Amputated Extremity

Figure 31.1 shows that changes in the dissolved oxygen concentration of a defined volume of buffer fluid within a closed chamber allows one to estimate the amount of oxygen transferred to the tissue from the buffer, but only if dead tissue is also tested for comparison.

The tissue oxygen consumption rate will vary according to the surrounding temperature [4] and freshness of the tissue. The latter parameter should, indirectly, indicate the tissue viability. The purpose of this study is to examine this concept as a method of assessing the capacity of the tissue to consume the oxygen and hence viability.

2 Methods

The experimental setup was developed in order to quantify the rate of oxygen transfer to tissue. The amount of oxygen transfer can be indirectly assessed by the change in dissolved oxygen concentration. Figure 31.2 shows the experimental setup. The volume-defined closed bath was made using a measuring cylinder and a oxygen sealing sheet with the oxygen diffusion capability of 60 [cc/m^2 day atm] (Saran Wrap®, Asahi kasei Home Products Corporation, Saitama Japan). A small

Fig. 31.1 Concept to estimate the amount of oxygen transfer to tissue

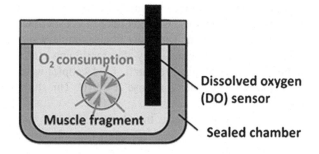

Fig. 31.2 Experimental setup to estimate the amount of oxygen transfer to tissue

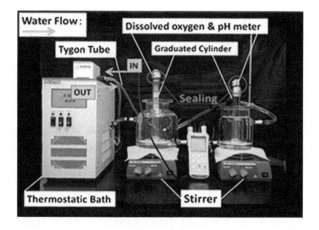

fragment of tissue, less than 5 g, is inserted in the chamber, and within the bath the temperature is controlled using a Thermostatic bath (SA-100, Sansyo Co. LTD, Japan). Within the temperature-controlled and volume-defined closed bath, the dissolved oxygen (DO) concentration in phosphate buffered saline (PBS) solution is monitored under temperature-controlled conditions. This PBS solution was prepared with ion-exchanged water acquired by the usage of ion-exchanged water machine (Pure port PP-101, Sibata scientific technology Ltd. Saitama Japan) or the PBS powder (162-19321, Wako Pure Chemical Industries, Ltd., Osaka Japan). Because the buffer volume is defined, the amount of oxygen transferred to the tissue can be assessed by the change in oxygen concentration. In this experiment, the DO sensor provided by Horiba (SS054, NaviH, Horiba Ltd, Japan) was used. As the sample tissue, rat skeletal muscle of the lower extremity (3 or 4.8 g) was used. Immediately after inserting the fresh and non-fresh tissue into the bath, the DO monitoring was started. During the experiment, a temperature of 4 or 37 °C was maintained. In this study, the definition of fresh tissue is rat skeletal muscle acquired within 30 min after death; the definition of non-fresh tissue is rat skeletal muscle acquired longer than 3 days after death. The duration of monitoring was

30 min. The test chamber volume was 100 or 50 cc. For experiments at 4 °C we used the 50 cc chamber, whereas for 37 °C experiments the 100 cc chamber was used. Each experiment with new tissue sample was performed five times. Fresh tissue was acquired immediately after sacrificing a rat, which had already been used in previous physiological research. 4.8 g samples were used for the experiments at 4 °C and 3.0 g samples for those at 37 °C. The study was approved by the local ethical committee.

After the experiments, the oxygen transfer rate into the tissue (OTRT) [mg/100 ml tissue/min] was derived by the following equation.

$$OTRT = (V/1000)*(100/W)*1/(Density*Time)$$

Where, V: chamber volume (50 or 100 ml) [ml], W: the weight of sample tissue (3 or 4.8 g) [g], Density: the muscle density (1.056) [g/cm^3], and Time: Time duration of DO monitoring test (30 min) [min].

And then, the tissue oxygen consumption rate (TOCR) [mg/100 ml tissue/min] can be derived by the difference of OTRT between the fresh and non-fresh tissue experiments like the following equation.

$$TOCR = (OTRT \ of \ fresh \ tissue \ experiment) \\ - (OTRT \ of \ non\text{-}fresh \ tissue \ experiment)$$

3 Result

Figure 31.3a, b present the experimental results separately for those at 4 °C and those at 37 °C showing the time course of the DO level. The white and black square plots represent the data from fresh and non-fresh tissue, respectively. And the grey triangles represent the negative control test data without tissue sample. Each plot in Fig. 31.3a represents the 4 °C averaged data of five experiments with the standard deviation. Each plot in the Fig. 31.3b represents the 37 °C averaged Fresh tissue data of seven experiments and the non-fresh tissue data of three experiments, with the negative control data without tissue sample of five experiments, respectively. All experiments show a continuous progressive decrease in the dissolved oxygen concentration as a function of time elapsed. In addition, the rate of decrease was greater with fresh tissue than with non-fresh tissue; for 5 °C over the first 10– 12 min, for 37 °C throughout the 30 min sampling. Figure 31.4 shows the monitored pH value which was obtained from the 37 °C test corresponding to Fig. 31.3b. The pH value fell compared with the unchanging negative control test values without tissue. The fall was the same for fresh and non-fresh tissues. Figure 31.5 shows the differential value of DO between the each sampling period.

Fig. 31.3 (a) Time series dissolved oxygen concentration change in PBS solution within closed vessel under the temperature of 4 °C. These data were acquired by using 50 ml chamber. (b) Time series dissolved oxygen concentration change in PBS solution within closed vessel under the temperature of 37 °C; this data is corresponding to the data in Fig. 31.4. These data were acquired by using 100 ml chamber

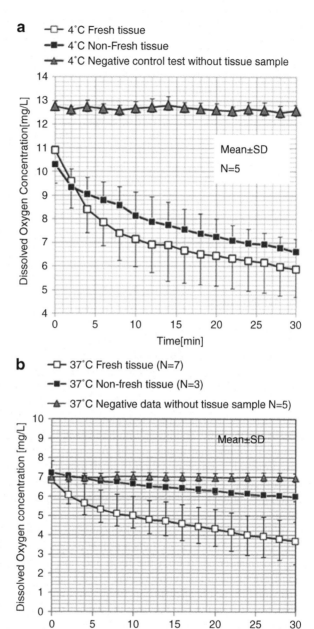

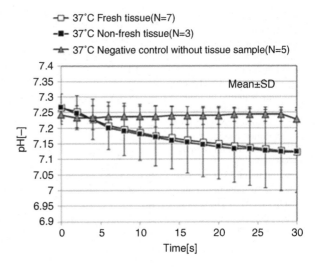

Fig. 31.4 Time series pH value change in PBS solution within closed vessel under the temperature of 37 °C; this plot data is corresponding to the data shown in Fig. 31.3b, each other

4 Discussion

The oxygen transfer would be driven by the oxygen concentration difference between the tissue inside and the surrounding buffer solution outside the tissue. Therefore the volume combination between the buffer solution and sample tissue would make an impact on the time series DO concentration change; therefore there is limitation when we compare the data between Fig. 31.3a and b. Even though there is a limitation for the interpretation, it is possible to compare the difference because of the same condition within the each figure (Fig. 31.3a, b) concerning the chamber volume and tissue weight. As shown in Fig. 31.4, the pH value slightly deteriorated by approximately 0.15 in 30 min under both fresh and non-fresh tissue experiments, with some deviation. This change would make slight impact upon the tissue metabolism; however we speculated them negligible for the diagnostics of the tissue viability.

Within the closed vessel, the DO concentration fell progressively, as shown in Fig. 31.3. The DO decrease indicates that oxygen was transferred to the tissue. The oxygen transferred to both the fresh and non-fresh tissue. We assumed that the non-fresh tissue did not consume the oxygen, and that the reason for the DO change was simply a result of diffusion from surrounding buffer solution into the tissue. Therefore, the difference between the data of fresh and non-fresh tissue supposedly represents the diffusion-normalized oxygen transfer data, i.e., the tissue oxygen consumption. The possible amount of oxygen consumption with the fresh and non-fresh tissue that was derived is shown in Tables 31.1 and 31.2, respectively: the tissue mass was converted into volume using the knowledge of muscle density of 1.056 [g/cm^3], as reported by Saito [5].

We speculated that the DO change in the non-fresh tissue experiment occurred due only to diffusion. On the other hand, the DO change in the fresh tissue

Fig. 31.5 (a) Time series differential value in DO during 2 min sampling time acquired by the experiment under the temperature of 4 °C. (b) Time series differential value in DO during 2 min sampling time acquired by the experiment under the temperature of 37 °C

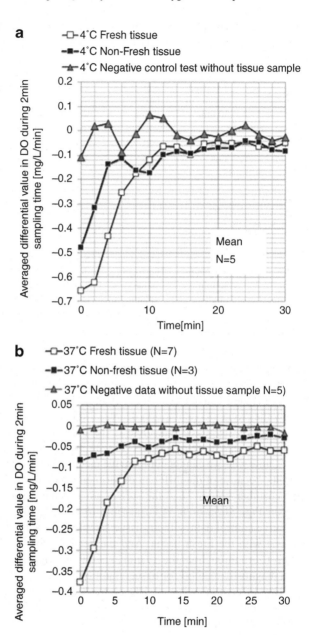

experiment includes both diffusion and oxygen consumption by vital fresh tissue. Therefore, the difference between the DO change of fresh and non-fresh tissue represents the oxygen consumption level by tissue. The oxygen consumption rates, based on these assumptions, at 37 and 4 °C of fresh tissue were 0.21 and 0.04 mg/ 100 ml tissue/min, respectively. An earlier study by Mottram [6] on human skeletal

Table 31.1 Experimental result with the fresh tissue (animal dead for less than 30 min); mass was converted into volume using knowledge of muscle density of 1.056 [g/cm^3], as reported by Saito [5]

Temperature	Decrease in dissolved oxygen in 30 min [mg/l]	Solvent phosphate buffered saline volume [ml]	Sample tissue weight [g]	Oxygen transfer rate into the tissue [mg/3 g tissue/ 30 min]	Oxygen transfer rate into the tissue [mg/100 ml tissue/ min]
37 °C	3.14	100	3.0	0.314	0.330
4 °C	5.02	50	4.8	0.157	0.165

Table 31.2 Experimental result with the non-fresh tissue (animal dead for longer than 3 days); mass was converted into volume using knowledge on muscle density of 1.056 [g/cm^3], as reported by Saito [5]

Temperature	Decrease in dissolved oxygen in 30 min [mg/l]	Solvent phosphate buffered saline volume [ml]	Sample tissue weight [g]	Oxygen transfer rate into the tissue [mg/3 g tissue/ 30 min]	Oxygen transfer rate into the tissue [mg/100 ml tissue/ min]
37 °C	1.12	100	3.0	0.112	0.118
4 °C	3.69	50	4.8	0.115	0.121

muscle oxygen consumption in another research team reported 0.273 mg 100 ml tissue/min, which is similar to our result but a little higher. Their experiment would have included conventional transfer in addition to diffusion transfer. Our results show the oxygen consumption decrease to be 21 % at 4 °C.

It can be suggested that there is the limitation with this study to evaluate the different temperature conditions (Fig. 31.3a, b). Metabolized more at 37 °C than at 4 °C, however, we cannot compare the metabolism level between them by using our results, because of the different experimental conditions such as the tissue amount, the chamber volume, and the surface area of the tissue, and so on. Even though, within Fig. 31.5a or b, it is possible to consider the tissue freshness level and negative control data without tissue samples. In each figure, the DO decrease was largest with the Fresh tissue. And the decrease for the negative control data was almost zero throughout the 30 min experiment, suggesting feasibility of our method as a diagnostic method for tissue viability.

The differential value of DO of fresh tissue tests, as shown in Fig. 31.5a, b, suggested the higher metabolism during first 10 min followed by lower metabolism during the monitoring.

The oxygen consumption rate will vary depending on the tissue freshness and also on the temperature. Additionally we calculated the temperature coefficient [7, 8] (Q10) using following equation in order to consider the temperature effect upon our results.

$$Q10 = (R2/R1)^\wedge(10/(T2 - T1))$$

Where, R1 and R2 are the rates at two temperatures (here we used TOCR), and T1 and T2, in this regard T2 > T1. The calculated Q10 was approximate 1.653. The similar data was found in other publication [8], which showed 1.56, as the data of Colorado potato beetle at temperature between 7 and 30 °C. Therefore our experimental data may be reasonable.

Anyway the temperature effect always gives the DO evaluation, even though tissue oxygen consumption seems highly related to tissue freshness, and can indirectly assess tissue viability. However it is necessary to fix the physical condition such as chamber volume and parameter to define the tissue sample amount such as weight, volume, surface area. Further study should optimize the experimental condition concerning these aspects.

5 Conclusions

We prototyped the experimental setup to evaluate tissue sample viability through measurement of the oxygen consumption rate. The level of oxygen transfer differed depending on the freshness of the tissue and the temperature. The tissue oxygen consumption seems to be highly related to tissue freshness, and can indirectly assess the tissue viability.

Acknowledgments A part of this paper was reported in the Master's thesis of S. Sawada [9] at Shibaura Institute of Technology, Japan (2013).

References

1. Levick JR (2010) An introduction to cardiovascular physiology, 5th edn. Hodder, Arnold and Hachette, London. ISBN 9780340942048
2. Hynes J, Floyd S, Soini AE, O'Connor R, Papkovsky DB (2003) Fluorescence-based cell viability screening assays using water-soluble oxygen probes. J Biomol Screen 8(3):264–272
3. Dmitriev RI, Papkovsky DB (2012) Optical probes and techniques for O2 measurement in live cells and tissue. Cell Mol Life Sci 69:2025–2039
4. Fry FEJ, Hart JS (1948) The relation of temperature to oxygen consumption in the goldfish. Biol Bull 94(1):66–77
5. Saito A (2003) Skeletal muscle structure. Rigakuryoho Kagaku 18(1):49–53
6. Mottram RF (1955) The oxygen consumption of human skeletal muscle in vivo. J Physiol 128 (2):268–276
7. Reyes AB, Pendergast JS, Yamazaki S (2008) Mammalian peripheral circadian oscillators are temperature compensated. J Biol Rhythms 23:95–98
8. Schmidt-Nielsen K (1997) Animal physiology adaptation and environment, 5th edn. Cambridge University Press, Cambridge, p 220
9. Sawada S (2014) Development of experimental setup to measure the tissue oxygen consumption. Master's thesis of Shibaura Institute of Technology (in Japanese)

Chapter 32
Monitoring Intracellular Oxygen Concentration: Implications for Hypoxia Studies and Real-Time Oxygen Monitoring

Michelle Potter, Luned Badder, Yvette Hoade, Iain G. Johnston, and Karl J. Morten

Abstract The metabolic properties of cancer cells have been widely accepted as a hallmark of cancer for a number of years and have shown to be of critical importance in tumour development. It is generally accepted that tumour cells exhibit a more glycolytic phenotype than normal cells. In this study, we investigate the bioenergetic phenotype of two widely used cancer cell lines, RD and U87MG, by monitoring intracellular oxygen concentrations using phosphorescent Pt-porphyrin based intracellular probes. Our study demonstrates that cancer cell lines do not always exhibit an exclusively glycolytic phenotype. RD demonstrates a reliance on oxidative phosphorylation whilst U87MG display a more glycolytic phenotype. Using the intracellular oxygen sensing probe we generate an immediate readout of intracellular oxygen levels, with the glycolytic lines reflecting the oxygen concentration of the environment, and cells with an oxidative phenotype having significantly lower levels of intracellular oxygen. Inhibition of oxygen consumption in lines with high oxygen consumption increases intracellular oxygen levels towards environmental levels. We conclude that the use of intracellular oxygen probes provides a quantitative assessment of intracellular oxygen levels, allowing the manipulation of cellular bioenergetics to be studied in real time.

Keywords Intracellular oxygen • Mitochondria • Oxidative phosphorylation • Oxygen probe • Aerobic glycolysis

M. Potter • Y. Hoade • K.J. Morten (✉)
Nuffield Department of Obstetrics and Gynaecology, University of Oxford, Oxford, UK
e-mail: Karl.morten@obs-gyn.ox.ac.uk

L. Badder
European Cancer Stem Cell Research Institute, University of Cardiff, Cardiff, UK

I.G. Johnston
Department of Mathematics, Imperial College London, London, UK

© Springer Science+Business Media, New York 2016
C.E. Elwell et al. (eds.), *Oxygen Transport to Tissue XXXVII*, Advances in Experimental Medicine and Biology 876, DOI 10.1007/978-1-4939-3023-4_32

257

1 Introduction

Oxygen has an important role to play in mammalian cell homeostasis, providing ATP for energy, assisting in enzymatic reactions (oxidation and hydroxylation) and generating reactive oxygen species (ROS), which are all key parameters in metabolism [1]. There are many physiological (and pathophysiological) implications for oxygen dynamics, especially in cancer and the adaptive response to hypoxia. Cells and tissues have different levels of intracellular and extracellular oxygen, with oxygen gradients existing across tissues. Cells and tissues are exposed to different levels of oxygen: most tissues have a partial pressure of oxygen (pO_2) of between 20 and 40 mmHg, whereas a large tumour can be severely hypoxic or anoxic with pO_2 levels <0.1 mmHg [2]. A frequent occurrence in most solid tumours is exposure of cells to levels of hypoxia equivalent to oxygen dissolved at less than 1 % in solution [3]. This hypoxia results in key adaptive responses, altering cell metabolism and making the cells resistant to chemotherapy and radiotherapy. With the development of new 3D culture systems, the ability to determine intracellular oxygen levels and model the in vivo situation will be of fundamental importance when investigating the effects of novel therapeutics to treat cancer or reduce ischaemic injury.

A recently-developed in vitro tool was utilised in this study to investigate the level of monolayer oxygenation. Monolayer oxygenation is a result of the balance between oxygen entering the cells by diffusion and oxygen consumed by physiological process. In most cells oxidative phosphorylation will be the major consumer of oxygen within the cell. MitoXpress®-Intra is an easy-to-use cell-penetrating phosphorescence-based oxygen sensing probe that allows the real-time quantitation of oxygen within the cell monolayer using a plate reader platform [4, 5]. We employed this new tool to examine the metabolism of two established cancer cell lines, RD (rhabdomyosarcoma) and U87MG (glioblastoma multiforme) under different applied oxygen concentrations.

2 Methods

RD, U87MG and 786-0 cells (ATCC) were seeded at 50,000 cells/well in black, clear bottomed 96 well plates (Nunc) in 150 µl Dulbecco's modified eagles medium (Invitrogen) containing 25 mM glucose and supplemented with 1 mM pyruvate, 2 mM glutamine, 10 % FBS and left overnight to adhere. Media was then removed and replaced with 150 µl fresh culture media containing 10 µg/ml MitoXpress-Intra probe. The cells were then returned to the incubator overnight to allow for probe internalisation. Before assaying, the media was removed and the cells were carefully washed thrice in PBS before 150 µl fresh assay media was added. The plate was read on a BMG Omega FluoStar plate reader with Ex 340 ± 50 nm and Em 650 ± 50 nm using dual delay time resolved measurements. The delay times were

30 and 70 µs. An atmospheric control unit (ACU) maintained the temperature at 37 °C and allowed us to control ambient oxygen levels. The raw data are expressed as TR-F intensity, requiring a conversion into phosphorescence lifetime values. Lifetime is calculated using the following equation: lifetime $\tau = (t_2 - t_1)/\ln (D_2/D_1)$ [t_i = delay times, D_i = measured intensity values], with each specific lifetime value corresponding to a specific oxygen concentration.

3 Results

To generate absolute values of probe fluorescence at different biological oxygen concentrations, a calibration curve for the MitoXpress-Intra was constructed using probe-containing antimycin-A treated cells. Prior to measurement, cells were treated with 1 µM antimycin-A in order to inhibit any mitochondrial respiration which would interfere with baseline probe signal. Atmospheric oxygen was set to 18 % and once the level of probe fluorescence reached a steady state, the oxygen in the reader was reduced in a stepwise manner as depicted in Fig. 32.1. Complete de-oxygenation was achieved by the addition of glucose oxidase to adherent cells in media. The average lifetime values of the non-respiring cell monolayer at each steady state were then plotted against their corresponding oxygen concentrations to produce the calibration curve depicted in Fig. 32.2. This calibration process facilitated the conversion of measured lifetime values into absolute percentage intracellular oxygen concentrations using the equation $[\% \ O_2] = A \exp(-B\tau)$, where the numerical values of the parameters $A \approx 370$ and $B \approx 0.1 \ \mu s^{-1}$ are derived from least-squares fitting the expression to the combined U87MG and RD data. We also computed 95 % confidence intervals on this fit using bootstrapping with the percentile method.

RD and U87MG cells, loaded with MitoXpress intra probe, were cultured in high glucose media and measured at different oxygen concentrations. The results show that, at 18 % ambient oxygen, RD cells contained intracellular O_2 levels of 13.2 ± 0.7 % (Fig. 32.3). The difference between ambient oxygen and intracellular oxygen concentrations is greater when ambient oxygen levels are dropped to more physiological levels. At 10 and 5 % ambient oxygen these metabolically proficient cells show a 50 % reduction in the level of intracellular oxygen (Fig. 32.3). In contrast U87MG cells, known for their glycolytic phenotype, were found to contain intracellular oxygen levels of 17.1 ± 0.8 % at 18 % ambient oxygen levels, more representative of ambient concentration (Fig. 32.3). In Fig. 32.4 we give an example of how the oxygen probes can be used to monitor drug effects in real time. 786-0 renal cancer cells pre-loaded with MitoXpress –Intra were incubated with different concentrations of the drug phenformin under an ambient concentration of 5 % oxygen. The cells show an initial drop in oxygen levels (increase in lifetime) as ambient oxygen levels change from 20 to 5 % followed by a complete inhibition of oxygen consumption in the phenformin treated samples (Fig. 32.4).

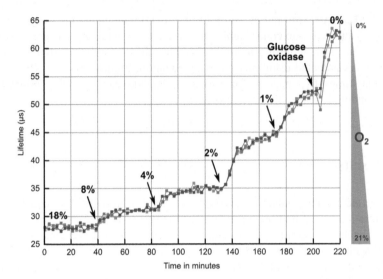

Fig. 32.1 Kinetic trace of the intracellular probe showing the response of a non-respiring cell monolayer (antimycin-A treated) to decreasing atmospheric oxygen concentrations (labelled). *Coloured traces* show three technical replicates

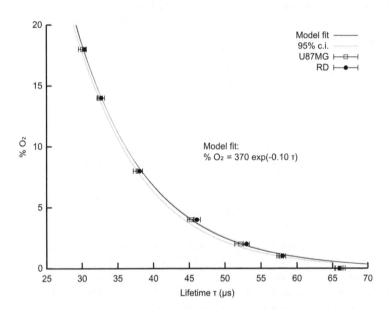

Fig. 32.2 Calibration curve, with 95 % confidence intervals, relating lifetime values to percentage ambient oxygen. *Values* represent the mean and standard deviation of 6 replicates. The equation gives an expression quantitatively linking lifetime measurements with O_2 levels, derived from fitting a model to these data points

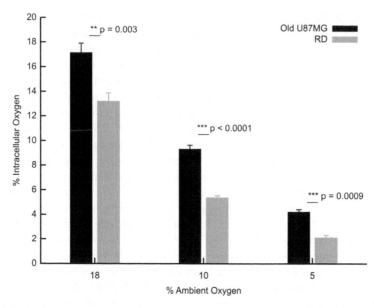

Fig. 32.3 At 18 % ambient oxygen RD cells were found to contain intracellular oxygen levels of 13.2 ± 0.7 %, thus suggesting that they are metabolically proficient cells. In contrast, U87MG cells, known for their glycolytic phenotype, had intracellular oxygen levels of 17.1 ± 0.8 % at 18 % ambient oxygen. ** -$p < 0.01$, *** – $p < 0.001$ using an unpaired t-test ($n = 3$)

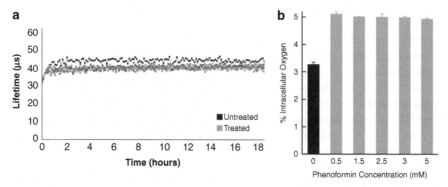

Fig. 32.4 Real-time measurement of the effect of phenformin treatment on intracellular oxygen levels on 786-0 renal carcinoma cells maintained in a BMG Omega Fluostar reader at 5 % O_2 and 5 % CO_2 (**a**) Lifetime measured over 18 h (**b**) Estimated intracellular oxygen levels from lifetime values., using a calibration curve derived from HEPG2 cell data

4   Discussion

A substantial body of evidence indicates that aerobic glycolysis, commonly known as the Warburg effect, is a hallmark of cancer [6]. Here we show that RD cells consume oxygen on high glucose media and do not rely on glycolysis for ATP production to the same extent as U87MG cells. This is also a feature of a large number of cancer cell lines (data not shown) indicating that not all cancer cells display this aerobic glycolysis 'hallmark' as previously suggested.

Our data also suggest that significant differences can exist between the oxygen concentration at the cell monolayer and the applied ambient oxygen concentration. This difference is not accounted for in standard in vitro analysis, where it is assumed the intracellular oxygen concentration reflects environmental levels (for example, assuming that an ambient oxygen level of 18 % implies an intracellular level of 18 %), consonant with previous reports that the rate of oxygen diffusion into the cells is faster than the oxygen consumption rate [7]. However, we observed that the aerobic RD cells consume and deplete intracellular oxygen to a dramatically greater extent than the glycolytic U87MG cells. The aerobically proficient RD cells, at ambient oxygen concentrations of 18, 10 and 5 %, have intracellular levels of 13.2, 5.4 and 2.1 %, respectively. In contrast to this are the glycolytic U87MG cells that displayed oxygen levels of 17.1, 9.3 and 4.2 %, more similar to ambient levels. The greater differences observed between the intracellular oxygen levels in the RD line compared to ambient are likely due a lower level of back diffusion from the environment as ambient oxygen levels are reduced. If mitochondrial respiration becomes limiting at lower levels of oxygen this difference may decrease. Interestingly our recent study in primary human bone cells showed very high rates of mitochondrial respiration even when ambient oxygen levels were reduced to 2 % [8].

The higher than expected value for the U87MG cells at 18 % oxygen with an error of 4 % is a property of the probe when measuring high non-physiological concentrations of oxygen. The probe is much more robust when working with oxygen levels in the physiological range <8 % O_2 with much smaller errors. Interestingly when the ambient oxygen levels were 2 %, the aerobic RD cells were almost anoxic (data not shown). Oxygen levels fluctuate and will drop significantly when glucose is used up during normal growth (as cells utilise mitochondrial substrates to generate ATP, thus consuming higher levels of oxygen). The intracellular oxygen monitoring system described here allows changes in oxygen levels to be monitored in real time. This monitoring can be either rapid (over minutes or hours, as with the mitochondrial inhibitor phenformin), or over several days (if the assays are run in a plate reader with an atmospheric control unit). The MitoXpress intra probe thus allows an insight into cellular oxygenation that has not previously been attainable. This new probe could be extremely useful as a tool to identify compounds or gene knockouts that modulate mitochondrial respiration in a high throughput format.

Acknowledgments MP and YH were supported by Williams Fund and LB by Luxcel Biosciences. We would also like to thank Dr James Hynes and Conn Carey (Luxcel Biosciences) and Jasmin Loo and Catherine Wark (BMG Labtech) for technical support.

References

1. Baynes JW (2009) Chapter 37 "oxygen and life". In: Baynes JW, Dominiczak MH (eds) Medical biochemistry, 3rd edn. MOSBY Elsevier, Philadelphia. http://www.us.elsevierhealth.com/biochemistry/medical-biochemistry-paperback/9780323053716/
2. Semenza GL (2007) Life with oxygen. Science 318:62–64
3. Vaupel P, Mayer A (2007) Hypoxia in cancer: significance and impact on clinical outcome. Cancer Metastasis Rev 26:225–239
4. Kondrashina AV, Papkovsky DB, Dmitriev RI (2013) Measurement of cell respiration and oxygenation in standard multichannel biochips using phosphorescent O2-sensitive probes. Analyst 138:4915–4921
5. Dmitriev RI, Papkovsky DB (2012) Optical probes and techniques for O2 measurement in live cells and tissue. Cell Mol Life Sci 69:2025–2039
6. Hanahan D, Weinberg RA (2011) Hallmarks of cancer: the next generation. Cell 144:646–674
7. Papandreou I, Cairns RA, Fontana L, Lim AL, Denko NC (2006) HIF-1 mediates adaptation to hypoxia by actively downregulating mitochondrial oxygen consumption. Cell Metab 3:187–197
8. Morten KJ, Badder L, Knowles HJ (2013) Differential regulation of HIF-mediated path ways increases mitochondrial metabolism and ATP production in hypoxic osteoclasts. J Pathol. doi:10.1002/path.4159

Chapter 33
Aging Effect on Post-recovery Hypofusion and Mortality Following Cardiac Arrest and Resuscitation in Rats

Kui Xu, Michelle A. Puchowicz, and Joseph C. LaManna

Abstract In this study we investigated the effect of aging on brain blood flow following transient global ischemia. Male Fisher rats (6 and 24 months old) underwent cardiac arrest (15 min) and resuscitation. Regional brain (cortex, hippocampus, brainstem and cerebellum) blood flow was measured in non-arrested rats and 1-h recovery rats using $[^{14}C]$ iodoantipyrene (IAP) autoradiography; the 4-day survival rate was determined in the two age groups. The pre-arrest baseline blood flows were similar in cortex, brainstem and cerebellum between the 6-month and the 24-month old rats; however, the baseline blood flow in hippocampus was significantly lower in the 24-month old group. At 1 h following cardiac arrest and resuscitation, both 6-month and 24-month groups had significantly lower blood flows in all regions than the pre-arrest baseline values; compared to the 6-month old group, the blood flow was significantly lower (about 40 % lower) in all regions in the 24-month old group. The 4-day survival rate for the 6-month old rats was 50 % (3/6) whereas none of the 24-month old rats (0/10) survived for 4 days. The data suggest that there is an increased vulnerability to brain ischemic-reperfusion injury in the aged rats; the degree of post-recovery hypoperfusion may contribute to the high mortality in the aged rats following cardiac arrest and resuscitation.

Keywords Transient global ischemia • Cerebral blood flow • Ischemia/reperfusion injury • Aged rat • Survival

K. Xu (✉) • J.C. LaManna
Department of Physiology and Biophysics, Case Western Reserve University, Cleveland, OH, USA
e-mail: kxx@case.edu

M.A. Puchowicz
Departments of Nutrition, Case Western Reserve University, Cleveland, OH, USA

© Springer Science+Business Media, New York 2016   265
C.E. Elwell et al. (eds.), *Oxygen Transport to Tissue XXXVII*, Advances in
Experimental Medicine and Biology 876, DOI 10.1007/978-1-4939-3023-4_33

1   Introduction

With aging, increased susceptibility to oxidative stress leads to increased incidence of cardiac arrest and poorer post-resuscitation outcome in elderly; about 80 % of patients with cardiac arrest are over 65 years old [1]. Most of the mortality and morbidity after initially successful resuscitation can be assigned to the immediate and delayed effects of reperfusion injury in the central nervous system. Reperfusion following global ischemia is often characterized by an initial hyperemia, followed by a period of hypoperfusion, where cerebral blood flow is reduced below pre-arrest baseline [2]. Despite return of spontaneous circulation (ROSC), persistent perfusion failure may contribute to brain ischemia and poor outcome. Emerging evidence supports the concept that the immediate post-arrest period is similar to the sepsis syndrome, with inflammation, endothelial dysfunction, and microcirculatory hypoperfusion [3]. Brain hypoperfusion may last for days following reperfusion, as demonstrated in our studies using rat model of cardiac arrest and resuscitation [4]. We have previously reported that in the aged rats, overall survival rate, hippocampal CA1 neuronal counts, and brain mitochondrial respiratory control ratio were significantly reduced compared to the younger rats following cardiac arrest and resuscitation [5].

   In this study we compared regional brain blood flow following cardiac arrest and resuscitation in 6-month and 24-month old rats to determine the aging effect on post-resuscitation perfusion of brain. Brain blood flow was measured in non-arrested rats and 1-h recovery rats using [^{14}C] iodoantipyrene (IAP) autoradiography; the 4-day survival rate was determined in the two age groups.

2   Methods

2.1   Animals and Induction of Transient Global Ischemia

The experimental protocol was approved by the Animal Care and Use Committee at Case Western Reserve University. Male Fisher rats (6 and 24 months old) were used in experiments. Transient global brain ischemia was achieved using a rat model of cardiac arrest and resuscitation as described previously [5, 6]. In brief, rats were anesthetized with isoflurane and cannulae were placed in femoral artery and external jugular vein. Cardiac arrest was induced in the conscious rat by the rapid sequential intra-atrial injection of D-tubocurare (0.3 mg) and ice-cold KCl solution (0.5 M; 0.12 ml/100 g of body weight). Resuscitation was initiated at 10 min after arrest. The animal was orotracheally intubated and ventilation was begun simultaneously with chest compressions and the intravenous administration of normal saline. Once a spontaneous heart beat returned, epinephrine (4–10 µg) was administered intravenously, the animal was considered to be resuscitated when mean blood pressure rose above 80 % of pre-arrest value. The duration of ischemia was about 15 min. Non-arrested rats went through the same surgical procedures except cardiac arrest. Brain blood flows were measured at 1 h following resuscitation; the 4-day survival rate was determined in the two age groups.

2.2 Measurement of Brain Blood Flow

Regional brain (cortex, hippocampus, brainstem and cerebellum) blood flow was measured using [^{14}C] iodoantipyrene (IAP) autoradiography as previously described [4]. In brief, the femoral artery catheter was attached to a withdrawal syringe pump set to a withdraw rate at 1.6 ml/min. A bolus of 25 μCi of [^{14}C] IAP was injected intra-arterially 3 s after the pump was started. The rat was decapitated and the pump stopped simultaneously 10 s later. The brain was quickly removed, frozen and stored at −80 °C. The reference arterial blood was collected and its radioactive content was determined. For autoradiography, each frozen brain was sectioned (20 μm) in a cryostat at the levels of atlas plate 13, 30 and 69 [7]. Brain sections and [^{14}C]-Micro-scale standards were placed on glass slides, covered with an autographic film and exposed for 21 days. The images were analyzed using a BIOQUANT image analysis system (R&M Biometries). Optical densities were converted to nCi per gram using standard curves generated from [^{14}C]-Micro-scale standards. The blood flow was calculated by the equation: Blood flow $(ml/g/min)$ = Tissue (nCi/g) × pump rate (ml/min)/Reference blood (nCi).

2.3 Statistical Analysis

Data are expressed as mean ± SD. Statistical analyses were performed using SPSS V20 for Windows. The comparison between any two groups was analyzed with a t-test, and significance was considered at the level of $p < 0.05$.

3 Results

3.1 Physiological Variables

Physiological variables were measured before cardiac arrest and 1 h post-resuscitation (Table 33.1) in the 6-month and 24-month old rats. At 1 h post-resuscitation, all rats had lower arterial blood pressure and pH, higher blood glucose and lactate compared to theirs pre-arrest baselines; there was no significant difference in any of the variables between the two age groups, though the plasma lactate was trended higher in the 24-month old group.

Table 33.1 Physiological variables before cardiac arrest and after resuscitation (mean ± SD)

		6-Month old (n = 6)	24-Month old (n = 5)
Body weight (g)		364 ± 21	360 ± 15
MAP (mmHg)	Pre	113 ± 4	113 ± 7
	1 h post	95 ± 12*	94 ± 7*
pH (unit)	Pre	7.41 ± 0.02	7.42 ± 0.01
	1 h post	7.36 ± 0.05*	7.35 ± 0.02*
PaO_2 (mmHg)	Pre	88 ± 7	88 ± 5
	1 h post	90 ± 7	90 ± 9
$PaCO_2$ (mmHg)	Pre	39 ± 3	37 ± 2
	1 h post	36 ± 3	35 ± 5
Hematocrit (%)	Pre	47 ± 4	47 ± 3
	1 h post	48 ± 3	46 ± 3
$Glucose_{plasma}$ (mM)	Pre	6.8 ± 1.5	8.4 ± 0.8
	1 h post	9.4 ± 1.6*	10.8 ± 2.3*
$Lactate_{plasma}$ (mM)	Pre	1.8 ± 0.5	1.9 ± 0.6
	1 h post	3.4 ± 0.7*	5.7 ± 2.4*

*Indicates significant difference compared to their corresponding pre-arrest values ($p < 0.05$, t-test)

3.2   Regional Brain Blood Flow

Brain blood flow was measured in 6-month and 24-month old rats before cardiac arrest and 1 h post-resuscitation, in the regions of cortex, hippocampus, brain stem and cerebellum (Fig. 33.1). The pre-arrest baseline blood flows were similar in cortex, brainstem and cerebellum in the 6-month and the 24-month old rats (n = 3 each) with slightly lower values in the aged group; however, the baseline blood flow in hippocampus was significantly lower in the 24-month old group. At 1 h post-resuscitation, both the 6-month and 24-month groups had significantly lower blood flows in all regions than the pre-arrest baseline values; compared to the 6-month old group, the blood flow was significantly lower (about 40 %) in all regions in the 24-month old group. For example, cortical blood flows (ml/g/min) were 0.44 ± 0.05 (n = 6) and 0.28 ± 0.06 (n = 5) in the 6-month and the 24-month old group.

3.3   Overall Survival

Overall survival rates were examined for 4 days following cardiac arrest and resuscitation in the 6-month and 24-month old groups (Table 33.2). The 4-day survival rate for the 6-month old rats was 50 % (3/6) whereas none of the 24-month old rats (0/10) survived for 4 days. The majority of the deaths of rats occurred within the first 2 days following resuscitation.

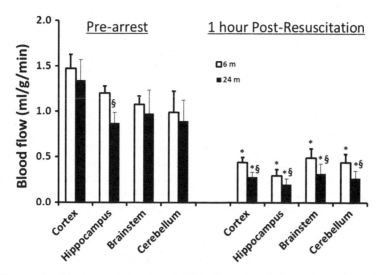

Fig. 33.1 Regional brain blood flow (ml/g/min) in 6-month and 24-month old rats. *Indicates significant difference (p < 0.05) compared to their corresponding pre-arrest values, § indicates significant difference (p < 0.05) compared to the 6-month old group

Table 33.2 Survival rates (4 days) after cardiac arrest and resuscitation

Age	Number of deaths				Survival rate
(Months)	<1 day	1–2 days	2–3 days	3–4 days	4 days/total
6	0	1	1	1	50 % (3/6)
24	4	3	2	1	0 (0/10)

4 Discussion

Ischemia/reperfusion injury is a complex biological phenomenon occurring after the return of blood to a tissue that has undergone ischemia. In cardiac arrest, the degree of damage depends mainly on the type and duration of the arrest as well as derangements in vital organs after reperfusion (the "post-resuscitation syndrome" [8]). Experimental studies reveal that during the arrest (no flow period) the metabolic rate of oxygen ($CMRO_2$) to be essentially zero. After resuscitation, the $CMRO_2$ recovers while at the same time there is multifocal absence of perfusion [9–11]. This is often followed by a short period of global cerebral hyperemia associated with a high $CMRO_2$ and then later cerebral hypoperfusion again. The mismatch between cerebral blood flow and the metabolic demand of the tissue (oxygen delivery/metabolism mismatch) may account for almost all of the injury associated with the no flow period and most of the subsequent injury associated with the post-resuscitation syndrome and mortality.

Recent studies showed decreased brain blood flows and regional heterogeneity with age [12]. Our findings, although obtained in a small number of animals, demonstrated the baseline of brain blood flow trended slightly lower in aged rats, with a statistically significant decrease in hippocampus. Being selectively vulnerable to neurodegenerative diseases, hippocampus might be more vulnerable to the change of neurovascular coupling during aging. In this study we found that the brain blood flows were decreased in both young and old rats during early recovery phase following cardiac arrest and resuscitation, however the degree of brain hypoperfusion was severer in the aged rats. This is probably due to the age-related changes in cerebral vasculature, endothelial function and the imbalance of an increase in free radical production and a decrease in defense mechanisms that exacerbates reperfusion injury in the aged brain.

In summary, our data suggest that there is an increased vulnerability to brain ischemic-reperfusion injury in the aged rats; the failure to restore brain blood flow may be one of the mechanisms contributes to the high mortality in the aged rats following cardiac arrest and resuscitation.

Acknowledgments This study was supported by NIH grant NINDS 1 R01 NS46074.

References

1. Zheng ZJ, Croft JB, Giles WH et al (2001) Sudden cardiac death in the United States, 1989 to 1998. Circulation 104:2158–2163
2. Gidday JM, Kim YB, Shah AR et al (1996) Adenosine transport inhibition ameliorates postischemic hypoperfusion in pigs. Brain Res 734:261–268
3. Adrie C, Adib-Conquy M, Laurent I et al (2002) Successful cardiopulmonary resuscitation after cardiac arrest as a "sepsis-like" syndrome. Circulation 106:562–568
4. Xu K, Puchowicz MA, Lust WD et al (2006) Adenosine treatment delays postischemic hippocampal CA1 loss after cardiac arrest and resuscitation in rats. Brain Res 1071:208–217
5. Xu K, Puchowicz MA, Sun X et al (2010) Decreased brainstem function following cardiac arrest and resuscitation in aged rat. Brain Res 1328:181–189
6. Xu K, LaManna JC (2009) The loss of hypoxic ventilatory responses following resuscitation after cardiac arrest in rats is associated with failure of long-term survival. Brain Res 1258:59–64
7. Palkovits M, Brownstein MJ (1988) Maps and guide to microdissection of rat brain. Elsevier, New York
8. Negovsky VA, Gurvitch AM (1995) Post-resuscitation disease--a new nosological entity. Its reality and significance. Resuscitation 30:23–27
9. Mortberg E, Cumming P, Wiklund L et al (2009) Cerebral metabolic rate of oxygen ($CMRO_2$) in pig brain determined by PET after resuscitation from cardiac arrest. Resuscitation 80:701–706
10. Oku K, Kuboyama K, Safar P et al (1994) Cerebral and systemic arteriovenous oxygen monitoring after cardiac arrest. Inadequate cerebral oxygen delivery. Resuscitation 27:141–152
11. Sterz F, Leonov Y, Safar P et al (1992) Multifocal cerebral blood flow by Xe-CT and global cerebral metabolism after prolonged cardiac arrest in dogs. Reperfusion with open-chest CPR or cardiopulmonary bypass. Resuscitation 24:27–47
12. Aanerud J, Borghammer P, Chakravarty MM et al (2012) Brain energy metabolism and blood flow differences in healthy aging. J Cereb Blood Flow Metab 32:1177–1187

Part VI
Brain Oxygenation

Chapter 34
Using fNIRS to Study Working Memory of Infants in Rural Africa

K. Begus, S. Lloyd-Fox, D. Halliday, M. Papademetriou, M.K. Darboe, A.M. Prentice, S.E. Moore, and C.E. Elwell

Abstract A pilot study was conducted to assess the feasibility of using fNIRS as an alternative to behavioral assessments of cognitive development with infants in rural Africa. We report preliminary results of a study looking at working memory in 12–16-month-olds and discuss the benefits and shortcomings for the potential future use of fNIRS to investigate the effects of nutritional insults and interventions in global health studies.

Keywords fNIRS • Working memory • Nutrition • Infants

K. Begus (✉)
Centre for Brain and Cognitive Development, Birkbeck College, University of London, London, UK
e-mail: k.begus@bbk.ac.uk

S. Lloyd-Fox
Centre for Brain and Cognitive Development, Birkbeck College, University of London, London, UK

Department of Medical Physics and Bioengineering, University College London, London, UK

D. Halliday • M. Papademetriou
Department of Medical Physics and Bioengineering, University College London, London, UK

M.K. Darboe
MRC International Nutrition Group, Keneba Field Station, Keneba, The Gambia

A.M. Prentice
MRC International Nutrition Group, Keneba Field Station, Keneba, The Gambia

MRC International Nutrition Group, London School of Hygiene and Tropical Medicine, London, UK

S.E. Moore
MRC International Nutrition Group, Keneba Field Station, Keneba, The Gambia

MRC Human Nutrition Research, Cambridge, UK

C.E. Elwell
Department of Medical Physics and Biomedical Engineering, University College London, London, UK

© Springer Science+Business Media, New York 2016
C.E. Elwell et al. (eds.), *Oxygen Transport to Tissue XXXVII*, Advances in Experimental Medicine and Biology 876, DOI 10.1007/978-1-4939-3023-4_34

273

1 Introduction

Inappropriate nutrition during fetal and early postnatal life can cause detrimental and persistent central nervous system alterations and deficits in behavioral functioning into childhood and adulthood [1–3]. These deficiencies are often only detected once the affected cognitive functions reach the point of observable behavior, usually during the second year of life or later, limiting the possibility of intervention at an earlier stage. Moreover, most assessments of cognitive development that can elucidate potential deficiencies are designed and validated for use in Western civilizations. Therefore, without substantial cultural adaptations and lengthy validation processes [4], these tests are often inappropriate for use in developing countries, where the majority of the population at risk for under nutrition live.

In this pilot project, we aimed to assess the feasibility of using fNIRS as an alternative to behavioral assessments of cognitive development of infants in their second year of life, in a resource-poor setting in rural west Africa. fNIRS is a non-invasive optical imaging technique, used to map the functional activation of the brain, by measuring absorption of near infrared light, as it travels from sources to detectors, placed on the surface of the head, through the underlying brain tissue [5]. In recent decades, it has been used extensively to study infant cognitive development [6], has proven to be a sensitive tool for highlighting early biomarkers of atypical brain development and function [7] and, in a previous phase of this project, we have provided evidence that fNIRS is a viable measurement tool in a resource-poor setting with young infants [8].

Two of the most reported deficits attributed to poor nutrition in infancy are suboptimal memory functioning and deficiencies in executive functions [9, 10]. By using fNIRS to directly map the cortical correlates of working memory in infants, we aimed to find a measure of cognitive development that would be less amenable to culture and that could be used to shed light on potential delays resulting from poor nutrition before these become apparent in behavior. We used an *object permanence* paradigm; a task testing the ability to create and hold a mental schema of an object in mind when it is no longer visible [11], tapping into both executive functions and working memory. A similar task was previously used in a longitudinal study of infants, directly relating the emergence of the behavioral ability to track an occluded object to the underlying brain activation as measured by fNIRS [11].

2 Method

2.1 Participants

Participants were identified from the West Kiang Demographic Surveillance System. All infants were born full term, with normal birth weight, as well as head circumferences no less than minus 3 z-scores against World Health Organisation (WHO) standards. Ethical approval was given by the joint Gambia Government— MRC Unit Ethics Committee, and written informed consent was obtained from all parents/care-givers prior to participation. Twenty-four 12–16-month-old infants were included in the final sample (mean age = 435.2 days, SD = 36.3). A further 15 infants participated but were excluded from the study due to an insufficient number of valid trials according to looking time measures (six infants), technical difficulties (no video recording, four infants), or due to tiredness/fussiness (five infants). This attrition rate is within the standard range for infant fNIRS studies [6].

2.2   Equipment and Procedure

Infants wore custom-built fNIRS headgear consisting of an array over the right hemisphere containing a total of 12 channels (source-detector separations at 2 cm, Fig. 34.1a), and were tested with the UCL optical topography system [12]. This system uses near-infrared light of two different wavelengths (780 and 850 nm) to make spectroscopic measurements of oxy (HbO_2) and deoxyhaemoglobin (HHb).

Once the fNIRS headgear was placed on an infant's head, they sat on their parent's lap in front of a screen. The parent was instructed to refrain from interacting with the infant during the stimuli presentation unless the infant became fussy or sought their attention.

2.3   Stimuli

Infants viewed videos in which an adult male actor (resident in The Gambia) picks up an object from the table in front of him, looks directly at the infant, vocalizes while holding the object (e.g. "Oooh!"), and then places the object into a box in

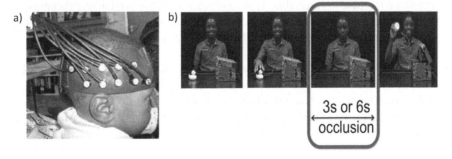

3s or 6s
occlusion

Fig. 34.1 (a) A Gambian infant wearing the custom-made headgear during the experiment. The headgear consists of an array of six sources and four detectors. (b) A sequence of video frames taken from the stimuli videos that the infants observed

front of him. While the object is occluded in the box (for 3 or 6 s, depending on condition), the actor is still and looking at the table. After the occlusion period, the actor retrieves the object from the box and again vocalizes, while holding the object and looking at the infant (Fig. 34.1b). Each experimental trial was preceded and followed by a baseline, which contained a display of static images of animals, infant faces, and landscapes, presented for random lengths of time.

2.4 Data Processing and Analysis

Changes in HbO_2 and HHb chromophore concentration (μmol) were calculated and used as hemodynamic indicators of neural activity [13]. Individual trials and channels were rejected from further analysis based on looking time measures and the quality of the signal, using artifact detection algorithms [6, 14]. Grand averaged time response curves of the hemodynamic responses (across all infants) for each channel were compiled. Either a significant increase in HbO_2 concentration or a significant decrease in HHb (but not concurrent increase or decrease in both measures) is commonly accepted as an indicator of cortical activation in infant work [6]. A time window was selected between 4 and 10 s post-stimulus onset for each trial. T-tests (two-way) were performed for each channel to analyze the activation during stimuli compared to baseline, and paired sample channel-by-channel t-tests (two-tailed) were performed to assess differences in activation between the two conditions.

3 Results

Figure 34.2a shows the time courses of the hemodynamic responses during the two trial types, in which the object was occluded for either 3 or 6 s. Channels with significant increase in HbO_2 (black) or significant decrease in HHb (grey) compared to baseline are marked in Fig. 34.2b. Comparisons between the two conditions as well as the results of t-test analyses for each condition separately are reported in Table 34.1.

By using a standardized scalp map of fNIRS channel locators to underlying anatomy [15], and the head measurements and photographs taken in the current study we estimated the location of activation in the condition with 3 s occlusion centered over the pSTS/TPJ region, and for the 6 s occlusion across a wider region of the cortex including the pSTS/TPJ, aSTG and IFG.

Object occluded for 3 seconds

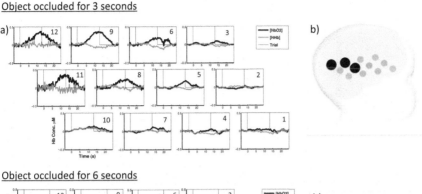

Object occluded for 6 seconds

Fig. 34.2 (**a**) Time courses of the haemodynamic responses for each channel, separately for the two trial types (3 or 6 s occlusion). (**b**) Channels with significant increase in HbO$_2$ (*black*) or significant decrease in HHb (*grey*) compared to baseline, for each trial type

4   Conclusions

We have successfully obtained high quality fNIRS data in 12–16-month-old infants in resource-poor settings of rural west Africa. Our results show differential neural activity when infants observed objects being hidden for 3 compared to 6 s. This cortical activation presumably reflects the infants sustaining a mental schema of the occluded objects in mind, potentially providing a sensitive measure of two of the cognitive functions that are most likely to be affected by poor nutrition—working memory and executive functioning.

Further work is currently underway to validate this measure by relating it to behavioral outcomes. We are also exploring novel ways of analyzing this data by looking at duration of sustained activation as opposed to maximum change [16], and are in the process of collecting data in an age-matched UK sample.

Establishing if fNIRS could be used to elucidate potential cognitive delays therefore requires much further work, however we have provided the first evidence that working memory can be measured in infants before they are able to express it behaviorally, even in a resource-poor setting such as rural Africa.

Table 34.1 Summary of t-test analyses for each channel, reporting statistical significance of changes in HbO_2 and HHb for each condition separately and the difference between the conditions (significant changes are marked in bold)

Ch.	3 s occlusion				6 s occlusion				6 s > 3 s occlusion			
	HbO_2		HHb		HbO_2		HHb		HbO_2		HHb	
	t	p	t	p	t	p	t	p	t	p	t	p
1	1.394	0.177	−0.353	0.727	0.042	0.967	**−1.942**	**0.064**	0.740	0.467	0.741	0.466
2	−1.160	0.258	1.124	0.272	−1.393	0.177	−0.160	0.874	0.238	0.814	1.164	0.256
3	0.022	0.983	0.347	0.731	−0.935	0.359	−0.527	0.604	0.638	0.530	0.736	0.469
4	−0.733	0.471	−1.551	0.135	0.486	0.632	**−3.483**	**0.002**	−1.185	0.248	1.118	0.275
5	1.014	0.321	0.317	0.754	**3.077**	**0.005**	−1.507	0.145	−1.633	0.116	1.709	0.101
6	1.427	0.167	−0.021	0.983	**3.489**	**0.001**	**−2.921**	**0.007**	−0.865	0.396	**1.763**	**0.091**
7	−0.105	0.916	0.179	0.859	0.495	0.625	**−1.919**	**0.067**	−0.609	0.548	**1.988**	**0.059**
8	**2.231**	**0.036**	1.628	0.118	**2.368**	**0.027**	−1.048	0.306	−0.166	0.870	**2.782**	**0.011**
9	**2.952**	**0.007**	−0.366	0.718	**3.447**	**0.002**	**−2.756**	**0.011**	0.255	0.801	1.233	0.231
10	1.512	0.144	2.509	0.019	0.063	0.950	−0.668	0.511	1.265	0.219	**2.455**	**0.022**
11	1.351	0.189	0.595	0.558	**1.860**	**0.076**	−0.137	0.891	0.099	0.922	1.073	0.294
12	**2.971**	**0.007**	1.172	0.253	**2.387**	**0.026**	**−1.830**	**0.080**	**1.826**	**0.081**	1.596	0.124

References

1. Victora C, Adair L, Fall C et al (2008) Maternal and child undernutrition: consequences for adult health and human capital. Lancet 371:340–357
2. Hackman D, Farah M (2009) Socioeconomic status and the developing brain. Trends Cogn Sci 13(2):65–73
3. Beard J (2003) Iron deficiency alters brain development and functioning. J Nutr 133:1468S–1472S
4. Gladstone MJ, Lancaster GA, Jones AP et al (2007) Can western developmental screening tools be modified for use in a rural Malawian setting? Arch Dis Child 93:23–29
5. Elwell C (1995) A practical users guide to near infrared spectroscopy. Hamamatsu Photonics, Hamamatsu
6. Lloyd-Fox S, Blasi A, Elwell CE (2010) Illuminating the developing brain: the past, present and future of functional near infrared spectroscopy. Neurosci Biobehav Rev 34:269–284
7. Lloyd-Fox S, Blasi A, Elwell CE et al (2013) Reduced neural sensitivity to social stimuli in infants at risk for autism. Proc R Soc B Biol Sci 280:20123026
8. Lloyd-Fox S, Papademetriou M, Darboe MK et al (2014) Functional near infrared spectroscopy (fNIRS) to assess cognitive function in infants in rural Africa. Sci Rep 4:4740
9. Lukowski AF, Koss M, Burden MJ et al (2010) Iron deficiency in infancy and neurocognitive functioning at 19 years: evidence of long-term deficits in executive function and recognition memory. Nutr Neurosci 13(2):54–70
10. Algarin C, Nelson CA, Peirano P et al (2013) Iron-deficiency anemia in infancy and poorer cognitive inhibitory control at 10 years. Dev Med Child Neurol 55:453–458
11. Baird AA, Kagan J, Gaudette T et al (2002) Frontal lobe activation during object permanence: data from near-infrared spectroscopy. Neuroimage 16(4):1120–1126
12. Everdell NL, Gibson AP, Tullis IDC et al (2005) A frequency multiplexed near-infrared topography system for imaging functional activation in the brain. Rev Sci Instrum 76:093705
13. Obrig H, Villringer A (2003) Beyond the visible—imaging the human brain with light. J Cereb Blood Flow Metab 23:1–18
14. Lloyd-Fox S, Blasi A, Volein A et al (2009) Social perception in infancy: a near infrared spectroscopy study. Child Dev 80:986–999
15. Lloyd-Fox S, Richards JE, Blasi A et al (2014) Co-registering fNIRS with underlying cortical areas in infants. Neurophotonics 1(2):025006
16. Coutts LV, Cooper CE, Elwell CE et al (2012) Time course of the haemodynamic response to visual stimulation in migraine, measured using near-infrared spectroscopy. Cephalalgia 32 (8):621–629

Chapter 35
Gender and Age Analyses of NIRS/STAI Pearson Correlation Coefficients at Resting State

T. Matsumoto, Y. Fuchita, K. Ichikawa, Y. Fukuda, N. Takemura, and K. Sakatani

Abstract According to the valence asymmetry hypothesis, the left/right asymmetry of PFC activity is correlated with specific emotional responses to mental stress and personality traits. In a previous study we measured spontaneous oscillation of oxy-Hb concentrations in the bilateral PFC *at rest* in normal adults employing two-channel portable NIRS and computed the *laterality index at rest* (LIR). We investigated the Pearson correlation coefficient between the LIR and anxiety levels evaluated by the State-Trait Anxiety Inventory (STAI) test. We found that subjects with right-dominant activity at rest showed higher STAI scores, while those with left dominant oxy-Hb changes at rest showed lower STAI scores such that the Pearson correlation coefficient between LIR and STAI was positive. This study performed Bootstrap analysis on the data and showed the following statistics of the target correlation coefficient: mean $= 0.4925$ and lower confidence limit $= 0.177$ with confidence level 0.05. Using the KS-test, we demonstrated that the correlation did not depend on age, whereas it did depend on gender.

Keywords NIRS • STAI • Pearson correlation coefficient • Bootstrap analysis • Kolmogorov-Smirnov test

1 Introduction

The valence asymmetry hypothesis [1–3], asserts that the left/right asymmetry of prefrontal cortex (PFC) activity is correlated with specific emotional responses to mental stress and personality traits. Electroencephalography (EEG) has demonstrated that subjects with greater relative left PFC activity shows more positive and

T. Matsumoto (✉) • Y. Fuchita • K. Ichikawa • Y. Fukuda
Department of Electrical Engineering and Bioscience, Waseda University, Tokyo, Japan
e-mail: takashi@matsumoto.elec.waseda.ac.jp

N. Takemura • K. Sakatani
Department of Electrical and Electronics Engineering, NEWCAT Research Institute, College of Engineering, Nihon University, Koriyama, Japan

© Springer Science+Business Media, New York 2016
C.E. Elwell et al. (eds.), *Oxygen Transport to Tissue XXXVII*, Advances in Experimental Medicine and Biology 876, DOI 10.1007/978-1-4939-3023-4_35

less negative dispositional mood than their right-dominant counterparts. In contrast, right frontally activated subjects respond more to negative affective challenges and less to positive affective challenges than their left dominant counterparts [4, 5].

Near-infrared spectroscopy (NIRS) is a noninvasive method for monitoring neuronal activity by measuring changes of oxyhemoglobin (oxy-Hb) and deoxyhemoglobin (deoxy-Hb) concentrations in cerebral vessels [6–8]. In [9], we proposed a new information criterion LIR (*Laterality Index at Rest*) defined in terms of oxy-Hb concentration changes from the right and left PFC at resting condition:

$$LIR = \frac{\sum\limits_{t \in analysis\ interval} ((\Delta oxyR_t - oxyR_{min}) - (\Delta oxyL_t - oxyL_{min}))}{\sum\limits_{t \in analysis\ interval} ((\Delta oxyR_t - oxyR_{min}) + (\Delta oxyL_t - oxyL_{min}))} \qquad (35.1)$$

where $\Delta oxyR_t$ and $\Delta oxyL_t$ stand for oxy-Hb concentration changes of the right and left PFC, whereas $\Delta oxyR_{min}$ and $\Delta oxyL_{min}$ are respective minimum values. Thus, LIR computes the degree of the right/left asymmetry of oxy-Hb concentration changes normalized between $[-1,+1]$. It is to be noted that LIR always converges, because the denominator is always non-negative. We demonstrated that there is a positive Pearson Correlation Coefficient between LIR and STAI-1 (State Anxiety of State-Trait Anxiety Inventory). A subject at rest is referred to as a subject is not performing any explicit task so that no task experimental design is necessary.

There were at least two issues for improvements over [9]:

1. The number of data as was small.
2. Parametric form of the probability distribution of Pearson Correlation Coefficient was difficult to obtain.

The purpose of this study is threefold:

(a) We will use bootstrap method [10] which is non-parametric in that no assumption is assumed about parametric form of the probability distribution of Pearson Correlation Coefficient and additional bootstrapped samples can be drawn. This could "robustify" the positive Pearson Correlation Coefficient.
(b) We will show that distribution of Pearson Correlation Coefficient does not depend on age by using Kolmogorov-Smirnov test.
(c) We will also show that distribution of Pearson Correlation Coefficient for male subjects appear different from that of female subjects, again by Kolmogorov-Smirnov test.

2 Material and Methods

2.1 Material

There were two groups consisting of young and aged subjects. The details are given in Table 35.1.

2.2 Experimental Protocol

The experimental protocol consists of the following:

1. STAI 1 questionnaire
2. Calibration
3. 1 min. preparation
4. 3 min. resting (analysis period)

The equipment used in this study was Pocket NIRS, Hamamatsu Photonics K.K., Japan for measurements of the concentration changes of oxy-Hb in the PFC. This system is wireless (Bluetooth®); such that the subject could move relatively freely. It is equipped with two channels where each uses light emitting diodes with wavelengths 735, 810, and 850 nm as light sources and one photo- detector. Two AAA batteries are used which can operate up to 8 h of continuous measurement for the two-probe operation. The sampling rate was 61.3 Hz. The concentration changes of hemoglobin were expressed in arbitrary units (a.u.). The NIRS probes were set symmetrically on the forehead with a flexible fixation pad, so that the midpoint between the emission and detection probes was 3 cm above the centers of the upper edges of the bilateral orbital sockets. The distance between the emitter and detector was set at 3 cm. This positioning is similar to positions Fp1 (left) and Fp2 (right) of the international electroencephalographic 10–20 system. Magnetic resonance imaging (MRI) confirmed that the emitter-detector was located over the dorsolateral and front polar areas of the PFC.

Table 35.1 Details of the subjects participated in the experiment

Gender/age	Male	Female	Total
Young (20–24)	6	13	19
Aged (60–69)	4	16	20
Total	10	29	39

2.3 Bootstrap Algorithm

One of the methods of "robustifying" the results under such circumstances is the bootstrap algorithm [6]. The bootstrap algorithm is non-parametric in that it does not assume any parametric form of a target probability distribution. It is one of the resampling methods *with replacements* such that many samples become available. Since it assumes no parametric form for target distributions, it is flexible for problems with unknown or difficult target distributions. It has been widely used in practical problems.

In order to explain the concept, consider Fig. 35.1 where the number of available data is n and each data is two dimensional consisting of LIR and STAI 1 from each subject. Bootstrap algorithm [10] draws B samples with replacement as demonstrated on the right hand side of the figure. The number of samples B in the experiment to be reported later, for instance, was 10,000. From those samples, we compute Pearson Correlation Coefficient between LIR and STAI 1 for each sample $i, i = 1, \ldots, B$, and draw histogram of the Pearson Correlation Coefficients. Given the histogram, one can compute confidence interval http://en.wikipedia.org/wiki/Statistical_hypothesis_testing.

2.4 Kolmogorov-Smirnov Test

The Kolmogorov-Smirnov test quantifies the distance between the empirical distribution functions of two datasets xi, $i = 1, \ldots, n$, yj , $j = 1, \ldots, m$. The empirical density and empirical distribution functions of the target random variables X and Y are defined as:

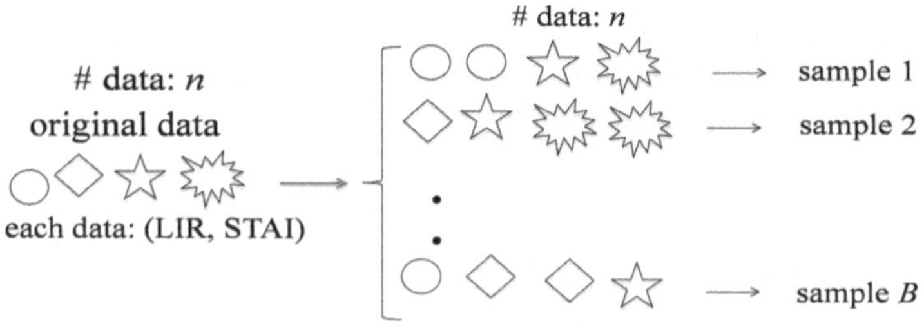

Fig. 35.1 A schematic of Bootstrap Algorithm

$$X_i(x) = \begin{cases} 1 & (x_i \leq x) \\ 0 & (x_i > x) \end{cases}, \quad F(x) := \frac{1}{n}\sum_{i=1}^{n}X_i(x)$$

$$Y_j(x) = \begin{cases} 1 & (y_j \leq x) \\ 0 & (y_j > x) \end{cases}, \quad G(x) := \frac{1}{m}\sum_{i=1}^{m}Y_j(x)$$

The test statistic is

$$D := \sup_{-\infty < x < \infty} |F(x) - G(x)| \qquad (35.2)$$

The null hypothesis here is that the two data come from the same distribution. It is to be rejected at level α if

$$\sqrt{\frac{-(n+m)log(\alpha/2)}{2nm}} < D \qquad (35.3)$$

otherwise, there is no reason to reject the null hypothesis.

3   Results

3.1   Bootstrap Histogram of the Pearson Correlation Coefficient

With the number of bootstrap samples 10,000, confidence level $\alpha = 0.05$, we obtained the following:

Mean $= 0.4925$
Upper confidence limit $= 0.738$
Lower confidence limit $= 0.177$

This indicates that the Pearson Correlation Coefficient between LIR and STAI-1 would be positive at least for the current dataset.

3.2   Kolmogorov-Smirnov Test for Pearson Correlation Coefficient Between LIR and STAI-1 Among Young and Aged Subjects

Recall D defined in (35.2). With confidence level $\alpha = 0.05$,

$$\sqrt{\frac{-(n+m)log(\alpha/2)}{2nm}} = 0.278 > D = 0.255$$

so that Pearson Correlation Coefficient *does not depend on age* at least among the present datasets.

3.3 Kolmogorov-Smirnov Test for Pearson Correlation Coefficient Between LIR and STAI-1 Between Male and Female Subjects

With confidence level $\alpha = 0.05$,

$$\sqrt{\frac{-(n+m)log(\alpha/2)}{2nm}} = 0.0136 < D = 0.0327$$

Therefore, Pearson Correlation Coefficient for male and female come from *different distributions* at least among the present datasets. One cannot, however, draw conclusions of how the two distributions are different at this stage.

4 Conclusions

We have attempted to "robustify" our previous result on the positivity of the LIR/STAI-1 Pearson Correlation Coefficient by using Bootstrap Algorithm. Since the algorithm is non-parametric and since the number of bootstrapped samples was reasonably large, the result was statistically robust. We have also used the Kolmogorov-Smirnov test together with the Bootstrap Algorithm to show that the probability distribution of the LIR/STAI-1 Pearson Correlation Coefficient does not appear to depend on age. Finally, we have demonstrated that the Kolmogorov-Smirnov test reveals that the Pearson Correlation Coefficient for male and female come from different distributions at least among the present datasets. One cannot, however, draw conclusions about how the two distributions are different at the moment.

Acknowledgments This research was supported in part by Grant-in-Aid for Exploratory Research (25560356).

References

1. Davidson RJ (1998) Affective style and affective disorders: perspectives from affective neuroscience. Cognit Emotion 12(3):307–330
2. Davidson RJ (1993) Cerebral asymmetry and emotion: conceptual and methodological conundrums. Cognit Emotion 7(1):115–138
3. Davidson RJ, Jackson DC, Kalin NH (2000) Emotion, plasticity, cortex, and regulation: perspectives from affective neuroscience. Psychol Bull 126(6):890–909
4. Tomarken AJ et al (1992) Psychometric properties of resting anterior EEG asymmetry: temporal stability and internal consistency. Psychophysiology 29(5):576–592
5. Wheeler RE, Davidson RJ, Tomarken AJ (1993) Frontal brain asymmetry and emotional reactivity: a biological substrate of affective style. Psychophysiology 30(1):82–89
6. Jöbsis FF (1977) Noninvasive, infrared monitoring of cerebral and myocardial oxygen sufficiency and circulatory parameters. Science 198(4323):1264–1267
7. Ferrari M et al (1992) Effects of graded hypotension on cerebral blood flow, blood volume, and mean transit time in dogs. Am J Physiol 262(6):H1908–H1914
8. Hoshi Y, Kobayashi N, Tamura M (2001) Interpretation of near-infrared spectroscopy signals: a study with a newly developed perfused rat brain model. J Appl Physiol 90(5):1657–1662
9. Ishikawa W, Sato M, Fukuda Y, Matsumoto T, Takemura N, Sakatani K (2014) Correlation between asymmetry of spontaneous oscillation of hemodynamic changes in the prefrontal cortex and anxiety levels: a near-infrared spectroscopy study. J Biomed Opt 19(2):027005
10. Efron B (1979) Bootstrap methods: another look at the jackknife. Ann Stat 7:1–26

Chapter 36
Effects of Cosmetic Therapy on Cognitive Function in Elderly Women Evaluated by Time-Resolved Spectroscopy Study

A. Machida, M. Shirato, M. Tanida, C. Kanemaru, S. Nagai, and K. Sakatani

Abstract With the rapid increase in dementia in developed countries, it is important to establish methods for maintaining or improving cognitive function in elderly people. To resolve such problems, we have been developing a cosmetic therapy (CT) program for elderly women. However, the mechanism and limitations of CT are not yet clear. In order to clarify these issues, we employed time-resolved spectroscopy (TRS) to evaluate the effect of CT on prefrontal cortex (PFC) activity in elderly females with various levels of cognitive impairment. Based on the Mini-Mental State Examination (MMSE) score, the subjects were classified into mild (mean MMSE score: 24.1 ± 3.8) and moderate (mean MMSE score: 10.3 ± 5.8) cognitive impairment (CI) groups ($p < 0.0001$). The mild CI group exhibited significantly larger baseline concentrations of oxy-Hb and t-Hb than the moderate CI group. CT significantly increased the baseline concentrations of oxy-Hb ($p < 0.002$) and t-Hb ($p < 0.0013$) in the left PFC in the mild CI group. In contrast, CT did not change the concentrations of oxy-Hb and t-Hb in the moderate CI group ($p > 0.05$). These results suggest that CT affects cognitive function by altering PFC activity in elderly women with mild CI, but not moderate CI.

Keywords Aging • NIRS • Prefrontal cortex • Stress • Cortisol

A. Machida • M. Shirato • M. Tanida
SHISEIDO Research Center, Yokohama, Japan

C. Kanemaru • S. Nagai
SHISEIDO Beauty Creation Research Center, Tokyo, Japan

K. Sakatani (✉)
Department of Electrical and Electronics Engineering, NEWCAT Research Institute, College of Engineering, 1 Nakagawara, Tokusada, Tamuramachi, Koriyama, Fukushima 963-8642, Japan

Department of Neurological Surgery, School of Medicine, Nihon University, Tokyo, Japan
e-mail: sakatani.kaoru@nihon-u.ac.jp

© Springer Science+Business Media, New York 2016
C.E. Elwell et al. (eds.), *Oxygen Transport to Tissue XXXVII*, Advances in Experimental Medicine and Biology 876, DOI 10.1007/978-1-4939-3023-4_36

1   Introduction

Neuropsychological disorders such as dementia and depression in the elderly are important issues in aging societies, and various non-pharmacologic therapies have been examined for maintaining a healthy physical and mental status. Females exhibit a higher rate of population aging and a higher incidence of dementia than males. Therefore, it is important to establish non-pharmacologic therapies suitable for aged females. For example, cosmetic therapy has recently received attention as an effective new method for improving cognitive function and QOL in aged females [1, 2]. However, the neurophysiological mechanism of cosmetic therapy is not yet clear. In the present study, we employed time-resolved spectroscopy (TRS) to evaluate the effect of cosmetic therapy on cerebral blood oxygenation (CBO) in the prefrontal cortex (PFC) in aged females with various levels of cognitive impairment.

2   Methods

2.1   Subjects

The subjects of this study were 61 elderly women (age: 82.2 ± 6.3 years). They had mild to moderate levels of dementia and were living in a nursing home. We evaluated cognitive impairment of the subjects employing the Mini-Mental State Examination (MMSE), which is the most commonly used examination for screening cognitive function, before cosmetic therapy. All subjects provided written informed consent as required by the Ethics Committee of Shiseido.

2.2   Cosmetic Therapy

We have developed a cosmetic therapy program which begins with deep breathing using fragrances and relaxing light exercise, followed by skin care and make-up. A seminar-style session was presented in which subjects followed the guidance of instructors (beauty therapists). Based on directions given by beauty therapists, skin care products were applied to their skin, and make-up was applied to their faces. Beauty therapists encouraged subjects to perform by themselves as much as possible. Before the end of the therapy, instructors and subjects were encouraged to comment to one another about the changes in their impression and appearance. The therapy lasted approximately 50 min, and the daily treatment was continued for 3 months.

2.3 Study on Cognitive Function

First, in order to assess the effects of the cosmetic therapy on cognitive function in aged women, we evaluated MMSE scores before and after the 3-month cosmetic therapy program and calculated the change of MMSE score in each subject. We compared the changes of MMSE scores in the cosmetic therapy group (n = 7, 92.3 ± 6.2 years) and the control group (n = 8, 86.7 ± 6.6 years). There was a significant difference in ages between the two groups.

2.4 Functional Study by TRS

In this study, we evaluated cognition in 61 aged women (82.5 ± 6.2 years). According to the MMSE scores, the subjects were classified into the mild cognitive impairment group (n = 32; mean score 24.1 ± 3.8) and the moderate cognitive impairment group (n = 29; mean MMSE score 10.3 ± 5.8, $p < 0.0001$). There was no significant difference in age between the mild (83.3 ± 6.3 years) and moderate (81.6 ± 6.0 years) cognitive impairment groups ($p > 0.05$).

First, we evaluated CBO in the PFC in a resting condition, employing TRS. Then, we evaluated the effect of cosmetic therapy on the CBO. It should be noted that, unlike continuous wave NIRS, TRS employs pico-second light pulses and the photon diffusion equation, and enables us to measure the absolute total hemoglobin (Hb) concentration at the resting condition [3]. Therefore, the TRS probe can be removed during cosmetic therapy. We used a TRS-20 system (Hamamatsu Photonics K.K, Hamamatsu, Japan), which has been used in several functional studies on normal adults [4, 5]. Details of this system have been described previously [4, 5]. Briefly, it consists of three pulsed laser diodes with different wavelengths (761, 791, and 836 nm) having a duration of 100 ps at a repetition frequency of 5 MHz, a photomultiplier tube, and a circuit for time-resolved measurement based on the time-correlated single photon counting method. The observed temporal profiles were fitted into the photon diffusion equation using the non-linear least-squares fitting method. The reduced scattering and absorption coefficients for the three wavelengths were calculated. The concentrations of oxy-Hb, deoxy-Hb, and total Hb (=oxy-Hb + deoxy-Hb; t-Hb) were then calculated using the least-squares method. The concentrations of Hb were expressed in μM. The distance between the emitter and detector was set at 3 cm.

3 Results

Figure 36.1 compares MMSE scores of the cosmetic therapy and control groups. Interestingly, the control group exhibited a decrease of MMSE scores after 3 months. In contrast, the cosmetic therapy group did not, and there was a significant difference between the MMSE scores of the two groups after 3 months (p < 0.05).

Employing TRS, we evaluated CBO in the PFC in the resting condition before and after cosmetic therapy. The mild cognitive impairment group exhibited significantly larger baseline concentrations of oxy-Hb and t-Hb than the moderate cognitive impairment group (Fig. 36.2). There were no significant differences in baseline concentrations of oxy-Hb and t-Hb between the right and left PFC.

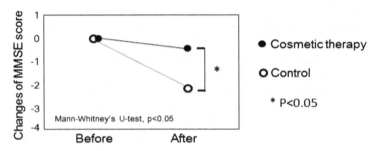

Fig. 36.1 Changes of MMSE scores after cosmetic therapy

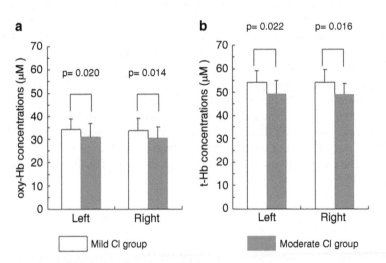

Fig. 36.2 Differences in baseline concentrations of oxy-Hb (**a**) and t-Hb (**b**) in the bilateral PFC between the mild and moderate cognitive impairment groups

Fig. 36.3 Effect of cosmetic therapy on baseline concentrations of oxy-Hb (**a**) and t-Hb (**b**) in the PFC in the mild cognitive impairment group

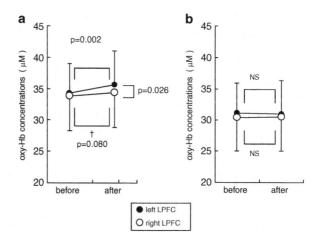

Daily cosmetic therapy (50 min) significantly increased the baseline concentrations of oxy-Hb ($p < 0.002$) and t-Hb ($p < 0.0013$) in the left PFC in the mild cognitive impairment group (Fig. 36.3). In contrast, the right PFC exhibited slight increases of oxy-Hb and t-Hb, but these were not statistically significant. In the moderate cognitive impairment group, the cosmetic therapy did not change the concentrations of oxy-Hb and t-Hb in the PFC ($p > 0.05$).

4   Discussion

The results of MMSE indicate that the aged women in the control group tended to show a decrease of cognitive function during 3 months; however, cosmetic therapy inhibited this decline. TRS demonstrated that cosmetic therapy increased the baseline concentrations of oxy-Hb and t-Hb in the PFC, suggesting that the therapy increased neuronal activity of the PFC at rest, since oxy-Hb and t-Hb reflect regional cerebral blood flow and blood volume, respectively [6, 7]. Considering that the PFC plays important roles in various higher brain functions, the effect of cosmetic therapy on aged women might be achieved through activation of the PFC.

Interestingly, the cosmetic therapy increased the baseline concentrations of oxy-Hb and t-Hb mainly in the left PFC, resulting in left-dominant PFC activity. It has been reported that the right PFC is dominant for negative emotions and the left PFC is dominant for positive emotions [8, 9]. These findings suggest that the cosmetic therapy induced positive emotions. Indeed, most of the subjects smiled after cosmetic therapy. It should be noted, however, that such increases of PFC activity induced by cosmetic therapy were observed only in the mild cognitive impairment group. These findings suggest a limitation in the effectiveness of cosmetic therapy on aged women.

The following limitations should be mentioned. Firstly, there was a significant difference in ages between the cosmetic and control groups in the study on cognitive function. In this study, we did not randomize the subjects, but classified the subjects into the two groups according to the floor on which they resided in the nursing home in order to avoid the possibility that interpersonal relationships might be adversely affected if subjects in different groups were mixed on one floor. It should be emphasized that, although the average age of the cosmetic group was older than that of the control group, the cosmetic group showed a higher MMSE score than the control group after the therapy. In addition, in the functional study by TRS, there was no significant difference in age between the mild and moderate cognitive impairment groups. Secondly, the physiological mechanism of the effect of cosmetic therapy is not clear. The cosmetic therapy had several components, including sensory stimulation with fragrances and skin care. It has been reported that sensory stimulation by aromatherapy and acupuncture improved cognitive function in dementia [10, 11]. We think that the effect of cosmetic therapy on cognitive function is holistic rather than being due to a single factor. Finally, we evaluated only the acute effects of cosmetic therapy on PFC activity in the functional study by TRS. Further studies are necessary to evaluate chronic effects of cosmetic therapy on PFC activity.

Acknowledgements This research was supported in part by Grants-in-Aid for Exploratory Research (25560356) and Strategic Research Foundation Grant-aided Project for Private Universities (S1411017) from the Ministry of Education, Culture, Sports, Sciences and Technology of Japan, and grants from Alpha Electron Co., Ltd. (Fukushima, Japan), Iing Co., Ltd. (Tokyo, Japan) and Southern Tohoku General Hospital (Fukushima, Japan).

References

1. Hayami M (1994) Applying cosmetic activates independent living, approaches in hospital Naruto Yamakami. Mon Total Care 4(11):27–34
2. Japan Society for Dementia Research (2009) Textbook for dementia. Chugai-Igakusha, Tokyo
3. Oda M, Nakano T, Suzuki A, Shimizu K, Hirano I, Shimomura F, Ohame E, Suzuki T, Yamashita Y (2000) Near infrared time-resolved spectroscopy system for tissue oxygenation monitor. SPIE 4160:204–210
4. Sakatani K, Tanida M, Hirao N, Takemura N (2014) Ginkobiloba extract improves working memory performance in middle-aged women: role of asymmetry of prefrontal cortex activity during a working memory task. Adv Exp Med Biol 812:295–301
5. Tanida M, Sakatani K, Tsujii T (2012) Relation between working memory performance and evoked cerebral blood oxygenation changes in the prefrontal cortex evaluated by quantitative time-resolved near-infrared spectroscopy. Neurol Res 34:114–119
6. Ferrari M, Wilson DA, Hanley DF, Traystma RJ (1992) Effects of graded hypotension on cerebral blood flow, blood volume, and mean transit time in dogs. Am J Physiol 262:1908–1914
7. Hoshi Y, Kobayashi N, Tamura M (2001) Interpretation of near-infrared spectroscopy signals: a study with a newly developed perfused rat brain model. J Appl Physiol 90:1657–1662
8. Hellige JB (1993) Hemispheric asymmetry. Harvard University Press, Cambridge

9. Ishikawa W, Sato M, Fukuda Y, Matsumoto T, Takemura N, Sakatani K (2014) Correlation between asymmetry of spontaneous oscillation of hemodynamic changes in the prefrontal cortex and anxiety levels: a near-infrared spectroscopy study. J Biomed Opt 19(2):027005
10. Thorgrimsen L, Spector A, Wiles A, Orrell M (2014) Aroma therapy for dementia. Cochrane Database Syst Rev 2:CD003150
11. Zeng BY, Salvage S, Jenner P (2013) Effect and mechanism of acupuncture on Alzheimer's disease. Int Rev Neurobiol 111:181–195

Chapter 37
Effects of Acupuncture on Anxiety Levels and Prefrontal Cortex Activity Measured by Near-Infrared Spectroscopy: A Pilot Study

K. Sakatani, M. Fujii, N. Takemura, and T. Hirayama

Abstract There is increasing evidence that acupuncture is useful in treating somatic and psychological disorders caused by stress; however, the physiological basis of the effect remains unclear. In the present study, we evaluated the effect of acupuncture on psychological conditions (i.e., anxiety) and prefrontal cortex (PFC) activity. We studied 10 patients with anxiety disorders and measured anxiety levels by means of the State-Trait Anxiety Inventory (STAI), including state anxiety (STAI-1) and trait anxiety (STAI-2). Employing a two-channel NIRS device, we measured oxy-Hb concentration in the bilateral PFC at rest, and evaluated asymmetry of the PFC activity by calculating the Laterality Index at Rest (LIR). The patients were treated by acupuncture at Yui Clinic in Osaka. The treatment significantly decreased the STAI-1 score ($p < 0.001$), but not the STAI-2 score ($p > 0.05$). The NIRS measurements indicated the presence of spontaneous oscillations of oxy-Hb in the bilateral PFC at rest before and after the treatment. Notably LIR decreased significantly in 7 out of the 10 subjects ($p < 0.01$), while 3 subjects showed an increasing tendency. The present pilot study indicates that acupuncture is effective in decreasing anxiety levels in patients with anxiety disorders. Our NIRS data suggest that acupuncture may alter the balance of PFC activity at rest, resulting in relaxation effects. Our NIRS data suggest that acupuncture changes the

---

K. Sakatani (✉)
Department of Electrical and Electronics Engineering, NEWCAT Research Institute, College of Engineering, 1 Nakagawara, Tokusada, Tamuramachi, Koriyama, Fukushima 963-8642, Japan

Department of Neurological Surgery, School of Medicine, Nihon University, Tokyo, Japan
e-mail: sakatani.kaoru@nihon-u.ac.jp

M. Fujii
Yui Acupuncture Clinic, Tokyo, Japan

N. Takemura
Department of Electrical and Electronics Engineering, NEWCAT Research Institute, College of Engineering, 1 Nakagawara, Tokusada, Tamuramachi, Koriyama, Fukushima 963-8642, Japan

T. Hirayama
Department of Neurological Surgery, School of Medicine, Nihon University, Tokyo, Japan

© Springer Science+Business Media, New York 2016
C.E. Elwell et al. (eds.), *Oxygen Transport to Tissue XXXVII*, Advances in
Experimental Medicine and Biology 876, DOI 10.1007/978-1-4939-3023-4_37

balance of PFC activity toward left-dominant, resulting in relaxation effects on the patients.

Keywords Acupuncture • Depression • NIRS • Prefrontal cortex • Stress

1   Introduction

It has been reported that acupuncture is useful in treating various neuropsychological diseases. For example, acupuncture has been used for the treatment of depression [1]. In addition, it was reported that acupuncture at certain points could reduce stress responses of the autonomic nervous system [2, 3]. However, the neurobiological basis of these effects is not yet clear.

The prefrontal cortex (PFC) plays an important role in emotion. Interestingly, left/right asymmetry of PFC activity is correlated with specific emotional responses to mental stress and personality traits [4–6]. For example, negative emotional stimuli increase relative right-sided PFC activation [6], whereas induced positive affective stimuli elicit an opposite pattern of asymmetric activation [7]. In addition, a functional near-infrared spectroscopy (NIRS) study found that subjects with right-dominant activity at rest showed higher scores in the State-Trait Anxiety Inventory (STAI), while those with left-dominant activity at rest showed lower STAI scores. These results suggest that NIRS-measured asymmetry in PFC activity at the resting state can predict emotional state. In the present study, we employed NIRS to examine the effect of acupuncture on psychological state (i.e., anxiety) and prefrontal cortex (PFC) activity at rest in patients with anxiety disorders.

2   Methods

2.1   Subjects

We studied 10 patients (male 1, female 9, mean age 41.8 ± 6.8 years) with anxiety disorders; anxiety levels were evaluated with the State-Trait Anxiety Inventory (STAI). The STAI assess state anxiety (STAI-1) and trait anxiety (STAI-2) separately; each type of anxiety has its own scale of 20 different questions that are scored. Scores range from 20 to 80, with higher scores correlating with greater anxiety. The patients were treated by acupuncture at Yui Acupuncture Clinic in Osaka. The following acupoints were used depending on the Traditional Chinese Medicine diagnosis of each patient: WHO-GV23, GV20, GV9, GV14, BL7, GV4, BL17, BL19, BL20, TE5, GB41, PC6, and SP4. The stimulation method was needle retention for 15 min. All subjects provided written informed consent as required by the Human Subjects Committee of Yui Acupuncture Clinic.

2.2 NIRS Measurements and Data Analysis

We employed NIRS to evaluate the asymmetry of PFC activity at rest, as in our previous study [8]. Briefly, oxy- and deoxy-Hb concentration changes in the bilateral PFC at rest were measured with a two-channel NIRS (PNIRS-10, Hamamatsu Photonics, Japan). The NIRS probes were set symmetrically on the forehead. The distance between the emitter and detector was set at 3 cm. Changes in t-Hb and oxy-Hb reflect changes in cerebral blood volume and flow, respectively [9, 10]

In order to analyze left/right asymmetry of PFC activity at rest, we calculated the laterality scores [8]. Consider

$$\Delta oxyR_{\min} = \min_{t \in analysis\ interval} \Delta oxyR_t \tag{37.1}$$

$$\Delta oxyL_{\min} = \min_{t \in analysis\ interval} \Delta oxyL_t \tag{37.2}$$

where $\Delta oxyR_t$ and $\Delta oxyL_t$ denote oxy-Hb concentration changes of the right and the left PFC. The quantities defined by Eqs. (37.1) and (37.2) are the variations with respect to their minimum values, so that they are always non-negative. Based on these quantities, we defined the *Laterality Index at Rest (LIR)* as follows:

$$LIR = \frac{\sum_{t \in analysis\ interval} ((\Delta oxyR_t - \Delta oxyR_{\min}) - (\Delta oxyL_t - \Delta oxyL_{\min}))}{\sum_{t \in analysis\ interval} ((\Delta oxyR_t - \Delta oxyR_{\min}) + (\Delta oxyL_t - \Delta oxyL_{\min}))} \tag{37.3}$$

The numerator of (37.3) consists of the difference between oxy-Hb concentration changes of the right and the left PFC summed over the analysis period (3 min). The index defined by Eq. (37.3) provides values in the range of $[-1, +1]$. A positive LIR indicates that the right PFC is more active at rest than the left PFC, while a negative LIR indicates that the left PFC is more active at rest than the right PFC. We previously found a positive correlation between LIR and STAI-1 (state anxiety) [8].

3 Results

All subjects exhibited fluctuations of oxy-Hb (reflecting neuronal activity) in the bilateral PFC at rest, both before and after acupuncture treatment. Acupuncture treatment significantly decreased the STAI-1 score from 50.8 ± 9.04 to 41.0 ± 10.1 ($F(1,8) = 15.9$, $p < 0.005$) (Fig. 37.1a). However, there was no significant effect on the STAI-2 score ($p > 0.05$).

The NIRS measurements indicated the presence of spontaneous oscillations of oxy-Hb in the bilateral PFC at rest before and after the treatment. Interestingly, LIR decreased in 7 out of the 10 subjects, indicating that the treatment changed the

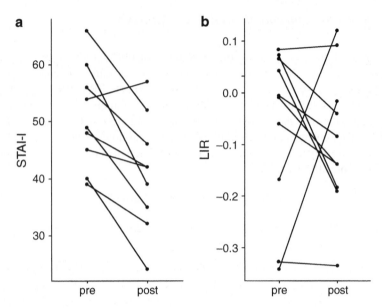

Fig. 37.1 Changes of STAI-1 score (**a**) and LIR (**b**) after acupuncture

balance of PFC activity toward left-dominant, at least in these subjects. The other three subjects showed an increasing tendency in LIR.

4 Conclusions

Acupuncture has been widely used for treatment of depression, especially over the last 10 years [1]. The present pilot study demonstrated that treatment by acupuncture significantly reduced state anxiety levels but not trait anxiety level. These results suggest that acupuncture is effective to decrease anxiety levels in patients with mood disorders. Indeed, a number of studies have indicated the effectiveness of acupuncture on depressive symptoms [11]. It should be noted, however, that a recent review concluded there was insufficient evidence of the effectiveness of acupuncture to treat depression, and that the studies indicating effectiveness contained methodological flaws [12].

Recently, our research group developed a new technique for objective assessment of mental stress levels by measuring fluctuation of oxy-Hb concentration in the bilateral PFC at rest, employing NIRS, and we used it to analyze left/right asymmetry of PFC activity by calculating the LIR [8]. In the present study, NIRS demonstrated that LIR decreased after acupuncture in 7 out of the 10 subjects, which indicates that the balance of PFC activity at rest changed toward the left side in those subjects. Notably, it was reported that LIR was positively correlated with STAI-1 (i.e., state anxiety) scores [8], which is consistent with the reported lateralization of emotion; the right hemisphere is dominant for negative emotions and the left hemisphere is dominant for positive emotions [13], although the lateralization of emotion is still an issue of debate.

Thus, it seems plausible that acupuncture changed the balance of PFC activity toward left-dominant, resulting in relaxation effects on the patients. However, it should be noted that acupuncture did not decrease LIR in 3 out of 10 patients. Further studies are necessary to clarify the physiological mechanism of acupuncture therapy.

The present pilot study has a number of limitations. First, it was not a randomized control study. Second, we did not measure changes of PFC activity during acupuncture since NIRS measurements during acupuncture could have influenced the putative treatment effects of acupuncture. Our previous study on normal adults revealed that acupuncture caused both activation and deactivation of the PFC [14]. Finally, the point of acupuncture stimulation varied among the subjects depending on the diagnosis according to Traditional Chinese Medicine, and the number of subjects was small. Nevertheless, our results suggest that further study in a larger group would be worthwhile.

Acknowledgements This research was supported in part by Grants-in-Aid for Exploratory Research (25560356) and Strategic Research Foundation Grant-aided Project for Private Universities (S1411017) from the Ministry of Education, Culture, Sports, Sciences and Technology of Japan, and grants from Alpha Electron Co., Ltd. (Fukushima, Japan), Iing Co., Ltd. (Tokyo, Japan) and Southern Tohoku General Hospital (Fukushima, Japan).

References

1. MacPherson H, Sinclair-Lian N, Thomas K (2006) Patients seeking care from acupuncture practitioners in the UK: a national survey. Complement Ther Med 14:20–30
2. Arai YC, Ushida T, Osuga T et al (2008) The effect of acupressure at the extra 1 point on subjective and autonomic responses to needle insertion. Anesth Analg 107:661–664
3. Middlekauff HR, Hui K, Yu JL et al (2002) Acupuncture inhibits sympathetic activation during mental stress in advanced heart failure patients. J Card Fail 8:399–406
4. Davidson RJ, Jackson DC, Kalin NH (2000) Emotion, plasticity, cortex, and regulation: perspectives from affective neuroscience. Psychol Bull 126:890–909
5. Canli T, Zhao Z, Desmond JE et al (2001) An fMRI study of personality influences on brain reactivity to emotional stimuli. Behav Neurosci 115:33–42
6. Fischer H, Andersson JL, Furmark T et al (2002) Right-sided human prefrontal brain activation during acquisition of conditioned fear. Emotion 2:233–241
7. Ahern GL, Schwartz GE (1985) Differential lateralization for positive and negative emotion in the human brain: EEG spectral analysis. Neuropsychologia 23:745–755
8. Ishikawa W, Sato M, Fukuda Y, Matsumoto T, Takemura N, Sakatani K (2014) Correlation between asymmetry of spontaneous oscillation of hemodynamic changes in the prefrontal cortex and anxiety levels: a near-infrared spectroscopy study. J Biomed Opt 19:027005
9. Ferrari M, Wilson DA, Hanley DF, Traystman RJ (1992) Effects of graded hypotension on cerebral blood flow, blood volume, and mean transit time in dogs. Am J Physiol 262:1908–1914

10. Hoshi Y, Kobayashi N, Tamura M (2001) Interpretation of near-infrared spectroscopy signals: a study with a newly developed perfused rat brain model. J Appl Physiol 90:1657–1662
11. Yoon-Hang K (2007) The effectiveness of acupuncture for treating depression: a review. Altern Complement Ther 13:129–131
12. Pilkington K (2010) Anxiety, depression and acupuncture: a review of the clinical research. Auton Neurosci 157:91–95
13. Hellige JB (1993) Hemispheric asymmetry. Harvard University Press, Cambridge
14. Sakatani K, Kitagawa T, Aoyama N, Sasaki M (2010) Effects of acupuncture on autonomic nervous function and prefrontal cortex activity. Adv Exp Med Biol 662:455–460

Chapter 38
Influence of Pleasant and Unpleasant Auditory Stimuli on Cerebral Blood Flow and Physiological Changes in Normal Subjects

Tomotaka Takeda, Michiyo Konno, Yoshiaki Kawakami, Yoshihiro Suzuki, Yoshiaki Kawano, Kazunori Nakajima, Takamitsu Ozawa, Keiichi Ishigami, Naohiro Takemura, and Kaoru Sakatani

Abstract The prefrontal cortex (PFC) plays an important role in emotion and emotional regulation. The valence asymmetry hypothesis, proposes that the left/right asymmetry of the PFC activity is correlated with specific emotional responses to stressors. However, this hypothesis still seems to leave room for clarifying neurophysiological mechanisms. The purpose of the present study was to investigate the effects of stimuli with positive and negative valence sounds (hereafter PS, NS) selected from the International Affective Digitized Sounds-2 on physiological and physiological responses, including PFC activity in normal participants. We studied the effect of both stimuli using 12 normal subjects (mean age 26.8 years) on cerebral blood oxygenation in the bilateral PFC by a multi-channel NIRS, alpha wave appearance rate in theta, alpha, beta by EEG, autonomic nervous function by heart rate, and emotional conditions by the State-Trait Anxiety Inventory (STAI) and the visual analogue scale (VAS). PS was selected over 7.00 and NS were fewer than 3.00 in the Pleasure values. Sounds were recorded during 3 s and reproduced at random using software. Every task session was designed in a block manner: seven rests with Brown Noise (30 s) and six tasks (30 s) blocks. All participants performed each session in random order with eyes closed. A paired Student's t-test was used for

T. Takeda (✉) • M. Konno • Y. Kawakami • Y. Suzuki • Y. Kawano • K. Nakajima • T. Ozawa • K. Ishigami
Department of Oral Health and Clinical Science, Division of Sports Dentistry, Tokyo Dental College, 2-9-18, Misaki, Thiyoda, Tokyo, 101-0061 Japan
e-mail: ttakeda@tdc.ac.jp

N. Takemura
Department of Electrical and Electronics Engineering, Nihon University, NEWCAT Institute, College of Engineering, Fukushima, Japan

K. Sakatani
Department of Electrical and Electronics Engineering, Nihon University, NEWCAT Institute, College of Engineering, Fukushima, Japan

School of Medicine, Department of Neurological Surgery, Nihon University, Tokyo, Japan

© Springer Science+Business Media, New York 2016   303
C.E. Elwell et al. (eds.), *Oxygen Transport to Tissue XXXVII*, Advances in
Experimental Medicine and Biology 876, DOI 10.1007/978-1-4939-3023-4_38

comparisons (P < 0.05). PFC activity showed increases bilaterally during both stimuli with a greater activation of the left side in PS and a tendency of more activation by
NS in the right PFC. Significantly greater alpha wave intensity was obtained in
PS. Heart rate tended to show smaller values in PS. The STAI level tended to show
smaller values in PS, and a significantly greater VAS score was obtained in PS which
indicated 'pleasant'. Despite the limitations of this study such as the low numbers of
the subjects, the present study indicated that PS provided pleasant psychological and
physiological responses and NS unpleasant responses. The PFC was activated bilaterally, implying a valence effect with the possibility of a dominant side.

Keywords Prefrontal cortex • Near-infrared spectroscopy • International Affective
Digitized Sounds-2 • Electroencephalogram • Stress

1   Introduction

Many people suffer from many kinds of stress in their life. Normally, a balance in
mental conditions is maintained between the activities of the sympathetic and
parasympathetic nervous systems. Chronic stress can disturb the balance and
cause stress-related health problems. When people are exposed to a stressor, the
brain initiates a stress response. This stress response is thought to be a healthy
defense mechanism. Many studies have investigated the effects of stress on: the
reaction of the circulation and central nervous system [1], and activity in the
prefrontal cortex (PFC) and the autonomic nervous system (ANS) [2, 3]. Even
relatively mild acute uncontrolled stress can cause a rapid stress response and
dramatic loss of prefrontal abilities [4].

The stress response involves activation of the PFC, which stimulates the
hypothalamic-pituitary-adrenal (HPA) axis and influences ANS, since neuronal
networks exist between the PFC and the neuroendocrine centers in the medial
hypothalamus, and the PFC has direct access to sympathetic and/or parasympathetic motor nuclei in brainstem and spinal cord [5]. The PFC will set the endocrine/
autonomic balance, depending on the emotional status [5]. The PFC is the region
that is most sensitive to the effects of stress exposure [4].

According to the valence asymmetry hypothesis, the left/right asymmetry of the
PFC activity is correlated with specific emotional responses to stressors and personality traits [6, 7]. The evidence supports a valence asymmetry hypothesis,
suggesting that the left hemisphere is specialized for the processing of positive
emotions, while the right hemisphere is specialized for the processing of negative
emotions [8, 9], including the HPA axis system [2]. However, this asymmetry
hypothesis has not received unequivocal support [10]. This hypothesis still seems
to leave room for clarifying neurophysiological mechanisms involved.

The International Affective Digitized Sounds-2 (IADS) is a standardized database of 167 naturally occurring sounds, and that is widely used in the study of
emotions. The IADS is part of a system for emotional assessment developed by the
Center for Emotion and Attention [11–13]. Studies that have used IADS stimuli

have revealed that auditory emotional stimuli activate the appetitive and defensive motivational systems similarly to the way that pictures do.

The purpose of the present study was to investigate the effects of stimuli with positive and negative valences selected from the IADS on psychological and physiological responses including the PFC activity measured by NIRS in normal participants.

2 Materials and Methods

A total of 12 healthy volunteers participated in the study (mean age 26.8 years). Participants were told to refrain from substances (e.g., coffee, etc. including caffeine) that could affect their nervous systems before and during the period of testing, and not to eat for 2 h before the test. They were also instructed to avoid excessive drinking and a lack of sleep the night before the test. In order to avoid the influence of environmental stress, the participants were seated in a comfortable chair in an air-conditioned room with temperature and humidity maintained at approximately 25 °C and 50 %, respectively. The study was conducted in accordance with the Principles of the Declaration of Helsinki, and the protocol was approved by the Ethics Committee of Tokyo Dental College (Ethical Clearance NO.436). Written informed consent was obtained from all participants.

After a 10-min rest, they then performed two auditory stimulation listening tasks by an earphone with eyes closed. Auditory tasks were positive and negative valence sounds selected from the IADS (hereafter PS, NS). PS tasks were selected over 7.0 and NS were fewer than 3.0 in the Pleasure value of IADS. Namely, PS was comfortable, and NA was uncomfortable. Sounds were recorded during 3 s and reproduced at random using software (Access Vision Co., Ltd., Japan). Each task session (in random order) was designed in a block manner; seven rests with a Smoothed Brown Noise (Technomind, Brazil) (30 s) and six tasks blocks with PS or NS (Fig. 38.1). The Noise was a low roar resembling a waterfall or heavy rainfall. It had more energy at lower frequencies. And it was used not too large in volume. Activity in the PFC was measured by a multi-channel NIRS; (OEG-16, Spectratech, Japan). The targets were placed at Fp1 and Fp2 in the international 10–20 system. Electroencephalogram (EEG) (Muse Brain System, Syscom, Japan) and heart rate (HR) (WristOx, NONIN, USA) were monitored simultaneously with the PFC activity. We used the 10-cm visual analogue scale (VAS) and the State-Trait

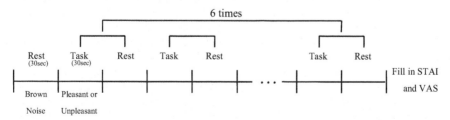

Fig. 38.1 Experimental protocol trial consisted of seven rests (Smoothed Brown Noise, 30 s) and six tasks blocks: Positive or Negative sounds from IADS. Two tasks were adopted at random

Anxiety Inventory (STAI) for psychological assessments. Statistical evaluations between PS and NS were performed using a paired Student's t-test (Excel Statistics, Microsoft Japan). A p-value of <0.05 was considered significant.

3   Results

Results are summarized in Table 38.1. PFC activity showed increases bilaterally during both stimuli with a greater activation of the left side in PS and a tendency of more activation by NS in the right PFC. Significantly greater alpha wave intensity

Table 38.1 Psychological and physiological assessment

		Stimulus or site	Mean (S.D.)	Statistics
Psychological assessment	STAI level (Bigger figure indicates "more unpleasant")	Positive sound	2.08 (0.79)	–
		Negative sound	2.41 (1.08)	
	VAS level (Smaller figure indicates "more unpleasant")	Positive sound	5.93 (0.96)	*
		Negative sound	4.02 (1.53)	
Physiological assessment	Heart rate (bpm)	Positive sound	62.46 (7.96)	–
		Negative sound	63.50 (8.59)	
	Alpha wave (%) (Appearance rate in theta, alpha, and beta)	Positive sound	47.88 (0.08)	*
		Negative sound	45.10 (0.07)	
	NIRS ⊿Oxy-Hb (a.u.)	Right PFC	Positive sound	0.06 (0.03)	–
			Negative sound	0.11 (0.08)	
		Left PFC	Positive sound	0.12 (0.09)	–
			Negative sound	0.11 (0.08)	
		Positive sound	Right PFC	Refer to the much column above	*
			Left PFC	Refer to the much column above	
		Negative sound	Right PFC	Refer to the much column above	–
			Left PFC	Refer to the much column above	

$*p < 0.05$

(appearance rate in theta, alpha, beta) was obtained in PS. HR tended to show smaller values in PS. The STAI level tended to show smaller values in PS, and a significantly greater VAS score was obtained in PS, indicating 'pleasant.'

4   Discussion

Psychological assessment of both VAS and STAI results showed 'more pleasant' in PS and 'more unpleasant' in NS. These results showed the effectiveness of IADS in the Japanese population with a different culture and environment than an American/English population. Also, selected positive and negative sound stimulus could induce stressed and relaxed conditions, respectively.

Physiological assessment of HR showed smaller values and greater alpha wave in PS. Stress can induce sympathetic excitation [14] and cause a rapid increase in HR [15]. EEG reflects neuronal activities of the human brain which can be directly affected by emotional states. Especially the alpha wave (8-13Hz) appearance in the awake EEG with an eye closed condition seemed to indicate a relaxed state [6]. The mean amplitude of the alpha wave was significantly reduced while performing mental arithmetic and/or listening to an unpleasant tone [16]. The amplitude of the alpha wave while listening to a siren was reduced [17]. These two physiological assessments are also consistent with the effectiveness of IADS in the psychological assessments.

The results of the four above-mentioned assessments indicate that PS induced 'pleasant' and NS 'unpleasant', which might cause responses in the brain. Indeed, PS and NS induced Oxy-Hb increases in the bilateral PFC. The emotional task-induced PFC activation could cause activation of the HPA axis and influence ANS, on the basis of networks between the PFC and the medial hypothalamus and ANS [5] Further, the greater activation of the left side in PS and a tendency of more activation by NS in the right PFC were obtained in the present study. These results are similar to a neuropsychological study that used auditory stimuli and revealed greater left-hemisphere frontotemporal activation in response to positively valenced auditory stimuli, while bilateral frontotemporal activation was observed in response to negatively valenced auditory stimuli [18] and might support the valence asymmetry hypothesis [6, 7] in which the left hemisphere is specialized for the processing of positive emotions and the right hemisphere is specialized for the processing of negative emotions [8, 9].

In summary, despite the limitations of this study such as the low numbers of the subjects, the present study indicated that positive sounds provided pleasant psychological and physiological responses and negative sounds unpleasant responses. The PFC was activated bilaterally, implying a valence effect with the possibility of a dominant side.

Acknowledgments This research was partly supported by Japan Science and Technology Agency, under Strategic Promotion of Innovative Research and Development Program, and a

Grant-in-Aid from the Ministry of Education, Culture, Sports, Sciences and Technology of Japan (Grant-in-Aid for Scientific Research 22592162, 25463025, and 25463024, Grant-in-Aid for Exploratory Research 25560356), and grants from Alpha Electron Co., Ltd. (Fukushima, Japan) and Iing Co., Ltd. (Tokyo, Japan).

References

1. Liu X, Iwanaga K, Koda S (2011) Circulatory and central nervous system responses to different types of mental stress. Ind Health 49:265–273
2. Tanida M, Katsuyama M, Sakatani K (2007) Relation between mental stress-induced prefrontal cortex activity and skin conditions: a near-infrared spectroscopy study. Brain Res 1184:210–216
3. Tanida M, Sakatani K, Takano R et al (2004) Relation between asymmetry of prefrontal cortex activities and the autonomic nervous system during a mental arithmetic task: near infrared spectroscopy study. Neurosci Lett 369:69–74
4. Arnsten AF (2009) Stress signalling pathways that impair prefrontal cortex structure and function. Nat Rev Neurosci 10:410–422
5. Buijs RM, Van Eden CG (2000) The integration of stress by the hypothalamus, amygdala and prefrontal cortex: balance between the autonomic nervous system and the neuroendocrine system. Prog Brain Res 126:117–132
6. Davidson RJ, Jackson DC, Kalin NH (2000) Emotion, plasticity, context, and regulation: perspectives from affective neuroscience. Psychol Bull 126:890–909
7. Ishikawa W, Sato M, Fukuda Y et al (2014) Correlation between asymmetry of spontaneous oscillation of hemodynamic changes in the prefrontal cortex and anxiety levels: a near-infrared spectroscopy study. J Biomed Opt 19:027005
8. Davidson RJ (2003) Affective neuroscience and psychophysiology: toward a synthesis. Psychophysiology 40:655–665
9. Coan JA, Allen JJ (2004) Frontal EEG asymmetry as a moderator and mediator of emotion. Biol Psychol 67:7–49
10. Royet JP, Zald D, Versace R et al (2000) Emotional responses to pleasant and unpleasant olfactory, visual, and auditory stimuli: a positron emission tomography study. J Neurosci 20:7752–7759
11. Soares AP, Pinheiro AP, Costa A et al (2013) Affective auditory stimuli: adaptation of the International Affective Digitized Sounds (IADS-2) for European Portuguese. Behav Res Methods 45:1168–1181
12. Stevenson RA, James TW (2008) Affective auditory stimuli: characterization of the International Affective Digitized Sounds (IADS) by discrete emotional categories. Behav Res Methods 40:315–321
13. Bradley MM, Lang PJ (2007) The International Affective Digitized Sounds (2nd Edn; IADS-2): affective ratings of sounds and instruction manual. Technical report B-3. University of Florida, NIH Center for the Study of Emotion and Attention, Gainesville
14. Van de Kar LD, Blair ML (1999) Forebrain pathways mediating stress-induced hormone secretion. Front Neuroendocrinol 20:1–48
15. Ritvanen T, Louhevaara V, Helin P et al (2006) Responses of the autonomic nervous system during periods of perceived high and low work stress in younger and older female teachers. Appl Ergon 37:311–318
16. Nishifuji S (2011) EEG recovery enhanced by acute aerobic exercise after performing mental task with listening to unpleasant sound. Conference proceedings: annual international conference of the IEEE Engineering in Medicine and Biology Society IEEE Engineering in Medicine and Biology Society Conference, vol 2011, pp 3837–3840

17. Horii A, Yamamura C, Katsumata T, Uchiyama A (2004) Physiological response to unpleasant sounds. J Int Soc Life Inform Sci 22:536–544
18. Altenmuller E, Schurmann K, Lim VK et al (2002) Hits to the left, flops to the right: different emotions during listening to music are reflected in cortical lateralisation patterns. Neuropsychologia 40:2242–2256

Chapter 39
Spontaneous Fluctuations of PO_2 in the Rabbit Somatosensory Cortex

Robert A. Linsenmeier, Daniil P. Aksenov, Holden M. Faber, Peter Makar, and Alice M. Wyrwicz

Abstract In many tissues, PO_2 fluctuates spontaneously with amplitudes of a few mmHg. Here we further characterized these oscillations. PO_2 recordings were made from the whisker barrel cortex of six rabbits with acutely or chronically placed polarographic electrodes. Measurements were made while rabbits were awake and while anesthetized with isoflurane, during air breathing, and during 100 % oxygen inspiration. In awake rabbits, 90 % of the power was between 0 and 20 cycles per minute (cpm), not uniformly distributed over this range, but with a peak frequently near 10 cpm. This was much slower than heart or respiratory rhythms and is similar to the frequency content observed in other tissues. During hyperoxia, total power was higher than during air-breathing, and the dominant frequencies tended to shift toward lower values (0–10 cpm). These observations suggest that at least the lower frequency fluctuations represent efforts by the circulation to regulate local PO_2.

R.A. Linsenmeier (✉)
Department of Biomedical Engineering, Northwestern University, 2145 Sheridan Road, Evanston, IL 60208-3107, USA

Department of Neurobiology, Northwestern University, Evanston, IL, USA

Department of Ophthalmology, Northwestern University, Chicago, IL, USA
e-mail: r-linsenmeier@northwestern.edu

D.P. Aksenov
NorthShore University HealthSystem, Evanston, IL, USA

H.M. Faber
Department of Biomedical Engineering, Northwestern University, 2145 Sheridan Road, Evanston, IL 60208-3107, USA

Department of Engineering Sciences and Applied Mathematics, Northwestern University, Evanston, IL, USA

P. Makar
Department of Biomedical Engineering, Northwestern University, 2145 Sheridan Road, Evanston, IL 60208-3107, USA

A.M. Wyrwicz
Department of Biomedical Engineering, Northwestern University, 2145 Sheridan Road, Evanston, IL 60208-3107, USA

NorthShore University HealthSystem, Evanston, IL, USA

© Springer Science+Business Media, New York 2016
C.E. Elwell et al. (eds.), *Oxygen Transport to Tissue XXXVII*, Advances in Experimental Medicine and Biology 876, DOI 10.1007/978-1-4939-3023-4_39

There were no consistent changes in total power during 0.5 or 1.5 % isoflurane anesthesia, but the power shifted to lower frequencies. Thus, both hyperoxia and anesthesia cause characteristic, but distinct, changes in spontaneous fluctuations. These PO_2 fluctuations may be caused by vasomotion, but other factors cannot be ruled out.

Keywords Oxygen • Cerebral cortex • Vasomotion • Rabbit • Anesthesia

1 Introduction

In the non-stimulated brain (e.g. [1, 2]) and retina [3], as well as other tissues (e.g. [4]), oxygen tension (PO_2) fluctuates spontaneously with amplitudes of a few mmHg. These are not correlated at different sites in the brain [1]. There are also spontaneous fluctuations of vessel tone (vasomotion), and it is generally believed that changes in vessel diameter cause flow changes that in turn change tissue PO_2 [2, 5]. While there are many examples of PO_2 fluctuations in the literature, there is rather little data on frequency content, or how it may change under different conditions. We have studied spontaneous PO_2 fluctuations in the whisker barrel cortex of awake rabbits during air breathing and hyperoxia, as well as comparing the awake and anesthetized state. By analyzing the power spectra of these oxygen signals under different conditions, we hoped to provide further data and provide a baseline for future studies of the relation between BOLD signals and tissue PO_2.

2 Methods

Experiments were approved by the Institutional Animal Care and Use Committee of NorthShore University HealthSystem. A pedestal was surgically implanted on the head of six adult female Dutch belted rabbits to carry electrodes, and rabbits recovered from surgery before recordings began. Rabbits were wrapped in a blanket during experimental episodes that lasted approximately 2 h, and did not exhibit any signs of stress during recordings. Further details of the preparation are available in [6].

Rabbits breathed either air or 100 % O_2, and they were either kept awake or anesthetized with 0.5 or 1.5 % isoflurane. The minimum alveolar concentration (MAC) for rabbits is 2.05 %. Insulated, 25 μm diameter Au/Ag alloy wires were plated with Au at the tip and chronically implanted into the whisker barrel cortex of three adult rabbits. In three other rabbits, electrodes were inserted into cortex through the head implant just before recording. Depth was not known precisely, but was within the gray matter. The electrodes were polarized at -0.7 V, and connected to an ammeter (Keithley 614) to record a current that was typically 3–10 nA during air breathing. The current was converted to voltage, notch and low-pass filtered (30 or 50 Hz), amplified, and digitized at 20 Hz. In chronic experiments it was not possible to calibrate the electrodes, so PO_2 values are approximate.

Files of data from 60 to 300 s were analyzed with Matlab programs that used the available Matlab algorithm to compute FFTs. Before computing power, the mean PO_2 was subtracted from each point in the data file. To approximately represent the amplitude of the fluctuations in each file, we divided the peak power in that file by the mean PO_2 that had been subtracted, resulting in an amplitude as a percentage of the mean. Recording sessions contained variable numbers of data files. Each recording session was weighted equally in grand averages shown in the figures by computing the average power spectrum for all data files of a particular type in a recording session (e.g. awake, air breathing), and then averaging across sessions. For comparing different conditions (Figs. 39.3, 39.4, and 39.5), we used only recording sessions in which we had matched sets of recordings.

3   Results

Irregular spontaneous fluctuations were observed in almost all recording sessions. The amplitude was 7.6 ± 1.5 % (mean and SEM) of the mean PO_2 during air breathing in awake animals.

Figure 39.1 shows representative examples of fluctuations from the conditions we studied after subtracting the mean PO_2. Also shown are the corresponding power spectra for each of those examples. In the oxygen breathing and anesthetized rabbit, the power in these examples was at lower frequencies than in the awake, air breathing animal. Each power spectrum is normalized by the total power between 0 and 30 cpm. Power beyond 30 cpm (0.5 hz) was generally very small. It was regarded as instrument noise and ignored. Individual FFTs did not exhibit characteristics that could be associated with a 1/f spectrum. Due to the normalization, these individual power spectra do not indicate whether the power was greater under one condition or another.

In order to average power spectra across trials, power was summed in 2 cpm bins (i.e. 0–2 cpm, 2–4 cpm, etc.) and plotted at the center points of those bands (i.e. 1, 3 cpm) as shown in Figs. 39.2, 39.3, 39.4, and 39.5. These spectra are normalized so that the total power between 0 and 30 cpm was 100 % for each condition. Figure 39.2 is the average power spectrum for $n = 16$ recording sessions (six rabbits) during air breathing. Ninety percent of the power was below 20 cpm and peaked near 10 cpm (0.17 Hz). There was no apparent difference between rabbits in which electrodes were implanted chronically and those in which they were inserted just before recordings. Figure 39.3 shows average power spectra for five recording sessions in which rabbits breathed both air and oxygen. More of the power shifted to lower frequencies during hyperoxia, although the peak at 10 cpm was still apparent. The amplitude of fluctuations was always higher during 100 % oxygen inspiration than during air breathing in the same recording session, by an average of 80 %.

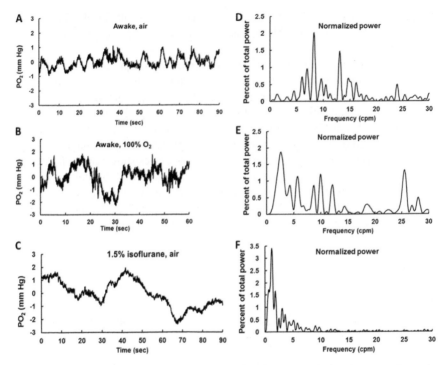

Fig. 39.1 Examples of PO₂ fluctuations after subtracting the mean PO₂, as well as the corresponding power spectrum for the awake, air breathing condition (**a, d**), the awake, 100 % O₂ condition (**b, e**), and the air breathing, 1.5 % isoflurane condition (**c, f**)

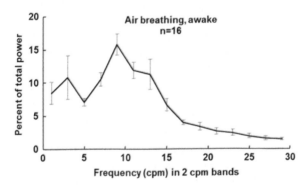

Fig. 39.2 Average power spectrum for n = 16 recording sessions (six rabbits) during air breathing. Spectra in Figs. 39.2, 39.3, 39.4, and 39.5 are normalized so that the total power between 0 and 30 cpm is 100 %. Each point in Figs. 39.2, 39.3, 39.4, and 39.5 collapses data from several frequencies in the original spectra, so each point contains more power than a point in Fig. 39.1. Error bars are SEM

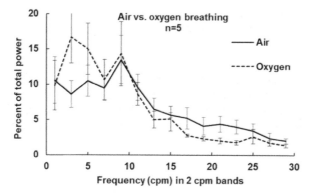

Fig. 39.3 Average power spectra for five recording sessions in which rabbits breathed both air and oxygen

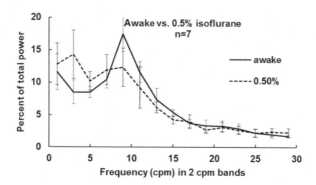

Fig. 39.4 Average normalized power spectra for recording sessions in which rabbits were both awake and anesthetized with 0.5 % isoflurane

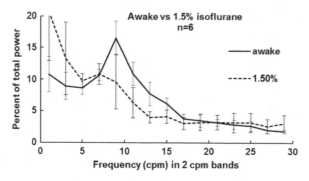

Fig. 39.5 Average normalized power spectra for recording sessions in which rabbits were both awake and anesthetized with 1.5 % isoflurane

Correspondingly, total power was always considerably greater during hyperoxia, by an average factor of 3.1 ($\pm$1.5 (SD)) (p = 0.038 compared to a ratio of 1).

Figures 39.4 and 39.5 show average normalized power spectra for recording sessions in which rabbits were both awake and anesthetized with 0.5 % or 1.5 % isoflurane. As in hyperoxia, power shifted to lower frequencies, but during anesthesia the peak at 10 cpm was blunted. While the spectrum shifted, there was no consistent increase or decrease in the amplitude or total power of fluctuations between air and either level of anesthesia. Baseline PO_2 also increased with anesthesia (not shown).

4   Discussion and Conclusions

As in previous work [1–4, 7], most of the power in spontaneous PO_2 fluctuations was below 20 cpm under all conditions. The peak during air breathing was just under 10 cpm here, similar to previous measurements in awake rabbit [1] and rat testes [4], and higher than observed in anesthetized cat retina (4 cpm [2]) or jejunum (3.4–5 cpm [7]). The peak frequency in rabbit brain is much lower than the rabbit's heart rate (around 200 cpm) or respiratory frequency (30–40 cpm), indicating that these fluctuations originate in the brain and not in the cardiovascular system in general. Fluctuations in different tissues do have somewhat different characteristics. There is a wider distribution of power in brain and retina than in testes and jejunum, where the oscillations are more regular. Differences could be due to differences in circulatory anatomy or regulation in different tissues, to the volume sampled by different oxygen electrodes, to anesthesia, or to other factors.

PO_2 fluctuations are likely to originate from arteriolar vasomotion, which occurs at similar frequencies, e.g. [2, 4, 5, 8]. It has been argued that vasomotion could be beneficial in enhancing tissue oxygen transport in hypoxic tissues [9, 10], although the purpose of fluctuations under normal conditions is not clear. Little attention has been paid to the possibility that factors apart from vasomotion may influence tissue PO_2 fluctuations. In the retina, where there is not much evidence of vasomotion, we argued that the reduction in red cell velocity resulting from leukocyte passage through the capillaries could play a role [3], and red cell clumping could have a similar effect. These have not been modeled.

During hyperoxia, the peak power shifted to lower frequencies, and the total power increased. Because brain blood flow decreases in hyperoxia to maintain a constant oxygen delivery, it is likely that the fluctuations reflect altered activity of the vascular smooth muscle. These changes are consistent with alterations in vasomotion in hamster muscle [8]. As far as we know, the effect of changes in blood pressure on the fluctuations has not been investigated.

We expected that anesthesia would disturb vascular regulation and depress the fluctuations. However, the fluctuations did not disappear, and total power did not change in a consistent way. Anesthesia, particularly at the higher level, did shift power to lower frequencies. It should be noted that this was still a modest level of anesthesia. The frequency shift suggests that there may be more than one mechanism responsible for fluctuations, and at least one is altered by anesthesia.

References

1. Manil J et al (1984) Properties of the spontaneous fluctuations in cortical oxygen pressure. Adv Exp Med Biol 169:231–239
2. Hudetz AG et al (1998) Spontaneous fluctuations in cerebral oxygen supply. An introduction. Adv Exp Med Biol 454:551–559
3. Braun RD, Linsenmeier RA, Yancey CM (1992) Spontaneous fluctuations in oxygen tension in the cat retina. Microvasc Res 44:73–84
4. Lysiak JJ, Nguyen QAT, Turner TT (2000) Fluctuations in rat testicular interstitial oxygen tensions are linked to testicular vasomotion: persistence after repair of torsion. Biol Reprod 63:1383–1389
5. Aalkjaer C, Boedtkjer D, Matchov V (2011) Vasomotion – what is currently thought? Acta Physiol 202:253–269
6. Aksenov D et al (2012) Effect of isoflurane on brain tissue oxygen tension and cerebral autoregulation in rabbits. Neurosci Lett 524(2):116–118
7. Hasibeder W, Germann R, Sparr H et al (1994) Vasomotion induces regular major oscillations in jejunal mucosal tissue oxygenation. Am J Physiol 266:G978–G986
8. Bertuglia S et al (1991) Hypoxia- or hyperoxia-induced changes in arteriolar vasomotion in skeletal muscle microcirculation. Am J Physiol 260:H362–H372
9. Goldman D, Popel AS (2001) A computational study of the effect of vasomotion on oxygen transport in capillary networks. J Theor Biol 209:189–199
10. Tsai AG, Intaglietta M (1993) Evidence of flowmotion induced changes in local tissue oxygen tension. Int J Microcirc Clin Exp 12:75–88

Chapter 40
Effect of the Antioxidant Supplement Pyrroloquinoline Quinone Disodium Salt (BioPQQ™) on Cognitive Functions

Yuji Itoh, Kyoko Hine, Hiroshi Miura, Tatsuo Uetake, Masahiko Nakano, Naohiro Takemura, and Kaoru Sakatani

Abstract Pyrroloquinoline quinone (PQQ) is a quinone compound first identified in 1979. It has been reported that rats fed a PQQ-supplemented diet showed better learning ability than controls, suggesting that PQQ may be useful for improving memory in humans. In the present study, a randomized, placebo-controlled, double-blinded study to examine the effect of PQQ disodium salt (BioPQQ™) on cognitive functions was conducted with 41 elderly healthy subjects. Subjects were orally given 20 mg of BioPQQ™ per day or placebo, for 12 weeks. For cognitive functions, selective attention by the Stroop and reverse Stroop test, and visual-spatial cognitive function by the laptop tablet Touch M, were evaluated. In the Stroop test, the change of Stroop interference ratios (SIs) for the PQQ group was significantly smaller than for the placebo group. In the Touch M test, the stratification analyses dividing each group into two groups showed that only in the lower group of the PQQ group (initial score < 70), did the score significantly increase. Measurements of physiological parameters indicated no abnormal blood or urinary adverse events, nor adverse internal or physical examination findings at any point in the study. The preliminary experiment using near-infrared spectrometry (NIRS) suggests that cerebral blood flow in the prefrontal cortex was increased by the

Y. Itoh • K. Hine • H. Miura
Department of Psychology, Keio University, Tokyo, Japan

T. Uetake
CX Medical Japan Co., Inc., Tokyo, Japan

M. Nakano (✉)
Niigata Research Laboratory, Mitsubishi Gas Chemical Co., Inc., Niigata, Japan
e-mail: masahiko-nakano@mgc.co.jp

N. Takemura
Department of Electrical and Electronics Engineering, Laboratory of Integrative Biomedical Engineering, College of Engineering, Nihon University, Tokyo, Japan

K. Sakatani
Department of Electrical and Electronics Engineering, Laboratory of Integrative Biomedical Engineering, College of Engineering, Nihon University, Tokyo, Japan

Department of Neurological Surgery, School of Medicine, Nihon University, Tokyo, Japan

© Springer Science+Business Media, New York 2016
C.E. Elwell et al. (eds.), *Oxygen Transport to Tissue XXXVII*, Advances in Experimental Medicine and Biology 876, DOI 10.1007/978-1-4939-3023-4_40

administration of PQQ. The results suggest that PQQ can prevent reduction of brain function in aged persons, especially in attention and working memory.

Keywords PQQ • Brain • Memory • Attention • Supplement

1 Introduction

Pyrroloquinoline quinone (PQQ) is a water-soluble quinone compound first identified in 1979 from bacteria as a cofactor for redox enzymes [1, 2]. It is contained in various everyday foods and beverages such as parsley, green tea or fermented soybeans [3]. PQQ is also detected in various human organs and tissues [4], and is especially high in human breast milk [5]. PQQ has anti-oxidative and mitochondrial biogenesis capacities [6]. Reduced PQQ has shown strong anti-oxidant capacity, 7.4x greater than ascorbic acid [7]. It was reported that PQQ inhibited cell death of cultured neuroblastoma in a dose-dependent manner [8], and increased expression of nerve growth factor (NGF) [9]. In addition, it has been demonstrated that rats fed a PQQ-supplemented diet showed better learning ability than controls at the early stage of the Morris water maze test [10]. These results suggest that PQQ is potentially effective for preventing neurodegeneration caused by oxidative stress, and may improve memory. However, the effect of PQQ on cognitive function in humans is unknown. If PQQ can maintain or improve cognitive function in the aged, PQQ may be useful for active and healthy aging. In the present study, we examined the effect of PQQ on cognitive function in the aged in a randomized, double-blinded, placebo-controlled trial. We evaluated cognitive functions employing the Stroop and reverse Stroop test for selective attention, and the laptop tablet Touch M for visual-spatial cognitive function. Our hypothesis was that lower interference ratio in the Stroop test or higher score in the Touch M can be achieved by the intake of PQQ.

2 Methods

2.1 Test Substance, Study Design and Subjects

The test substance was provided in the form of a two-piece hard capsule containing 10 mg of PQQ disodium salt (BioPQQ™) manufactured by Mitsubishi Gas Chemical Co., Inc. (Tokyo, Japan) or placebo. Each capsule was ingested with a cup of water once a day after breakfast for 12 weeks. The dose of PQQ disodium salt (20 mg/day) was based on our previous animal study [10], which was far beyond the estimated daily intake of PQQ from everyday foods and beverages (0.01–0.4 mg/day).

The study was a randomized, placebo-controlled, double-blinded two-group parallel study, conducted in compliance with the Declaration of Helsinki (2008) and The Ethical Guideline of Epidemiological Study (issued in 2004 by the Ministry of Education, Culture, Sports, Science and Technology, and the Ministry of Health, Labor and Welfare). The study protocol (#1112-05-001MGC03352) was approved by the ethical committee of Nihon University Itabashi Hospital (Tokyo, Japan) on 17th April 2012 (Approved #24-1). The 42 volunteers were randomly divided into two groups, the placebo group ($n = 21$) and the PQQ group ($n = 21$).

All participants underwent internal and physical examinations, peripheral blood test and urinalysis test at baseline (before intake, 0 weeks) and at 12 weeks.

2.2   Evaluations

The Stroop and reverse Stroop test examine selective attention ability, which are frequently used to examine individual differences in cognitive ability. Participants took a group version of the Stroop and reverse Stroop test [11] at 0 and 12 weeks. In each test, participants were required to choose and check one color name or patch that corresponded to a sample color name or ink color from five alternatives. Low Stroop interference ratios (SIs) and reverse Stroop interference ratios (RIs) are considered to indicate high attention ability and ratios are higher in elderly persons as a result of reduced attention ability. Previous study showed the range of mean SIs and RIs of each generation covering 7–92 years old was from 8.20 to 40.9 and -10.89 to 23.30, respectively [12]. We regard SIs as more important, as RIs might be affected by sensory sensitivity for color, which is sometimes low in elderly persons [12]. Statistical analysis methods are shown with the results below.

Touch M is a simple evaluation system for visual-spatial cognitive function utilizing a laptop tablet with a 12-in. size screen originally developed by Yamato Sangyo Ltd. (Yokohama, Japan) [13]. The score of Touch M (from 0 to 100 points) is calculated from the reproduction fidelity of the order of the targets' appearance, which is taken to reflect visual-spatial cognitive function, working memory and action procedure. Studies to date have shown that the score decreases according to aging after the 60-year age point has been reached. A score below 70 points can be regarded to reflect a decline in brain functions [13].

3   Results

3.1   Subjects

One subject in the placebo group dropped out for personal reasons. A total of 41 volunteers completed the study. The placebo group contained 20 subjects

(7 male, 13 female) with a mean age of 58.4 ± 5.2 (*SD*) years. The PQQ group contained 21 subjects (7 male, 14 female) with a mean age of 58.6 ± 5.1 (*SD*) years.

Measurements of physiological parameters indicated no abnormal blood or urinary adverse events, nor adverse internal or physical examination findings at any point in the study.

3.2 Stroop and Reverse Stroop Test

The SIs and RIs were calculated at 0 and 12 weeks for each participant, according to Watanabe et al [12]. The changes in the SIs were 0.4 for the PQQ group (8.0 for 0 weeks and 8.4 for 12 weeks) and 3.1 for the placebo group (8.9 and 12.0, respectively). A one factor ANOVA revealed no significant difference between the groups ($F(1,39) = 0.72$, $p > 0.10$). Here, F stands for variance ratio likewise below. One sample t-test did not reject the hypothesis that the average change for both groups was 0 ($t(40) = 1.10$, $p > 0.10$). This fact suggests that no effect was caused by practice of the test. The changes in the RIs were 2.2 for the PQQ group (9.0 for 0 weeks and 11.2 for 12 weeks) and -2.9 for the placebo group (12.9 and 10.0, respectively). The difference in the changes in RIs between the groups was calculated as significant ($F(1,39) = 4.34$, $p < 0.05$). Although one-sample t-tests revealed that the changes between 0 and 12 weeks were non-significant for both groups ($p > 0.10$), these results seem to show that the placebo, not the test substance, has an effect in improving attention ability. However, the difference between the two groups might be an artifact that can be attributed to the performances in the 0 weeks test that were slightly higher for the placebo group than the PQQ group ($F(1, 39) = 3.34$, $p < 0.01$) by one factor ANOVA. An ANCOVA for the RIs changes with the group as a factor and the 0 weeks RIs as a covariant revealed no effects of group, nor the interaction between group and 0 weeks RIs, but an effect of 0 weeks RIs.

Each of the interference ratios was calculated with performances in two of the subtests that were susceptible to noise. As the SIs is the main target of the analyses and discussion, we conducted an additional analysis removing the outliers. Three participants (one from the PQQ group and two from the placebo group) whose 0 weeks or 12 weeks SIs were more than 2 *SD* from the mean SIs (for all participants) were removed. The change for the PQQ group (-0.9, 9.0 for 0 weeks and 8.1 for 12 weeks) was significantly smaller ($F(1,36) = 5.41$, $p < 0.05$) than for the placebo group (5.2, 8.7 for 0 weeks and 14.1 for 12 weeks) (Fig. 40.1). A one-sample t-test showed significant change in the SIs for the placebo group ($t(17) = 2.41$, $p < 0.05$) whereas no significant change was detected for the PQQ group ($t(19) = -0.59$, $p > 0.10$).

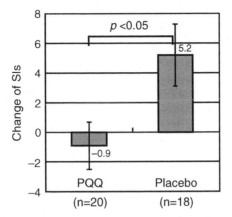

Fig. 40.1 Mean changes of the Stroop interference ratios (SIs) from 0 to 12 weeks. *Error bars* indicate *SEs*. *p* value of paired *t*-test, comparison between 0 and 12 weeks

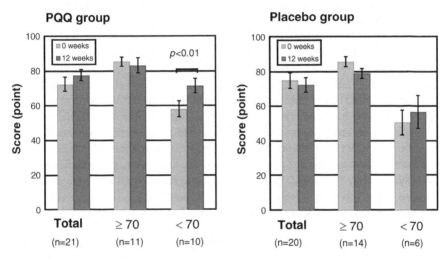

Fig. 40.2 Mean values of the Touch M score. *Error bars* indicate *SEs*. *p* value of paired *t*-test, comparison between 0 and 12 weeks. Stratification analyses were performed, dividing each group into two groups at the threshold of an initial score of 70 points

3.3 Touch M

Touch M results were analyzed by two-tail paired *t*-test with setting a significant level at 5 %. In the PQQ group, the mean score had a tendency to increase from 72.4 at 0 weeks to 77.6 at 12 weeks ($t(20) = 1.68$, $p \cong 0.10$), while in the placebo group the score did not significantly differ between 0 weeks (74.8) and 12 weeks (72.2) ($t(19) = 0.85$, $p > 0.10$) (Fig. 40.2). Stratification analyses were performed, dividing each group into two groups at the threshold of an initial score of 70 points.

In the PQQ group, the mean score of higher group (initial score $\geq$ 70points) did not differ between 0-weeks (85.4) and 12 weeks (83.2) ($t(10) = 0.55$, $p > 0.10$); however, that of the lower group (initial score <70 points) significantly increased from 58.1 at 0 weeks to 71.5 at 12 weeks ($t(9) = 3.88$, $p < 0.01$). In the placebo group, the mean score of the higher group (initial score $\geq$ 70 points) decreased from 85.4 at 0 weeks to 78.9 at 12 weeks ($t(13) = 1.59$, $p < 0.10$), and that of the lower group (initial score < 70 points) did not differ between 0 weeks (50.2) and 12 weeks (56.3) ($t(5) = 1.29$, $p > 0.10$).

4 Discussion

The change in SIs in the Stroop test within the PQQ group was significantly lower than that within the placebo group when the three outliers were omitted, and the Touch M score significantly increased only in the PQQ group with the lower initial score of < 70 points after 12 weeks.

It was expected that the SIs and RIs would decrease between 0 weeks and 12 weeks only for the PQQ group if the 12-week intake of the substance improves participants' cognitive ability. We did not observe such a decrease in either the SIs or the RIs. These data show no effects of PQQ on cognitive ability. However, it is premature to conclude that PQQ has no effect on cognitive ability, because the Stroop test measures only limited aspects of cognitive ability and the indices are not very stable.

There are several reasons to focus on the additional analysis of the SIs with exclusion of outliers: the SIs seem to be a better index than the RIs because changes of RIs might be canceled out by changes in color sensitivity, and both SIs and RIs are susceptible to noise [12], so removal of outliers is desirable. In the additional analysis, there was a significant difference between changes in the SIs at 0 weeks and 12 weeks for the PQQ and placebo groups. However, the change for the PQQ group was not significant, while the SIs increased significantly from 0 weeks to 12 weeks for the placebo group. It seems that unknown factors increased the SIs for both the PQQ and placebo groups and the improvement of cognitive ability by PQQ canceled out the increase of the SIs for the PQQ group. Hakoda and Sasaki [11] observed an increase in SIs by repetition with a similar group version of the Stroop and reverse Stroop test. Although the reason for the increase was not clear, this observation might support the discussion above.

PQQ is an anti-oxidant compound. As such compounds often cause improvement of brain functions, a similar effect was expected with PQQ [14]. Moreover, our preliminary testing on NIRS during a working memory task suggested that the blood flow detected as cerebral blood oxygenation may influence such improvement. The results of this study suggest that PQQ can prevent a reduction of brain functions in aged persons, especially in attention and working memory. Further study on the mechanism involved is needed.

Acknowledgments   This research was supported in part by a Grant-in-Aid from the Ministry of Education, Culture, Sports, Science and Technology of Japan (Grant-in-Aid for Exploratory Research 25560356), and grants from Alpha Electron Co., Ltd. (Fukushima, Japan) and Iing Co., Ltd. (Tokyo, Japan).

References

1. Salisbury SA, Forrest HS, Cruse WBT et al (1979) A novel coenzyme from bacterial primary alcohol dehydrogenase. Nature 280:843–844
2. Duine JA, Frank J, Van Zeeland JK (1979) Glucose dehydrogenase from acinetobacter calcoaceticus: a quinoprotein. FEBS Lett 108:443–446
3. Kumazawa T, Sato K, Seno H et al (1995) Levels of pyrroloquinoline quinone in various foods. Biochem J 307:331–333
4. Kumazawa T, Seno H, Urakami T et al (1992) Trace levels of pyrroloquinoline quinone in human and rat samples detected by gas chromatography/mass spectrometry. Biochim Biophys Acta 1156:62–66
5. Mitchell AE, Johnes AD, Mercer RS et al (1999) Characterization of pyrroloquinoline quinone amino acid derivatives by electrospray ionization mass spectrometry and detection in human milk. Anal Biochem 269:317–325
6. Rucker R, Chowanadisai W, Nakano M (2009) Potential physiological importance of pyrroloquinoline quinone. Altern Med Rev 14:268–277
7. Mukai K, Ouchi A, Nakano M (2011) Kinetic study of the quenching reaction of singlet oxygen by pyrroloquinolinequinole ($PQQH_2$, a reduced form of pyrroloquinolinequinone) in micellar solution. J Agric Food Chem 59:1705–1712
8. Nunome K, Miyazaki S, Nakano M et al (2008) Pyrroloquinoline quinone prevents oxidative stress-induced neuronal death probably through changes in oxidative status of DJ-1. Biol Pharm Bull 31(7):1321–1326
9. Yamaguchi K, Sasano A, Urakami T et al (1993) Stimulation of nerve growth factor production by pyrroloquinoline quinone and its derivatives in vitro and in vivo. Biosci Biotechnol Biochem 57:1231–1233
10. Ohwada K, Takeda H, Yamazaki M et al (2008) Pyrroloquinoline quinone (PQQ) prevents cognitive deficit caused by oxidative stress in rats. J Clin Biochem Nutr 42:29–34
11. Hakoda Y, Sasaki M (1990) Group version of the Stroop and reverse-Stroop test –the effects of reaction mode, order and practice. Jpn J Educ Psychol 38:389–394 (Japanese)
12. Watanabe M, Hakoda Y, Matsumoto A (2011) Group version of the Stroop and reverse-Stroop test: an asymmetric development trait in two kinds of interference. Psychol Res Kusyu Univ 12:41–50
13. Hayashi Y, Kijima T, Satou K et al (2011) Examination of the evaluation method of visual-spatial cognitive function using the touch screen device. Jpn J Geriatr Psychiatry 22 (4):439–447 (Japanese)
14. Clausen J, Nielsen SA, Kristensen M (1989) Biochemical and clinical effects on an antioxidative supplementation of geriatric patients. A double blind study. Biol Trace Elem Res 20:135–151

Chapter 41
Estimation of Skin Blood Flow Artefacts in NIRS Signals During a Verbal Fluency Task

Akitoshi Seiyama, Kotona Higaki, Nao Takeuchi, Masahiro Uehara, and Naoko Takayama

Abstract The aim of this study was to clarify effects of skin (scalp) blood flow on functional near infrared spectroscopy (fNIRS) during a verbal fluency task. In the present study, to estimate the influence of skin blood flow on fNIRS signals, we conducted examinations on 19 healthy volunteers (39.9 ± 13.1 years, 11 male and 8 female subjects). We simultaneously measured the fNIRS signals, skin blood flow (i.e., flow, velocity, and number of red blood cells [RBC]), and pulse wave rates using a multimodal fNIRS system. We found that the effects of skin blood flow, measured by the degree of interference of the flow, velocity, and number of RBCs, and pulse wave rates, on NIRS signals varied considerably across subjects. Further, by using the above physiological parameters, we evaluated application of the independent component analysis algorithm proposed by Molgedey and Schuster (MS-ICA) to remove skin blood flow artefacts from fNIRS signals.

Keywords Functional near infrared spectroscopy • Verbal fluency task • Independent component analysis • Pearson's correlation coefficient • Coefficient of spatial uniformity

A. Seiyama (✉) • M. Uehara
Division of Medical Devices for Diagnoses, Human Health Sciences, Graduate School of Medicine, Kyoto University, Kyoto, Japan
e-mail: aseiyama@hs.med.kyoto-u.ac.jp

K. Higaki • N. Takeuchi
Division of Medical Devices for Diagnoses, Human Health Sciences, Graduate School of Medicine, Kyoto University, Kyoto, Japan

Laboratory of Clinical Examination, Takayama Medical Clinic, Medical Corporation Taihoukai, Takayama, Japan

N. Takayama
Laboratory of Clinical Examination, Takayama Medical Clinic, Medical Corporation Taihoukai, Takayama, Japan

© Springer Science+Business Media, New York 2016
C.E. Elwell et al. (eds.), *Oxygen Transport to Tissue XXXVII*, Advances in Experimental Medicine and Biology 876, DOI 10.1007/978-1-4939-3023-4_41

1   Introduction

A verbal fluency task (VFT) is a neuropsychological test that is widely used to assess cognitive function in the frontal cortex and as an aided differential diagnosis of psychiatric disorders [1–5]. However, several recent reports have cautioned that the skin (scalp) blood flow in the frontal cortex considerably influences functional near infrared spectroscopy (fNIRS) signals during a VFT [6, 7]. On the other hand, independent component analyses (ICA) have been recently used to evaluate and/or extract a task-related neural response from the fNIRS signals for various tasks including cognitive and motor tasks [8–12]. In the present study, to estimate the influence of skin blood flow on fNIRS signals during a VFT and to evaluate the successful application of ICA, we performed simultaneous measurements of the fNIRS signals, skin blood flow (i.e., flow, velocity, and number of red blood cells [RBC]) and pulse wave rates using a multimodal fNIRS system [13]. Here, we discuss the successful application of the ICA algorithm proposed by Molgedey and Schuster (MS-ICA) [14] to remove skin blood flow artefacts from fNIRS signals during the VFT experiments.

2   Methods

2.1   Experimental Design

Nineteen right-handed healthy volunteers (11 males, 8 females, age 39.9 ± 13.1 years) participated in this study. Before beginning the experiments, written informed consent was obtained from all the participants. The Ethics Committee of Kyoto University approved the study.

A VFT consisted of a 30-s pre-task control period, 60-s task period, and 70-s post-task control period in a single trial for each subject [1–5]. During control periods, participants were requested to repeatedly verbalise five Japanese vowels (/ a/, /i/, /u/, /e/, and /o/). During the task period, participants were requested to verbalise as many words as possible that began with a Japanese character shown on a computer screen every 20 s (thus, 3 characters per task period). The characters included /ko/, /ka/, /si/, /ta/, /ki/, or /ha/.

The multimodal fNIRS system was used [11]. Simultaneous measurements of fNIRS (SMARTNIRS, Shimadzu, Japan), Laser Doppler Flowmeter (LDF) (FLO-C1, Omega Flow, Japan), and pulse oxymeter (OLV-3100, Nihon Kohden) were performed at 24-ms sampling intervals. The fNIRS system is a CW mode, and three wave lengths, i.e., 780, 805 and 830 nm, are used to measure changes in the concentration of oxygenated ([oxy-Hb]) and deoxygenated hemoglobin ([deoxy-Hb]) according to the Modified Beer–Lambert law. The source probe emits light sequentially in order to avoid cross-talk noise. Sampling time is adjusted to 25 ms. Each channel is supposed to measure the changes at the mid-point between the two probes at 20–30 mm deep under the scalp [15]. Optical probes (2×7 arrays) of the fNIRS system were attached on the forehead with a distance of 3 cm between them. This alignment enabled us to measure 19 regions, which covered the Brodmann's areas 9 and 46 (dorsolateral region), 10 (front polar region), and 45 (ventrolateral region). The illuminating light guide in the fourth column two lines was positioned at Fpz according to the international 10–20 system.

One LDF probe was attached to the left (or right) temple [centred between the corner of left (or right) eye and the base of left (or right) ear]. The FLO-C1 instrument enabled us to measure the velocity, and the number of RBCs simultaneously in a tissue, and the mass flow of RBCs was then calculated as the product of RBC number and RBC velocity. To estimate the absolute values, the following calibration values were used. For the estimation of RBC number, the mean optical path length is 1 mm, collision number (m) of laser light to RBC is 1, scattering area of erythrocytes is 4.5 μm^2 at 780 nm, and RBC density is $2.2 \times 10^4/mm^3$. For the estimation of RBC velocity, 6.2 krad/s of the mean angular frequency (1 kHz) of the power spectrum corresponded to 2.9 mm/s of RBC velocity ($m \approx 1$). Our instrument ensured a linear relationship between RBC velocity and LDF signal change up to 12 kHz. Thus, our instrument ensured measurements of only the skin or the scalp blood flow parameters and a linear relationship between RBC velocity and LDF signal change up to 12 kHz [16].

2.2 Analyses of Data

We applied the MS-ICA, with time delays between 0 and 0.48 s with steps of 0.024 s, to 19-channels of fNIRS signals and used the following two criteria to remove skin blood flow effects and reconstruct the fNIRS signals [8].

1. Correlation coefficient between fNIRS and LDF signals, where Pearson's correlation coefficient 'r' was converted to a Zr value in accordance with Fisher's Z-transformation expression $Zr = (1/2)*\ln\{(1+r)/(1-r)\}$ for statistical analyses.
2. Coefficient of spatial uniformity ($CSU(j) = |<aj>/\sigma j|$), where $<aj>$ and σj are the mean and standard deviation of each column from the mixing matrix and j is the column number. Thus, when the signal distributes in a wide area, the value of CSU will be large [8].

To use the above criteria, we adopted the following assumptions; (1) changes in skin blood flow parameters take place uniformly over the measurement area, and changes in the blood flow and the number of RBCs measured by the LDF reflect changes in [oxy-Hb] and [total-Hb] in the skin or scalp, respectively. Therefore, we applied the above criteria independently on changes in [oxy-Hb] and [total-Hb], because fNIRS gives changes in [oxy-Hb] and [deoxy-Hb], and change in [total-Hb] is given as sum of [oxy-Hb] and [deoxy-Hb]. Thus, calculation of two, here oxy-Hb and total-Hb, of the three parameters could give another parameter (deoxy-Hb).

After removal of components (those having 'r' ≥ 0.4000 and CSU ≥ 1.50, see Discussion section), fNIRS signals were reconstructed. The pre-task baseline was determined as the mean over a 10-s period just prior to the task period, and the post-task baseline was determined as the mean over the last 10 s of the 70-s post-task period. For the baseline correction, linear fitting was performed using the data obtained from the two baselines. The moving average method with 4.44 s window was adopted to remove high frequency artefacts from the data.

3   Results

Figure 41.1 shows one example of application of MS-ICA for removal of skin blood flow effects. It should be noted, this type of changes in fNIRS signal is rare in our present experiments (one of 19 subjects). Thus, although further identification of the skin blood flow in terms of changes in [oxy-Hb] and [deoxy-Hb] (or [total-Hb]) is required for strict evaluation of the present method in this case, before ICA

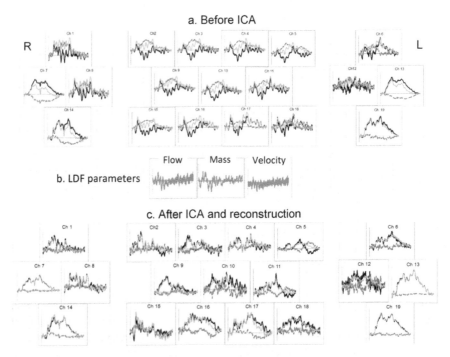

Fig. 41.1 One example of a successful application of the MS-ICA for removal of skin blood flow effects. *Top* (**a**) and *bottom* (**c**) panels denote 19-channel fNIRS signals before and after the MS-ICA reconstruction (arbitrary units). *Middle panel* (**b**) is the LDF signals, where flow, mass, and velocity denote the flow, number, and velocity of RBCs (arbitrary units), respectively. The *solid black*, *broken grey*, and *solid grey lines* denote oxy-Hb, deoxy-Hb, and total-Hb, respectively

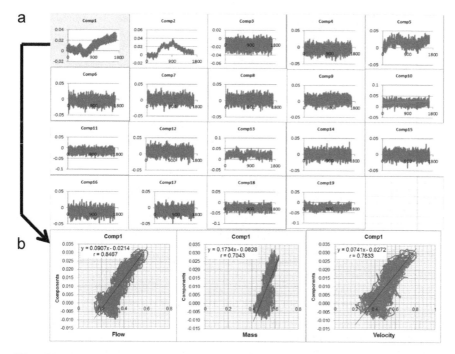

Fig. 41.2 One example for the removal of the MS-ICA-component based on the correlation coefficient with LDF parameters. (**a**) 19-MS-ICA components. (**b**) Correlation coefficients between component 1 and LDF parameters (r > 0.7000). Here, only component 1 was removed. The original fNIRS data are shown in Fig. 41.1a

(Fig. 41.1a), this subject shows strong correlation between oxy-Hb and LDF parameters (Figs. 41.1b and 41.2b).

Figure 41.2 shows a removed component for oxy-Hb and its correlation to LDF parameters. In this example, only component 1 shows strong correlation with each LDF parameters, and no components showed CSU ≥ 1.50 (data not shown).

Figure 41.3 shows the removed components for total-Hb and their coefficients of spatial uniformity (CSU). Components 1 and 2 show strong correlation with LDF parameters, and components 1, 5, and 6 show CSU ≥ 1.50 (see bottom of panel).

Table 41.1 summarises the MS-ICA application for 19 subjects. Ten of nineteen subjects show successful application of the MS-ICA with the criteria, r ≥ 0.4000 and CSU ≥ 1.50. Here, we judged the successful application of the MS-ICA based on the reductions of correlation coefficients between fNIRS signals and LDF parameters after removal of components having 'r' > 0.4000 and CSU > 1.50.

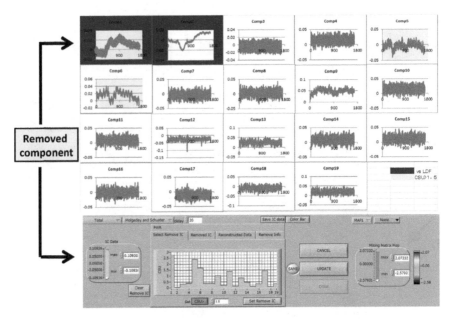

Fig. 41.3 One example for removal of the MS-ICA-component based on the CSU. Components 1, 5, and 6 showing CSU $\geq$ 1.50 were removed. Further, components 1 and 2 were removed based on the correlation coefficient with LDF parameters. The original and reconstructed fNIRS data are shown in Fig. 41.1a and c, respectively

4 Discussion

In the present study, we examined the potential of the MS-ICA to remove effects of scalp blood flow in fNIRS signals in the forehead regions during a VFT. We found that the effects of the skin blood flow, measured by the degree of interference of the flow, velocity, and number of RBCs, and pulse wave rates, on fNIRS signals varied considerably across subjects, but the effect of the pulse wave rate (Zr correlation $= 0.2011 \pm 0.1812$) was relatively small as compared with the LDF parameters (see Table 41.1). In 11 of 19 subjects, successful applications of the MS-ICA were observed (Table 41.1). The result suggests that the following two criteria are useful for successful application of the MS-ICA to remove effects of scalp blood flow in fNIRS signals: (1) correlation coefficient between fNIRS and LDF signals ($r \geq 0.4000$) and (2) coefficient of spatial uniformity (CSU ≥ 1.50).

Although further evaluations are required to use the ICA algorithm to remove skin blood flow from fNIRS signals, the MS-ICA has a potential to remove effects of scalp blood flow from fNIRS signals and extract actual task-related neural activation.

Table 41.1 Summary of the MS-ICA application for removal of skin blood flow effects

Subject	Before ICA(F-Zr correlation)			After ICA(F-Zr correlation)		
	Flow	Mass	Velocity	Flow	Mass	Velocity
1	0.3768	0.1935	0.3283	0.3807	0.3854	0.2626
2	0.0490	0.3607	0.1243	0.0783	0.3367	0.0862
3	0.5623	0.0808	0.6424	0.4506	0.0548	0.4091
4	0.0661	0.0000	0.0283	–	–	–
5	0.2237	0.1538	0.1223	–	–	–
6	0.5148	0.5453	0.1794	0.2035	0.1814	0.1104
7	0.5660	0.1104	0.6667	0.3101	0.0084	0.3976
8	0.3739	0.2799	0.1247	0.3044	0.2186	0.0095
9	0.4200	0.2035	0.1923	0.0814	0.0839	0.1048
10	0.3289	0.0224	0.3125	0.0245	0.0200	0.0095
11	0.1831	0.0558	0.0077	0.0400	0.0413	0.0100
12	0.8195	0.3846	0.6725	0.1331	0.0387	0.1067
13	0.3209	0.3214	0.2814	–	–	–
14	0.3182	0.0332	0.4365	–	–	–
15	0.0265	0.0316	0.0100	–	–	–
16	0.2716	0.4222	0.0679	0.1292	0.2346	0.0687
17	1.0113	0.6266	0.6872	0.0400	0.3791	0.0756
18	0.6249	0.7725	0.5432	0.1934	0.0539	0.1893
19	0.4797	0.5452	0.0100	0.3976	0.4169	0.0200
Mean	0.3967	0.2707	0.2862	0.1299	0.1586	0.0731
SD	0.2541	0.2305	0.2490	0.1227	0.1626	0.0615
r >0.4	6/19	5/19	6/19	2/11	1/11	1/11

'r' denotes Pearson's correlation coefficient. '–' denotes no effects of ICA application

Acknowledgments This study was supported in part by grants-in-aid from the Ministry of Education, Science and Culture of Japan.

References

1. Suto T, Fukuda M, Ito M et al (2004) Multichannel near-infrared spectroscopy in depression and schizophrenia: cognitive brain activation study. Biol Psychiatry 55:501–511
2. Kameyama M, Fukuda M, Yamagishi Y et al (2006) Frontal lobe function in bipolar disorder: a multichannel near-infrared spectroscopy study. Neuroimage 29:172–184
3. Takizawa R, Kasai K, Kawakubo Y et al (2008) Reduced frontopolar activation during verbal fluency task in schizophrenia: a multi-channel near-infrared spectroscopy study. Schizophr Res 99:250–262
4. Kinou M, Takizawa R, Marumo K et al (2013) Differential spatiotemporal characteristics of the prefrontal hemodynamic response and their association with functional impairment in schizophrenia and major depression. Schizophr Res 150(2–3):459–467
5. Takizawa R, Fukuda M, Kawasaki S et al (2014) Neuroimaging-aided differential diagnosis of the depressive state. Neuroimage 85(Pt 1):498–507

6. Takahashi T, Takikawa Y, Kawagoe R et al (2011) Influence of skin blood flow on near-infrared spectroscopy signals measured on the forehead during a verbal fluency task. Neuroimage 57:991–1002
7. Kirilina E, Jelzow A, Heine A et al (2012) The physiological origin of task-evoked systemic artifacts in functional near infrared spectroscopy. Neuroimage 61:70–81
8. Kohno S, Miyai I, Seiyama A et al (2007) Removal of the skin blood flow artifact in functional near-infrared spectroscopic imaging data through independent component analysis. J Biomed Opt 12(6):062111
9. Katura T, Sato H, Fuchino Y et al (2008) Extracting task-related activation components from optical topography measurement using independent components analysis. J Biomed Opt 13:054008
10. Markham J, White BR, Zeff BW et al (2009) Blind identification of evoked human brain activity with independent component analysis of optical data. Hum Brain Mapp 30:2382–2392
11. Patel S, Katura T, Maki A et al (2011) Quantification of systemic interference in optical topography data during frontal lobe and motor cortex activation: an independent component analysis. Adv Exp Med Biol 915:45–51
12. Funane T, Atsumori H, Katura T et al (2014) Quantitative evaluation of deep and shallow tissue layers' contribution to fNIRS signal using multi-distance optodes and independent component analysis. Neuroimage 85:150–165
13. Seiyama A, Sasaki Y, Takatsuki A et al (2012) Effects of the autonomic nervous system on functional neuroimaging: analyses based on the vector autoregressive model. Adv Exp Med Biol 737:77–82
14. Molgedey L, Schuster HG (1994) Separation of a mixture of independent signals using time delayed correlations. Phys Rev Lett 72:3634–3637
15. Shimodera S, Imai Y, Kamimura N et al (2012) Mapping hypofrontality during letter fluency task in schizophrenia: a multi-channel near-infrared spectroscopy study. Schizophr Res 136:63–69
16. Kashima S, Ono Y, Sohda A et al (1994) Separate measurement of two components of blood flow velocity in tissue by dynamic light scattering method. Jpn J Appl Phys 33:2123–2127

Chapter 42
Effect of Transcranial Direct Current Stimulation over the Primary Motor Cortex on Cerebral Blood Flow: A Time Course Study Using Near-infrared Spectroscopy

Haruna Takai, Atsuhiro Tsubaki, Kazuhiro Sugawara, Shota Miyaguchi, Keiichi Oyanagi, Takuya Matsumoto, Hideaki Onishi, and Noriaki Yamamoto

Abstract Transcranial direct current stimulation (tDCS) is a noninvasive brain stimulation technique that is applied during stroke rehabilitation. The purpose of this study was to examine diachronic intracranial hemodynamic changes using near-infrared spectroscopy (NIRS) during tDCS applied to the primary motor cortex (M1). Seven healthy volunteers were tested during real stimulation (anodal and cathodal) and during sham stimulation. Stimulation lasted 20 min and NIRS data were collected for about 23 min including the baseline. NIRS probe holders were positioned over the entire contralateral sensory motor area. Compared to the sham condition, both anodal and cathodal stimulation resulted in significantly lower oxyhemoglobin (O_2Hb) concentrations in the contralateral premotor cortex (PMC), supplementary motor area (SMA), and M1 ($p < 0.01$). Particularly in the SMA, the O_2Hb concentration during anodal stimulation was significantly lower than that during the sham condition ($p < 0.01$), while the O_2Hb concentration during cathodal stimulation was lower than that during anodal stimulation ($p < 0.01$). In addition, in the primary sensory cortex, the O_2Hb concentration during anodal stimulation was significantly higher than the concentrations during both cathodal stimulation and the sham condition ($p < 0.05$). The factor of time did not demonstrate significant differences. These results suggest that both anodal and cathodal tDCS cause widespread changes in cerebral blood flow, not only in the area immediately under the electrode, but also in other areas of the cortex.

H. Takai (✉) • A. Tsubaki • K. Sugawara • S. Miyaguchi • T. Matsumoto • H. Onishi
Institute for Human Movement and Medical Sciences, Niigata University of Health and Welfare, 1398 Shimami-cho, Kita-ku, Niigata-shi, Niigata 950-3198, Japan
e-mail: hpm13005@nuhw.ac.jp

K. Oyanagi
Kobe City Medical Center General Hospital, 2-2-1, Minatojimaminamimachi, Chuo-ku, Kobe, Hyogo 650-0047, Japan

N. Yamamoto
Niigata Rehabilitation Hospital, 761, Kizaki, Kita-ku, Niigata-shi, Niigata 950-3304, Japan

© Springer Science+Business Media, New York 2016
C.E. Elwell et al. (eds.), *Oxygen Transport to Tissue XXXVII*, Advances in Experimental Medicine and Biology 876, DOI 10.1007/978-1-4939-3023-4_42

Keywords Transcranial direct current stimulation • Near-infrared spectroscopy • Oxyhemoglobin • Primary motor area • Time course study

1 Introduction

Transcranial direct current stimulation (tDCS) is a noninvasive brain stimulation technique that is currently being investigated as a neuromodulation therapy for a range of conditions including stroke. Although its mechanism of action is currently unclear, it is well known that the nature of the neural modulations depends on the tDCS polarity, which can modulate corticomotor excitability [1]. As an evaluation of the effects of tDCS, various functional brain imaging studies have already been performed. In a study by Merzagora et al. [2], anodal stimulation induced a significant increase in the oxyhemoglobin (O_2Hb) concentration under the electrode after tDCS. Similarly, in a study using functional magnetic resonance imaging (fMRI), anodal tDCS increased the cortical excitability of not only the underlying primary motor cortex (M1), but also of the bilateral supplemental motor area (SMA) and contralateral premotor cortex (PMC), which have strong functional connectivity with M1 [3]. In addition, it has been shown that the effects of tDCS are observed only with an adaptation time above a certain level [4]. However, these studies used stimulation durations that were shorter than the stimulation times used during actual treatments. Furthermore, while many studies on the after-effects of tDCS exist, studies investigating changes over the duration of tDCS adaptation are limited. Therefore, the purpose of this study was to examine diachronic intracranial hemodynamic changes using near-infrared spectroscopy (NIRS) during tDCS applied to the M1.

2 Methods

Seven healthy volunteers (mean ± standard deviation [SD]; age: 22.8 ± 2.0 years; height: 168.0 ± 8.8 cm; weight: 61.6 ± 5.4 kg) participated in this study. All subjects were free of any known cardiovascular or respiratory diseases and were not taking any medications. Each subject received verbal and written explanations of the study objectives, measurement techniques, risks, and benefits associated with the investigation. The study was approved by the Ethics Committee of Niigata University of Health and Welfare (17487-140509) and conformed to the standards set by the Declaration of Helsinki.

Subjects were seated in a chair in a quiet room where they then received the stimulation. They were tested during two real stimulation conditions (anodal or cathodal) and during a sham stimulation condition. Stimulation lasted 20 min and NIRS data were collected for about 23 min including during the 3 min baseline period before stimulation onset.

2.1 tDCS

tDCS was delivered by a direct current stimulator (Eldith, NeuroConn GmbH, Germany) with a pair of saline-soaked surface sponge electrodes (5×7 cm, 35 cm^2). For the anodal tDCS condition, the anodal and cathodal electrodes were placed over the right M1 and contralateral supraorbital region, respectively, while the arrangement was reversed for the cathodal tDCS condition. A constant current with an intensity of 1.0 mA was applied, with ramp up prior to the first tDCS phase and ramp down after the tDCS phase for several seconds. Three conditions (anodal tDCS condition: anodal electrode placed on the right M1; cathodal tDCS condition: cathodal electrode placed on the right M1; sham condition: stimulation occurred only for the first 30 s) were tested at random on separate days.

2.2 NIRS

A multichannel NIRS imaging system (OMM-3000; Shimadzu Co., Kyoto, Japan) with three wavelengths (780, 805, and 830 nm) was used to perform the measurements. NIRS optodes, consisting of 12 light-source fibers and 12 detectors providing 34-channel simultaneous recording, were set in the double density setting by joining two 3×8 multichannel probe holders positioned over the left PMC, SMA, left M1, and left primary sensory cortex (S1). The measurement channel included all 34 channels. The Cz position of the International 10–20 system was used to ensure consistent optode placement among all subjects. To remove any interference effects due to hair at the region of interest, we pushed the hair that was located under the sensors aside using an earpick to be able to expose the scalp.

2.3 Mean Arterial Pressure (MAP), Heart Rate (HR), and Skin Blood Flow (SBF)

MAP, which may influence the NIRS signal [5], and HR were recorded by volume clamping the finger pulse with finger photoplethysmography (Finometer; Finapres Medical Systems, Amsterdam, The Netherlands) on the left middle finger. SBF changes, which may also influence the NIRS signal [6], were measured at the forehead using a laser Doppler blood flow meter (Omegaflow FLO-CI; Omegawave Inc., Osaka, Japan). These analog data were converted into digital data with an A/D converter (PowerLab; AD Instruments, Australia) at a 4000-Hz sampling rate.

2.4 Analysis

The average MAP, HR, and SBF, as well as the average O_2Hb in each area were expressed as changes from the average of the 3 min baseline period acquired before stimulation onset. These parameters were averaged every 30 s. Differences were assessed with two-way analyses of variance (ANOVAs) (IBM SPSS, Armonk, NY, USA), with significance set at $p < 0.05$.

3 Results

3.1 MAP, HR, and SBF

The two-way ANOVAs for MAP, HR, and SBF did not show any significant differences (time × stimulation [anodal, cathodal, sham]).

3.2 Regional O_2Hb Changes

Compared to the sham condition, both anodal and cathodal stimulation resulted in significantly lower O_2Hb concentrations in the PMC, SMA, and M1 ($p < 0.01$). Particularly, in the SMA, the O_2Hb concentration during anodal stimulation was significantly lower than that during the sham condition ($p < 0.01$), while the O_2Hb concentration during cathodal stimulation was lower than that during anodal stimulation ($p < 0.01$). In addition, in the S1, the O_2Hb concentration during anodal stimulation was significantly higher than the concentrations during both cathodal stimulation and the sham condition ($p < 0.05$). The factor of time did not demonstrate significant differences (Fig. 42.1).

4 Discussion

One of the main findings of this study was that, compared to the sham condition, both anodal and cathodal tDCS stimulation over M1 resulted in significantly lower O_2Hb concentrations throughout the contralateral motor cortex with the exception of the S1. Another important finding was that the O_2Hb concentration during anodal stimulation was significantly higher than the concentrations during both cathodal stimulation and the sham condition, specifically in the S1.

Regarding the decrease in O_2Hb in the entire motor area, except the S1, a similar change that did not depend on polarity was observed in a precedent study [1]. Polanía et al. [7] reported that two directional types of tDCS electric current

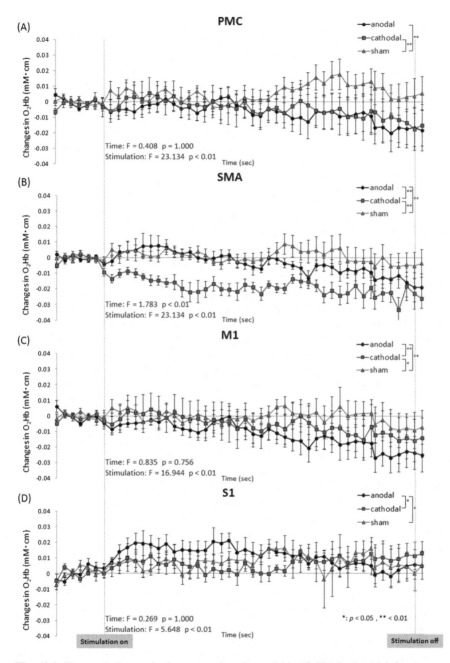

Fig. 42.1 Temporal changes in the averaged oxyhemoglobin (O₂Hb) in the (**a**) left premotor cortex (PMC), (**b**) supplemental motor area (SMA), (**c**) left primary motor cortex (M1), and (**d**) left primary sensory cortex (S1). *p < 0.05, **p < 0.01

changed the functional connections between different parts of the human brain; therefore, cerebral blood flow might be distributed to the reinforcement region. This suggests the possibility that O_2Hb decreased in the region where the connection was not strengthened. Although we did not measure the brain blood flow in this study, the fact that the MAP did not change during stimulation may support this hypothesis.

Another possibility that does not depend on the polarity of tDCS is that the electricity applied to the brain may directly influence the diameter of blood vessels. Spontaneous firing changes resulting from alterations in the neuronal resting membrane potential are thought to underlie the effects of tDCS [8]. An astrocytic function that coordinates with neurotransmitters and ions outside of the cell may participate in changing this neuronal resting membrane potential. Ruohonen and Karhu [9] suggested that theoretical calculations indicated that tDCS can affect the glial transmembrane potential. In addition, the astrocytes function to regulate the cerebral blood flow by expanding the blood vessels. If astrocytic depolarization occurred by tDCS in this study, then it may have influenced the results. While the astrocyte responses for polarity are unclear, the neuromodulatory effect of tDCS could be secondary to changes in astrocytic activity.

On the other hand, it was expected that sensory input (pain and itching) during the anodal stimulation may be the cause of the increased O_2Hb concentration, which was specific for the S1 during the anodal condition. However, the side effects of pain and itching rarely occur during tDCS, and no study has reported their influence on brain activity [10, 11]. Thus, it is difficult to draw conclusions based only on our result.

In conclusion, these results suggest that both anodal and cathodal tDCS cause widespread changes in cerebral blood flow, not only in the areas immediately under the electrode, but also in other areas of the cortex. Furthermore, anodal stimulation of M1 results in O_2Hb changes, specifically in the S1.

Acknowledgments This study was supported by a Grant-in-Aid for Challenging Exploratory Research 24650335 from the Japan Society for the Promotion of Science.

References

1. Nitsche MA, Paulus W (2000) Excitability changes induced in the human motor cortex by weak transcranial direct current stimulation. J Physiol 527(3):633–639
2. Merzagora AC, Foffani G, Panyavin I et al (2010) Prefrontal hemodynamic changes produced by anodal direct current stimulation. Neuroimage 49(3):2304–2310
3. Jang SH, Ahn SH, Byun WM et al (2009) The effect of transcranial direct current stimulation on the cortical activation by motor task in the human brain: an fMRI study. Neurosci Lett 460 (2):117–120
4. Kwon YH, Ko MH, Ahn SH et al (2008) Primary motor cortex activation by transcranial direct current stimulation in the human brain. Neurosci Lett 435(1):56–59
5. Franceschini MA, Joseph DK, Huppert TJ et al (2006) Diffuse optical imaging of the whole head. J Biomed Opt 11(5):054007

6. Kohno S, Miyai I, Seiyama A et al (2007) Removal of the skin blood flow artifact in functional near-infrared spectroscopic imaging data through independent component analysis. J Biomed Opt 12(6):062111

7. Polanía R, Paulus W, Antal A, Nitsche MA (2011) Introducing graph theory to track for neuroplastic alterations in the resting human brain: a transcranial direct current stimulation study. Neuroimage 54(3):2287–2296

8. Bikson M, Inoue M, Akiyama H et al (2004) Effects of uniform extracellular DC electric fields on excitability in rat hippocampal slices in vitro. J Physiol 557(Pt 1):175–190

9. Ruohonen J, Karhu J (2012) tDCS possibly stimulates glial cells. Clin Neurophysiol 123 (10):2006–2009

10. Nitsche MA, Liebetanz O, Lang N et al (2003) Safety criteria for transcranial direct current stimulation (tDCS) in humans. Clin Neurophysiol 114(11):2220–2222

11. Nitsche MA, Niehaus L, Hoffmann KT et al (2004) MRI study of human brain exposed to weak direct current stimulation of the frontal cortex. Clin Neurophysiol 115:2419–2423

Chapter 43
Relationships Between Gum-Chewing and Stress

Michiyo Konno, Tomotaka Takeda, Yoshiaki Kawakami, Yoshihiro Suzuki, Yoshiaki Kawano, Kazunori Nakajima, Takamitsu Ozawa, Keiichi Ishigami, Naohiro Takemura, and Kaoru Sakatani

Abstract Studies have shown that chewing is thought to affect stress modification in humans. Also, studies in animals have demonstrated that active chewing of a wooden stick during immobilization stress ameliorates the stress-impaired synaptic plasticity and prevents stress-induced noradrenaline release in the amygdala. On the other hand, studies have suggested that the right prefrontal cortex (PFC) dominates the regulation of the stress response system, including the hypothalamic-pituitary-adrenal (HPA) axis. The International Affective Digitized Sounds-2 (IADS) is widely used in the study of emotions and neuropsychological research. Therefore, in this study, the effects of gum-chewing on physiological and psychological (including PFC activity measured by NIRS) responses to a negative stimulus selected from the IADS were measured and analyzed. The study design was approved by the Ethics Committee of Tokyo Dental College (No. 436).

We studied 11 normal adults using: cerebral blood oxygenation in the right medial PFC by multi-channel NIRS; alpha wave intensity by EEG; autonomic nervous function by heart rate; and emotional conditions by the State-Trait Anxiety Inventory (STAI) test and the 100-mm visual analogue scale (VAS). Auditory stimuli selected were fewer than 3.00 in Pleasure value. Sounds were recorded in 3 s and reproduced at random using software. Every task session was designed in a block manner; seven rests: Brown Noise (30 s) and six task blocks: auditory stimuli or auditory stimuli with gum-chewing (30 s). During the test, the participants' eyes

M. Konno (✉) • T. Takeda • Y. Kawakami • Y. Suzuki • Y. Kawano • K. Nakajima • T. Ozawa • K. Ishigami
Department of Oral Health and Clinical Science, Division of Sports Dentistry, Tokyo Dental College, 2-9-18, Misaki, Chiyoda 101-0061, Japan
e-mail: mkonno@tdc.ac.jp

N. Takemura
Department of Electrical and Electronics Engineering, Nihon University, NEWCAT Institute, College of Engineering, Fukushima, Japan

K. Sakatani
Department of Electrical and Electronics Engineering, Nihon University, NEWCAT Institute, College of Engineering, Fukushima, Japan

Department of Neurological Surgery, School of Medicine, Nihon University, Tokyo, Japan

© Springer Science+Business Media, New York 2016
C.E. Elwell et al. (eds.), *Oxygen Transport to Tissue XXXVII*, Advances in Experimental Medicine and Biology 876, DOI 10.1007/978-1-4939-3023-4_43

were closed. Paired Student's t-test was used for the comparison (P < 0.05). Gum-chewing showed a significantly greater activation in the PFC, alpha wave appearance rate and HR. Gum-chewing also showed a significantly higher VAS score and a smaller STAI level indicating 'pleasant'. Gum-chewing affected physiological and psychological responses including PFC activity. This PFC activation change might influence the HPA axis and ANS activities. In summary, within the limitations of this study, the findings suggest that gum-chewing reduced stress-related responses. Gum-chewing might have a possible effect on stress coping.

Keywords Prefrontal cortex • Near-infrared spectroscopy • International Affective Digitized Sounds-2 • Electroencephalogram • Gum-chewing

1 Introduction

The basic function of mastication is to make food soft, smaller, and to mix it with enzymes in saliva for swallowing and digestion. However, many people chew gum for relaxation and concentration. Studies have shown that gum-chewing is thought to affect both physical and psychological stress modification in humans [1–4]. Also, studies in animal have demonstrated that active chewing of a wooden stick during immobilization stress, ameliorates the stress-impaired synaptic plasticity [5]. However, the effect of chewing gum on stress has not received unequivocal support and the neurophysiological mechanisms involved are unclear.

The prefrontal cortex (PFC) is the most sensitive to the effects of stress exposure [6]. The stress response involves activation of the PFC which stimulates the hypothalamic-pituitary-adrenal (HPA) axis and influences Autonomic nervous system (ANS), since neuronal networks exist between the PFC and the neuroendocrine centers in the medial hypothalamus [7], and the PFC has direct access to sympathetic and/or parasympathetic motor nuclei in brainstem and spinal cord. The PFC will set the endocrine/autonomic balance, depending on the emotional status [7]. The left hemisphere is specialized for the processing of positive emotions, while the right hemisphere is specialized for the processing of negative emotions [8].

The International Affective Digitized Sounds-2 (IADS) is a standardized database of 167 naturally occurring sounds, which is widely used in the study of emotions. The IADS is part of a system for emotional assessment developed by the Center for Emotion and Attention [9, 10]. Studies using IADS stimuli have revealed that auditory emotional stimuli activate the appetitive and defensive motivational systems similar to the way that pictures do.

Therefore, in this study, the effects of gum-chewing on a negative stimulus selected from the IADS on the right medial PFC activity using NIRS and other physiological responses, were measured and analyzed.

2 Materials and Methods

A total of 11 healthy volunteers participated in the study (mean age, 26.8 ± 1.66 years, male: female $= 9$: 2). Participants were told to refrain from substances (e.g., coffee etc. including caffeine) that could affect their nervous system before and during the period of testing, and not to eat for 2 h before the test. They were also instructed to avoid excessive drinking and lack of sleep the night before the test. In order to avoid the influence of environmental stress, the participants were seated in a comfortable chair in an air-conditioned room with temperature and humidity maintained at approximately 25 °C and 50 %, respectively. The study was conducted in accordance with the Principles of the Declaration of Helsinki, and the protocol was approved by the Ethics Committee of Tokyo Dental College (Ethical Clearance NO.436). Written informed consent was obtained from all participants. After a 10-min rest, they then performed an auditory stimulation task with eyes closed, negative sounds (NS) selected from the IADS, the NS were fewer than 3.0 in Pleasure value, in random order (Fig. 43.1). Activity in the right PFC was measured by a multi-channel NIRS; (OEG-16, Spectratech, Japan). The source-detector distance of the NIRS device is 30 mm. The locations of the shells were determined on the international 10–20 system. The most inferior channel was located at Fp2. The region of interest was placed at medial right PFC [8]. Electroencephalogram (EEG) (Muse Brain System, Syscom, Japan) and heart rate (HR) (WristOx, NONIN, USA) were monitored simultaneously with PFC activity (Fig. 43.2). The self-rated psychological measurement was taken using a 100-mm visual analogue scale (VAS) [11] and we used the State-Trait Anxiety Inventory (STAI) [12] to assess psychological assessments. We analyzed the Oxy-Hb values during task averaged across three channels on the medial right PFC. Alpha wave (8–13 Hz) appearance rates in theta, alpha and beta waves were calculated. Statistical evaluations between NS and NS with Gum-chewing, were performed using a paired Student's t-test (Excel Statistics, Microsoft Japan). A p-value of <0.05 was considered significant.

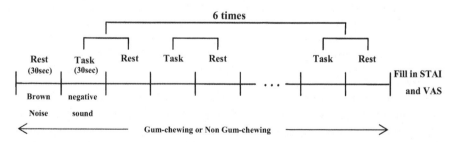

Fig. 43.1 Experimental protocol trial consisted of seven rests (Smoothed Brown Noise, 30 s) and six blocks of task: negative sounds from IADS

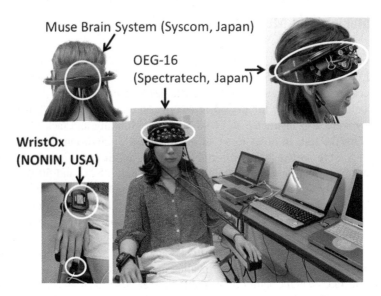

Fig. 43.2 PFC activity was measured by a multi-channel NIRS. The region of interest was placed at medial right PFC across three channels. EEG and HR were monitored simultaneously with NIRS

3 Results

Results are summarized in Table 43.1. PFC activity show a significantly greater activation with gum-chewing in the right PFC (NS = 0.045 ± 0.032, NS with Gum = 0.084 ± 0.055 µmol/l). A significantly greater alpha wave appearance rate (NS = 44.00 ± 0.06, NS with Gum = 47.10 ± 0.08 %) and HR (NS = 61.91 ± 8.57, NS with Gum = 67.83 ± 7.26 bpm) was obtained in gum-chewing. The STAI level tended to show smaller values in gum-chewing (NS = 2.55 ± 1.04, NS with Gum = 2.45 ± 0.82), and a significantly higher VAS score was obtained in gum-chewing indicating 'pleasant' (NS = 40.36 ± 15.95, NS with Gum = 54.80 ± 15.82).

4 Discussion

Judged mainly by the psychological results, the subjects felt discomfort during NS, which could cause stress responses in the brain. Indeed, NS increased Oxy-Hb in the right medial PFC, indicating that NS induced neural activation of the PFC. This is in line with previous findings used loud noise [16]. The unpleasant task-induced PFC activation could cause activation of the HPA axis and influence ANS, on the basis of networks between the PFC and the medial hypothalamus and ANS [7].

Table 43.1 Psychological and physiological assessment

			Stimuli of site	Mean (S.D.)	Statistics
Psychological assessment	STAI level (bigger figure indicates "more unpleasant")		Negative sound	2.55 (1.04)	
			Negative sound + Gum	2.45 (0.82)	
	VAS level (smaller figure indicates "more unpleasant")		Negative sound	40.36 (15.95)	*
			Negative sound + Gum	54.80 (15.82)	
Physical assessment	Heart rate (bpm)		Negative sound	61.91 (8.57)	*
			Negative sound + Gum	67.83 (7.26)	
	Alpha wave (%) (appearance rate in theta, alpha, and beta waves)		Negative sound	44.00 (0.06)	*
			Negative sound + Gum	47.10 (0.08)	
	NIRS $\varDelta$Oxy-Hb (a.u.) Approximately stabilized 15 s date in task—8 s of pre rest	Right PFC	Negative sound	0.045 (0.032)	*
			Negative sound + Gum	0.084 (0.055)	

*$p < 0.05$

Gum-chewing reduced the unpleasant psychological feeling and increased the alpha wave appearance rate, indicating the subjects' unpleasant feeling with coping. Gum-chewing increased HR and decreased PFC activity. This PFC activation change might influence the HPA axis and ANS activities. These gum-chewing effects in stress responses are also consistent with those of previous reports regarding the relationship between gum-chewing and stress [1–5, 16].

EEG reflects neuronal activities of a human brain that can be directly affected by emotional states. Especially, alpha wave appearance in the awaked EEG with an eyes-closed condition seems to indicate a relaxed state [13]. Listening to an unpleasant tone [14] and a loud siren reduced the alpha wave [15]. The effect of the increased alpha wave by gum-chewing might indicate a reduction of the uncomfortable mood.

Gum-chewing itself increased HR [16]. In the present study, NS with gum-chewing increased HR. The HR increase could relate to PFC blood flow. The blood supply change might have an influence on emotional control.

Gum-chewing reduced negative psychological scores on the VAS and STAI, indicating relief of mental stress.

Two possible interpretations for this Gum-chewing in the subjects being distracted from the unpleasant sounds was: Gum-chewing or some other activity may shield the organism from the external stressor (i.e., the noise) through attention distraction and reduces the experience of stress. Gum-chewing generates internal noise within the auditory system which may itself partially mask the impact of external noise or distract attention away from external noise [4].

The NIRS technique has some shortcomings; NIRS can detect only in surface areas of the brain, NIRS measurement has possible confounders such as skin blood flow, respiration and blood pressure, NIRS has relatively low spatial resolution. Despite its shortcomings, NIRS is becoming increasingly useful in neuroscience.

In summary, within the limitations of this study, the findings suggest that gum-chewing reduced stress-related responses, and there was a gum-chewing-induced change in the level of PFC activity. Gum-chewing might have a possible effect on stress coping [1–4].

Acknowledgments This research was partly supported by Japan Science and Technology Agency, under Strategic Promotion of Innovative Research and Development Program, and a Grant-in-Aid from the Ministry of Education, Culture, Sports, Sciences and Technology of Japan (Grant-in-Aid for Scientific Research 22592162, 25463025, and 25463024, Grant-in-Aid for Exploratory Research 25560356), and grants from Alpha Electron Co., Ltd. (Fukushima, Japan) and Iing Co., Ltd. (Tokyo, Japan).

References

1. Scholey A, Haskell C, Robertson B et al (2009) Chewing gum alleviates negative mood and reduces cortisol during acute laboratory psychological stress. Physiol Behav 97(3–4):304–312
2. Smith AP, Chaplin K, Wadsworth E (2012) Chewing gum, occupational stress, work performance and wellbeing. An intervention study. Appetite 58(3):1083–1086
3. Kamiya K, Fumoto M, Kikuchi H et al (2010) Prolonged gum chewing evokes activation of the ventral part of prefrontal cortex and suppression of nociceptive responses: involvement of the serotonergic system. J Med Dent Sci 57(1):35–43
4. Yu H, Chen X, Liu J et al (2013) Gum chewing inhibits the sensory processing and the propagation of stress-related information in a brain network. PLoS One 8(4):e57111
5. Ono Y, Yamamoto T, Kubo KY et al (2010) Occlusion and brain function: mastication as a prevention of cognitive dysfunction. J Oral Rehabil 37(8):624–640
6. Arnsten AF (2009) Stress signalling pathways that impair prefrontal cortex structure and function. Nat Rev Neurosci 10(6):410–422
7. Buijs RM, Van Eden CG (2000) The integration of stress by the hypothalamus, amygdala and prefrontal cortex: balance between the autonomic nervous system and the neuroendocrine system. Prog Brain Res 126:117–132
8. Coan JA, Allen JJ (2004) Frontal EEG asymmetry as a moderator and mediator of emotion. Biol Psychol 67(1–2):7–49
9. Soares AP, Pinheiro AP, Costa A et al (2013) Affective auditory stimuli: adaptation of the International Affective Digitized Sounds (IADS-2) for European Portuguese. Behav Res Methods 45(4):1168–1181

10. Stevenson RA, James TW (2008) Affective auditory stimuli: characterization of the International Affective Digitized Sounds (IADS) by discrete emotional categories. Behav Res Methods 40(1):315–321
11. Folstein MF, Luria R (1973) Reliability, validity, and clinical application of the Visual Analogue Mood Scale. Psychol Med 3(4):479–486
12. Demirbas H, Ilhan IO, Dogan YB et al (2011) Assessment of the mode of anger expression in alcohol dependent male inpatients. Alcohol Alcohol 46(5):542–546
13. Davidson RJ, Jackson DC, Kalin NH et al (2000) Emotion, plasticity, context, and regulation: perspectives from affective neuroscience. Psychol Bull 126(6):890–909
14. Nishifuji S (2011) EEG recovery enhanced by acute aerobic exercise after performing mental task with listening to unpleasant sound. Conf Proc IEEE Eng Med Biol Soc 2011:3837–3840
15. Horii A, Yamamura C, Katsumata T (2004) Physiological response to unpleasant sounds. J Int Soc Life Inform Sci 22(2):536–544
16. Suzuki M, Ishiyama I, Takiguchi T et al (1994) Effects of gum hardness on the response of common carotid blood flow volume, oxygen uptake, heart rate and blood pressure to gum-chewing. J Masticat Health Soc 4(1):51–62

Chapter 44
Effects of Anodal High-Definition Transcranial Direct Current Stimulation on Bilateral Sensorimotor Cortex Activation During Sequential Finger Movements: An fNIRS Study

Makii Muthalib, Pierre Besson, John Rothwell, Tomas Ward, and Stephane Perrey

Abstract Transcranial direct current stimulation (tDCS) is a non-invasive electrical brain stimulation technique that can modulate cortical neuronal excitability and activity. This study utilized functional near infrared spectroscopy (fNIRS) neuroimaging to determine the effects of anodal high-definition (HD)-tDCS on bilateral sensorimotor cortex (SMC) activation. Before (Pre), during (Online), and after (Offline) anodal HD-tDCS (2 mA, 20 min) targeting the left SMC, eight healthy subjects performed a simple finger sequence (SFS) task with their right or left hand in an alternating blocked design (30-s rest and 30-s SFS task, repeated five times). In order to determine the level of bilateral SMC activation during the SFS task, an Oxymon MkIII fNIRS system was used to measure from the left and right SMC, changes in oxygenated (O_2Hb) and deoxygenated (HHb) haemoglobin concentration values. The fNIRS data suggests a finding that compared to the Pre condition both the "Online" and "Offline" anodal HD-tDCS conditions induced a significant reduction in bilateral SMC activation (i.e., smaller decrease in HHb) for a similar motor output (i.e., SFS tap rate). These findings could be related to anodal HD-tDCS inducing a greater efficiency of neuronal transmission in the bilateral SMC to perform the same SFS task.

Keywords Functional near-infrared spectroscopy • tDCS • Neuroplasticity • Neuromodulation • Sensorimotor cortex

M. Muthalib (✉) • P. Besson • S. Perrey
Movement to Health (M2H) Laboratory, EuroMov, University of Montpellier, Montpellier, France
e-mail: makii.muthalib@gmail.com; makii.muthalib@univ-montp1.fr

J. Rothwell
Institute of Neurology, University College London, London, UK

T. Ward
Department of Electronic Engineering, National University of Ireland, Maynooth, Ireland

© Springer Science+Business Media, New York 2016
C.E. Elwell et al. (eds.), *Oxygen Transport to Tissue XXXVII*, Advances in Experimental Medicine and Biology 876, DOI 10.1007/978-1-4939-3023-4_44

1 Introduction

Transcranial direct current stimulation (tDCS) is a non-invasive electrical brain stimulation technique that applies mild (1–2 mA) direct currents over time (10–20 min) via the scalp to increase (anodal tDCS) or decrease (cathodal tDCS) cortical neuronal excitability [1]. The subsequent increase in spontaneous neuronal firing rates (during "Online" tDCS), coupled with synaptic neuroplasticity ("Online" and after "Offline" tDCS), contribute to anodal tDCS effects of increasing cortical excitability [1].

In order to increase sensorimotor cortex (SMC) excitability, tDCS is conventionally applied using two large (~35 cm^2) rubber-sponge electrodes with the anode electrode placed on a target region (i.e., SMC) and return electrode on the contralateral supraorbital region or non-cephalic region [2]. High-definition (HD)-tDCS is a recent approach that uses arrays of small EEG size (~3 cm^2) electrodes whose configuration can be optimized for more focal targeting of cortical regions determined using computational modeling of current flows between the electrodes [3]. Recently, anodal HD-tDCS (2 mA, 20 min) targeting the SMC (via a 4×1 electrode montage) was shown to induce Offline increases in resting corticospinal excitability assessed using transcranial magnetic stimulation [3]. However, it is not clear how Online and Offline anodal HD-tDCS modulates SMC activation during performance of a motor task.

An indirect marker of motor task-related SMC activation is the subsequent increase in the regional cortical blood flow and oxygenation (i.e., neurovascular coupling), which can be assessed using functional near infrared spectroscopy (fNIRS) neuroimaging [4]. fNIRS measures several physiological parameters related to cortical blood flow and oxygenation including measurements of changes in oxygenated (O$_2$Hb) and deoxygenated (HHb) haemoglobin concentration values [5]. Therefore, the aim of this study was to utilize fNIRS neuroimaging to measure bilateral SMC activation during a simple finger sequence (SFS) task in order to determine the Online and Offline effects of anodal HD-tDCS targeting the left SMC.

2 Methods

2.1 Subjects

Eight healthy subjects 30.4 ± 10.6 years (mean ± SD) participated in the study. All subjects were right handed as determined by the Edinburgh handedness questionnaire [6]. All subjects had no known health problems (e.g. metabolic or neuromuscular disorders) or any upper extremity muscle or joint injuries. The study conformed to the recommendations of the local Human Research Ethics Committee in accordance with the Declaration of Helsinki.

2.2 Protocol

Before (Pre), at 10 min during (Online), and 3 min after (Offline) anodal HD-tDCS (2 mA, 20 min) targeting the left SMC, subjects performed a self-paced SFS task (i.e., sequential tapping of the index, middle, ring and fourth finger against the thumb) with their right or left hand in an alternating blocked design (30-s rest and 30-s SFS task, repeated five times for each hand). Prior to the start of the experiment, subjects were familiarised with the SFS task in order to maintain a consistent rate of finger sequence taps (between 2 and 3 Hz), which was confirmed prior to the start of the Pre condition. The number of finger sequence taps was counted by the experimenter during each of the experimental SFS task blocks.

2.3 Experimental Setup

tDCS A Startim® tDCS system (Neuroelectrics, Spain) was used to deliver constant direct currents to the left SMC via a 4×1 anodal HD-tDCS electrode montage (active anode electrode at the centre surrounded by four return electrodes each at a distance of ~3.5 cm from the active electrode) [3]. The five electrodes (3.14 cm^2 AgCl electrodes) were secured on the scalp in the adjacent 10-10 EEG electrode system positions (anode: C3, and 4 return electrodes: FC1, FC5, CP1, CP5) using conductive paste (Ten20®, Weaver and Company, USA) and held in place using a specially designed synthetic cap to hold the HD-tDCS electrodes and fNIRS probes on the head (see Fig. 44.1 for layout).

fNIRS A continuous wave multi-channel Oxymon MkIII fNIRS system (Artinis Medical Systems, The Netherlands) was used to measure changes in bilateral SMC O_2Hb and HHb concentration values during the SFS task. Four receiver (avalanche photodiode) and 12 transmitter (pulsed laser diode) probes were placed in the synthetic cap to obtain 16 channels (each channel represented by a receiver-transmitter combination separated by ~3 cm) primarily covering the left (eight channels) and right (eight channels) SMC regions (see Fig. 44.1 for locations of the 16 channels). Two wavelengths (856 and 781 nm) per channel were used at a sampling rate of 10 Hz.

The changes in O_2Hb and HHb concentration values (expressed in µM), calculated according to a modified Beer-Lambert Law and including an age-dependent constant differential pathlength factor ($4.99 + 0.067 * \text{Age}^{0.814}$) [7], were transferred from the fNIRS system to a personal computer. During the data collection procedure, the time course of changes in O_2Hb and HHb concentration values were displayed in real time, and the signal quality and absence of movement artefacts were verified.

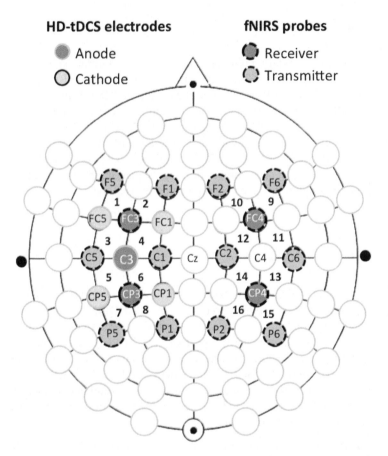

Fig. 44.1 Locations of the HD-tDCS electrodes and fNIRS probes on a 10-10 EEG electrode system layout. Each of the 16 fNIRS channels are represented by a receiver-transmitter combination

2.4 Data Analysis

The time course of changes in O_2Hb and HHb concentration values for each of the 16 channels were first low-pass filtered at 0.1 Hz to attenuate cardiac signal, respiration, and Mayer-wave systemic oscillations [5]. The time course of changes in O_2Hb and HHb concentration values for each SFS task block (30-s duration) were then normalized using the mean of the O_2Hb and HHb values measured during the last 5 s of the 30-s rest period preceding each SFS task block. These were then sample-to-sample averaged (i.e., 10 samples/s) the O_2Hb and HHb time course values over the 5 SFS task blocks, yielding one average O_2Hb and HHb time course for each subject.

In order to locate the channel to represent the level of SMC activation during the SFS task period for each subject, we first selected in the Pre condition one channel on the left and right SMC (i.e., the channel corresponding to one of the four channels located adjacent to the C3 and C4 electrode positions; see Fig. 44.1) showing peak and consistent SFS task-related haemodynamic responses (i.e., increase in O_2Hb and decrease in HHb), and then used the same channels for analysis in the Online and Offline conditions. Following the selection of the left and right SMC channel, we computed first the individual subject O_2Hb maximum (O_2Hb_{max}) and the HHb minimum (HHb_{min}) values from the left and right SMC and then group averaged these values for the Pre, Online and Offline conditions.

2.5   Statistical Analysis

For statistical analysis of the fNIRS dependent variables (O_2Hb_{max} and HHb_{min}), a Condition (Pre, Online, Offline) x Hemisphere (Left SMC, Right SMC) x Hand (Left, Right) repeated measures ANOVA was used. If a significant main or inter-action effect was evident, then post-hoc Tukey's HSD (honestly significant differ-ence) tests were performed. For statistical analysis of the behavioural dependent variable (SFS tap rate), a Condition (Pre, Online, Offline) $\times$ Hand (Left and Right) repeated measures ANOVA was used. Significance was set at $P \leq 0.05$. Data are presented as mean $\pm$ SD.

3   Results

The behavioural results indicated that subjects were able to perform the SFS task at a consistent rate (2.51 ± 0.32 Hz) with their right and left hand with no significant difference over the three conditions.

Figure 44.2 shows a typical time course of changes in O_2Hb and HHb from the left and right SMC during the right hand SFS task. Before anodal HD-tDCS (i.e., Pre condition), the right and left hand SFS task induced a cortical haemodynamic response (i.e., increase in O_2Hb and decrease in HHb) in the bilateral SMC, with a greater response in the contralateral hemisphere to the hand performing the task (see Fig. 44.2 for right hand SFS task).

Table 44.1 shows the group average O_2Hb_{max} and HHb_{min} values from the left and right SMC during the SFS task for the Pre, Online and Offline conditions. The ANOVA showed no significant Condition x Hemisphere x Hand interaction, but a significant ($p < 0.001$) Condition x Hemisphere interaction effect for both O_2Hb_{max} and HHb_{min}. The post-hoc showed that for the right SMC (i.e., unstimulated hemisphere), although there was no significant difference in O_2Hb_{max} over the three conditions, there was a significantly ($p < 0.001$) smaller HHb_{min} in both the Online and Offline conditions compared to Pre. For the left SMC (i.e., stimulated hemisphere), O_2Hb_{max} significantly ($p < 0.001$) increased in the Online condition compared to Pre, but returned to Pre levels in the Offline condition. In contrast,

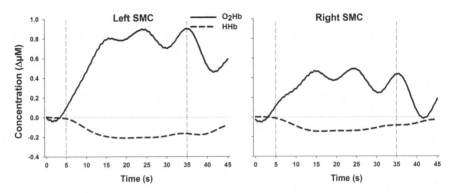

Fig. 44.2 Typical task-related changes in oxygenated (O_2Hb) and deoxygenated (HHb) haemoglobin concentrations in the *left* (Left SMC) and *right* (Right SMC) sensorimotor cortex during the right hand simple finger sequence task in the Pre condition. *Dashed vertical lines* denote the start and end of the 30-s SFS task period

there was a significantly ($p < 0.001$) smaller HHb_{min} during both the Online and Offline conditions compared to Pre.

4 Discussion

The main new finding of this study was of a significant reduction in bilateral SMC activation (based on smaller HHb_{min}) for a similar motor behaviour (i.e., SFS tap rate) in the Online and Offline conditions compared to the Pre condition.

Although O_2Hb_{max} increased significantly only in the Online condition for the stimulated left SMC, we consider that changes in O_2Hb were likely contaminated by anodal HD-tDCS induced local skin blood flow changes in the vicinity of the HD-tDCS electrodes. In contrast, changes in HHb are considered less affected by skin blood flow changes [8] and we found less variability in HHb responses during the five blocks of the SFS task than with O_2Hb responses. Therefore, we suggest that HHb may be a more reliable marker of HD-tDCS induced effects on task-related cortical activation.

The present study findings of smaller bilateral SMC HHb_{min} values during the SFS task in the Online and Offline conditions compared to Pre could be related to a greater efficiency of neuronal transmission [9] in the bilateral SMC (i.e., less synaptic input for the same neuronal output) that reduced SFS task-induced regional blood flow and thus produced smaller changes in fNIRS-derived HHb in the bilateral SMC. Furthermore, since the effect of anodal HD-tDCS on SFS task-related SMC activation was similar for both the Online and Offline conditions, it seems that synaptic neuroplastic modifications are necessary to induce these motor task-related reductions in SMC activation.

Table 44.1 Group mean ($\pm$SD) oxygenated haemoglobin maximum (O_2Hb_{max}) and deoxygenated haemoglobin minimum (HHb_{min}) concentration values from the left (Left SMC) and right (Right SMC) sensorimotor cortex during the simple finger sequence task performed before (Pre), during (Online) and after (Offline) anodal HD-tDCS

	Left SMC			Right SMC		
	Pre	Online	Offline	Pre	Online	Offline
O_2Hb_{max} ($\Delta\mu M$)	0.83 ± 0.28	$0.99 \pm 0.29^*$	0.83 ± 0.26	0.78 ± 0.38	0.76 ± 0.49	0.77 ± 0.40
HHb_{min} ($\Delta\mu M$)	-0.38 ± 0.14	$-0.33 \pm 0.11^*$	$-0.27 \pm 0.09^*$	-0.34 ± 0.14	$-0.29 \pm 0.17^*$	$-0.28 \pm 0.11^*$

*:$p < 0.001$; significantly different from Pre

In the present study, despite the attempt at focal stimulation to the left SMC by anodal 4×1 HD-tDCS, the effects on motor task-related cortical activation were bilateral, probably because intervening in one part of a distributed neural network system has effects on many nodes in the system [10]. It should also be noted that we found the same effect on bilateral SMC activation during SFS movements with the left and right hand. Although it would have been more expected to have observed a difference in ipsilateral and contralateral SMC activation between the left and right hand, evidence exists from recent studies which demonstrate that unilateral tDCS of the SMC can have bilateral effects [11, 12]. For example, Hendy et al. [12] have found bilateral changes in activation and muscle strength after anodal unilateral tDCS, and Roy et al. [12] found wide bihemispheric effects on EEG of unilateral HD-tDCS.

5 Conclusion

This preliminary study has shown for the first time that both Online and Offline anodal HD-tDCS reduced bilateral SMC activation to perform a sequential finger movement task. These positive initial results justify further research efforts to optimize the effects and enhance our understanding of the neurophysiological mechanisms of HD-tDCS-induced neuroplastic modifications.

References

1. Stagg CJ, Nitsche MA (2011) Physiological basis of transcranial direct current stimulation. Neuroscientist 17(1):37–53
2. Muthalib M, Kan B, Nosaka K, Perrey S (2013) Effects of transcranial direct current stimulation of the motor cortex on prefrontal cortex activation during a neuromuscular fatigue task: a fNIRS study. Adv Exp Med Biol 789:73–79
3. Kuo HI, Bikson M, Datta A, Minhas P, Paulus W, Kuo MF, Nitsche MA (2012) Comparing cortical plasticity induced by conventional and high-definition 4 x 1 ring tDCS: a neurophysiological study. Brain Stimul 6(4):644–648
4. Muthalib M, Anwar AR, Perrey S, Dat M, Galka A, Wolff S, Heute U, Deuschl G, Raethjen J, Muthuraman M (2013) Multimodal integration of fNIRS, fMRI and EEG neuroimaging. Clin Neurophysiol 124(10):2060–2062
5. Basso Moro S, Bisconti S, Muthalib M, Spezialetti M, Cutini S, Ferrari M, Placidi G, Quaresima V (2014) A semi-immersive virtual reality incremental swing balance task activates prefrontal cortex: a functional near-infrared spectroscopy study. Neuroimage 85 (Pt 1):451–460
6. Oldfield RC (1971) The assessment and analysis of handedness: the Edinburgh inventory. Neuropsychologia 9(1):97–113
7. Duncan A, Meek JH, Clemence M, Elwell CE, Tyszczuk L, Cope M, Delpy DT (1995) Optical pathlength measurements on adult head, calf and forearm and the head of the newborn infant using phase resolved optical spectroscopy. Phys Med Biol 40(2):295–304

8. Kirilina E, Jelzow A, Heine A, Niessing M, Wabnitz H, Bruhl R, Ittermann B, Jacobs AM, Tachtsidis I (2012) The physiological origin of task-evoked systemic artefacts in functional near infrared spectroscopy. Neuroimage 61(1):70–81
9. Holland R, Leff AP, Josephs O, Galea JM, Desikan M, Price CJ, Rothwell JC, Crinion J (2011) Speech facilitation by left inferior frontal cortex stimulation. Curr Biol 21(16):1403–1407
10. Lang N, Siebner HR, Ward NS, Lee L, Nitsche MA, Paulus W, Rothwell JC, Lemon RN, Frackowiak RS (2005) How does transcranial DC stimulation of the primary motor cortex alter regional neuronal activity in the human brain? Eur J Neurosci 22(2):495–504
11. Hendy AM, Kidgell DJ (2014) Anodal-tDCS applied during unilateral strength training increases strength and corticospinal excitability in the untrained homologous muscle. Exp Brain Res. doi:10.1007/s00221-014-4016-8
12. Roy A, Baxter B, He B (2014) High-definition transcranial direct current stimulation induces both acute and persistent changes in broadband cortical synchronization: a simultaneous tDCS-EEG study. IEEE Trans Biomed Eng 61(7):1967–1978

Part VII
Multimodal Imaging

Chapter 45
Towards Human Oxygen Images with Electron Paramagnetic Resonance Imaging

Boris Epel, Gage Redler, Victor Tormyshev, and Howard J. Halpern

Abstract Electron paramagnetic resonance imaging (EPRI) has been used to noninvasively provide 3D images of absolute oxygen concentration (pO_2) in small animals. These oxygen images are well resolved both spatially (~1 mm) and in pO_2 (1–3 mmHg). EPRI preclinical images of pO_2 have demonstrated extremely promising results for various applications investigating oxygen related physiologic and biologic processes as well as the dependence of various disease states on pO_2, such as the role of hypoxia in cancer.

Recent developments have been made that help to progress EPRI towards the eventual goal of human application. For example, a bimodal crossed-wire surface coil has been developed. Very preliminary tests demonstrated a 20 dB isolation between transmit and receive for this coil, with an anticipated additional 20 dB achievable. This could potentially be used to image local pO_2 in human subjects with superficial tumors with EPRI. Local excitation and detection will reduce the specific absorption rate limitations on images and eliminate any possible power deposition concerns. Additionally, a large 9 mT EPRI magnet has been constructed which can fit and provide static main and gradient fields for imaging local anatomy in an entire human. One potential obstacle that must be overcome in order to use EPRI to image humans is the approved use of the requisite EPRI spin probe imaging agent (trityl). While nontoxic, EPRI trityl spin probes have been injected intravenously when imaging small animals, and require relatively high total body injection doses that would not be suitable for human imaging applications. Work has been done demonstrating the alternative use of intratumoral (IT) injections, which can

B. Epel • G. Redler • H.J. Halpern (✉)
Center for EPR Imaging In Vivo Physiology, Chicago, IL 60637, USA

Department of Radiation and Cellular Oncology, University of Chicago, Chicago, IL 60637, USA
e-mail: h-halpern@uchicago.edu

V. Tormyshev
Center for EPR Imaging In Vivo Physiology, Chicago, IL 60637, USA

Novosibirsk Institute of Organic Chemistry and the University of Novosibirsk, Novosibirsk, Russia

© Springer Science+Business Media, New York 2016
C.E. Elwell et al. (eds.), *Oxygen Transport to Tissue XXXVII*, Advances in
Experimental Medicine and Biology 876, DOI 10.1007/978-1-4939-3023-4_45

reduce the amount of trityl required for imaging by a factor of 2000- relative to a whole body intravenous injection.

The development of a large magnet that can accommodate human subjects, the design of a surface coil for imaging of superficial pO_2, and the reduction of required spin probe using IT injections all are crucial steps towards the eventual use of EPRI to image pO_2 in human subjects. In the future this can help investigate the oxygenation status of superficial tumors (e.g., breast tumors). The ability to image pO_2 in humans has many other potential applications to diseases such as peripheral vascular disease, heart disease, and stroke.

Keywords Radiation therapy • Oxygen guided therapy • Oxygen imaging • EPR imaging

1   Introduction

EPR oxygen images have been shown to reproduce the ability of both the Eppendorf electrode and the more recent Oxylite quenching by molecular oxygen (O_2) of the decay of fluorescence excited by a short optical pulse of light [1]. However, as images, they provide much more information. The images provide an inventory of locations within a tumor of the subregions where O_2 is reduced: hypoxic subvolumes with fractions of its image voxels less than a threshold value of pO_2 less than a certain value, e.g., 10 mmHg, in this case referred to as the hypoxic fraction (HF) less than 10 mmHg (HF10). This is accomplished by infusing intravenously (IV) in mice, a nontoxic spin probe carrying an unpaired electron prepared in a very low magnetic field, 9 milliTesla (mT) and subject to linear field gradients. The rate at which the longitudinal magnetization of an unpaired spin relaxes from an excitation provided by a short (50 ns) pulse of 250 MHz radiofrequency is nearly absolutely proportional to the local concentration of O_2 through Heisenberg spin exchange with one of either of the unpaired O_2 electron spins [2]. Small animal experiments provide a proof of principle that EPR O_2 images can direct local therapies such as radiation to resistant portions of tumors, hypoxic subregions lacking O_2, that can be a major source of therapeutic failure [3].

The frequencies at which these experiments have been carried out are those used for a 6 T whole body MRI. This suggests that EPR technology can be applied to human subjects to enhance local radiation therapy. In this paper we suggest that the initial investigation of EPR O_2 imaging in the enhancement of radiation therapy will be in the derivation of local images, characterizing the oxygen physiology of localized tumors. Dealing with localized cancers with localized images is a natural starting point for the technology to minimize the dose of spin probe provided to human subjects and the applied specific (power) absorption rate (SAR).

2 Methods

Local EPR oxygen images provide near absolute measures of the pO_2 in each of the approximately 1 mm^3 voxels in the image. This is enabled by suffusing relevant tissue by the extracellular OX063d$_{24}$ trityl [2], whose spin lattice relaxation rates (R_1) report the average local oxygen concentration. Preparation of the trityl electron spins is accomplished with a low main magnetic field, 9 mT, with an excitation frequency of 250 MHz [4]. For the work at our center, EPR imaging is accomplished with fixed stepped magnetic field gradients, currently possible with our large imaging system, capable of accommodating human size samples. At each gradient angle, the electron spins are subjected to the sequence of $\pi - \pi/2 - \pi$ pulses, referred to as Inversion Recovery with Electron Spin Echo (IRESE) readout. This allows acquisition of magnetization decay profiles representing the superposition of the decays of the trityl electrons shifted in planes perpendicular to the direction of the magnetic field gradient. The Fourier transform of these profiles in turn provides a frequency display of the projection which, with a complete set of gradient angles for each temporal decay, can provide an image of the magnetization amplitude in each voxel, reconstructed using filtered back projection. A sequence of eight T delays allows the reconstruction of the voxel spin lattice relaxation rate, calibrated absolutely proportional to voxel pO_2. Confounding variation by trityl concentration or viscosity is less than the error of the voxel measurement, 2 mmHg [2]. To give anatomic meaning to the oxygen images, they must be registered with images that provide high anatomic resolution, such as T_2-weighted MRI. A particular advantage of defining anatomy using MRI (Fig. 45.1B) is the increase in free, slowly relaxing water within tumor tissue. Registration with EPR pO_2 images allows definition of the tumor boundary within the pO_2 image, providing a full inventory of the tumor pO_2 values, many of which show steep gradients, as much as 50 mmHg/mm (Fig. 45.1C).

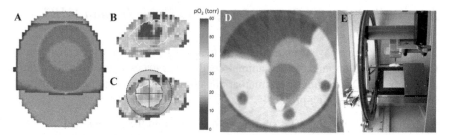

Fig. 45.1 (**A**) General tumor radiation (**B**) White contour is defined by MRI T_2 weighted image. (**C**) Boost to hypoxic tumor (dotted red region) or well oxygenated shell (darkened shell). (**D**) CT scan of tumor with hypoxic boost. (**E**) Mouse IMRT radiator

2.1   Animal Imaging and Radiation Treatment

In mice, the entire animal is injected, IV, with the 1475 D spin probe OX63d$_{24}$ trityl or a dose of approximately 50 mg/25 g animal, 2 g/kg [2]. Tumors are imaged with the IRESE sequence delivering approximately 0.06 W to a 2 mL tumor bearing leg (30 W/kg) in a 21 G animal, in a 19 mm diameter by 15 mm resistively Q-spoiled (from 100 to 14) loop gap resonator, with tumor temperature measurements showing no measurable increase within 0.5 °C. This SAR with minimal temperature rise argues for the correct reduction in SAR limit for limited portions of the animal radiated.

2.2   Mouse Intensity Modulated Radiation Therapy (IMRT)

IMRT to a mouse is enabled using an XRAD225 animal irradiator capable of delivering 5 Gy/min to small radiation fields configured to deliver spherical radiation volumes to tumor regions of diameters from 4 to 10.25 mm in 1.25 mm increments to cover either hypoxic or well oxygenated tumor subregions of similar volumes. Larger fields are given to whole tumor regions as part of a treatment course. The diameter of the tumor volume to be boosted is determined by choosing that which minimizes the distance between the curve shown in Fig 45.2B and the upper left value of the graph with 100 % hypoxic voxels and 0 % well oxygenated voxels. Tumors with dose boosts in this fashion are expected to be cured more frequently than those boosted with a shell of dose of similar volume to well oxygenated tumor voxels shown in the lower panel of Fig 45.1C. Data are currently being acquired to validate this hypothesis.

2.3   Techniques for Oxygen Imaging to Guide Human IMRT

Although 381 mice have been treated with the trityl OX063 or its deuterated analog OX063d$_{24}$, with no unexplained animal deaths (e.g. excessive anesthesia), initial approaches to O$_2$ imaging for human application should involve two major modifications of our imaging protocol:

1. Reduction of the total trityl spin probe dose given to the subject, and
2. Reduction of the whole subject specific absorption rate (SAR) or radiofrequency power deposition.

Reduction of total trityl spin probe dose can be accomplished using intratumoral spin probe injection. As an example of this in mouse tumors, we show in Fig. 45.3 the pO$_2$ distributions from the same tumors imaged via IV injection (Fig. 45.3A) and then, after a period of approximately 1 h, when the IV trityl has substantially

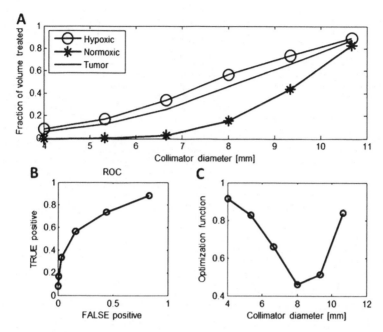

Fig. 45.2 (**A**) Plot of fraction of hypoxic or normoxic * cells vs spherical boost diameter. (**B**) Plot of hypoxic cell fraction (True Positive) vs fraction of well oxygenated cells for each boost volume diameter. (**C**) Distance to *upper left corner* in (**B**) whose minimum is optimum

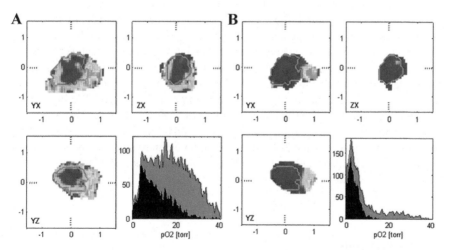

Fig. 45.3 Images and histograms of pO_2 in mouse leg (*blue histogram*) bearing tumor (*red contour, red histogram*) (**A**) after IV injection and (**B**) after IT injection of trityl spin probe

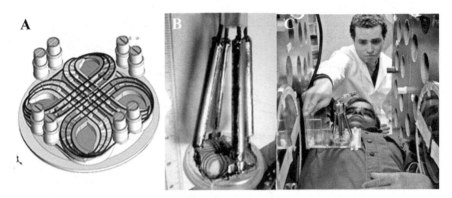

Fig. 45.4 (**A**) Scheme of bimodal surface coil with geometric mode decoupling; (**B**) Construction of bimodal surface coil; (**C**) placement of surface coil

cleared, more trityl is injected via intratumoral (IT) injection. The apparent pO_2 distribution from the IT injection is similar to the IV (Fig. 45.3B, red histograms), also seen in the additional half dozen animals in which this was tried. Early tumor stage for human tumors is often the case with maximum tumor linear dimension ≤ 2.0 cm (~4 mL volume). The next tumor size categorized by the American Joint Commission (AJC) for tumor staging involves tumors with dimension >2.0 and ≤ 5.0 cm in most cases (up to ~ 63 mL volume) so that an average tumor size can be taken as ~ 15 mL. For such tumor size and IT injection, the reduction in the subject dose for a 70 kg human subject is a factor of 5000. This translates to a subject dose of order 10s of nM rather than 100s of μM.

Local tumor excitation and detection using surface coils reduces the whole subject SAR. This is recognized in the SAR specifications for MRI where the maximum local SAR is relaxed with reduced fraction of the subject radiated. https://www.aapm.org/meetings/02AM/pdf/8356-48054.pdf. The local excitation and detection system shown in Fig. 45.4 provides without any optimization isolation between the excitation power and detection signal of 23 dB. We have found that with greater than 40 dB isolation eliminates virtually all of the baseline signal that requires off resonance subtraction. This should be easily achievable. Modelling of the radiofrequency sensitivity of the bimodal excitation/detection surface coils indicates good B_1 distributions to depths of 2–3 cm, making all T1 size tumors and many T2 size tumors accessible.

3   Conclusions

This combination of local trityl injection and local excitation and local detection reduces the risk of toxicity for the initial use of both spin probe and instrument power deposition to extremely low. Nanomolar concentrations of tri-acid spin probe, with extracellular distribution and rapid renal excretion and no murine toxicity argues that this substance approaches human tissue concentrations low enough to be below the threshold for concern.

Acknowledgments Supported by NIH grants P41 EB002034 and R01 CA98575.

References

1. Elas M, Bell R, Hleihel D et al (2008) Electron paramagnetic resonance oxygen image hypoxic fraction plus radiation dose strongly correlates with tumor cure in FSa fibrosarcomas. Int J Radiat Oncol Biol Phys 71:542–549
2. Epel B, Bowman MK, Mailer C, Halpern HJ (2014) Absolute oxygen R1e imaging in vivo with pulse electron paramagnetic resonance. Magn Reson Med 72:362–368
3. Overgaard J (2007) Hypoxic radiosensitization: adored and ignored. J Clin Oncol 25:4066–4074
4. Epel B, Sundramoorthy SV, Mailer C, Halpern HJ (2008) A versatile high speed 250-MHz pulse imager for biomedical applications. Concepts Magn Reson Part B Magn Reson Eng 33B:163–176

Chapter 46
Evaluation of Brown Adipose Tissue Using Near-Infrared Time-Resolved Spectroscopy

Shinsuke Nirengi, Takeshi Yoneshiro, Takeshi Saiki, Sayuri Aita, Mami Matsushita, Hiroki Sugie, Masayuki Saito, and Takafumi Hamaoka

Abstract Human brown adipose tissue (BAT) activity (SUV_{max}) has been typically evaluated by ^{18}F-fluorodeoxy glucose (FDG)–positron emission tomography (PET) combined with computed tomography (CT). In this study, the objective was to detect human BAT by near-infrared time-resolved spectroscopy (NIR_{TRS}), a noninvasive and simple method for measuring total hemoglobin concentration [total-Hb] and reduced scattering coefficient ($\mu s'$) in the tissue. The [total-Hb] in the supraclavicular region of the BAT (+) ($SUV_{max} \geq 2.0$) group was 95.0 ± 28.2 µM (mean +/− SD), which was significantly higher than that of the BAT (−) ($SUV_{max} < 2.0$) group (52.0 ± 14.8 µM), but not in other regions apart from the BAT deposits. The $\mu s'$ in the supraclavicular region of the BAT (+) group was 8.4 ± 1.7 cm^{-1}, which was significantly higher than that of BAT (−) group (4.3 ± 1.0 cm^{-1}), but not in other regions. The area under the receiver operating characteristic curve closest to (0, 1) for

S. Nirengi
Division of Preventive Medicine, Clinical Research Institute, National Hospital Organization Kyoto Medical Center, 1-1 Mukaihata-cho, Fukakusa, Fushimi-ku, Kyoto, Kyoto 612-8555, Japan

T. Yoneshiro
Department of Biomedical Sciences, Graduate School of Veterinary Medicine, Hokkaido University, Sapporo, Japan

T. Saiki
Graduate School of Sport and Health Science, Ritsumeikan University, Kusatsu, Japan

S. Aita
Department of Food and Nutrition, Hakodate Junior College, Hakodate, Japan

M. Matsushita
Department of Nutrition, School of Nursing and Nutrition, Tenshi College, Sapporo, Japan

H. Sugie
LSI Sapporo Clinic, Sapporo, Japan

M. Saito
Hokkaido University, Sapporo, Japan

T. Hamaoka (✉)
Department of Sports Medicine for Health Promotion, Tokyo Medical University, Shinjyuku, Japan
e-mail: kyp02504@nifty.com

© Springer Science+Business Media, New York 2016
C.E. Elwell et al. (eds.), *Oxygen Transport to Tissue XXXVII*, Advances in Experimental Medicine and Biology 876, DOI 10.1007/978-1-4939-3023-4_46

[total-Hb] and $\mu s'$ to discriminate BAT (+) from BAT (−) was 72.5 μM and 6.3 cm^{-1}, respectively. The sensitivity, specificity, and accuracy for both parameters were 87.5, 100, and 93.3 %, respectively. Our novel NIR$_{TRS}$ method is noninvasive, simple, and inexpensive compared with FDG-PET/CT, and is reliable for detecting human BAT.

Keywords ^{18}F-fluorodeoxy glucose • Positron emission tomography • Computed tomography • Total hemoglobin concentration • Scattering coefficient

1 Introduction

Brown adipose tissue (BAT) regulates whole-body energy expenditure and adiposity, and attracts attention as countermeasures of overweight [1]. Recent studies, using ^{18}F-fluorodeoxy glucose (FDG)–positron emission tomography (PET) combined with computed tomography (CT) scanning, demonstrated that, in humans, BAT exists in adults [2]. However, FDG-PET/CT has serious limitations, including the enormous cost of the device and radiation exposure. In addition, a 2-h cold exposure is needed to reveal BAT [1, 2].

Near-infrared time-resolved spectroscopy (NIR$_{TRS}$) is a method for determining optical properties such as absorption (μ_a) and reduced scattering coefficients ($\mu s'$), and [total-Hb] which are respective indices of tissue vasculature and mitochondria content [3, 4], as an absolute value in a noninvasive, simple, inexpensive approach [3]. Since BAT has abundant capillaries and mitochondria compared with white adipose tissue (WAT), NIR$_{TRS}$ is expected to be able to detect BAT. Thus, the purpose of this study was to detect BAT using NIR$_{TRS}$ in 15 healthy subjects.

2 Methods

2.1 Methods and Procedures

Fifteen healthy male subjects aged 20–29 years participated in this study. Subjects provided informed, written consent. The protocol was approved by the institutional reviewer board of Tenshi College and Ritsumeikan University. The body weight and body fat were determined using a bioelectric impedance method (HBF-361; Omron Healthcare, Kyoto, Japan). All measurements were performed in winter.

2.2 FDG-PET/CT

After 2-h cold exposure at 19 °C, BAT activity was evaluated by FDG-PET/CT (Aquiduo; Toshiba Medical Systems, Otawara, Japan) [1, 2]. The BAT activity in the supraclavicular region was quantified by calculating the maximal standardized uptake value (SUV_{max}), defined as the radioactivity per milliliter within the region of interest divided by the injected dose in MBq/g body weight. For dividing subjects into BAT-positive [BAT (+)] and BAT-negative [BAT (−)] groups, a cutoff value of 2.0 [1, 2] was applied.

2.3 NIR_{TRS}

The [total-Hb] and $\mu s'$ were measured using NIR_{TRS} (TRS-20; Hamamatsu Photonics K.K., Hamamatsu, Japan) for 5 min at 27 °C by placing the probes on the skin in the supraclavicular region (right side, n = 15; left side, n = 15) adjacent to the BAT deposits and also in the subclavicular (right side, n = 15; left side, n = 14) and deltoid muscle (right side, n = 14; left side, n = 14) regions which were apart from the BAT deposits. The distance between the emitter and detector was set at 30 mm. The NIR_{TRS} system provided data every 10 s.

2.4 Statistical Analyses

Data are expressed as mean ± SD. Bland-Altman analysis was done for comparing NIR_{TRS} parameters between the right and left side probes. Comparisons between the two groups were performed using the Student's t-test. Receiver operating characteristic (ROC) curves were constructed by plotting true-positive rates (sensitivity) against false-positive rates (1—specificity) to compare discriminatory accuracy for survival and event-free survival for [total-Hb] and $\mu s'$. The Pearson product–moment correlation analysis was used to determine the relationship between the NIR_{TRS} index and SUV_{max}. Values were considered to be statistically significant if P was < 0.05. All statistical analyses were performed using SPSS v19.

3 Results

In the supraclavicular region, we found eight subjects to be BAT (+) and seven to be BAT (−). The height, body weight, and body fat were 174.1 ± 5.8 cm, 62.0 ± 5.2 kg, 13.4 ± 1.9 % in the BAT (+) group and 173.3 ± 6.5 cm, 67.4 ± 8.5 kg, 16.2 ± 4.8 % in the BAT (−) group, respectively. No significant differences were observed in these anthropometric characteristics between the BAT (+) and BAT (−) groups.

There were no significant differences in the NIR_{TRS} parameters between the left and right side probes in all regions. The [total-Hb] in the supraclavicular region of the BAT (+) group was 95.0 ± 28.2 μM, which was significantly higher than that of the BAT (−) group (52.0 ± 14.8 μM) (Fig. 46.1a), but not in other regions (Fig. 46.1b, c). The $\mu s'$ in the supraclavicular region of the BAT (+) group was

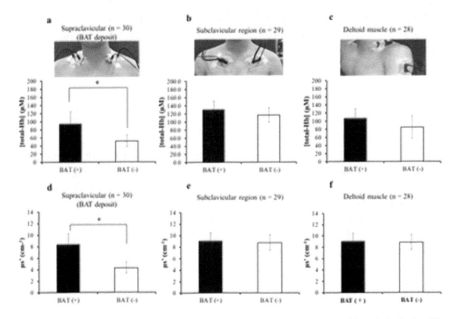

Fig. 46.1 (**a–c**) The total hemoglobin ([total-Hb]) in the supraclavicular (**a**), subclavicular (**b**), and deltoid (**c**) regions of the BAT-positive (BAT [+]) and BAT-negative (BAT [−]) groups. (**d–f**) The reduced scattering coefficient ($\mu s'$) in the supraclavicular (**a**), subclavicular (**b**), and deltoid (**c**) regions of the BAT (+) and BAT (−) groups. *(p < 0.01): significantly different from BAT (−) group

8.4 ± 1.7 cm^{-1}, which was significantly higher than that of the BAT (−) group (4.3 ± 1.0 cm^{-1}) (Fig. 46.1d), but not in other regions (Fig. 46.1e, f).

The area under the ROC curve closest to (0, 1) for [total-Hb] was 72.5 μM. If the cut-off value was set to 72.5 μM, the drawning diagnosis sensitivity was 87.5 %, with a specificity of 100 %, a positive predictive value (PPV) of 100 %, a negative predictive value (NPV) of 88.0 %, and an accuracy of 93.3 % (Fig. 46.2a). The area under the ROC curve closest (0, 1) for $\mu s'$ was 6.3 cm^{-1}. If this was set as the cut-off value, the drawning diagnosis sensitivity was 87.5 %, with a specificity of 100 %, a PPV of 100 %, an NPV of 88.0 %, and an accuracy of 93.3 % (Fig. 46.2b).

The [total-Hb] in the supraclavicular region was significantly correlated to the SUV_{max} ($r = 0.69$, $p < 0.01$) only in the BAT (+) group. There was no significant relationship between $\mu s'$ and SUV_{max} in the both groups.

4   Discussion

The major finding of this study was that we were able to detect human BAT by using NIR_{TRS} parameters such as [total-Hb] and $\mu s'$ with high sensitivity, specificity, and accuracy.

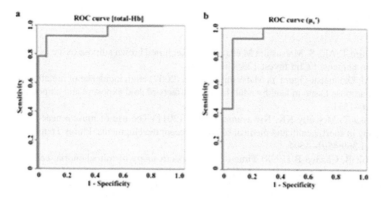

Fig. 46.2 (**a**) Receiver operating characteristic (ROC) curves at the total hemoglobin ([total-Hb]). The area under the ROC curve closest to (0, 1) for [total-Hb] was selected, and the ROC curve was plotted, if the cut-off value of [total-Hb] was set to 72.5 μM. (**b**) ROC curves at the reduced scattering coefficient (μs'). The area under ROC curve closet (0, 1) for μs' was selected, and the ROC curve was plotted, if the cut-off value of μs' was set to 6.3 cm^{-1}

When FDG-PET/CT is used for detecting BAT activity, it is a prerequisite for subjects to be acutely exposed to cold, because glucose/FDG uptake into BAT is activated by cold exposure [1, 2]. Indeed, no FDG uptake is observed under warm conditions even in subjects having BAT [2]. In contrast, [total-Hb] and μs' obtained from NIR_{TRS} are respective indices of vascularity and mitochondria content, which could be insensitive to acute temperature change. Thus, our NIR_{TRS} method is capable of detecting BAT at room temperature.

There is one previous study which evaluated human BAT using continuous wave NIRS [5]. The previous study evaluated only changes in oxygen saturation relative to deltoid muscle with cold exposure. However, that study did not find a significant difference ($p = 0.08$) in NIRS parameters between the BAT (+) and BAT (−) groups. Thus, ours is the first study which distinguished the BAT (+) group from the BAT (−) group using quantitative NIR_{TRS} without cold exposure.

Adiposity greatly influences the NIR signal intensity from the target tissue [3]. A correction curve is available for the influence of a fat layer thickness against the underlying muscle tissue from the results of simulation and in vivo tests [3]. However, in this study, since there is no difference in body fat between BAT (+) and BAT (−) groups, we did not have to consider the effect of the difference in adiposity on the NIRS parameters in these subjects. To enhance the reliability of this method, further validation is necessary in a wider range of individuals with varying adiposity.

5 Conclusion

In conclusion, NIR_{TRS} is able to detect human BAT. NIR_{TRS} is noninvasive, simple, inexpensive and, more importantly, it can be conducted without radiation exposure compared with FDG-PET/CT.

References

1. Yoneshiro T, Aita S, Matsushita M et al (2013) Recruited brown adipose tissue as an antiobesity agent in humans. J Clin Invest 123:3404–3408
2. Saito M, Okamatsu-Ogura Y, Matsushita M et al (2009) High incidence of metabolically active brown adipose tissue in healthy adult humans: effects of cold exposure and adiposity. Diabetes 58:1526–1531
3. Hamaoka T, McCully KK, Niwayama M et al (2011) The use of muscle near-infrared spectroscopy in sport, health and medical sciences: recent developments. Philos Trans A Math Phys Eng Sci 369:4591–4604
4. Beauvoit B, Chance B (1998) Time-resolved spectroscopy of mitochondria, cells and tissues under normal and pathological conditions. Mol Cell Biochem 184:445–455
5. Muzik O, Mangner TJ, Leonard WR et al (2013) ^{15}O PET measurement of blood flow and oxygen consumption in cold-activated human brown fat. J Nucl Med 54:523–531

Chapter 47
Near-Infrared Image Reconstruction of Newborns' Brains: Robustness to Perturbations of the Source/Detector Location

L. Ahnen, M. Wolf, C. Hagmann, and S. Sanchez

Abstract The brain of preterm infants is the most vulnerable organ and can be severely injured by cerebral ischemia. We are working on a near-infrared imager to early detect cerebral ischemia. During imaging of the brain, movements of the newborn infants are inevitable and the near-infrared sensor has to be able to function on irregular geometries. Our aim is to determine the robustness of the near-infrared image reconstruction to small variations of the source and detector locations. In analytical and numerical simulations, the error estimations for a homogeneous medium agree well. The worst case estimates of errors in reduced scattering and absorption coefficient for distances of r = 40 mm are acceptable for a single source-detector pair. The optical properties of an inhomogeneity representing an ischemia are reconstructed correctly within a homogeneous medium, if the error in placement is random.

Keywords Near-infrared imaging • Optical properties • Stability • Reconstruction • Optoacoustic imaging

1 Introduction

At the moment it is not possible to detect cerebral ischemia early and to thus prevent brain injury in preterm infants. Therefore we are developing a novel multimodal hybrid diagnostic imager, combining ultrasound, opto-acoustic [1, 2] and near-infrared imaging (NIRI) [3, 4] to quantitatively image anatomical and functional information (oxygen saturation) of the brain in real-time. The novel multimodal instrument will be applied at the bedside to non-invasively image the brain of a newborn infant.

L. Ahnen (✉) • M. Wolf • S. Sanchez
Biomedical Optics Research Laboratory, University Hospital Zurich, Zurich, Switzerland
e-mail: linda.ahnen@usz.ch

C. Hagmann
Division of Neonatology, University Hospital Zurich, Zurich, Switzerland

© Springer Science+Business Media, New York 2016
C.E. Elwell et al. (eds.), *Oxygen Transport to Tissue XXXVII*, Advances in
Experimental Medicine and Biology 876, DOI 10.1007/978-1-4939-3023-4_47

Near-infrared imaging functions properly in well-defined geometries, where analytical calculations can be performed. For measurements on irregular geometries, the errors in positioning have to be taken into account. The near-infrared sources and detectors will be placed in a circle around the opto-acoustic imaging probe. During imaging, movements of the newborn infants are inevitable and the near-infrared sensor has to function properly for the varying geometries of the newborns' heads. The stability of the NIRI sensor placement introduces an uncertainty in the measurement. The aim is to estimate the error in optical properties in reconstructed images in dependence of small deviations of the source and detector locations.

2 Methods

The error in the optical properties caused by variations in source/detector positioning is estimated first for a homogeneous material and then for an inhomogeneity representing an ischemic lesion within the bulk medium. For the homogeneous medium, this error is assessed analytically and numerically and for the inhomogeneity only numerically. The error depends on the optical properties, which have been measured in neonates [5].

Considering an infinite homogeneous medium (Fig. 47.1a), the diffusion approximation yields

$$G(r) = \frac{1}{4\pi D |r|} e^{ik_0 |r|}, \quad \text{with } k_0 = \sqrt{-\frac{\mu_a}{D} + \frac{i\omega}{Dc_0}} \tag{47.1}$$

Where k_0 is the wavenumber, $D = 1/(3\mu_s')$ is the diffusion coefficient, c_0 the speed of light in the medium, ω the frequency, r is the distance between source and detector, μ_a the absorption coefficient and μ_s' the reduced scatteringcoefficient.

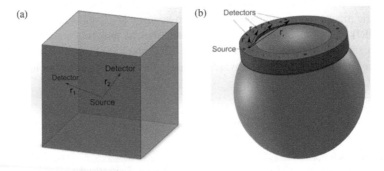

Fig. 47.1 (a) Infinite homogeneous medium. (b) Spherical homogeneous medium

The wavenumber can be decomposed into its real and imaginary part. With an approximation for small frequencies [6] the absorption and reduced scatteringcoefficients for normalized intensities become

$$\mu'_s \cong \frac{2c_0\varphi}{3\omega r^2}\ln\left(\frac{1}{Ar}\right), \ \mu_a \cong \frac{\omega}{2c_0\varphi}\ln\left(\frac{1}{Ar}\right) \tag{47.2}$$

Where φ is the phase and A is the amplitude.

The error in absorption and reduced scatteringcoefficient can now be calculated as

$$\varepsilon(\mu'_s) = \left(\frac{2\mu'_s}{r} + \frac{1}{r^2}\sqrt{\frac{\mu'_s}{3\mu_a}}\right)\varepsilon(r) \tag{47.3}$$

$$\varepsilon(\mu_a) = \frac{1}{r^2}\sqrt{\frac{\mu_a}{3\mu'_s}}\varepsilon(r) \tag{47.4}$$

For the numerical estimation of the error in a homogeneous medium the measurement setup displayed in Fig. 47.1b was applied. This is the future NIRI measurement setup, where sources and detectors will be placed around the OA sensor on the preterm infant's head. To simulate the measured amplitude and phase forward simulations are performed with NIRFAST [7, 8], a finite element software to simulate near-infrared light propagation in tissue. Various source-detector distances and absorption and reduced scatteringcoefficients are simulated.

The geometric setup shown in Fig. 47.4a is employed to numerically estimate the error in the absorption coefficient of an inhomogeneity within a homogeneous medium ($\mu_a = 0.005\,\text{mm}^{-1}$ and $\mu'_s = 0.35\,\text{mm}^{-1}$).

3 Results

Figure 47.2 illustrates the analytically estimated errors in (a) absorption and (b) reduced scatteringcoefficients for different errors in source detector distances. The source-detector distances on preterm infants' heads are assumed to be in the range of $r = 40$ mm. The achievable accuracy in source detector placement is assumed to be 1 mm. The analytically estimated errors for small frequencies are then $\varepsilon(\mu_a, r = 40mm, \mu'_s = 0.35mm^{-1}, \varepsilon(r) = 1mm) = 1\%$ and $\varepsilon(\mu'_s, r = 40mm, \mu_a = 0.005mm^{-1}, \varepsilon(r) = 1mm) = 6\%$.

The contour plot of the simulated logarithmic amplitude is shown in Fig. 47.3 for several (a) absorption and (b) reduced scatteringcoefficients. Considering the setup shown in Fig. 47.1b, the numerically determined errors (for a source detector distance of $r = 40$ mm and an error in positioning of 1 mm) are $\varepsilon(\mu'_s, r = 40mm, \mu_a \approx 0.005mm^{-1}, \varepsilon(r) = 1mm) \approx 4\%$ and $\varepsilon(\mu_a, r = 40mm, \mu'_s \approx 0.35mm^{-1}, \varepsilon(r) = 1mm) \approx 7\%$.

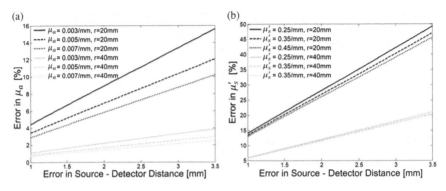

Fig. 47.2 Analytical error estimation: (**a**) error in absorption coefficient $\varepsilon(\mu_a)$ for fixed $\mu_s'=0.35\ \mathrm{mm}^{-1}$ (**b**) error in reduced scatteringcoefficient $\varepsilon(\mu_s')$ for fixed $\mu_a=0.005\ \mathrm{mm}^{-1}$

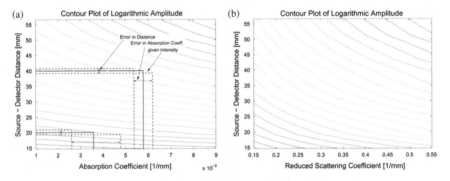

Fig. 47.3 Numerical error estimation: Contour plot of the logarithmic simulated amplitude for a single source-detector pair. (**a**) For different absorption coefficients and source-detector distances at fixed $\mu_s'=0.35\ \mathrm{mm}^{-1}$. (**b**) For different reduced scatteringcoefficient and source-detector distances at fixed $\mu_a=0.005\ \mathrm{mm}^{-1}$. The plots show how an error in the source-detector distance for a given intensity relates to an error in the optical properties for a homogeneous medium

For the inhomogeneity within a homogeneous medium the contour plot of the logarithmic amplitude (Fig. 47.4b) shows a high sensitivity for small perturbations in the source-detector distance. However simulations with several source-detector pairs tend to cancel the errors: By adding a second pair, the minimization problem, considering only two unknown areas, the bulk and the inhomogeneity becomes $\chi^2(\mu_a \pm \varepsilon(\mu_a)) = \left(lnA_M^{(1)}(\mu_a, r + \varepsilon(r)) - lnA_C^{(1)}(\mu_a \pm \varepsilon(r), r)\right)^2 + \left(lnA_M^{(2)}(\mu_a, r - \varepsilon(r)) - ln\ A_C^{(2)}(\mu_a \pm \varepsilon(r), r)\right)^2 \rightarrow min$ with the measured amplitude A_M and the calculated amplitude A_C. This leads to $\chi^2(\mu_a \pm \varepsilon(\mu_a\ r, A)) \approx 0,441*10^{-3}$ and $\chi^2(\mu_a) \approx 0,221*10^{-3}$, i.e. the correct absorption coefficient is reconstructed.

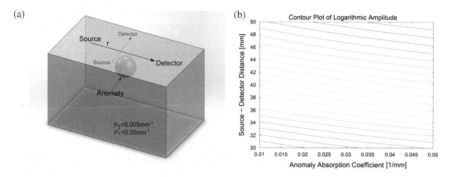

Fig. 47.4 (**a**) Geometry with inhomogeneity employed for simulations. (**b**) Contour plot of logarithmic simulated amplitude for different absorption coefficients and source-detector distances of a single source-detector pair, considering the geometry on the *left*

4 Conclusions

The development of a novel multimodal diagnostic imager faces various challenges. One major challenges the adaptation of the setup to a universal clinical application. Smaller source-detector distances are more sensitive to variations in the source and detector position than larger ones. Analytical and numerical errors for a homogeneous medium agree well and yield acceptable worst case errors for a single source-detector pair. The error in optical properties of an inhomogeneity within a homogeneous bulk material is sensitive to positioning perturbations for a single source-detector pair. However, the error cancels and the optical properties of the inhomogeneity are reconstructed correctly if the positioning error is random.

Acknowledgement This work was supported by the KFSP Tumor Oxygenation of the University of Zurich and SwissTransMed.

References

1. Beard P (2011) Biomedical photoacoustic imaging. Interface Focus 1:602–631. doi:10.1098/rsfs.2011.0028
2. Xu M, Wang LV (2006) Photoacoustic imaging in biomedicine. Rev Sci Instrum 77(4):041101
3. Arridge SR (1999) Optical tomography in medical imaging. Inverse Prob 15:R41–R93
4. Hebden JC, Austin T (2007) Optical tomography of the neonatal brain. Eur Radiol 17(11):2926–2933, Epub 2007/05/03
5. Arri SJ, Muehlemann T, Biallas M et al (2011) Precision of cerebral oxygenation and hemoglobin concentration measurements in neonates measured by near-infrared spectroscopy. J Biomed Opt 16(4):047005–047005
6. Lorenzo JR (2012) Principles of diffuse light propagation: light propagation in tissues with applications in biology and medicine. World Scientific, Singapore

7. Dehghani H, Eames ME, Yalavarthy PK et al (2009) Near infrared optical tomography using NIRFAST: algorithm for numerical model and image reconstruction. Commun Numer Methods Eng 25:711–732
8. Jermyn M, Ghadyani H, Mastanduno MA et al (2013) Fast segmentation and high-quality three-dimensional volume mesh creation from medical images for diffuse optical tomography. J Biomed Opt 18(8):086007. doi:10.1117/1.JBO.18.8.086007

Part VIII
Optical Techniques

Chapter 48
Simultaneous Measurement of Dissolved Oxygen Pressure and Oxyhemoglobin Spectra in Solution

Maritoni Litorja and Jeeseong C. Hwang

Abstract The measurement of the spatial distribution of oxygen saturation (sO_2) in superficial tissues using optical reflectance imaging has been useful in the clinical venue especially in temporally demanding applications such as monitoring tissue oxygenation during surgery. The measurement is based on relative spectrometry of oxy- and deoxyhemoglobin in tissues. We titrated deoxyhemoglobin with oxygen gas and simultaneously measured the dissolved oxygen pressure and the visible absorbance spectra to verify spectral shapes at different saturations. sO_2 values derived from the measured pO_2 are compared to those derived from the hemoglobin spectra at various stages of oxygenation.

Keywords Oximetry • Calibration • Oxyhemoglobin • sO_2 • Deoxyhemoglobin

1 The Need for an In Vitro Standard

Reflectance imaging oximetry uses a camera to report oxygen saturation (sO_2) of superficial tissues. It has been especially useful in visualizing oxygenation distribution during surgery [1]. The typical camera has a responsivity range from 400 nm to about 950 nm region, coincident with the spectral region in which the hemoglobin molecule exhibits significant absorbance changes as it transitions from the deoxygenated to oxygenated species and vice versa due to excitation of the porphyrin ring electrons. The relative spectrometry of the oxy- and deoxyhemoglobins is the measurement basis for many optical oximeters in use, including the pulse oximeter and the clinical laboratory hemoximeter.

Instrumentally, there are many variations to the reflectance oximeter in optical design, spatial resolution, spectral regions and bandwidths used. It measures oxygenation in different tissues with different optical reflectance and scattering properties. It is useful to have an in vitro calibrator for device development and

M. Litorja (✉) • J.C. Hwang
National Institute of Standards and Technology, Gaithersburg, MD, USA
e-mail: maritoni.litorja@nist.gov

© Springer Science+Business Media, New York 2016
C.E. Elwell et al. (eds.), *Oxygen Transport to Tissue XXXVII*, Advances in Experimental Medicine and Biology 876, DOI 10.1007/978-1-4939-3023-4_48

proficiency testing since it is not practical in all cases to generate a calibration curve for the whole saturation range from human subjects [2]. Tissue phantom calibrators are prepared, consisting of blood with other absorbers such as ink to mimic melanin content, and optical scatterers such as Intralipid, titanium dioxide or polystyrene spheres [3]. Spectral absorbance from fully oxygenated hemoglobin is used as the calibration point for $sO_2 = 1$; fully deoxygenated hemoglobin using an oxygen consumer, such as yeast, is used as the calibration point for $sO_2 = 0$. These two points are used to define a calibration line assuming a linear combination of the absorbance spectra of the two endpoints for intermediate values [4, 5].

2   Calibration Methods of Related Devices

Any in vivo reflectance oximeter, whether measuring a single spatial point or in two dimensions, has to be calibrated to have an sO_2 scale. It is relevant to look at the calibration procedures used in blood oximetry, which measures the same analytes. Blood oximetry is a standardized clinical laboratory technique, with established measurement protocols, reporting guidelines, and consensus standards [6]. Arterial, venous or capillary blood from a patient is drawn into an evacuated container and submitted to the clinical laboratory for hemoglobin sO_2 determination, along with other quantities. The clinical laboratory will analyze for sO_2 using either a blood gas analyzer or a hemoximeter, also known as CO-oximeter or cuvette oximeter as it was originally developed to measure carboxyhemoglobin [7].

In blood gas analysis, the oxygen from the patient blood hemoglobin is released into solution by converting all hemoglobins into cyanohemoglobin and the dissolved oxygen measured electrochemically. The primary calibration for the blood gas analyzer uses tonometered blood [8]. For quality control, the analyzer is checked using a reference standard gas mixture which is traceable to the international system of units (SI). These are sold in tanks labelled as blood gas mixtures and contain 5% carbon dioxide, 21% oxygen and the balance is nitrogen. It also measures pH and pCO_2 and the sO_2 is calculated from the measured pO_2 using conversion equations [9].

In contrast, a hemoximeter determines sO_2 spectrophotometrically. Absorbances of the sample are measured at multiple wavelengths to determine the concentration of oxy-, deoxy-, carboxy-, met- and sulf-hemoglobin using multicomponent analysis. The hemoglobins are then converted to cyanohemoglobin to determine the total hemoglobin concentration from its absorbance, based on an external calibrator measured on site, specific to the instrument. The determination of the various hemoglobin derivatives are based on predetermined spectra of the pure derivatives (parent spectra) which are stored, along with the multicomponent analysis software, in the instrument's memory. There are no published guidelines for the primary calibration of hemoximeters for oxygen saturation as there is for blood gas analyzers, so it is assumed that the two-point calibration commonly mentioned in the literature and described above, is practiced. While each instrument manufacturer conducts a primary calibration, this is typically not published and is part of the proprietary instrument analysis algorithm.

3 Preliminary Results of Simultaneous Spectra at Dissolved Oxygen Measurement

It is important to know accurately the sO_2 values of the calibration points especially if there are only two points used. Towards this purpose, oxygen gas titration of available hemoglobin was undertaken to verify hemoglobin spectral shapes at various degrees of saturation. A commercially available luminescence-lifetime-based oxygen pressure probe was chosen for convenience and speed. This type of oxygen pressure probe can be calibrated with a reference gas [10].

A gas mixing station for oxygen, carbon dioxide and nitrogen (Qubit Systems, Kingston, ON, Canada)[1] was set up for the titration experiment. Human hemoglobin A_0 (H0267, Sigma Aldrich, St. Louis, MO, USA) dry powder was dissolved in deoxygenated buffer solution inside a spectrophotometer cuvette with a stirring bar and a sealed top with inlets and outlets for the gas and solution mixtures. The gas mixture was humidified using deoxygenated buffer prior to entry into the solution. This was introduced into the 0.5 mL hemoglobin solution (5 mg/mL) through a 0.78 mm diameter Teflon tubing at a flowrate of less than 1 mL/min. This was low enough such that evaporation of the solution was not significant over the duration of the measurement. Visible spectra of the solution were collected every 10 s using a fiber-optic array spectrometer with 0.33 nm bandwidth (Avantes, Broomfield, CO). Partial pressure of dissolved oxygen in the solution was synchronously collected using a luminescence-lifetime-based probe (Neofox, Ocean Optics, Dunedin, FL, USA). The dissolved oxygen probe was placed just above the spectrophotometer beam path such that it measured oxygen partial pressure at a point close to spectrometer beam path without interfering in the spectroscopy. Prior to O_2 addition, the solution was titrated with sodium dithionite until the pressure probe registered zero and the hemoglobin spectrum matched that of published deoxyhemoglobin [11]. Several methods of deoxygenation were tried and the sodium dithionite titration worked most effectively. When the deoxygenation was complete, the oxygen titration was started. A gas mixture of 5 % carbon dioxide, 5 % oxygen and balance of nitrogen for a total headspace pressure of approximately 100 Torr was introduced into the solution. The delivery gas tubing was in contact with the solution such that the gas mixture flowed into the stirred solution instead of simply allowing the gas to diffuse into the solution as that was simply too slow to be practical. The gas flow was controlled by electronic metering valves via the

---

[1] Identification of commercial equipment is for information purposes only. It does not imply recommendation or endorsement by NIST, nor does it imply that the equipment identified is necessarily the best available for the purpose.

Fig. 48.1 The cuvette for oxygen titration of hemoglobin in solution has a gastight top with feedthroughs for the premixed gas from the mixer (*A*) and auxiliary tubing (*B*) through which solutions are introduced. The dissolved oxygen probe (*C*) is situated just above the light source (*D*) and spectrometer (*E*) beam path

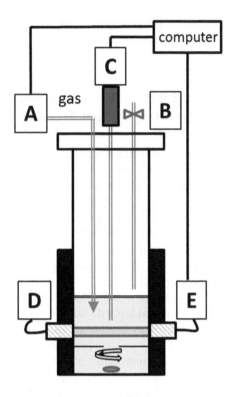

software. Thus, the oxygen partial pressure could be held at specific amounts allowing the oxygen in the headspace and solution to equilibrate for several minutes while spectra of the hemoglobin solution at specific partial pressures of oxygen were collected. In this experiment, it is assumed that the oxygen partitions between the hemoglobin and the buffer and the system is in equilibrium during the measurement. For each measured pO_2 value, a corresponding hemoglobin spectrum was collected. Figure 48.1 shows an illustration of the measurement setup.

Figure 48.2 shows the measurement results from the titration experiment. Each pO_2 value (abscissa), corrected for the experimental conditions used here (20 °C, pH 7.0), is plotted against sO_2 values derived from the pO_2 (diamonds) according to the Severinghaus equations [12]. The sO_2 was then determined from the hemoglobin spectrum at each corresponding pO_2 value using partial least squares method of multicomponent spectral analysis (Grams Software, ThermoScientific Inc, US). Spectral points from 520 to 600 nm were used. The spectrum of the prepared deoxyhemoglobin as described above, was used as the parent spectrum for fully deoxygenated hemoglobin ($sO_2 = 0$). The spectrum at the end of the measurement, when no changes were detected with further increase in pO_2, was used as the parent spectrum for fully oxygenated hemoglobin ($sO_2 = 1$). The sO_2 from this calibration model is used to generate Curve A in Fig. 48.2. The sO_2 values from Curve A vary

Fig. 48.2   sO$_2$ calculated from each measured pO$_2$ (*diamonds*) and sO$_2$ calculated from the spectra (Curves A and B) using two-component spectral analysis. Both A and B use the same deoxygenated model spectra but with sO$_2 = 0$ and sO$_2 = 0.25$ respectively, as the starting values

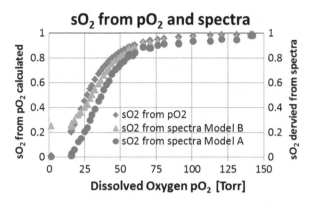

significantly from the corresponding sO$_2$ derived from the pO$_2$, especially at the lower sO$_2$ values. It was observed during the titration experiment that the fully deoxygenated spectrum persisted as some oxygen was added even as the dissolved pO$_2$ values increased to what is estimated as sO$_2 = 0.20$. The same spectral data were recalculated to generate Curve B but instead of assuming that the parent deoxygenated spectrum is fully deoxygenated, it is given a value of sO$_2 = 0.25$. The sO$_2$ values in Curve B are closer to the pO$_2$-derived sO$_2$ values.

Spectra-derived measurement of sO$_2$ has been implicitly taken to be "true" or correct sO$_2$ values, compared to those calculated from pO$_2$. Our preliminary results show that it cannot be assumed that the absorbance change scales linearly with the degree of saturation, or that the two endpoints are truly absolute. The observed persistence of the deoxygenated spectral shape with the addition of oxygen gas at the low saturation end can be due to instrumental and physical reasons, such as slow diffusion of the oxygen at the low concentration or poor sensitivity of the dissolved oxygen probe at low pO$_2$. The sample may also contain some contaminants although since we are only probing the reactive or functional hemoglobin, it likely does not play a significant role. Also plausible is that hemoglobin exhibits deoxygenated spectral shape even as some oxygen binding already occurs. Hemoglobin is known to exist, in a simplified model, in two conformations, which could have different absorptivities. The measured absorbance would then be a mixture of the two and their proportions vary according to the extent of oxygenation. It exists in the taut or T state when fully deoxygenated. It changes to the R or relaxed state upon binding with molecular oxygen. The switch, from T to R due to a rearrangement of several bonds and the motion of the protein subunits, allows hemoglobin to bind oxygen cooperatively, and is a subject of mechanistic studies [13]. This conformational change has a low probability of happening with only one oxygen molecule bound. The hemoglobin molecule therefore, could have greater than zero saturation and not exhibit observable change in the visible absorbance spectrum.

4 Summary and Future Work

We titrated deoxyhemoglobin in solution with O_2 and simultaneously measured the dissolved oxygen (unbound O_2) and the spectra (bound O_2). This is analogous to measurements with a blood gas analyzer and a clinical hemoximeter. Our preliminary results show that the accuracy of $sO_2 = 0$ calibration endpoint needs to be further verified. The experiment needs to be repeated with more data points at the lower pO_2 levels and eliminate any instrumental concerns. Preferably, an independent technique that is sensitive to the presence of the first dioxygen binding, such as Raman spectroscopy, will be used to verify the lack of absorbance change during the first oxygen binding. If the spectrum of deoxygenated hemoglobin does not reproducibly correlate to zero saturation, then it should not be used as a calibration point.

References

1. Tracy CR, Terrell JD, Francis RP et al (2010) Characterization of renal ischemia using DLP hyperspectral imaging: a pilot study comparing artery-only occlusion versus artery and vein occlusion. J Endourol 24:321–325
2. CLSI HS3-A (2005) Pulse oximetry: approved guideline. Clinical Laboratory Standards Institute, Wayne
3. Pogue BW, Patterson MS (2006) Review of tissue simulating phantoms for optical spectroscopy, imaging and dosimetry. J Biomed Opt 11(4):041102-1–16
4. Zwart A, Kwant A, Oeseburg A et al (1982) Oxygen dissociation curves for whole blood, recorded with an instrument that continuously measures pO_2, and SO_2 independently at constant t, CO_2 and pH. Clin Chem 28:1287–1292
5. Nahas GG (1951) Spectrophotometric determination of hemoglobin and oxyhemoglobin in hemolyzed blood. Science 113:723–725
6. NCCLS C25-A (1997) Fractional oxyhemoglobin oxygen content and saturation and related quantities in blood: terminology, measurement and reporting: approved guideline. NCCLS, Wayne
7. Moran RF (1993) Laboratory assessment of oxygenation. J Int Fed Clin Chem 5:170–182
8. International Federal of Clinical Chemistry (1989) IFCC document stage 3, draft 1 dated 1989. An approved IFCC recommendation. IFCC method (1988) for tonometry of blood: reference materials for pCO_2 and pO_2. Clin Chim Acta 185:S17–S24
9. Breuer H-WM, Groeben H, Breuer J et al (1989) Oxygen saturation calculation procedures: a critical analysis of six equations for the determination of oxygen saturation. Intensive Care Med 15:385–389
10. Thomas PC, Halter M, Tona A et al (2009) A Noninvasive thin film sensor for monitoring oxygen tension during in vitro cell culture. Anal Chem 81:9239–9246
11. Zijlstra WG, Buursma A, van Assendelft OW (2000) Visible and near infrared absorption spectra of human and animal hemoglobin. VSP BV, Netherlands, pp 57–59
12. Severinghaus JW (1979) Simple, accurate equations for human blood O_2 dissociation computations. J Appl Physiol 46:599–602
13. Wyman J, Gill SJ (1990) Binding and linkage: functional chemistry of biological macromolecules. University Science, Mill Valley, p 123

Chapter 49
Local Measurement of Flap Oxygen Saturation: An Application of Visible Light Spectroscopy

Nassim Nasseri, Stefan Kleiser, Sascha Reidt, and Martin Wolf

Abstract The aim was to develop and test a new device (OxyVLS) to measure tissue oxygen saturation by visible light spectroscopy independently of the optical pathlength and scattering. Its local applicability provides the possibility of real time application in flap reconstruction surgery. We tested OxyVLS in a liquid phantom with optical properties similar to human tissue. Our results were in good agreement with a conventional near infrared spectroscopy device.

Keywords Visible light spectroscopy • Flap oxygenation • Oxygenation in liquid phantom

1 Introduction

Non-invasive techniques with the ability to locally determine flap oxygenation are important to increase the rate of flap salvage. Conventional assessment of flaps in the operation theatre consists of clinical observation, examination of capillary refilling, temperature, and pinprick testing [1–3]. Although these methods are widely applied in flap reconstruction surgery, they are qualitative and highly dependent on the surgeon's level of experience. Recently, near-infrared spectroscopy (NIRS) was recognized as a reliable tool for flap viability assessment [4–6] and may be able to close this gap. However, NIRS measures relatively large

N. Nasseri (✉) • S. Kleiser
Biomedical Optics Research Laboratory, Department of Neonatology, University Hospital Zurich, University of Zurich, Zurich, Switzerland

Institute of Biomedical Engineering, ETH Zurich, Zurich, Switzerland
e-mail: nassim.nasseri@usz.ch

S. Reidt
Institute of Biomedical Engineering, ETH Zurich, Zurich, Switzerland

M. Wolf
Biomedical Optics Research Laboratory, Department of Neonatology, University Hospital Zurich, University of Zurich, Zurich, Switzerland

© Springer Science+Business Media, New York 2016
C.E. Elwell et al. (eds.), *Oxygen Transport to Tissue XXXVII*, Advances in Experimental Medicine and Biology 876, DOI 10.1007/978-1-4939-3023-4_49

volumes of deep tissue, but for flap salvage an instrument which resolves more localized tissue would be of interest. The instrument needs to provide rapid and real time readings of the tissue oxygen saturation (StO_2) values. Such a system has a wide range of other applications: from intraoperative monitoring of cerebral blood oxygenation [7] to oxygen saturation measurement in human and animal skin transplants [8].

Hemoglobin, the main absorber of blood, has a specific spectrum in the range of visible light ($\lambda = 520$–600 nm). There are two easily distinguishable local maxima at $\lambda_1 = 542$ nm and $\lambda_2 = 577$ nm when it is 100 % oxygenated and just one local maximum at $\lambda = 556$ nm when it is 100 % deoxygenated [9, 10]. The distance between the two maxima is an indicator of the oxygen saturation of hemoglobin [11]. This distance is independent of scattering and optical pathlength [11].

The aim is to develop a new device (OxyVLS) based on visible light spectroscopy and to test its functionality in a liquid phantom with optical properties similar to human tissue.

2 Material and Methods

2.1 Material

The Maya2000 Pro (Ocean Optics) spectrometer is applied to capture the attenuation spectrum of illuminated tissue. The spectrometer was combined with a tungsten halogen source (wavelength range: 360–2000 nm, power: 7 W) and a 400 μm reflection probe with 2 mm distance between light emission and detection. Reference data were acquired by the conventional frequency-domain NIRS device Imagent (ISS). We applied a self-calibrating probe [12] with distances of 1.5/2.5 and 1.7/2.7 cm and wavelengths of 692, 808 and 840 nm.

The liquid phantom consisted of human blood, saline, SMOFlipid, and glucose. SMOFlipid is a lipid emulsion, which scatters light [13]. The human blood mainly absorbs light. De-oxygenation of blood was achieved by adding yeast [14]. The sensor of the Imagent was sealed in a milky translucent plastic bag to protect it from moisture.

2.2 Method

We calculated the hemoglobin oxygen saturation of the phantom by measuring the wavelength interval between the two maxima in the optical attenuation spectrum of the phantom. Figure 49.1 shows the spectrum of hemoglobin (0–100 % oxygen saturation). As it is clear in Fig. 49.1 the distance between the two maxima in the spectrum of hemoglobin decreases as it gets de-oxygenated. In 1973, Wodick and

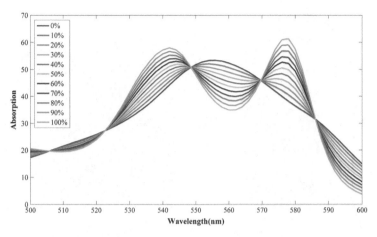

Fig. 49.1 Absorption spectrum of hemoglobin with varying oxygen saturation. Data from [8]

Luebbers proved the independence of this distance from the optical pathlength [11]. Their method is briefly outlined in Eqs. (49.1)–(49.3). Equations (49.1)–(49.3) show how maxima in the optical attenuation spectrum of the phantom represent maxima in the absorption spectrum of pure hemoglobin.

$$A_{phantom}(\lambda) = H\big(A_{hemoglobin}(\lambda)\big), \tag{49.1}$$

$$\frac{d\big(A_{phantom}(\lambda)\big)}{d\lambda}\big|_{\lambda_0} = \frac{\partial H}{\partial A_{hemoglobin}} \times \frac{\partial A_{hemoglobin}}{\partial \lambda}\big|_{\lambda_0}, \tag{49.2}$$

$$\frac{d\big(A_{phantom}(\lambda)\big)}{d\lambda}\big|_{\lambda_0} = 0 \overset{H \text{ is strictly monotonic}}{\Longrightarrow} \frac{\partial A_{hemoglobin}}{\partial \lambda}\big|_{\lambda_0} = 0 \tag{49.3}$$

Where $A_{phantom}$ is the optical attenuation spectrum of the liquid phantom, $A_{hemoglobin}$ is the absorption spectrum of hemoglobin, and $H(x)$ is the unknown and non-linear transfer function of the phantom. Hemoglobin is assumed to be the main absorber of light in the phantom. Equations (49.1)–(49.3) show even though the transfer function of the phantom, $H(x)$, is assumed to be non-linear and unknown, it does not alter the locations of the maxima in the optical attenuation spectrum of the phantom compared to pure hemoglobin. Thus the optical attenuation spectrum from the phantom has local maximum if and only if there is local maximum at the same wavelength in the absorption spectrum of hemoglobin. There is no assumption made on $H(x)$, except from it being strictly monotonic. Furthermore, the distance between the maxima depends on the StO_2 of the phantom based on empirical results [8].

Initially the phantom (Fig. 49.2) was 100 % oxygenated due to its direct contact with room air. De-oxygenation started when we added dry yeast to the phantom. The temperature of the phantom was kept at 30 °C. To re-oxygenate we pumped

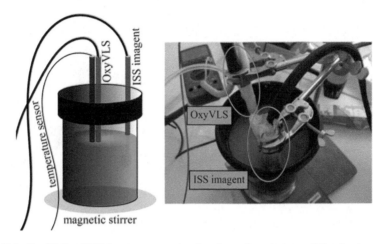

Fig. 49.2 OxyVLS and ISS imagent measuring the attenuation spectrum of the phantom

room air into the phantom. In addition, we injected either SMOFlipid or blood to change the optical properties of the phantom. In order to avoid any O_2-gradient due to contact to the room air, the phantom was constantly stirred by a magnetic stirrer. During 4 h, we measured 8 cycles of oxygenation/de-oxygenation.

The spectrometer was set to record at least 3 samples per second. Later on we down sampled to 1 spectrum per 10 s (without any averaging). After low pass filtering the spectrum, we fitted a spline interpolation in the range of 520 and 600 nm and calculated the distance between the two local maxima of the fitted signal. We created a reference database which mapped the distance between the two peaks in the spectrum of hemoglobin to its oxygen saturation based on the data from [9] and compared the spectra we recorded to the reference database we created. With the current algorithm we were not able to detect the maxima in the spectrum of hemoglobin with <30 % StO_2. For $StO_2 < 30$ % we returned a value of 30 %.

The Imagent StO_2 was computed by applying the equations from [12], attenuation coefficients measured by the instrument, and assuming 98 % water content in the phantom.

3 Results

Figure 49.3 shows that the two maxima approach each other while hemoglobin deoxygenates. Figure 49.4 shows multiple cycles of oxygenation and de-oxygenation in the range 0–90 % recorded by both devices. OxyVLS reacts similarly to changes in the oxygenation of the phantom as the ISS Imagent, although there is no perfect agreement, the agreement is good.

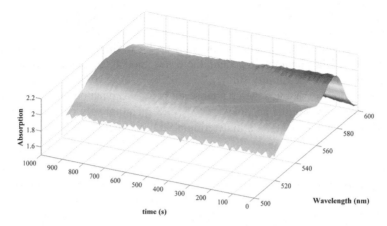

Fig. 49.3 Two maxima in the optical attenuation spectrum of hemoglobin approach each other as oxygenation decreases (*to the left*)

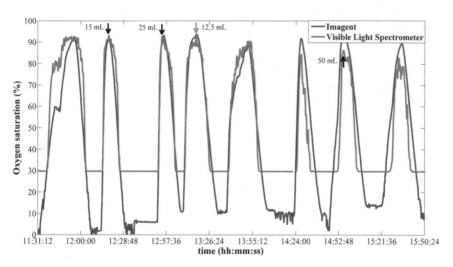

Fig. 49.4 The phantom oxygen saturation measured by OxyVLS compared to ISS Imagent. During the experiment, the light scattering was increased by adding SMOFlipid (*black arrows*) and the absorption by adding blood (*red arrow*)

4   Discussion

The distance between the maxima in the spectrum of the phantom depended on the StO_2, which is in agreement with the literature [9]. However, for $StO_2 < 30\%$ it was impossible to detect two separate maxima by the current method. For oxygen saturation $>30\%$ OxyVLS was reacting similarly to changes of phantom oxygenation as the Imagent, a commercial NIRS device applied widely in research. This comparison is not valid within the time period in which we bubble air into the phantom (phantom gets oxygenated), because air bubbles disturb the measurement by creating local inhomogeneity. Differences between the StO_2 of the OxyVLS and Imagent during the deoxygenating cycles may be due to slight inhomogeneity of the oxygenation of the phantom. In general there was good agreement between the two methods within this feasibility study.

In the future we plan to develop a more sophisticated phantom without inhomogeneities to determine the accuracy of OxyVLS quantitatively compared to standard methods and an algorithm to determine StO_2 also below 30 %.

5 Conclusion

We developed a system based on visible light spectroscopy with the ability of measuring StO_2. The main features of this device are that it measures independently of optical pathlength and scattering; it has a rapid response time, and is able to measure StO_2 locally and superficially. This method has a high potential for flap reconstruction surgery or other applications.

Acknowledgments This work is dedicated to D. Lübbers and R. Wodick, who had the original idea. We gratefully acknowledge funding by the KFSP Tumor Oxygenation of the University of Zurich. In addition, this work was scientifically evaluated by the SNSF, partially financed by the Swiss Confederation and funded by Nano-Tera.ch.

References

1. Disa JJ, Cordeiro PG, Hidalgo DA (1999) Efficacy of conventional monitoring techniques in free tissue transfer: an 11-year experience in 750 consecutive cases. Plast Reconstr Surg 104 (1):97–101
2. Gardiner MD, Nanchahal J (2010) Strategies to ensure success of microvascular free tissue transfer. J Plast Reconstr Aesthet Surg 63(9):665–673
3. Keller A (2007) Non-invasive tissue oximetry for flap monitoring: an initial study. J Reconstr Microsurg 23:189–198
4. Irwin MS, Thorniley MS, Doré CJ, Green CJ (1995) Near infrared spectroscopy: a non-invasive monitor of perfusion and oxygenation within the microcirculation of limbs and flaps. Br J Plast Surg 48:14–22
5. Thorniley MS, Sinclair JS, Barnett NJ, Shurey CB, Green CJ (1998) The use of near infrared spectroscopy for assessing flap viability during reconstructive surgery. Br J Plast Surg 51:218–226
6. Repez A, Oroszy D, Arnez ZM (2008) Continuous postoperative monitoring of cutaneous free flaps using near infrared spectroscopy. J Plast Reconstr Aesthet Surg 61(1):71–77
7. Hoshino T, Katayama Y, Sakatani K et al (2006) Intracranial monitoring of cerebral blood oxygenation and hemodynamics during extracranial-intracranial bypass surgery by a newly developed visible light spectroscopy system. Surg Neurol 65(6):569–576
8. Figulla HR, Austermann KH, Luebbers DW (1981) A new non-invasive technique for measuring O_2 saturation of hemoglobin using wavelength-distances and its application to human and animal skin transplants. Adv Physiol Sci 25:317–318

9. Zijlstra WG, Buursma A, van Assendelft OW (2000) Visible and near infrared absorption spectra of human and animal hemoglobin. VSP, Utrecht. ISBN 90-6764-317-3

10. Hoelzle F, Loeffelbein DJ, Nolte D, Wolff KD (2006) Free flap monitoring using simultaneous non-invasive laser Doppler flowmetry and tissue spectrophotometry. J Craniomaxillofac 34:25–33

11. Wodick R, Luebbers DW (1973) Quantitative Analyse von Reflexionsspektren und anderen Spektren mit inhomogenen Lichtwegen an Mehrkomponentensystemen mit Hilfe der Queranalyse, I. Z Physiol Chem 354:903–915

12. Hueber DM, Fantini S, Cerussi AE et al (1999) New optical probe designs for absolute (self-calibrating) NIR tissue hemoglobin measurements. Optical Tomography and Spectroscopy of Tissue III 3597, pp 618–631

13. Ninni PD, Martelli F, Zaccanti G (2011) Intralipid: towards a diffusive reference standard for optical tissue phantoms. Phys Med Biol 56(2):21–28

14. Suzuki S, Takasaki S, Ozaki T et al (1999) A tissue oxygenation monitor using NIR spatially resolved spectroscopy. Optical Tomography and Spectroscopy of Tissue III 3597, pp 582–592

Chapter 50
Hemodynamic Measurements of the Human Adult Head in Transmittance Mode by Near-Infrared Time-Resolved Spectroscopy

Hiroaki Suzuki, Motoki Oda, Etsuko Ohmae, Toshihiko Suzuki, Daisuke Yamashita, Kenji Yoshimoto, Shu Homma, and Yutaka Yamashita

Abstract Using a near-infrared time-resolved spectroscopy (TRS) system, we measured the human head in transmittance mode to obtain the optical properties and the hemodynamic changes of deep brain tissues in seven healthy adult volunteers during hyperventilation. For six out of seven volunteers, we obtained the optical signals with sufficient intensity within 10 sec. of sampling. We confirmed that it is possible to non-invasively measure the hemodynamic changes of the human head during hyperventilation, even in the transmittance measurements by the developed TRS system. These results showed that the level of deoxygenated hemoglobin was significantly increased, and the level of oxygenated and total hemoglobin and tissue oxygen saturation were also significantly decreased during hyperventilation. We expect that this TRS technique will be applied to clinical applications for measuring deep brain tissues and deep biological organs.

Keywords Near-infrared spectroscopy • Time-resolved spectroscopy • Deep brain tissues • Hemodynamics • Tissue oxygen saturation

1 Introduction

Near-infrared time-resolved spectroscopy (TRS) is an effective method for quantifying absorption coefficients, reduced scattering coefficients, and hemodynamics of human tissues [1–3]. In the majority of near-infrared spectroscopy (NIRS) studies, the optical source and optical detector are placed a few centimeters apart in reflection mode because of the limited detected light intensity. In our previous study [4], to measure the optical properties and hemodynamics of the human adult head in transmittance mode using NIRS equipment, we developed a highly sensitive TRS system and an algorithm using the wavelength-dependence of a reduced

H. Suzuki (✉) • M. Oda • E. Ohmae • T. Suzuki • D. Yamashita • K. Yoshimoto • S. Homma • Y. Yamashita
Central Research Laboratory, Hamamatsu Photonics K.K., Hamamatsu, Japan
e-mail: hiro-su@crl.hpk.co.jp

© Springer Science+Business Media, New York 2016
C.E. Elwell et al. (eds.), *Oxygen Transport to Tissue XXXVII*, Advances in Experimental Medicine and Biology 876, DOI 10.1007/978-1-4939-3023-4_50

scattering coefficient. The improved fitting method using the wavelength-dependence of the reduced scattering coefficient has been previously reported [5]. This was more robust than conventional techniques, especially at a low signal-to-noise ratio. Using this TRS system, we measured the human adult head in transmittance mode to determine absorption coefficients and hemodynamics for 50 healthy adult volunteers during the resting state. We also confirmed that the TRS system could non-invasively measure the hemodynamics of the human adult head in transmittance mode. However, the sampling required an extended time period of 104 ± 64 s for this transmittance measurements because the detected light intensity was low.

In the present study, to improve the sensitivity of the TRS system and investigate the hemodynamic changes of the human head during hyperventilation, we also developed high-power light source modules emitting light pulses at three different wavelengths (762, 801, and 834 nm). We then performed simultaneous measurements of the human head in transmittance mode using the developed TRS system and of the left forehead in reflectance mode using the conventional TRS system during hyperventilation.

2 Materials and Methods

2.1 Subjects

To measure the absorption coefficients and hemodynamics of the human head in transmittance mode and the left forehead in reflectance mode during hyperventilation, seven healthy male subjects were studied. Informed consent was obtained from all volunteers before the experiments.

2.2 Three Wavelength Time-Resolved Spectroscopy (TRS) System

A photograph and schematic diagram of the TRS system are displayed in Fig. 50.1. The TRS system consists of three-wavelengths of high-power light sources called Nanosecond Light Pulsers (NLPs; 762, 801, and 834 nm), a photomultiplier tube (PMT) with a GaAs photocathode for single-photon counting, a TRS circuit that used the time-correlated single-photon counting (TCSPC) method to measure the temporal profile function of the sample, and a detection and an irradiation optical bundle fibers with a numerical aperture (N.A.) of 0.56 and 0.29, respectively. We placed a thin diffuser panel in front of the detection optical fiber during the instrumental response measurements [6]. Each NLP generates a light pulse with full-width at half maximum (FWHM) of approximately 1.8–1.9 ns, pulse rate of 5 MHz, and an average output power of approximately 6 mW. Comparing with the

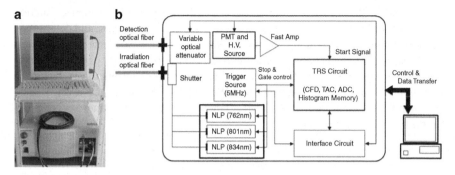

Fig. 50.1 (a) Photograph of the developed TRS system, and (b) Block diagram of the TRS system

conventional TRS system [7], the sensitivity of our developed TRS system increased approximately 100 times by increase in fiber N.A. and average output power, so it may measure the hemodynamics of deeper brain tissues.

2.3 Data Analysis

The TRS system can provide the temporal profiles in reflectance mode and transmittance mode. For the TRS data analysis, we used the solution [3] of a semi-infinite homogeneous model for the reflectance mode, and an infinite homogeneous slab for the transmittance mode. The solution $R(\rho,t)$ for the reflectance mode, and the solution $T(d, \rho, t)$ for the transmittance mode, are expressed by a function of the fiber distance ρ, the thickness of slab d, and time t, as shown in Eqs. (50.1) and (50.2), respectively.

$$R(\rho,t) = (4\pi Dc)^{-\frac{3}{2}} z_0 t^{-\frac{5}{2}} \exp(-\mu_a ct)\exp\left(-\frac{\rho^2 + z_0^2}{4Dct}\right), \tag{50.1}$$

$$T(\rho,d,t) = (4\pi Dc)^{-\frac{3}{2}} t^{-\frac{5}{2}} \exp(-\mu_a ct)\exp\left(-\frac{\rho^2}{4Dct}\right)\left\{(d-z_0)\exp\left[-\frac{(d-z_0)^2}{4Dct}\right]\right.$$

$$\left. -(d+z_0)\exp\left[-\frac{(d+z_0)^2}{4Dct}\right] + (3d-z_0)\exp\left[-\frac{(3d-z_0)^2}{4Dct}\right]\right.$$

$$\left. -(3d+z_0)\exp\left[-\frac{(3d+z_0)^2}{4Dct}\right]\right\}, \tag{50.2}$$

where c is the speed of the light in the medium and $D = (3\mu_s')^{-1}$, $z_0 = (\mu_s')^{-1}$. Using the non-linear least squares method based on the Levenberg–Marquardt method [8], we fitted Eqs. (50.1) or (50.2) that convolved the instrumental response function into the observed temporal profiles obtained from the TRS systems, and determined the absorption coefficient μ_a and the reduced scattering coefficient μ_s' at each wavelength. The oxygenated hemoglobin concentration (HbO_2; µM) and deoxygenated hemoglobin concentration (Hb; µM) were then determined by referring to the hemoglobin spectroscopic data [8]. We can then calculate the level of total hemoglobin concentration (tHb [µM] $= HbO_2 + Hb$) and the tissue oxygen saturation ($SO_2[\%] = HbO_2/tHb$).

2.4   Protocol

As shown Fig. 50.2, the irradiation and the detection optical fibers of each TRS system were attached to the human heads. In the transmittance measurements of the human head, the right ear canal was irradiated with three-wavelengths of pulsed light, and the photons passing through the human head were collected at the left ear canal. We then assumed a slab thickness of $d = 9$ cm ($\rho = 0$ cm). The absorption coefficient is minimally affected by the difference between d and real distance. In the reflectance measurement of the left forehead, we set a fiber distance of $\rho = 4$ cm. In the condition as described above, we performed the transmittance and the reflectance measurements for 5 min (2 min: resting state, 1 min: hyperventilation, 2 min: resting state) for the seven healthy volunteers at the same time. We have preliminarily confirmed the interferences between the two methods are negligible.

Fig. 50.2 Schematic of the position of the irradiation and the detection of optical fibers for the transmittance measurement of the human head using the developed TRS system, and for the reflectance measurement of the left forehead using the conventional TRS system

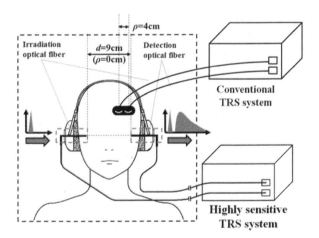

3 Results

For the transmittance measurement of the human head, we could obtain the optical signals with sufficient intensity in less than 10 sec. sampling time for six out of seven volunteers (mean age, 33.2 ± 3.6; range, 29–38 years). To improve the measurement accuracy [5], we used the ratio of the reduced scattering coefficient of $R_{762nm}:R_{801nm}:R_{834nm} = 1.022:1.000:0.985$, in which values were obtained from the mean value of the human head measurement results (data not shown) in the transmittance mode of the developed TRS system, using the conventional algorithm. Mean values $\pm$ standard deviation ($n = 6$) at the points of hyperventilation task start and hyperventilation task peak (defined as maximum point of absolute change from task start) for the human head in transmittance mode and the left forehead in reflectance mode are summarized in Table 50.1. These results show that Hb concentration measured in both modes significantly increased during hyperventilation. Additionally, HbO_2, tHb, SO_2 measured in both modes significantly decreased during hyperventilation. Figure 50.3 shows an example of changes in HbO_2, Hb and SO_2 when using the transmittance measurement of the human head and when using the reflectance measurement of the left forehead. We confirmed that both modes of the TRS system could detect the hemodynamic changes due to hyperventilation. Comparing the changes in HbO_2, Hb and SO_2 shown in Fig. 50.3, those of the human head measurement in transmittance mode were smaller than those of change of the left forehead measurement.

4 Discussions

Using our developed TRS system in six volunteers, we could determine the absorption coefficient, hemoglobin concentrations (HbO_2, Hb, tHb), and the SO_2 of the human head in transmittance mode during hyperventilation. Hb and HbO_2, tHb, SO_2 were significantly increased and decreased, respectively, during hyperventilation. These results reflect a reduction in the cerebral blood flow caused by hypocapnia during hyperventilation. Also, the hemodynamic changes of human forehead during hyperventilation and hypercapnia have been reported [9]. Therefore, we will investigate the hemodynamic changes during hypoventilation by this transmittance measurement.

Changes in HbO_2, Hb and SO_2 when using the transmittance measurement of the human head were smaller than those when using the reflection measurement of the left forehead (Fig. 50.3). We consider that these results might be caused by the employed homogeneous model. Actually, we used an infinite homogeneous slab for the transmittance mode to analyse TRS data. However, between the irradiation and the detection optical fibers in the transmittance measurements, various body parts such as the eardrum, middle ear, inner ear, pons, and medulla oblongata are included in the temporal profile data. Therefore, to eliminate the influence of

Table 50.1 Mean values ± SD (n = 6) at the points of hyperventilation (HV) task start and HV task peak for the human head in transmittance mode and the left forehead in reflectance mode

	Hemoglobin concentration of the human head in transmittance mode [μM]				Hemoglobin concentration of the left forehead in reflectance mode [μM]			
	HbO$_2$	Hb	tHb	SO$_2$ [%]	HbO$_2$	Hb	tHb	SO$_2$ [%]
HV task start	31.4 ± 2.4	15.0 ± 1.3	46.4 ± 3.3	67.7 ± 1.6	47.7 ± 4.0	23.8 ± 2.5	71.5 ± 5.2	66.7 ± 2.6
HV task peak	30.2 ± 2.2	16.1 ± 1.6	45.7 ± 3.4	65.4 ± 1.5	45.6 ± 5.8	25.5 ± 2.7	70.3 ± 6.3	64.2 ± 4.1
Hemodynamic change	−1.1 ± 0.8	+1.1 ± 0.5	−0.7 ± 0.5	−2.2 ± 0.8	−2.1 ± 2.4	+1.6 ± 0.8	−1.2 ± 1.4	−2.4 ± 2.0
Paired t-test	**$P < 0.01$	**$P < 0.01$	**$P < 0.01$	**$P < 0.01$	*$P < 0.05$	**$P < 0.01$	*$P < 0.05$	*$P < 0.05$

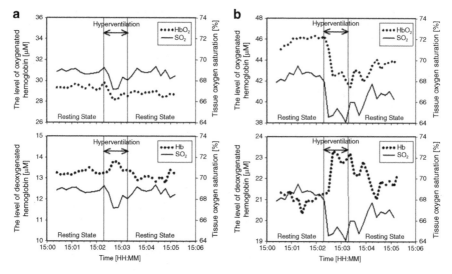

Fig. 50.3 Example of the measurement result of oxygenated hemoglobin (HbO₂), deoxygenated hemoglobin (Hb) and tissue oxygen saturation (SO₂) during hyperventilation. (**a**) The transmittance measurement of the human head using the developed TRS system. (**b**) The reflectance measurement of the left forehead using a conventional TRS system

heterogeneity of the human head, it is necessary to develop a new model that simulates the complicated structure of living tissue. Additionally, it is necessary to determine the specific regional cerebral tissues that contribute to the temporal profiles. We expect that this TRS technique will be applied to clinical applications for measuring deep brain tissues and deep biological organs.

References

1. Chance B, Nioka S, Kent J et al (1988) Time-resolved spectroscopy of hemoglobin and myoglobin in resting and ischemic muscle. Anal Biochem 174:698–707
2. Delpy DT, Cope M, van der Zee P et al (1988) Estimation of optical pathlength through tissue from direct time of flight measurement. Phys Med Biol 33:1433–1442
3. Patterson MS, Chance B, Wilson BC (1989) Time resolved reflectance and transmittance for the non-invasive measurement of tissue optical properties. Appl Opt 28:2331–2336
4. Suzuki H, Oda M, Yamaki E et al (2014) Hemodynamic measurements in deep brain tissues of humans by near-infrared time-resolved spectroscopy. Proc SPIE 8928–23
5. Andrea CD, Spinelli L, Bassi A et al (2006) Time-resolved spectrally constrained method for the quantification of chromophore concentrations and scattering parameters in diffusing media. Opt Express 14:1888–1898
6. Liebert A, Wabnitz H, Grosenick D et al (2003) Fiber dispersion in time domain measurements compromising the accuracy of determination of optical properties of strongly scattering media. J Biomed Opt 8:512–516

7. Ohmae E, Ouchi Y, Oda M et al (2006) Cerebral hemodynamics evaluation by near-infrared time-resolved spectroscopy: correlation with simultaneous positron emission tomography measurements. Neuroimage 29:697–705
8. Suzuki K, Yamashita Y, Ohta K et al (1994) Quantitative measurement of optical parameters in the breast using time-resolved spectroscopy. Phantom and preliminary in vivo results. Invest Radiol 29:410–414
9. Smielewski P, Kirkpatrick P, Minhas P et al (1995) Can cerebrovascular reactivity be measured with near-infrared spectroscopy? Stroke 26(12):2285–2292

Chapter 51
Non-invasive Monitoring of Hepatic Oxygenation Using Time-Resolved Spectroscopy

Tomotsugu Yasuda, Keiji Yamaguchi, Takahiro Futatsuki,
Hiroaki Furubeppu, Mayumi Nakahara, Tomohiro Eguchi,
Shotaro Miyamoto, Yutaro Madokoro, Shinsaku Terada,
Kentaro Nakamura, Hiroki Harada, Taniguchi Junichiro,
Kosuke Yanagimoto, and Yasuyuki Kakihana

Abstract The aim of the present study was to investigate whether changes in hepatic oxygenation can be detected by time-resolved spectroscopy (TRS) placed on the skin surface above the liver. METHODS: With approval of the local Hospital Ethics Committee and informed consent, six healthy volunteers aged 28.8 (25–36) years, and five patients with chronic renal failure aged 70.6 (58–81) years were studied. In six healthy volunteers, following echography, TRS (TRS-10, Hamamatsu Photonics K.K., Hamamatsu, Japan) probes consisting of a near-infrared light (at 760, 800, 835 nm) emitter and a receiver optode, were placed 4 cm apart on the abdominal skin surface above the liver or at least 10 cm distant from the liver. In five patients with chronic renal failure, following echography, TRS probes were placed 4 cm apart on the skin surface above the liver during hemodialysis (HD). RESULTS: In six healthy volunteers, the values of abdominal total hemoglobin concentration (tHb) were significantly higher in the liver area than in the other area (80.6 ± 26.81 vs 44.6 ± 23.1 µM, p $= 0.0017$), while the value of abdominal SO_2 in the liver area was nearly the same as that in the other area (71.5 ± 3.6 vs 73.6 ± 4.6 %, p $= 0.19$). The values of mean optical pathlength and scattering coefficient ($\mu's$) at 800 nm in the liver area were significantly different from those in the other area (21.3 ± 4.9 vs 29.2 ± 5 cm, p $= 0.0004$, and 7.97 ± 1.14 vs 9.02 ± 0.51 cm^{-1}, p $= 0.015$). One of five patients with chronic renal failure complained of severe abdominal pain during HD, and abdominal SO_2 decreased from 53 to 22 %; however, pain relief occurred following cessation of HD, and SO_2 recovered to the baseline level. CONCLUSIONS: Our data suggest that the optical properties of the liver may be measured by the TRS placed on the skin surface, and

T. Yasuda (✉) • K. Yamaguchi • T. Futatsuki • H. Furubeppu • M. Nakahara • T. Eguchi •
S. Miyamoto • Y. Madokoro • S. Terada • K. Nakamura • H. Harada • T. Junichiro •
K. Yanagimoto • Y. Kakihana
Department of Emergency and Intensive Care Medicine, Kagoshima University Graduate
School of Medical and Dental Sciences, Kagoshima, Japan
e-mail: yasutomo@m.kufm.kagoshima-u.ac.jp

© Springer Science+Business Media, New York 2016   407
C.E. Elwell et al. (eds.), *Oxygen Transport to Tissue XXXVII*, Advances in
Experimental Medicine and Biology 876, DOI 10.1007/978-1-4939-3023-4_51

the hepatic oxygenation may act as a non-invasive monitoring for early detection of intestinal ischemia.

Keywords Near-infrared spectroscopy • Time-resolved spectroscopy • Splanchnic ischemia • Liver • Tissue oxygenation

1 Introduction

Near-infrared spectroscopy (NIRS) is useful for continuously and non-invasively assessing the oxygenation state in the brain [1]. At present it is not clear whether NIRS provides reliable values on the tissue oxygenation of various organs (e.g. the liver as a corresponding site in the splanchnic region, which reacts very sensitively to hemodynamic instability). The aim of the present study was to investigate whether in adults the hemoglobin concentration in the liver and the changes in hepatic oxygenation can be detected by time-resolved spectroscopy (TRS: a new method of NIRS) placed on the skin above the liver surface.

2 Materials and Methods

With approval of the local Hospital Ethics Committee and informed consent, six healthy volunteers aged 28.8 (25–36) years in protocol A, and five patients with chronic renal failure (CRF) aged 70.6 (58–81) years in protocol B, were studied (Table 51.1).

Protocol A: In healthy volunteers, following echography, TRS probes consisting of an emitter optode and a receiver optode, were placed 4 cm apart on the abdominal skin above the liver surface. After monitoring the liver area, probes were removed to the other (intestine) area at least 10 cm distant from the liver surface and monitored. The values of mean optical pathlength, scattering coefficient ($\mu's$), absorption coefficient ($\mu'a$), total hemoglobin (tHb) concentration, and oxygen saturation (SO_2) in the liver area were compared with those in the intestine area.

Table 51.1 Characteristics of the five patients in the Protocol A (*DM* diabetes mellitus, *ASO* arteriosclerosis obliterans)

Case patient number	Age (years)	Sex	Complication
1	72	M	DM, ASO
2	58	M	DM, ASO
3	70	M	ASO
4	73	M	Polycystic kidney, postoperative internal iliac aneurysm and sigmoid colon cancer
5	81	M	ASO, polycystic kidney and postoperative pharyngeal cancer

Protocol B: In five patients with CRF, following echography, TRS probes were placed 4 cm apart on the skin above the liver surface, and the Hb and SO_2 in the liver area was continuously monitored during hemodialysis (HD). The TRS system (TRS-10, Hamamatsu Photonics K.K., Hamamatsu, Japan) uses a time-correlate single photon counting (TCPC) method to measure the temporal profile of the detected photons [2]. The system consists of a three-wavelength light pulser as the light source, which generates light pulses with a peak power of 60 mW, pulse width of 100 ps, pulse rate of 5 MHz, and an average power of 30 μW. The system measures the intensity of light in a time domain and enables analysis of the data with the time domain photo diffusion equation.

3 Results

In protocol A, the values of tHb were significantly higher in the liver area than in the intestine area (80.6 ± 26.81 vs 44.6 ± 23.1 M, p = 0.0017) (Figs. 51.1 and 51.2), while the value of SO_2 in the liver was nearly the same as that of the intestine area (71.5 ± 3.6 vs 73.6 ± 4.6 %, p = 0.19). The values of mean optical pathlength, absorption coefficient and scattering coefficient at 800 nm in the liver area were significantly different from those in the intestine area (21.3 ± 4.9 vs 29.2 ± 5 cm, p = 0.0004, 0.181 ± 0.058 vs 0.105 ± 0.049 cm^{-1}, p = 0.0021 and 7.97 ± 1.14 vs 9.02 ± 0.51 cm^{-1}, p = 0.015) (Fig. 51.2). Optical parameters at 760 and 835 nm are shown in Table 51.2.

In protocol B, one of five patients (case patient no. 5), with CRF and arteriosclerosis obliterans (ASO), complained of severe abdominal pain during HD, and the value of SO_2 in the liver area decreased from 53 to 22 %; however, pain relief occurred following cessation of HD and fluid resuscitation, and SO_2 recovered to the baseline level (Fig. 51.3).

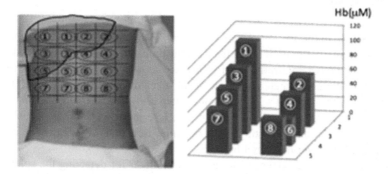

Fig. 51.1 tHb concentration measured transcutaneously by TRS. *Left picture*: the numbers indicate the positions where the TRS probes were placed. *Right-hand graph* indicates the tHb concentration measured by TRS on the corresponding number position

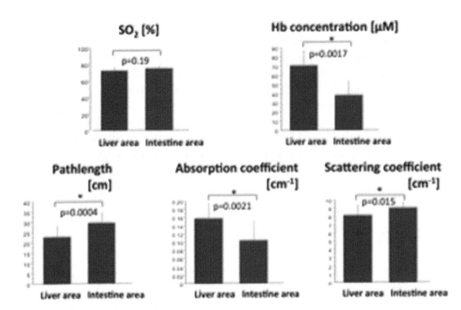

Fig. 51.2 Data from the liver and intestine measured by TRS

Table 51.2 Optical parameters at 760 and 835 nm from the liver and intestine measured by TRS

		Liver area	Intestine area	p value
Pathlength (cm)	760 nm	22.2 ± 1.92	30.1 ± 4.96	<0.0001
	835 nm	20.35 ± 1.91	27.6 ± 4.43	<0.0001
Absorption coefficient (cm^{-1})	760 nm	0.165 ± 0.018	0.105 ± 0.046	0.0005
	835 nm	0.182 ± 0.012	0.121 ± 0.050	0.0006
Scattering coefficient (cm^{-1})	760 nm	8.47 ± 0.70	9.62 ± 0.61	0.0008
	835 nm	7.86 ± 0.97	9.19 ± 0.51	0.0014

4   Discussion

In the present study, we investigated whether TRS can transcutaneously measure the tissue oxygen saturation of the liver in real-time and non-invasively in adults as a clinical hemodynamic monitor of global tissue oxygenation. The main finding was that the tissue oxygen saturation of the liver measured by TRS had a good correlation with a patient's hemodynamic state and clinical manifestation during HD. Our study was unique in identifying the liver position transcutaneously with tHb concentration and detecting the splanchnic ischemia with the tissue oxygen saturation of the liver measured by TRS.

We chose the tissue oxygen saturation of the liver for detecting the splanchnic ischemia during HD. During reduced cardiac output with reduced intestinal perfusion, decreased portal vein oxygen saturation and increased oxygen extraction rate in the liver results in a decrease of liver tissue oxygenation [3].

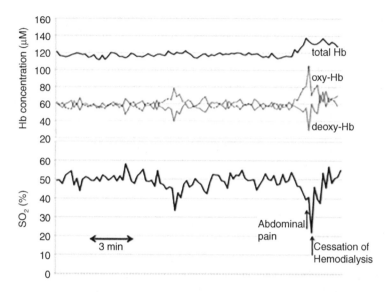

Fig. 51.3 Hepatic oxygenation and Hb concentration measured by TRS during hemodialysis (case patient no. 5)

Several reports have described the transcutaneous measurement of tissue oxygen index of the liver by means of NIRS [4–6]. First we investigated whether the TRS can identify the liver transcutaneously. By using abdominal echography simultaneously, we compared the data measured by TRS with the anatomical structure of the liver. The liver is rich in blood flow and, therefore, has a high concentration of hemoglobin. But the previous NIRS method cannot measure hemoglobin concentration and can only measure oxygen saturation. However a newly-developed TRS can measure hemoglobin concentration of the tissue as previously reported [1] and has the capability to identify the hemoglobin-rich liver position. The results of protocol A suggest that the liver can be identified by the TRS placed on the skin above the liver area.

However, patients with ASO have been reported to predispose to complain of the abdominal pain during HD, which suggests the occurrence of ischemia on the splanchnic region [7]. The results of protocol B suggest that continuous monitoring at the liver area by TRS can detect the ischemic insults of splanchnic organ in real-time and non-invasively. However, data by TRS above the liver area contain the signals of subcutaneous tissue such as fat and muscle. Therefore, further studies are required to evaluate the identification of the liver correctly, and early detection of splanchnic ischemia by TRS.

TRS can measure not only the absorption coefficient but also scattering coefficient in real time. These absorption and scattering coefficients are expected to construct the histological structure [8]. Analyzing the images by echography and the absorption and scattering coefficient may have the capability of detecting the changes of histological structure. However, it is structural changes of mitochondria that most contribute to changes of scattering coefficient in the human body. A previous study reported that severe cell injury causes mitochondria swelling in about 3 h [9]. The structural changes of mitochondria seem to be useful for not only assessing cell injury, but also assessing therapeutic value. Therefore, continuous measurement of scattering coefficient with TRS is considered to be able to detect the structural changes of mitochondria in real-time and is expected to detect the viability of biological tissue.

Our study has some limitations. We used comparatively young subjects in protocol A. The fat layer is normally thicker in elderly subjects which influences the measured oxygenation values. So in further study, we need to use elderly subjects in protocol A.

5 Conclusions

TRS can measure the hemoglobin concentration of the liver and identify the liver position and might be able to identify the liver and detect splanchnic ischemia in real-time and non-invasively. To confirm the usefulness of TRS in detecting splanchnic ischemia, we need further investigations in animal and in clinical setting.

References

1. Kakihana Y, Okayama N, Matsunaga A et al (2012) Cerebral monitoring using near-infrared time-resolved spectroscopy and postoperative cognitive dysfunction. Adv Exp Med Biol 737: 19–24
2. Oda M, Yamashita Y, Nishimura G et al (1996) A simple and novel algorithm for time-resolved multiwavelength oximetry. Phys Med Biol 41:551–562
3. Rhee P, Langdale L, Mock C et al (1997) Near-infrared spectroscopy: continuous measurement of cytochrome oxidation during hemorrhagic shock. Crit Care Med 25:166–170
4. Schulz G, Weiss M, Bauersfeld U et al (2002) Liver tissue oxygenation as measured by near-infrared spectroscopy in the critically ill child in correlation with central venous oxygen saturation. Intensive Care Med 28:184–189
5. Weiss M, Schulz G, Fasnacht M et al (2002) Transcutaneously measured near-infrared spectroscopic liver tissue oxygenation does not correlate with hepatic venous oxygenation in children. Can J Anaesth 49:824–829
6. Teller J, Wolf M, Keel M et al (2000) Can near infrared spectroscopy of the liver monitor tissue oxygenation? Eur J Pediatr 159:549
7. Lee T-C, Wang H-P, Chiu H-M et al (2010) Male gender and renal dysfunction are predictors of adverse outcome in nonpostoperative ischemic colitis patients. J Clin Gastroenterol 44: e96–e100
8. Suzuki K, Yamashita Y, Ohta K et al (1994) Quantitative measurement of optical parameters in the breast using time-resolved spectroscopy. Phantom and preliminary in vivo results. Invest Radiol 29:410–414
9. Lifshitz J, Janmey PA, McIntosh TK (2006) Photon correlation spectroscopy of brain mitochondrial populations: application to traumatic brain injury. Exp Neurol 197:318–329

Chapter 52
Comparison of Near-Infrared Oximeters in a Liquid Optical Phantom with Varying Intralipid and Blood Content

S. Kleiser, S. Hyttel-Sorensen, G. Greisen, and M. Wolf

Abstract The interpretation of cerebral tissue oxygen saturation values (StO_2) in clinical settings is currently complicated by the use of different near-infrared spectrophotometry (NIRS) devices producing different StO_2 values for the same oxygenation due to differences in the algorithms and technical aspects. The aim was to investigate the effect of changes in scattering and absorption on the StO_2 of different NIRS devices in a liquid optical phantom. We compared three continuous-wave (CW) with a frequency domain (FD) NIRS device. Responsiveness to oxygenation changes was only slightly altered by different intralipid (IL) concentrations. However, alterations in haematocrit (htc) showed a strong effect: increased htc led to a 20–35 % increased response of all CW devices compared to the FD device, probably due to differences in algorithms regarding the water concentration.

Keywords Near-infrared spectroscopy • Instrumentation • Comparison • Water correction • Liquid phantom

1 Introduction

In the last decades, NIRS has evolved to a valuable tool to measure and monitor StO_2 in research and various clinical applications [1]. StO_2 gives insight into the local balance of oxygen supply and demand. This new information could be clinically useful. Especially for critically ill patients such as preterm children who are often suffering from unstable cerebral oxygen supply and lack of cerebral auto-regulation, this information may be of value for tailoring clinical management [2].

S. Kleiser (✉) • M. Wolf
Biomedical Optics Research Laboratory, Department of Neonatology, University Hospital Zurich, University of Zurich, Zurich 8091, Switzerland
e-mail: stefan.kleiser@usz.ch

S. Hyttel-Sorensen • G. Greisen
Department of Neonatology, National University Hospital, Rigshospitalet, Copenhagen 2100, Denmark

© Springer Science+Business Media, New York 2016   413
C.E. Elwell et al. (eds.), *Oxygen Transport to Tissue XXXVII*, Advances in Experimental Medicine and Biology 876, DOI 10.1007/978-1-4939-3023-4_52

Despite this potential, routine use is being hampered by the fact, that different devices give different readings [3]. It is necessary to quantify these differences in order to transfer findings from one device to another. Testing devices on human subjects is problematic due to a lack of a reference method, inter-subject variations, poor precision, and physiological fluctuations in oxygenation [4–6]. Furthermore, analysis of dynamic response is limited because large variations of oxygenation can only be induced non-invasively and safely by cuff occlusion in limbs, but not in the brain.

Recently, we compared multiple NIRS devices under controlled conditions in a liquid phantom with scattering and absorption similar to neonatal brain tissue [7]. Oxygenation was changed by a membrane oxygenator, which allowed a comparison over the whole saturation range. The aim of the present study was to improve the experimental procedure and to additionally investigate mixtures with different blood and IL content to examine how NIRS devices compare under variable conditions.

2   Methods

Three CW NIRS devices (Somanetics INVOS 5100 adult sensor, Hamamatsu NIRO 300 and OxyPrem, an in-house built prototype) and one FD NIRS device (ISS OxiplexTS) were compared. The OxiplexTS was calibrated on a solid phantom with known optical properties before the first experiment. The sensors were fixed to the rim of a black bucket (see Fig. 52.1a). The NIRO 300 sensor was wrapped in thin plastic foil for moisture protection, whereas all other sensors were directly immersed into the liquid. The top of the bucket was covered with thin plastic foil and dark cloth to reduce entrance of room air and ambient light.

The main ingredients of the phantom were 6 l of phosphate buffered saline (pH 7.4, pre-heated to 37 °C), 4 g of baker's yeast, a variable amount of Fresenius Kabi Intralipid (IL) 200 mg/ml and human whole blood. The liquid was permanently mixed with a magnetic stirrer. Temperature of the phantom was stabilized and monitored by a MTRE Criticool temperature controller with its heat exchange mat wrapped around the bucket. Oxygenation was changed similarly to a previous experiment [8]: yeast continuously metabolized glucose and O_2, thus caused a steady decrease of phantom oxygenation. Reoxygenation was achieved by bubbling pure O_2. Important events were marked on all devices within a few seconds.

Two experiments were conducted, both with the same initial composition. In the first experiment, starting from 0.5 %, IL content was increased to 1.5 % in two steps while haematocrit (htc) was kept constant at 1 %. This resulted in changes of the scattering coefficient, which is essentially determined by the IL concentration [9]. In a second experiment, htc was raised from 1 to 2 % by adding blood while IL concentration was kept at 0.5 %. This mainly changed absorption and not scattering.

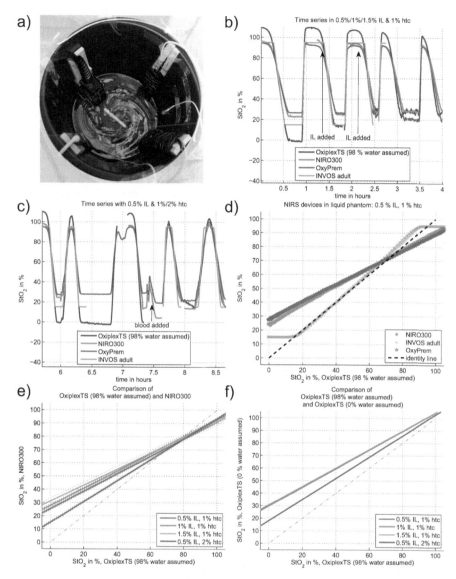

Fig. 52.1 (**a**) Bucket with sensors mounted. Clock-wise from the *top left*: ISS OxiplexTS, OxyPrem, Hamamatsu NIRO300, Somanetics INVOS adult sensor. (**b**) Time series showing readings from all instruments during experiment 1 with constant htc and increasing IL concentration. (**c**) Time series showing readings from all instruments during experiment 2 with constant IL and increasing htc. (**d**) Data points of NIRO 300, INVOS adult sensor and OxyPrem in 0.5 % IL and 1 % htc compared to ISS OxiplexTS with 98 % water assumption. (**e**) Relationship of NIRO 300 and ISS OxiplexTS with 98 % water assumption for different IL concentrations and htc. (**f**) Relationship of ISS OxiplexTS with 0 and 98 % water assumption for different IL concentrations and htc. For (**e**) and (**f**) *dotted lines* show the 95 % confidence interval of the linear fits

OxyPrem StO_2 was computed from raw data using a self-calibrating algorithm [10] assuming 0 % water, scattering of 0.5 % IL (interpolated from [9]) and applying absorption coefficients from [11] (weighted with LED emission spectra). OxiplexTS StO_2 was computed from raw data using the ISS Software OxiTS (version 3.1.1.0). Absorption coefficients were applied from the manual (version 3.1) and as suggested by the manufacturer, 0.01 cm^{-1} were subtracted at 692 nm as background absorption. All light paths were included in the analysis. Two StO_2 datasets were calculated: one accounting for the presumably true water content of 98 % (ISS98) and another without water correction (ISS0).

StO_2, timestamps and events from all devices were imported into Matlab. Time shifts between device clocks were determined and removed by searching the maximum of the event cross-correlation. Subsequently, data were interpolated (piecewise cubic) to a common time base and down-sampled to the slowest sampling rate of any of the devices (~0.2 Hz). For all comparisons and linear fits between the values of any two devices, only data during periods of decreasing oxygenation were used and only if both devices were in the StO_2 range of 40–85 %.

3   Results

The yeast induced a steady decline in oxygenation until all devices abruptly reached a lower steady state at the same time. Adding oxygen similarly led to an upper steady state quite synchronized repeatedly in all devices (Fig. 52.1b and c).

During repeated desaturations in mixtures of 0.5 % IL and 1 % htc the INVOS adult sensor was in good agreement with the ISS98 between 15 and 95 %, whereas both OxyPrem and NIRO 300 were much less responsive to the change in oxygenation with similar values at 70 %, but an overreading of as much as 25 % at the lowest StO_2 (Fig. 52.1d).

The stepwise increase in IL led to a stepwise increase in the StO_2 at lower steady state by the ISS98 and to a much smaller degree by the NIRO 300 (Fig. 52.1b). This response was not visible for the OxyPrem. The INVOS signal was at all times clipped at <15 and >95 %.

Increasing IL did not impact the association between ISS98 and NIRO 300, but doubling the htc increased the steepness of the curve by a factor of 1.19 and reduced the overreading at the lowest StO_2 to approximately 15 % (Fig. 52.1e). For the OxyPrem and the INVOS adult sensor, the steepness increased by a factor of 1.25 and 1.35, respectively. ISS98 vs. ISS0 showed an overreading at the lowest StO_2 of 30 % in the mixture with 0.5 % IL and 1 % htc which reduced to 19 % StO_2 when doubling htc, whereas steepness increased by a factor of 1.14 (Fig. 52.1f). The linear correlation coefficient between ISS98 and all CW devices was $R^2 > 0.99$ in the StO_2 range 40–85 % for all mixtures.

4 Discussion and Conclusion

In the present liquid phantom study changes in IL concentration only had minor impact on the pair-wise device relations, whereas an increase in htc resulted in a 'steeper' response of all CW devices compared to the FD device (ISS98).

The absolute value of scattering theoretically cancels out when calculating StO_2 and only the wavelength dependence remains, which was shown to be quite stable across different human tissues [12]. Consequently, it is not surprising that the responsiveness of the StO_2 was independent of changes in IL concentration.

NIRO 300 and ISS0 behaved remarkably similarly to the increase in htc, which seems related to both [8] not taking water absorption into account. The overreading by 25–30 % by OxyPrem, NIRO 300 and ISS0 at 1 % htc compared to ISS98 StO_2 at 0 % could be caused by water absorption at longer wavelengths (>800 nm). This contributes more significantly to the total absorption when htc is low, thus causing an overestimation of mainly oxyhemoglobin concentration. This effect is dependent on wavelengths employed by the device. This can explain the observed overreading at the lowest levels of StO_2 as well as the lowered responsiveness to oxygenation changes. Metz et al. [13] reported only a small influence of water assumption on absolute StO_2 values measured on the neonatal head (high StO_2) and a high influence on the adult forearm (low StO_2), which agrees well with our findings (Fig. 52.1f); they observed higher variability when accounting for water, which can be explained by the increased responsiveness also observed in this study.

Although the design of the phantom aimed at reducing the possibilities of inhomogeneity in the phantom by eliminating the distinct outlet tube in [7], the presence of oxygenation gradients from top to bottom as well as a gradient in yeast concentration cannot be entirely excluded. Sensors were placed at slightly different heights in the phantom. Therefore, all absolute results have to be treated with care, as sensors could potentially have 'seen' different true oxygenations and changes thereof. However, in the two experiments shown here, sensor positions were unchanged, thus the relative changes caused by changing mixtures should not be affected by these possible gradients.

In mixtures with increased IL or htc, the lower steady-state StO_2 for ISS98 did not reach 0 %. Also, the ISS98 reached maximum StO_2 well above 100 %. An O_2 gradient could explain these observations. However, we do not expect a possible gradient to affect the current analysis which is looking only at relative changes of pair-wise relations, because we only considered values in the range 40–85 % showing linear relations ($R^2 > 0.99$) and it seems unlikely that a potential gradient was different at different phantom compositions.

In comparison with [7], the OxyPrem vs INVOS adult sensor relation was different in the present experiments. Possibly different sensor heights in combination with an O_2 gradient could explain both changes in relative sensitivity as well as absolute numbers. Furthermore, OxyPrem measured horizontally whereas all other sensors measured vertically in the direction of the gradient, which could also lead to a difference. Responsiveness of NIRO 300 vs INVOS adult sensor was approximately

0.59 at 1 % htc and 0.51 at 2 % htc, which is in line with 0.53 at 1.5 % htc reported in [7]. This would be reasonable even if there was a saturation gradient, as they were mounted at the same height.

In conclusion, we have employed an optical phantom with variable scattering and blood content to investigate the effect of such changes on the steepness of the pair-wise relation of three CW NIRS oximeters and one FD NIRS oximeter which measured scattering and allowed to make user-defined corrections for water content of the phantom. Findings were that scattering changes had a minor influence on the devices, but changes in htc led to different responsiveness to oxygenation changes presumably due to the water assumption made in the algorithms.

Acknowledgments This work was scientifically evaluated by the SNSF, partially financed by the Swiss Confederation and funded by Nano-Tera.ch as well the Danish Council for Strategic Research (SafeBoosC project), which the authors would like to gratefully acknowledge.

References

1. Wolf M et al (2012) A review of near infrared spectroscopy for term and preterm newborns. J Near Infrared Spectrosc 20(1):43–55
2. Greisen G, Leung T, Wolf M (2011) Has the time come to use near-infrared spectroscopy as a routine clinical tool in preterm infants undergoing intensive care? Philos Transact A Math Phys Eng Sci 369(1955):4440–4451
3. Sorensen LC, Leung TS, Greisen G (2008) Comparison of cerebral oxygen saturation in premature infants by near-infrared spatially resolved spectroscopy: observations on probe-dependent bias. J Biomed Opt 13(6):064013
4. Bickler PE, Feiner JR, Rollins MD (2013) Factors affecting the performance of 5 cerebral oximeters during hypoxia in healthy volunteers. Anesth Analg 117(4):813–823
5. Hessel TW, Hyttel-Sorensen S, Greisen G (2014) Cerebral oxygenation after birth – a comparison of INVOS((R)) and FORE-SIGHT near-infrared spectroscopy oximeters. Acta Paediatr 103(5):488–493
6. Hyttel-Sorensen S, Hessel TW, Greisen G (2014) Peripheral tissue oximetry: comparing three commercial near-infrared spectroscopy oximeters on the forearm. J Clin Monit Comput 28(2): 149–155
7. Hyttel-Sorensen S et al (2013) Calibration of a prototype NIRS oximeter against two commercial devices on a blood-lipid phantom. Biomed Opt Express 4(9):1662–1672
8. Suzuki S et al (1999) A tissue oxygenation monitor using NIR spatially resolved spectroscopy. In: Proceedings of the optical tomography and spectroscopy of tissue III, 3597, pp 582–592
9. Ninni PD, Martelli F, Zaccanti G (2011) Intralipid: towards a diffusive reference standard for optical tissue phantoms. Phys Med Biol 56(2):N21–N28
10. Hueber DM et al (1999) New optical probe designs for absolute (self-calibrating) NIR tissue hemoglobin measurements. In: Proceedings of the SPIE, optical tomography and spectroscopy of tissue III, 3597, pp 618–631
11. Matcher SJ et al (1995) Performance comparison of several published tissue near-infrared spectroscopy algorithms. Anal Biochem 227(1):54–68
12. Matcher SJ et al (1995) Absolute quantification methods in tissue near-infrared spectroscopy. In: Proceedings of the SPIE, San Jose, pp 486–495
13. Metz AJ et al (2013) The effect of basic assumptions on the tissue oxygen saturation value of near infrared spectroscopy. Adv Exp Med Biol 765:169–175

Chapter 53
Improvement of Speckle Contrast Image Processing by an Efficient Algorithm

A. Steimers, W. Farnung, and M. Kohl-Bareis

Abstract We demonstrate an efficient algorithm for the temporal and spatial based calculation of speckle contrast for the imaging of blood flow by laser speckle contrast analysis (LASCA). It reduces the numerical complexity of necessary calculations, facilitates a multi-core and many-core implementation of the speckle analysis and enables an independence of temporal or spatial resolution and SNR. The new algorithm was evaluated for both spatial and temporal based analysis of speckle patterns with different image sizes and amounts of recruited pixels as sequential, multi-core and many-core code.

Keywords Laser speckle contrast analysis • Speckle algorithm • Perfusion imaging • Blood flow • Parallel processing

1 Introduction

Laser speckle contrast analysis (LASCA) is widely used as a tool for the assessment of blood flow of brain or pathological tissue. Quantification of blood flow changes allows the exploration of the mechanics of neurovascular coupling and regulation in normal and pathological brain [1]. The calculation of speckle contrast can be based either on a spatial or temporal analysis of the speckle pattern acquired with a digital camera. In each case it is essential that the spatial or temporal resolution of the speckle contrast images has to be balanced with the signal to noise ratio (SNR) [2]. Using a moving sub-window avoids this disadvantage but requires more arithmetic calculations which sets limits for a real time analysis. To solve this problem a new efficient numerical algorithm for the spatial and temporal based calculation of speckle contrast was developed which reduces the numerical complexity and is easy to implement as multi-core and many-core code.

A. Steimers (✉) • W. Farnung • M. Kohl-Bareis
RheinAhrCampus Remagen, University of Applied Sciences Koblenz, Remagen, Germany
e-mail: steimers@rheinahrcampus.de

© Springer Science+Business Media, New York 2016
C.E. Elwell et al. (eds.), *Oxygen Transport to Tissue XXXVII*, Advances in
Experimental Medicine and Biology 876, DOI 10.1007/978-1-4939-3023-4_53

2 Method

When illuminating a surface with coherent laser light it appears granular to the observer. This effect is commonly known as speckle effect. A scattering of the laser light by moving particles like blood cells results in fluctuations of the speckle pattern. If these fluctuations are observed by an integrating system like a camera with fixed exposure time the pattern appears blurred, and therefore this blurring correlates with the blood flow.

Based on a pattern produced under ideal conditions of e.g. a perfectly diffusing surface with a Gaussian distribution of the surface height, path differences due to the roughness of the surface which is much smaller than the coherence length and the scattering area and an absolutely coherent, monochromatic laser illumination, there is an exponentially decreasing intensity distribution when assuming a Gaussian statistic. A characteristic of this distribution within a specified area of N pixels is a standard deviation σ proportional to the mean intensity <I>. In practise, the standard deviation of the speckle pattern is often smaller than the mean intensity and therefore the speckle contrast defined as $SC = \sigma/<I>$ is reduced. Moving particles do result in blurred speckle images whereas a completely stable object maximizes the speckle contrast [3].

In case of a spatial based calculation of speckle contrast, the N pixels are recruited in a sub-window of size $w \times w$ of a full speckle image, whereas in case of a temporal based calculation they are recruited from the same single pixel as a vector with size N in temporal direction.

Increasing the number of pixels N does increase the robustness of the statistics but reduces either the spatial or the temporal resolution of the contrast image.

The standard algorithm to calculate the contrast out of N pixels is:

$$SC = \frac{\sigma}{\langle I \rangle} = \frac{\sqrt{\frac{1}{N-1}\sum_{i=1}^{N}(I_i - \langle I \rangle)^2}}{\langle I \rangle} \tag{53.1}$$

I_i is the intensity of a pixel and the mean intensity is calculated as: $\langle I \rangle = \frac{1}{N}\sum_{i=1}^{N} I_i$.

3 Optimized Speckle Algorithm

Equation (53.1) can be rewritten to find a more computationally efficient expression. To this end the displacement law of variance

$$\sum_{i=1}^{N}(I_i - \langle I \rangle)^2 = \left(\sum_{i=1}^{N} I_i^2\right) - \frac{1}{N}\left(\sum_{i=1}^{N} I_i\right)^2 \qquad (53.2)$$

allows restating Eq. (53.1) as:

$$SC = \frac{\sqrt{\frac{1}{N-1}\left[\left(\sum_{i=1}^{N} I_i^2\right) - \frac{1}{N}\left(\sum_{i=1}^{N} I_i\right)^2\right]}}{\frac{1}{N}\sum_{i=1}^{N} I_i} \qquad (53.3)$$

This can be rewritten as:

$$SC = \sqrt{\frac{N^2}{N-1} \cdot \frac{\sum_{i=1}^{N} I_i^2}{\left(\sum_{i=1}^{N} I_i\right)^2} - \frac{N}{N-1}} \qquad (53.4)$$

This improved algorithm reduces the numerical complexity of the speckle contrast calculation significantly and allows a much more efficient implementation in a multi-core and many-core code.

Furthermore this algorithm allows a more efficient calculation of the intensity and squared intensity sums of the recruited pixels using a technique based on a moving sum calculation. For this, vectors are shifted stepwise over the speckle image in temporal or spatial direction and a new vector sum is calculated at every step (see Fig. 53.1). By using this technique only the calculation of the first stripe dependents on the number of recruited pixel $N = w \times w$ whereas the calculation of the other sums and square sums is independent of N. In this implementation the (spatial or temporal) resolution is independent of the number of recruited pixels N and therefore the SNR has not to be balanced with the resolution of the speckle contrast images.

Table 53.1 lists the computational intensity of all arithmetic calculations for the speckle contrast calculations. When calculated as described, the computation time is independent of N. Furthermore, the speckle contrast image has the same nominal resolution as the full raw speckle image when calculated by the spatial based algorithm, except for two small stripes at the edges of the image. The overall benefit of the optimized algorithm is difficult to exactly predict as it depends on the architecture of the CPU.

1. sum up stripes of first w columns   2. moving sum over columns

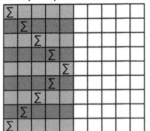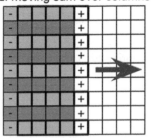

3. sum up stripes of first w rows   4. moving sum over rows
   of the col sum picture   of the col sum picture

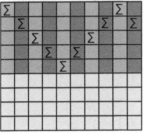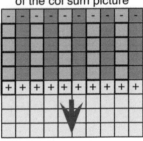

Fig. 53.1 Realization of an efficient moving sum algorithm for the spatial based calculation of the speckle contrast. First, in each row the sum is calculated for the vector of the first w elements. Second, these vectors are shifted by 1 pixel, i.e. the first element subtracted and one element added to give a new vector sum. This calculation is repeated until the last column element. Finally the vector sums are saved in a new matrix and the first two steps are repeated in the other dimension of the matrix

Table 53.1 Number of arithmetic calculations for both speckle contrast calculations. M is the resolution of the full speckle image, i.e.: 1024×768 pixels, and N the number of pixels in a spatial sub-window (generally 7×7 pixels) or in a time array (generally 25–49 pixels)

Arithmetic calculation	Speckle-contrast calculations		
	Base [based on Eq. (53.1)]	Optimized [based on Eq. (53.4)]	Benefit
Additions	$2 \times (M \times N)$	$2 \times M$	$1/N$
Subtractions	$N \times M$	$3 \times M$	$3/N$
Multiplications	$N \times M$	$3 \times M$	$3/N$
Divisions	$3 \times M$	M	$1/3$
Roots	M	M	1

4 Results

All algorithms were implemented as sequential and multi-core code and compiled by Visual Studio 2008 Professional Edition (Microsoft Corp., USA) using the open multi-processing library 2.0 (openMP) for parallelisation The used workstation included an Intel i7 950 processor with four physical cores that can be addressed as eight virtual cores due to simultaneous multi threading (SMT).

Both algorithms [Eqs. (53.1) and (53.4)] in their sequential and parallelized implementation were executed and the average computation time for a single execution was measured for different resolutions of raw data images and numbers of recruited pixels for speckle contrast calculation. The resolutions of raw data speckle images were: 500×500, 512×384, 752×480, 782×582, 1000×1000, 1024×768, 1296×966, 1392×1040, 1626×1236, 2000×2000, 3000×3000 and 4000×4000 pixels. The window size for spatial based contrast calculation was 5×5, 6×6, 7×7 and 8×8 pixels and the buffer size for temporal based contrast calculation was 25, 36, 49 and 64 frames.

Figure 53.2 shows the results of these measurements for all resolutions of speckle images for all four window sizes when based on spatial mode. In Fig. 53.3 the same finding can be drawn for temporal speckle analysis. Again the gain in computation speed by using the optimized algorithm can be seen. Figure 53.4 shows the comparison of the computation times of the spatial based and temporal based as a function of window size for an image with a resolution M of 1000×1000.

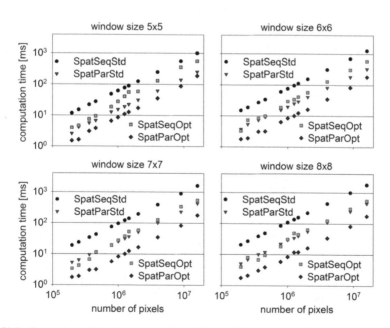

Fig. 53.2 Comparison of the computation time of the spatial based algorithm for the window sizes of 5×5 pixels (*upper, left*), 6×6 pixels (*upper, right*), 7×7 pixels (*lower, left*) and 8×8 pixels (*lower, right*) as a function of pixel number M of the whole raw image. The standard algorithm [Eq. (53.1), Std.] and the optimized variant [Eq. (53.4), Opt.] is compared, with implementation as sequential (Seq.) and parallelized (Par.) code

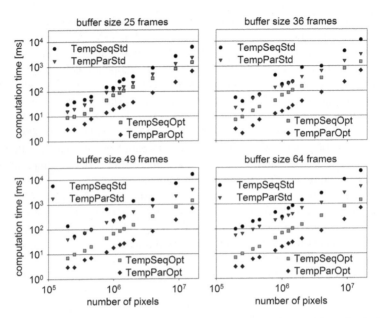

Fig. 53.3 Comparison of the computation times of the temporal based algorithm for the buffer sizes of 25 pixels (*upper, left*), 36 pixels (*upper, right*), 49 pixels (*lower, left*) and 64 pixels (*lower, right*) as a function of pixel number M of the whole raw image. The standard algorithm [Eq. (53.1), Std.] and the optimized variant [Eq. (53.4), Opt.] is compared, with implementation as sequential (Seq.) and parallelized (Par.) code

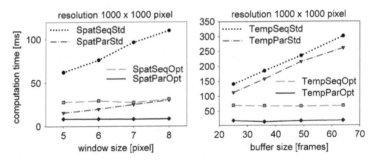

Fig. 53.4 Comparison of the computation times of the spatial based (*left*, as function of widow size) and temporal based (*right*, as function of buffer size) as a function of window size for an image of 1000 × 1000 pixels. The standard algorithm [Eq. (53.1), Std.] and the optimized variant [Eq. (53.4), Opt.] is compared, with implementation as sequential (Seq.) and parallelized (Par.) code

On average, the code is faster by a factor of 4.9 for the optimized compared to the standard algorithm. When adding the multi-core feature, the overall improvement is a factor of 15.5.

Furthermore a many-core code was implemented using CUDA (NVIDIA Corp., USA) and tested for images with a resolution of 1024×768 pixels and a window size of 8×8 pixels on a NVIDIA GTX 480 GPU, resulting in an improvement of a factor of 22.5 compared to the standard algorithm.

5   Discussion

In conclusion an efficient algorithm for the spatial and temporal based calculation of speckle contrast was demonstrated. This efficient algorithm reduces the numerical complexity of necessary calculations, facilitates a parallelized implementation of the speckle analysis and enables an independence of temporal or spatial resolution and SNR. Furthermore this algorithm was evaluated for both spatial and temporal based analysis of speckle patterns with different image sizes and amounts of recruited pixels as sequential multi-core and many-core code.

References

1. Dunn AK, Bolay H, Moskowitz MA et al (2001) Dynamic imaging of cerebral blood flow using laser speckle. J Cereb Blood Flow Metab 21:195–201
2. Thompson O, Andrews M, Hirst E (2011) Correction for spatial averaging in laser speckle contrast analysis. Biomed Opt Express 2(4):1021–1029
3. Goodman JW (1984) Laser speckle and related topics, vol 9. Springer, Berlin, pp 9–75

Chapter 54
A Clinical Tissue Oximeter Using NIR Time-Resolved Spectroscopy

Shin-ichi Fujisaka, Takeo Ozaki, Tsuyoshi Suzuki, Tsuyoshi Kamada, Ken Kitazawa, Mitsunori Nishizawa, Akira Takahashi, and Susumu Suzuki

Abstract The tNIRS-1, a new clinical tissue oximeter using NIR time-resolved spectroscopy (TRS), has been developed. The tNIRS-1 measures oxygenated, deoxygenated and total hemoglobin and oxygen saturation in living tissues. Two-channel TRS measurements are obtained using pulsed laser diodes (LD) at three wavelengths, multi-pixel photon counters (MPPC) for light detection, and time-to-digital converters (TDC) for time-of-flight photon measurements. Incorporating advanced semiconductor devices helped to make the design of this small-size, low-cost and low-power TRS instrument possible. In order to evaluate the correctness and reproducibility of measurement data obtained with the tNIRS-1, a study using blood phantoms and healthy volunteers was conducted to compare data obtained from a conventional SRS device and data from an earlier TRS system designed for research purposes. The results of the study confirmed the correctness and reproducibility of measurement data obtained with the tNIRS-1. Clinical evaluations conducted in several hospitals demonstrated a high level of usability in clinical situations and confirmed the efficacy of measurement data obtained with the tNIRS-1.

Keywords Time-resolved spectroscopy • NIRS • Tissue oximeter • Multi-pixel photon counter • Medical device

1 Introduction

In the nearly 40 years since tissue near infrared spectroscopy (NIRS) was first described, demonstrated and developed by F. Jobsis [1], it has become an established tool to monitor the oxygenation of brain tissue and other tissues. During these years, several types of NIRS machines based on such methods as modified-Beer Lambert (MBL) [2], spatially-resolved spectroscopy (SRS) [3], and time-

S. Fujisaka (✉) • T. Ozaki • T. Suzuki • T. Kamada • K. Kitazawa • M. Nishizawa •
A. Takahashi • S. Suzuki
Systems Division, Hamamatsu Photonics K.K., Hamamatsu, Japan
e-mail: fujisaka@sys.hpk.co.jp

© Springer Science+Business Media, New York 2016
C.E. Elwell et al. (eds.), *Oxygen Transport to Tissue XXXVII*, Advances in Experimental Medicine and Biology 876, DOI 10.1007/978-1-4939-3023-4_54

resolved spectroscopy (TRS) [4], were developed. Since devices based upon the
MBL and SRS methods employ continuous wave (CW) light and have advantages
of cost and practicability, such devices have long been used in clinical settings as
approved medical devices [5, 6]. On the other hand, the TRS method is well-known
for its superiority over the SRS and MBL methods with respect to correctness and
reproducibility of the measured data [7]. However, until now, devices used for TRS
measurements have been bulky and expensive, and they required sophisticated
technologies to handle the picosecond-speed signals. Therefore, although a TRS
device has long been sought for clinical bedside measurements, the use of TRS
oximeters has been limited to research fields. Our purpose was to develop a small-
size, low-cost and easily-operated TRS tissue oximeter suitable for use in clinical
environments. This paper describes the technical features of the instrument, the
results of measurements from blood phantoms and healthy volunteers, and clinical
data obtained during cardiac surgery to replace a descending aorta.

2 Theory of Measurement

Propagation of photons in highly scattering media can be described by the diffusion
approximation [8]. For a semi-infinite half-space geometry, the solution of the
photon diffusion equation for the impulse (δ-function) input is expressed by
Eq. (54.1) [8].

$$R\left(\rho, t, \mu_a, \mu_s'\right) = (4\pi Dc)^{-\frac{3}{2}} \frac{1}{\mu_s'} t^{-\frac{5}{2}} \exp(-\mu_a ct) \exp\left(-\frac{\rho^2 + \mu_s'^{-2}}{4Dct}\right) \qquad (54.1)$$

where R is the intensity of reflected light, ρ and t are the distance and time from the
impulse input, respectively, μ_a and μ_s' are the absorption and reduced scattering
coefficients, respectively, $D = 1/3(\mu_a + \mu_s')$ is the diffusion coefficient and c is the
velocity of light in the medium.

In TRS measurements, the instrument response function, IRF(t), which represents the response waveform of the instrument itself, is measured in advance with a
proper optical filter and is stored in the instrument. The time response waveform
including the measured tissue, M(ρ, t), is theoretically expressed by the convolution
of IRF and R as shown in Eq. (54.2).

$$M(\rho, t) = IRF(t) * R\left(\rho, t, \mu_a, \mu_s'\right) \qquad (54.2)$$

Therefore, with the conventional iteration method, the μa and $\mu s'$ which give the
least square error between M(ρ,t) and IRF(t)*R(ρ, t, μ_a, μ_s') are determined as those
of the measured tissue. With the tNIRS-1, the absorption coefficients at three
wavelengths, $\mu_a(\lambda_1)$, $\mu_a(\lambda_2)$ and $\mu_a(\lambda_3)$, are measured, and the concentration of
oxygenated hemoglobin (O$_2$Hb) and deoxygenated hemoglobin (HHb) are

calculated by the conventional method [7]. The tissue total hemoglobin (tHb) and oxygen saturation (StO_2) are derived from the equations, $tHb = O_2Hb + HHb$ and $StO_2 = (O_2Hb/tHb) \times 100$, respectively.

3   Design of the Instrument

The design of the instrument is shown schematically in Fig. 54.1. Temperature controlled pulsed LDs (typical wavelengths: 755, 816 and 850 nm, repetition rate: 9 MHz), which are chosen from the products available, are sequentially irradiated to the skin through the emission fibers. Light transmitted through the tissue is collected by the detection fibers and detected by the MPPCs through the variable neutral density filters (NDFs), which enables a dynamic range of the TRS measurement up to 3 orders of magnitude. The pulse width of the IRF of the tNIRS-1 is approximately 1.5 ns. In the time period between LD trigger and the photon detection by MPPCs, TDC measures the time of flight (TOF) of the transmitted light with an accuracy of 27 ps. The time response waveforms are created as a histogram of TOF every 5 s.

Figure 54.2 shows the external appearance of the instrument. The size is 292 mm (W) × 207 mm (D) × 291 mm (H) and the weight is 7.5 kg. The user operates the system using a touch screen. The system has sufficient battery back-up power to operate the system for 30 min. The probe consists of fiber bundle cables that are placed in a pad containing optical prisms that is secured in a rubber holder and attached to the patient. The patient pad is disposable and may be easily attached or detached from the patient, enabling it to be left in place on the patient between measurements. In this way, intermittent measurements with high reproducibility are

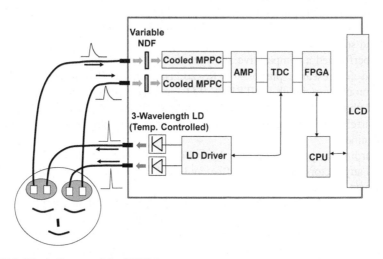

Fig. 54.1 Block diagram of the tNIRS-1

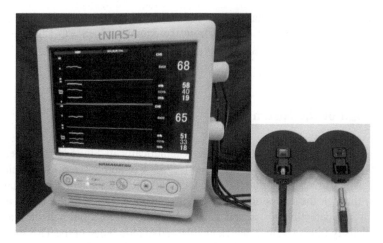

Fig. 54.2 The tNIRS-1 device and the probe

possible without restricting patient movement when measurements are not being made.

4   Measurements made for Instrument Evaluation

The correctness of tHb measured with the tNIRS-1 was evaluated using blood phantoms [5]. While adding human blood with known hemoglobin content measured by Stat Profile pHOx (Nova Biomedical Co., USA) into intralipid solutions (0.6 and 1.0 %) using a particular step, the tHb was recorded, and the hemoglobin concentrations in the phantoms and tHb were compared. As shown in Fig. 54.3a, the tHb values correlate well with the hemoglobin concentration in both of the intralipid solutions.

We performed a comparative study with the tNIRS-1 and another TRS system designed for research purposes (TRS-20, Hamamatsu Photonics [7]). Oxygen saturation measured with the TRS-20 (SO_2) and the tNIRS-1 was compared using a blood phantom (hematocrit; 1.1 %, intralipid; 0.7 %). As shown in Fig. 54.3b, good correlation between oxygen saturation measured with TRS-20 and tNIRS-1 was obtained.

Regarding the y-intercepts of -3.5 in Fig. 54.3a and -6.6 in Fig. 54.3b, we did not find any factors of systemic offset and think it comes from a limitation of the experiment accuracy.

We performed comparison studies on the reproducibility of measured data with a conventional SRS device (NIRO-200NX, Hamamatsu Photonics). With 8 healthy subjects (A–H), reproducibility of the data was evaluated by (i) repeating the attachment and detachment of the probe 8 times at the same position on the forehead and (ii) repeating the measurements at the 6 different positions on the

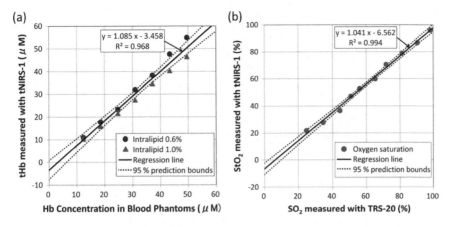

Fig. 54.3 (**a**) Hb concentration in blood phantoms and tHb measured with tNIRS-1, (**b**) Oxygen saturation measured with TRS-20 and tNIRS-1

Table 54.1 Average of standard deviations of measured oxygen saturation among eight subjects

Measurements \ Results		Average of SDs of measured oxygen saturation among 8 subjects	
		NIRO-200NX	tNIRS-1
(i) 8 times at the same position	R0	2.0 %	1.0 %
(ii) 6 different positions	R2 R3 R1 L1 L2 L3	5.1 %	2.1 %

forehead. Each measurement was carried out for 1 min, and the mean values over this time were recorded. The results of (i) and (ii) are shown in Table 54.1. In measurement (i), standard deviations of TOI (NIRO-200NX) and StO_2 (tNIRS-1) were 2.0 and 1.0 %, respectively, and in the measurement (ii), these were 5.1 and 2.1 %.

The results show that reproducibility of the measured data with the tNIRS-1 is higher than that with the NIRO-200NX, particularly in cases of measurements at different positions. Since the SRS method tends to be susceptible to the geometry or curvature of measured parts, the reproducibility of data was much improved with the TRS method.

Figure 54.4 shows an example of clinical data during cardiac surgery (descending aorta replacement with the circulating blood temperature of 18 °C). The probes were set on the forehead of the patient. In each stage of the surgery, expected data changes such as the decrease in StO_2 and tHb during circulatory arrest were observed.

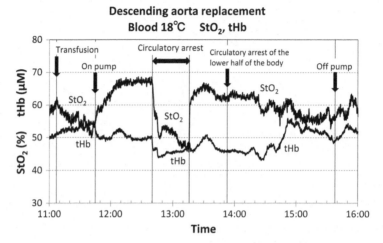

Fig. 54.4 An example of clinical data (By courtesy of Dr. Kenji Yoshitani, Department of Anesthesiology, National Cerebral and Cardiovascular Center, Japan)

In all the clinical evaluations performed thus far, the tNIRS-1 has demonstrated a high degree of usability, and the efficacy of the measured data in terms of correctness and reproducibility has been confirmed.

5   Conclusions

The tNIRS-1, a new clinical tissue oximeter using the TRS method, has been developed. A small size and low power design has been realized by employing advanced semiconductor devices such as MPPC and TDC. Expected performances were confirmed by evaluating measurements with blood phantoms, healthy volunteers and clinical measurements. The tNIRS-1 has been approved for use as a medical device in Japan. The applications to the US and EU countries are also intended.

References

1. Jobsis FF (1977) Non-invasive infrared monitoring of cerebral myocardial oxygen sufficiency and circulatory parameters. Science 198:1264–1267
2. Cope M, Delpy DT (1988) A system for long-term measurement of cerebral blood and tissue oxygenation in newborn infants by near-infrared transillumination. Med Biol Eng Comput 26: 289–294
3. Matcher SJ, Kirkpatrick P, Nahid M et al (1993) Absolute quantification methods in tissue near infrared spectroscopy. Proc SPIE 2389:486–495

4. Miwa M, Ueda Y, Chance B (1995) Development of time resolved spectroscopy system for quantitative non-invasive tissue measurement. Proc SPIE 2389:142–149
5. Suzuki S, Takasaki S, Ozaki T et al (1999) A tissue oxygenation monitor using NIR spatially resolved spectroscopy. Proc SPIE 3597:582–592
6. Roberts KW, Crnkowic AP, Linneman LJ (1998) Near infrared spectroscopy detects critical cerebral hypoxia during carotid endarterectomy in awake patients. Anesthesiology 89:A934
7. Ohmae E, Oda M, Suzuki T et al (2007) Clinical evaluation of time-resolved spectroscopy by measuring cerebral hemodynamics during cardiopulmonary bypass surgery. J Biomed Opt 12: 062112
8. Patterson MS, Chance B, Wilson BC (1989) Time resolved reflectance and transmittance for the noninvasive measurement of tissue optical properties. Appl Opt 28:2331–2336

Chapter 55
Validation of a Hybrid Microwave-Optical Monitor to Investigate Thermal Provocation in the Microvasculature

Allann Al-Armaghany, Kenneth Tong, David Highton, and Terence S. Leung

Abstract We have previously developed a hybrid microwave-optical system to monitor microvascular changes in response to thermal provocation in muscle. The hybrid probe is capable of inducing deep heat from the skin surface using mild microwaves (1–3 W) and raises the tissue temperature by a few degrees Celsius. This causes vasodilation and the subsequent increase in blood volume is detected by the hybrid probe using near infrared spectroscopy. The hybrid probe is also equipped with a skin cooling system which lowers the skin temperature while allowing microwaves to warm up deeper tissues. The hybrid system can be used to assess the condition of the vasculature in response to thermal stimulation. In this validation study, thermal imaging has been used to assess the temperature distribution on the surface of phantoms and human calf, following microwave warming. The results show that the hybrid system is capable of changing the skin temperature with a combination of microwave warming and skin cooling. It can also detect thermal responses in terms of changes of oxy/deoxy-hemoglobin concentrations.

Keywords Near infrared spectroscopy • Thermoregulation • Vasodilation • Muscle • Microwave applicator

A. Al-Armaghany • K. Tong
Department of Electronic and Electrical Engineering, University College London, London, UK

D. Highton
Neurocritical Care, University College London Hospitals, Queen Square, London, UK

T.S. Leung (✉)
Department of Medical Physics and Biomedical Engineering, University College London, London, UK
e-mail: t.leung@ucl.ac.uk

© Springer Science+Business Media, New York 2016   435
C.E. Elwell et al. (eds.), *Oxygen Transport to Tissue XXXVII*, Advances in Experimental Medicine and Biology 876, DOI 10.1007/978-1-4939-3023-4_55

1 Introduction

Thermal provocation has been used in combination with skin blood flow measurement to assess a number of conditions including chronic spinal cord injury [1], digital obstructive arterial disease (DOAD) [2] and transplanted free flaps [3]. In these studies, the region of interest is warmed by placing it into a thermal box [2], immersing it into warm water [3] or using a heating element directly on the skin [1], and the subsequent changes in blood flow are measured by a laser Doppler flowmeter. Healthy skin will respond by vasodilation leading to an increase in blood flow, while suboptimal or pathological skin may initiate a much smaller increase. It is believed that there are at least two factors which may affect the magnitude of the thermal response: (1) the obstruction of the vasculature as in the case of DOAD [2], and (2) the disruption of the sympathetic pathways as in the case of spinal cord injury [1] and transplanted free flaps [3].

Although current studies focus on the skin, one natural progression of this research is to investigate the thermal response of the muscle below the skin and see whether similar observations can be found. This work aims to develop a device to facilitate such investigations. In our previous publication, we described the development of a hybrid microwave-optical system to monitor thermal provocation in muscle [4]. The hybrid probe has integrated a number of devices including (1) a microwave patch applicator to induce deep heat, (2) a NIRS (near infrared spectroscopy) optical probe to monitor changes in oxy/deoxy/total-hemoglobin concentration ($\Delta[HbO_2]/\Delta[HHb]/\Delta[HbT]$) in the tissue, (3) a temperature sensor to monitor skin temperature, (4) a skin cooling system to minimize skin temperature, and (5) a laser Doppler probe to monitor skin blood flow. Since the NIRS and laser Doppler probes interfere with each other in the current design, they cannot be used simultaneously. The whole system has been built in-house.

In this paper, thermal imaging has been adopted to investigate the temperature distribution in phantoms and human calf, following microwave warming, with and without skin cooling. The hybrid probe remains largely the same as the previous design and its details can be found in the previous publication [4]. The current design has a small modification on the location of the temperature sensor to improve the feedback on skin cooling.

2 Methods

2.1 Phantom Study

A cylindrical tissue mimicking phantom was made which consisted of three layers (thicknesses): skin (2 mm), fat (10 mm) and muscle (40 mm) layers, with a diameter of 120 mm. The ingredients for the phantom included deionized water, gelatin, paraffin, safflower oil, detergent, n-propanol, formaldehyde and p-tolic acid

[5]. The proportion of the ingredients was adjusted for the three layers separately so that their dielectric properties matched those found in real human tissues. The relative permittivity/electrical conductivity (S/m) were measured to be 37.7/1.6, 5.6/0.2 and 54.7/2.2 for the skin, fat and muscle layers, respectively.

To establish the amount and depth of heat that the hybrid probe can induce, it was placed on top of the phantom and 3 W microwaves were used to warm up the phantom. After 5 min, the microwave system was switched off and the hybrid probe was removed to expose the skin phantom's surface so that a thermal image can be taken with an infrared thermal camera (FLIR i7). Afterwards, the skin and fat phantom layers were also removed to expose the top of the fat phantom layer for thermal images to be taken. The experiment was conducted twice, one with skin cooling and one without.

2.2   In Vivo Study

The in vivo study has been approved by the UCL Ethics Committee. Three adult subjects were involved in this study. The subject sat in a comfortable chair in a quiet room. The hybrid probe was placed securely on the calf of the subject's leg. The placement of the hybrid probe was judged satisfactory if the corresponding NIRS signals ($\Delta[HbO_2]/\Delta[HHb]$) had good signal quality. The skin temperature and the NIRS signals typically settled to a stable level after approximately 10 min. The monitoring then started in the resting state, lasting for 5–7 min, after which the microwave system was switched on for a period of time. Afterwards, the monitoring continued for another 20 min during the recovery state.

To establish how the combination of microwave warming and skin cooling affected the results, a series of experiments were conducted for three cases, i.e., case I: microwave warming without skin cooling, case II: microwave warming with skin over cooling, and case III: microwave warming with moderate skin cooling. In case I, 1 W microwaves were used for 60 min without skin cooling. In case II, 1 W microwaves were used for 60 min, this time with skin cooling to 1 °C below the initial temperature (over cooling). In case III, 3 W microwaves were used for 15 min, with skin cooling to 0.2 °C below the initial temperature. In cases I, II and III, the experiments were repeated again but this time the hybrid probe was removed immediately after the microwave system was switched off to allow thermal images to be taken.

3 Results

3.1 Phantom Results

Figure 55.1a, b show the thermal images taken on the skin and muscle phantoms after 5 min of microwave warming without skin cooling. It can be seen that for the skin phantom, the temperature at the central cursor location was raised to 28.6 °C (from 20 °C) while the temperature of the muscle phantom raised to 25.1 °C, confirming the capability of the microwave applicator to induce deep heat. With the skin cooling system switched on, the skin phantom temperature only raised slightly to 21.5 °C (from 19 °C) as shown in Fig. 55.1c. The muscle phantom temperature, however, was raised to 24.4 °C as depicted in Fig. 55.1d, confirming the capability of the skin cooling system to reduce superficial temperature and that of the microwave applicator to induce deep heat.

3.2 In Vivo Results

Figure 55.2a and b show case I (microwave warming without skin cooling) results including the $\Delta[HbO_2]/\Delta[HHb]/\Delta[HbT]$ signals and the skin temperature during the experiment. Figure 55.2c depicts the thermal image of the skin surface immediately after 60 min of microwave warming in a second experiment. It can be seen that the skin temperature increased rapidly at the beginning and then settled at 34.7 °C approximately 24 min into microwave warming. In the meantime, both $\Delta[HbO_2]$ and $\Delta[HbT]$ increased significantly, showing that the blood volume increase was mainly due to the dilation of arterial vessels, while the slight decrease in $\Delta[HHb]$ indicated an increased blood flow.

Figure 55.3 shows case II (microwave warming with skin over cooling) results. The skin cooling system lowered the skin temperature to ~1 °C below the initial temperature, as shown in Fig. 55.3b. The corresponding $\Delta[HbO_2]$ and $\Delta[HbT]$ in Fig. 55.3 decreased gradually while $\Delta[HHb]$ remained relatively stable. These

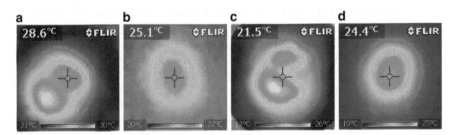

Fig. 55.1 Thermal images of phantom during microwave exposure (**a**) skin phantom without cooling, (**b**) muscle phantom without cooling (**c**) skin phantom with cooling (**d**) muscle phantom with cooling

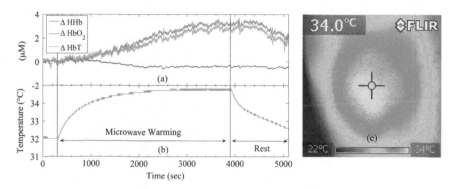

Fig. 55.2 Case I: with microwave warming without cooling (**a**) Measured NIRS signals, (**b**) Measured skin temperature (**c**) Skin thermal image immediately after warming

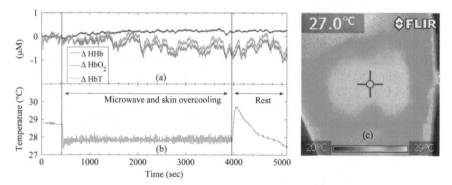

Fig. 55.3 Case II: microwave warming with skin over cooling (1 °C below initial temperature) (**a**) Measured NIRS signals, (**b**) Measured skin temperature (**c**) Skin thermal image

behaviors are consistent with vasoconstriction in the arterial vessels which is what one would expect when human tissue is cooled. The thermal image in Fig. 55.3c also confirms the lowering of skin temperature. It is noted from Fig. 55.3b that when both the microwave and skin cooling systems were switched off, the skin temperature rose significantly before it gradually went down again, possibly due to a combination of rewarming of the skin and the diffusion of heat from the microwave applicator back to the skin.

Figure 55.4 shows case III (microwave warming with moderate skin cooling) results. The microwave power was 3 W higher than those used in the previous two cases. The warming duration was therefore shortened to 15 min to avoid over warming the muscle. Figure 55.4b shows that the skin cooling system kept the skin temperature at ~0.2 °C below the initial temperature. Figure 55.4a shows that after the microwaves were turned on, $\Delta[HbO_2]$ and $\Delta[HbT]$ initially decreased for approximately 3 min, followed by an increase in $\Delta[HbO_2]$ and a decrease in

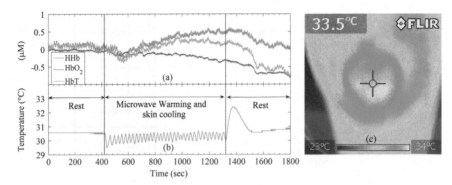

Fig. 55.4 Case III: microwave warming at 3 W with skin moderate cooling (0.2 °C below initial temperature) (**a**) Measured NIRS signals, (**b**) Measured skin temperature (**c**) Skin thermal image

Δ[HHb] over the next 12 min. In comparison to case I, the increase of Δ[HbO$_2$] in case III was much smaller. After the microwaves were switched off, the skin temperature increased momentarily and then dropped, as in the case of case II. As for Δ[HbO$_2$], this gradually decreased back to the initial level while Δ[HHb] continued to decrease. It is noted that although the skin temperature was kept at below the initial temperature of 30.5 °C as shown in Fig. 55.4b, the thermal image in Fig. 55.4c shows that the temperature distribution on the skin was not evenly distributed. The skin temperature near the optical probe had in fact increased, causing an increase in Δ[HbO$_2$] and a decrease in Δ[HHb] on the skin.

4 Discussions and Conclusions

This study has shown that by combing microwave deep warming and skin cooling, the skin temperature can be adjusted accordingly. In order to investigate the thermal provocation of muscle microvasculature, it would be necessary to be able to warm preferentially the deeper region while maintaining the skin temperature relatively constant so that any changes in the NIRS signals could be related to the muscle rather than to the skin. This preferential warming can potentially be achieved by optimizing the balance between microwave warming and skin cooling. However, the thermal imaging results suggest that the current design of the hybrid probe needs to be modified to achieve this. The main technical problem, as revealed by the thermal images, is that the skin temperature distribution is not uniform and the skin temperature indicated by the temperature sensor is only valid for its immediate neighboring region. As a result, the skin cooling only aims to maintain the skin temperature at a certain level for a small area under the temperature sensor, rather than for the whole region under the hybrid probe. To tackle this problem, a more thorough and representative temperature monitoring system needs to be introduced and a more effective configuration of cooling the skin at the right place would also be necessary. The initial aim of this work is to develop a device which facilitates investigations into thermal provocation in deep tissues. As for the accompanied physiology, it will be left to be investigated in the next phase of this work in a larger scale study involving many more subjects.

Acknowledgement This work was partly funded by EPSRC (Grant Code EP/G005036/1).

References

1. Nicotra A, Asahina M, Mathias CJ (2004) Skin vasodilator response to local heating in human chronic spinal cord injury. Eur J Neurol 11:835–837
2. Mahe G, Liedl DA, McCarter C, Shepherd R, Gloviczki P, McPhail IR et al (2014) Digital obstructive arterial disease can be detected by laser Doppler measurements with high sensitivity and specificity. J Vasc Surg 59:1051–1057.e1
3. Rahmanian-Schwarz A, Schiefer JL, Amr A, Rothenberger J, Schaller HE, Hirt B (2011) Thermoregulatory response of anterolateral thigh flap compared with latissimus dorsi myocutaneous flap: an evaluation of flaps cutaneous flow and velocity due to thermal stress. Microsurgery 31:650–654
4. Al-Armaghany A, Tong K, Leung TS (2014) Development of a hybrid microwave-optical thermoregulation monitor for the muscle. Adv Exp Med Biol 812:347–353
5. Lazebnik M, Madsen EL, Frank GR, Hagness SC (2005) Tissue-mimicking phantom materials for narrowband and ultrawideband microwave applications. Phys Med Biol 50:4245–4258

Part IX
Blood Substitutes

Chapter 56
Possibilities of Using Fetal Hemoglobin as a Platform for Producing Hemoglobin-Based Oxygen Carriers (HBOCs)

Khuanpiroon Ratanasopa, Tommy Cedervall, and Leif Bülow

Abstract The expression levels of fetal hemoglobin (HbF) in bacterial recombinant systems are higher compared with normal adult hemoglobin (HbA). However, heme disorientation in globins are often observed in recombinant production processes, both for HbA and HbF, although the degree of heme oriental disorder is much lower for HbF. In addition, the heme disorientation can be converted to a normal conformation by an oxidation-reduction process. A chromatographic cleaning process involving a strong anion exchanger can be utilized to remove such unstable and nondesirable forms of Hb.

Keywords Heme • *E. coli* • Hemoglobin purification • Circular dichroism • Autooxidation

1 Introduction

When designing next generation hemoglobin-based oxygen carriers (HBOCs), multiple issues need to be considered. These include hemoglobin (Hb) stability, toxicity management, possibilities of scaling-up production and efficiency in oxygen delivery. The biological source of the Hb moiety is also critical in the preparation of blood substitutes. The development of HBOCs has to date largely been focused either on adult human hemoglobin (HbA) originating from outdated human blood, or bovine Hb [1]. These Hb molecules need to be formulated to generate an effective product, e.g. involving chemical cross-linking, polymerization or PEGylation to improve its stability [2]. Such modifications alter the physical and biological properties of Hb and also influence its final degradation. In addition, cell-

K. Ratanasopa • L. Bülow (✉)
Pure and Applied Biochemistry, Department of Chemistry, Lund University, 221 00 Lund, Sweden
e-mail: Leif.Bulow@tbiokem.lth.se

T. Cedervall
Biochemistry and Structural Biology, Department of Chemistry, Lund University, 221 00 Lund, Sweden

© Springer Science+Business Media, New York 2016
C.E. Elwell et al. (eds.), *Oxygen Transport to Tissue XXXVII*, Advances in Experimental Medicine and Biology 876, DOI 10.1007/978-1-4939-3023-4_56

445

free Hb is toxic outside the protective environment of the erythrocytes. Such toxicity issues can be linked to sequestering of NO and complex oxidative side-reactions of the Hb molecule [3]. This in turn may result in damages to surrounding tissues, proteins, nucleic acids and lipids. The acute-phase plasma protein hapto-globin (Hp) therefore has a protective role and binds rapidly and avidly to free Hb. The formation of this Hp-Hb complex effectively reduces and eliminates several of the intrinsic toxic effects of Hb [4, 5].

In this study, the potential of exploring HbF for HBOC applications has been examined. HbF is the main oxygen transport protein in the fetus during the last 7 months of development in the uterus and in the newborn until roughly 6 months of age. Functionally, HbF differs from HbA in that it is able to bind oxygen with higher affinity, giving the developing fetus better access to oxygen from the mother. In adults, HbF production can be reactivated pharmacologically, which is useful in the treatment of several Hb disorders, such as sickle-cell disease. Whereas HbA is composed of two alpha and two beta subunits, HbF is harbouring two alpha and two gamma subunits, commonly denoted as $\alpha_2\gamma_2$ [6]. This has consequences on the stability of the protein and on the radical formation reactions linked to HbF. Interestingly, Hp is not present in the fetal blood stream indicating that the toxicity issues may be less for HbF compared with HbA. However, the availability of HbF needs to be secured and recombinant means appear particularly attractive. The possibilities for producing HbF in a cost-effective way and linked to powerful downstream processing must therefore be critically considered. In this communi-cation, we have compared the recombinant production of HbA with HbF. We have found that Hbs produced by bacterial expression system partly involve heme disorientation. HbF was found to have less heme orientation disorders, and the irregular forms of Hb can be removed by a chromatographic step embracing a strong anion exchanger.

2 Methods

In order to produce recombinant HbF and HbA, respectively, the genes were codon optimised for *Escherichia coli* and chemically synthesised (Epoch Biolabs, USA). They were subsequently inserted into the pET-Duet-1 expression plasmid (Novagen) and expressed in *E. coli* BL21 (DE3) [7]. The bacteria were disrupted by sonication and the soluble protein fraction was purified firstly by cation exchange chromatography (CM Sepharose) equilibrated with 10 mM NaP buffer pH 6.0. The Hb protein was eluted by a linear gradient ranging up to 70 mM NaP buffer pH 7.2. This fraction was collected and the buffer was changed to 20 mM Tris–HCl buffer pH 8.3 using a Sephadex G-25 desalting column (GE Healthcare). The concentrated protein sample was then applied to a strong anion exchanger (Q Sepharose HP, GE Healthcare) equilibrated with the same buffer. The Hb proteins could be eluted with a linear gradient ranging up to 50 mM NaP buffer pH 7.2 supplemented with 100 mM NaCl, The Hb containing fractions were collected and concentrated using a Vivaspin column with 30,000 MWCO (Sartorius). The concentrated sample was rapidly frozen in liquid nitrogen and stored at $-80\ ^{\circ}C$ until further characterization. During handling, the Hb proteins were regularly kept in the CO-form. To prepare the oxy-form, samples were first converted to the ferric form by adding a 1.5 M excess of potassium ferriccyanide. The extra potassium salt was removed by desalting on a Sephadex G-25 column. A few grains of sodium dithonite were subsequently added to the Hb protein. The sample was then again passed through a Sephadex G-25 column and the oxy-form fraction was collected. The concentration of Hb was determined spectroscopically using a Cary 60 spectrophotometer (Agilent). The structural measurements were carried out with a Jasco J-815 Circular Dichroism (CD) spectropolarimeter at $20\ ^{\circ}C$.

3 Results

The comparison between HbA and HbF was initiated by examining their bacterial expression behavior. By using the pET-Duet1 plasmid/*E. coli* BL21 as expression system, soluble proteins of both HbA and HbF could easily be obtained. However, despite their close structural similarities, the expression patterns differed largely for these two proteins. The aeration of the cultivation proved to be especially critical. This can most likely be associated with the stronger oxygen affinity of HbF compared with HbA. After inducing the cultures with IPTG, HbA thus required a modest shaking (60 rpm) while the HbF cultures needed a vigorous shaking (150 rpm) to improve expression levels. Under these optimised conditions HbA and HbF were expressed at protein levels corresponding to approximately 10 and 15 mg/L culture, respectively.

HbA and HbF were purified from crude cellular extracts using two chromatographic steps. Ion exchange chromatography (IEX) is a very effective modality in the downstream processing of these proteins. In the first step, a cation IEX (CM Sepharose) was used to remove the bulk of the host proteins. In this standard procedure for protein isolation, both proteins behaved identically and no difference in their chromatographic performances was observed. Most often, a gel filtration step concludes the purification. However, the HbA containing fractions exhibited a reduced stability compared with HbF and also compared with HbA purified from blood. Several alternative measures were screened to identify the reason for this detrimental property of the HbA samples. A second ion exchanger proved to be very useful. The collected fractions were therefore subsequently applied on a strong anion exchanger (Q Sepharose). Interestingly, HbF behaved as a single protein species, while HbA could easily be separated into two well-defined fractions (Fig. 56.1). Absorption spectra (over the range 300–700 nm) were therefore collected to identify the individual Hb fractions. In the CO-form, the purified samples behaved identical to those of native Hb isolated from blood samples, exhibiting peaks at 419, 539 and 569 nm. HbA peak one and HbF generated well-defined

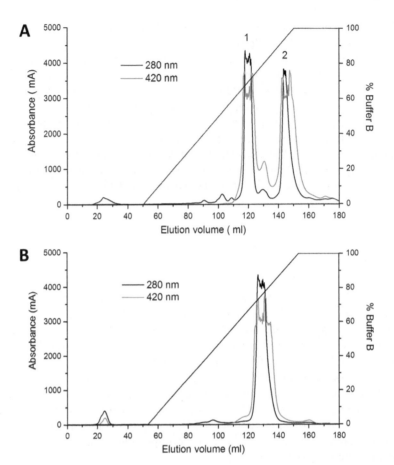

Fig. 56.1 Elution profiles of recombinant Hbs on a Q Sepharose column (16/10), flow rate 2.5 ml/ min, and using a linear gradient of buffer A (20 mM Tris–HCl at pH 8.3) mixed with buffer B (50 mM NaP at pH 7.2 and 100 mM NaCl) from 0 to 100 %. (a) HbA, (b) HbF

spectroscopic patterns characteristic for native Hb also in the ferric form. However, HbA peak 2 produced different results. The ferric form of peak 2 resulted in peaks at 409 and 535 nm, and a shoulder near 565 nm. Moreover, the ferric form of peak 2 was not stable and precipitated rapidly as observed from the baseline shifted (Fig. 56.2a, red line). Adding a few grains of sodium dithionite to this fraction resulted in deoxyHb spectra similar to those of hexacoordinated Hb [8, 9]. The same phenomenon did not happen when the deoxyHbA was prepared from the oxy form. This confirms that hemichrome formation occurs in the oxidation step. To further elucidate the structural composition of the obtained Hb fractions, CD spectra were collected both in the CO- and oxy-forms of the individual proteins.

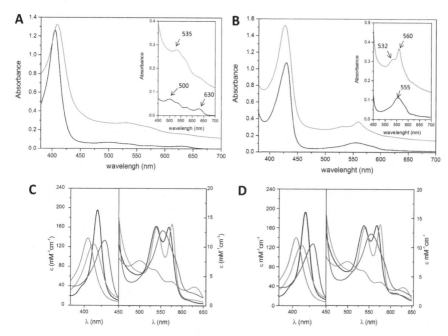

Fig. 56.2 Absorption spectra of the ferric and deoxy form of recombinant Hbs. (**a**) Absorption spectra of ferric Hb; *black line*: HbA peak 1; *red line*: HbA peak 2. (**b**) Absorption spectra of deoxyHb; *black*; HbA peak 1, *red*; HbA peak 2. The main differences in the spectra between the two Hbs are indicated by *arrows*. (**c**) and (**d**) is the spectra of four different forms of wtHbA peak 1 and wtHbF respectively: CO form; *black*, oxy form; *red*, deoxy form; *blue*, and ferric form; *magenta*

Far-UV CD spectra of recombinant Hbs showed a typical α helical structure with only minor differences from native Hbs. Differences between rHb and native Hb were observed in CD spectra (Fig. 56.3). When analysed in their CO-form, the recombinant Hb proteins exhibited significant differences in the obtained spectra compared with the native counterparts isolated from blood samples. However, when converted to the oxy-form, the samples from recombinant HbF and HbA peak 1, generated results which are very close or identical with the native proteins.

Autoxidation studies of the oxyhemoglobin fractions at 37 °C revealed an unstable character of the HbA fraction originating from the second peak. As shown in Fig. 56.4, the absorbance at the Soret region (415 nm) decreased over time and peaks at the Q band region (541 and 575 nm) were gradually diminished until they hardly were observed (after 12 h of experiment). These phenomena indicate a bleaching of the heme group in the globin pocket. In addition, a baseline

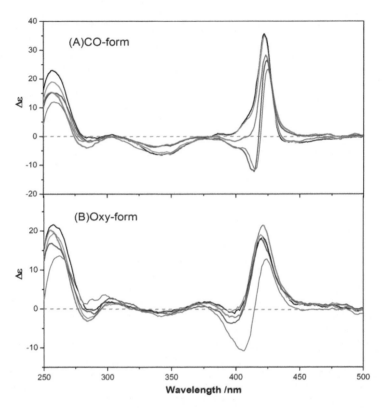

Fig. 56.3 CD spectra of native HbA (*black*), native HbF (*red*), recombinant HbA peak 1 (*blue*), recombinant HbA peak 2 (*magenta*), and recombinant HbF (*olive*) in the CO- (**a**) and oxy-forms (**b**). CD spectra (60 µM in heme) in 50 mM sodium phosphate buffer pH 7.2 were measured in a cell with a light path of 2 mm with scanning speed 50 nm/min, and three scans were averaged

shift resulting from significant light scattering was also observed due to protein precipitation in this second fraction. It is also worth noting that the final product of autoxidation after 24 h indicated the occurrence of hemichrome. This result was even more pronounced when small amounts of H_2O_2 were added (data not shown). H_2O_2 is frequently generated during oxidative side reactions of Hb. The first collected peak of HbA as well as the HbF collected fractions exhibited only a low autooxidative activity, at levels associated with the native proteins. The autoxidation rates were thus estimated to be 0.134 and 0.085 h^{-1} for rHbA peak 1 and rHbF, respectively.

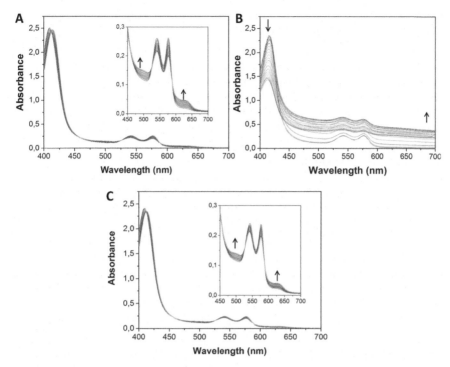

Fig. 56.4 Autooxidation study of recombinant Hbs, (**a**) HbA peak 1, (**b**) HbA peak 2, (**c**) HbF. The experiment was performed in 20 mM NaP buffer pH 7.4 at 37 °C. The spectra were recorded every 30 min for 12 h

4 Discussion

The proper insertion of the heme group into the globin in recombinant expression systems has previously been examined by a few other groups [6, 10]. It has also been suggested that the incorrect heme insertion into globins can be converted into a normal orientation after an oxidation-reduction procedure. A reversed heme orientation also occurs naturally in, for instance, Yellowfin tuna Mb [11] and even in human HbA [12, 13]. It has previously been observed that HbA expressed in *E. coli* can result in several different protein populations. Fractions containing reversed heme orientation generated different CD spectra. Particularly the HbA protein seems to contain a substantial fraction of heme disorder, as detected primarily by the negative band at the Soret region. Production of HbF resulted in a much more well-defined protein. This observation is, however, inconsistent with a previous report, pointing out that HbF is structurally more disordered in terms of heme orientation than HbA [14]. However, the formation of a proper Hb protein is highly dependent on the expression system used and this needs to be carefully optimized. HbA and HbF thus require different cultivation conditions to achieve high and functional protein levels. In addition, the fractions that contained reversed heme in HbF and HbA (peak 1) could be converted to a normal heme by an oxidation-reduction procedure. This can be monitored from the CD spectra of the oxy-form, which was prepared from the ferric from. The heme orientation of the unstable fraction of HbA (peak 2) could not be converted to a normal and functional protein. Instead, trying to convert this Hb to the ferric from yielded a bis-His (hemichrome) species. As a result, production of HbA resulted in a lower final yield (40–50 % less) compared with production of HbF. The reason for this phenomenon is still not fully known. However, it has been found that bacterial endotoxins (lipopolysaccharides, LPS) can bind to the hydrophobic patches on Hb and enhance the oxidation reaction, producing met-Hb and hemichrome as final oxidation products [15, 16]. The long hydrophobic moieties of fatty acid in LPS might be responsible for this conversion. LPS are released during the cell disruption process. As a consequence, such interactions can lead to conformational changes of the protein, which in turn then becomes more prone to unusual oxidation processes. Irregular oxidation process resulting in hemichrome as a final oxidation product has earlier been observed in Antarctic fish Hb [17]. It has been suggested that such hemichrome formation may have physiological roles since a significantly enhanced peroxidase activity has been observed [18]. However, in this study it has been shown that this form of human Hb is unstable and easily precipitates during autoxidation and other chemical oxidation processes. Therefore, the removal of this unstable form occurring during bacterial production process is essential. This observation is particularly applicable to field of basic Hb research, where recombinant Hb molecules have been used extensively to probe complex folding patterns or explain redox activity mechanisms. In addition, for HBOC purposes, recombinant Hb needs to be produced on a very large scale with fully functional properties. When produced in recombinant systems, HbF is more stable than HbA and it can be produced at higher levels, and therefore it must be considered as a more useful starting material for further HBOC development.

References

1. Kim HW, Greenburg AG (eds) (2013) Hemoglobin-based oxygen carriers as red cell substitutes and oxygen therapeutics. Springer, Berlin, 741 pp
2. Winslow RM (ed) (2006) Blood substitutes. Academic, London, 548 pp
3. Olsson MG, Allhorn M, Bülow L et al (2012) Pathological conditions involving extracellular hemoglobin: molecular mechanisms, clinical significance, and novel therapeutic opportunities for α(1)-microglobulin. Antioxid Redox Signal 17:813–846
4. Ratanasopa K, Chakane S, Ilyas M et al (2013) Trapping of human hemoglobin by haptoglobin: molecular mechanisms and clinical applications. Antioxid Redox Signal 18:2364–2374
5. Alayash AI, Andersen CB, Moestrup SK et al (2013) Haptoglobin: the hemoglobin detoxifier in plasma. Trends Biotechnol 31:2–3
6. Shen TJ, Ho NT, Zou M et al (1997) Production of human normal adult and fetal hemoglobins in *Escherichia coli*. Protein Eng 10:1085–1097

7. Reeder BJ, Grey M, Silaghi-Dumitrescu RL et al (2008) Tyrosine residues as redox cofactors in human hemoglobin: implications for engineering nontoxic blood substitutes. J Biol Chem 283:30780–30787

8. Dewilde S, Kiger L, Burmester T et al (2001) Biochemical characterization and ligand binding properties of neuroglobin, a novel member of the globin family. J Biol Chem 276: 38949–38955

9. Sawai H, Kawada N, Yoshizato K et al (2003) Characterization of the heme environmental structure of cytoglobin, a fourth globin in humans. Biochemistry 42:5133–5142

10. Shen TJ, Ho NT, Simplaceanu V et al (1993) Production of unmodified human adult hemoglobin in Escherichia coli. Proc Natl Acad Sci U S A 90:8108–8112

11. Levy MJ, La Mar GN, Jue T et al (1985) Proton NMR study of yellowfin tuna myoglobin in whole muscle and solution. Evidence for functional metastable protein forms involving heme orientational disorder. J Biol Chem 260:13694–13698

12. La Mar GN, Yamamoto Y, Jue T et al (1985) Proton NMR characterization of metastable and equilibrium heme orientational heterogeneity in reconstituted and native human hemoglobin. Biochemistry 24:3826–3831

13. Nagai M, Nagai Y, Aki Y et al (2007) Effect of reversed heme orientation on circular dichroism and cooperative oxygen binding of human adult hemoglobin. Biochemistry 47: 517–525

14. Yamamoto Y, Nagaoka T (1998) A 1H NMR comparative study of human adult and fetal hemoglobins. FEBS Lett 424:169–172

15. Kaca W, Roth RI, Levin J (1994) Hemoglobin, a newly recognized lipopolysaccharide (LPS)-binding protein that enhances LPS biological activity. J Biol Chem 269:25078–25084

16. Kaca W, Roth RI, Vandegriff KD et al (1995) Effects of bacterial endotoxin on human cross-linked and native hemoglobins. Biochemistry 34:11176–11185

17. Vitagliano L, Bonomi G, Riccio A et al (2004) The oxidation process of Antarctic fish hemoglobins. Eur J Biochem 271:1651–1659

18. Vergara A, Franzese M, Merlino A et al (2009) Correlation between hemichrome stability and the root effect in tetrameric hemoglobins. Biophys J 97:866–874

Chapter 57
The βLys66Tyr Variant of Human Hemoglobin as a Component of a Blood Substitute

R.S. Silkstone, G. Silkstone, J.A. Baath, B. Rajagopal, P. Nicholls, B.J. Reeder, L. Ronda, L. Bulow, and C.E. Cooper

Abstract It has been proposed that introducing tyrosine residues into human hemoglobin (e.g. βPhe41Tyr) may be able to reduce the toxicity of the ferryl heme species in extracellular hemoglobin-based oxygen carriers (HBOC) by facilitating long-range electron transfer from endogenous and exogenous antioxidants. Surface-exposed residues lying close to the solvent exposed heme edge may be good candidates for mutations. We therefore studied the properties of the βLys66Tyr mutation. Hydrogen peroxide (H_2O_2) was added to generate the ferryl protein. The ferryl state in βLys66Tyr was more rapidly reduced to ferric (met) by ascorbate than recombinant wild type (^{r}wt) or βPhe41Tyr. However, βLys66Tyr suffered more heme and globin damage following H_2O_2 addition as measured by UV/visible spectroscopy and HPLC analysis. βLys66Tyr differed notably from the ^{r}wt protein in other ways. In the ferrous state the βLys66Tyr forms oxy, CO, and NO bound heme complexes similar to ^{r}wt. However, the kinetics of CO binding to the mutant was faster than ^{r}wt, suggesting a more open heme crevice. In the ferric (met) form the typical met Hb acid-alkaline transition (H_2O to ^-OH) appeared absent in the mutant protein. A biphasicity of cyanide binding was also evident. Expression in *E. coli* of the βLys66Tyr mutant was lower than the ^{r}wt protein, and purification included significant protein heterogeneity. Whilst, βLys66Tyr and ^{r}wt autoxidised (oxy to met) at similar rates, the oxygen p50 for βLys66Tyr was very low. Therefore, despite the apparent introduction of a new electron transfer pathway in the βLys66Tyr mutant, the heterogeneity, and susceptibility to oxidative damage argue against this mutant as a suitable starting material for a HBOC.

R.S. Silkstone • G. Silkstone • B. Rajagopal • P. Nicholls • B.J. Reeder • C.E. Cooper (✉)
School of Biological Sciences, University of Essex, Colchester, UK
e-mail: ccooper@essex.ac.uk

J.A. Baath • L. Bulow
Department of Pure and Applied Biochemistry, Lund University, Lund, Sweden

L. Ronda
Department of Biochemistry and Molecular Biology, Universita' degli Studi di Parma, Parma, Italy

© Springer Science+Business Media, New York 2016
C.E. Elwell et al. (eds.), *Oxygen Transport to Tissue XXXVII*, Advances in Experimental Medicine and Biology 876, DOI 10.1007/978-1-4939-3023-4_57

Keywords Hemoglobin • Blood substitute • HBOC • Mutation • Oxidative stress

1   Introduction

The highly oxidised and redox active ferryl state ($Fe^{4+}=O^{2-}$) is suggested to be responsible for part of the observed toxicity of cell free hemoglobin (Hb) [1]. As part of an on-going synthetic biology project to decrease the toxicity of extra-cellular Hb, we have been characterizing mutations of hemoglobin that have the potential to decrease this oxidative toxicity. A promising pathway is the intro-duction of redox-active tyrosine (Tyr) residues that facilitate the reduction of ferryl heme [2]. For example, by introducing a new Tyr residue into the β-subunit of Hb (βPhe41Tyr), in the same position as already existed in the α-subunit (αTyr42), we showed enhanced reduction of ferryl to met Hb [3]. However, the reduction kinetics displayed were not as fast as seen when tyrosines were introduced in an *Aplysia* myoglobin model system [3]. It is possible that the introduced Tyr in the β subunit was too close to that in the α subunit, enabling electron transfer cross-talk between the two residues and limiting their effectiveness. We therefore attempted to engi-neer a new mutation on the opposite side of the heme. In this paper we discuss the properties of this mutant: βLys66Tyr.

2   Methods

Recombinant proteins were expressed in a pETDuet system in an *E. coli* BL21DE3 host and then purified by ion exchange chromatography and gel filtration as described previously [2]. The proteins studied were recombinant wild type (wt), and the βLys66Tyr and βLys66Phe mutants. Proteins were purified as the ferrous CO com-plex in 70 mM sodium phosphate buffer, pH 7.2 from which the different liganded and redox states were prepared [2]: ferric met (Fe^{3+}); ferrous deoxy (Fe^{2+}); ferrous oxy (Fe^{2+}-O_2); and ferryl ($Fe^{4+}=O^{2-}$). Optical and EPR characterisations of the different states of the proteins was as in [2]. Note that all concentrations in this paper are expressed in molar heme equivalents, not molar hemoglobin protein.

For pH titrations, met Hb (at 25 °C) was titrated to ~5.5 (+ HCl) followed by small stepwise additions of alkali (+ NaOH) until a pH of ~9.5 was reached (increments of ~0.25 in pH). Back titrations to neutral pH confirmed the reversibility of the reaction. The pH change was monitored optically using the wavelength pairs 425–405 nm (wt and βLys66Phe) and 425–410 nm (βLys66Tyr). The data were then fitted to the Henderson-Hasselbach equation and a pK_{app} value obtained (average of n = 3).

In order to measure the rate of ferryl reduction, Met Hb (10 μM) was reacted with H_2O_2 (~60–80 μM) at pH 7.4, at 25 °C for 10–15 min. Catalase enzyme (~20 nM) was then added to remove excess H_2O_2. The reductant ascorbate was added at varying concentrations (0–800 μM). The conversion of ferryl back to met was monitored optically in the visible and Soret regions until the reaction had finished. The time course (425–406 nm) was fitted to a double exponential function minimizing the least squares using the Microsoft Excel Solver program. Each set of rate constants resulting from reduction of α-Hb or β-Hb was then plotted as a function of ascorbate concentration, and this profile was fitted to a double rectangular hyperbola [2].

To measure the tendency for the proteins to undergo oxidative damage, 50 μM protein was treated with excess H_2O_2 (5 and 50 fold excess) and the samples incubated at room temperature for ~12 h, followed by reverse phase HPLC. Heme and protein degradation were monitored at 280 nm. Protein autoxidation (oxy to met conversion) was followed optically (577–630 nm) for 10 μM protein and the data fitted to single exponential. Oxygen affinity (p50) measurements and co-operativity (n) values were carried out as described previously [4], using a Hayashi reducing system at 15 °C to minimize autoxidation. The rates of CO and NO binding were measured in a stopped-flow spectrophotometer. The ferrous proteins (~5 μM) were mixed with CO of varying concentrations from 0 to 250 μM and monitored optically. Time courses (419–426 nm) were plotted and fitted to single exponential fits. The resulting k_{obs} (s^{-1}) values were plotted against the CO concentrations and straight lines of best fit made through these points. From the slopes of these straight lines and their intercepts, rate constants (k_{on}, $M^{-1} s^{-1}$) for CO binding and CO off rates (k_{off}, s^{-1}) were calculated. Cyanide (CN^-) binding to the met protein was measured optically at 424 nm Hb samples were ~2–10 μM in heme, and [CN^-] was varied.

3   Results

The initial step of purification involves the binding of the protein at pH 6.0 (low salt) to a cation exchange resin on a column followed by elution at pH 7.2 (high salt). The βLys66Tyr mutant showed differences compared to wt. For wt protein ~95 % bound at pH 6.0 and this was all eluted at pH 7.2. For the βLys66Tyr protein only ~10 % bound at pH 6.0 and this was all eluted at pH 7.2. The large fraction of this mutant that did not bind was characterised by cation exchange chromatography on a pH gradient and found to contain at least three different binding fractions. All the different binding fractions of this mutant behaved similarly by simple spectroscopic comparison (met, oxy, deoxy, and CO optical spectra at pH 7). All further studies in this paper refer to the fraction that behaved similarly to wt, i.e., bound at pH 6 and eluted at 7.2.

For the met forms of the proteins (at pH 7) the wt recombinant protein showed the same peaks as hemoglobin purified from a human subject, i.e., 405, 500, 540, 568, and 630 nm. However, the βLys66Tyr showed peaks at 410, 535, 570 and a diminished peak at 630 nm. The virtual absence of the 630 nm band in the mutant suggested an absence of high spin ferric iron. This was confirmed by low temperature EPR spectroscopy (results not shown). For oxy, deoxy and CO-bound forms (pH 7), both wt and the K66Y mutant showed the characteristic peaks expected (430, 555, and 580 nm for deoxy; 417, 540, and 577 nm for oxy and 421, 537, and 570 for CO-bound).

For native Human met Hb, a typical high-spin to low-spin transition is observed on increasing the pH as the distal heme ligand changes from weakly bound H_2O to hydroxide ligation, with a $pK_{app} = \sim8.0$. The wt recombinant protein on titration from low to high pH behaved spectrally very similarly to the native protein (high- to low-spin transition) with a $pK_{app} = 7.70$. The βLys66Phe mutant behaved spectrally similarly to the native and recombinant wt proteins, with a high to low-spin transition on increasing the pH ($pK_{app} = 8.05$). The βLys66Tyr mutant (already starting with a low-spin heme ligand) had a similar pH transition to a different low spin form ($pK_{app} = 8.0$). EPR analysis of the wt and βLys66Tyr samples at the low and high pH confirmed the spin states assigned by optical spectroscopy (data not shown).

To ascertain whether a new intra-protein electron transfer pathway had been introduced, the rate constants for ferryl reduction as a function of [ascorbate] were measured and compared to the wt protein (Fig. 57.1). In the alpha subunit, the K_D for the high affinity pathways (assigned to electron transfer via α-Tyr42) was similar in the wt and the βLys66Phe mutant (30 μM). A slight increase in affinity was seen in βLys66Tyr (6 μM). However, this was offset by a lower K_D for the low affinity pathway (electron transfer directly to the heme group): wt (169 μM); βLys66Phe (171 μM); βLys66Tyr (621 μM). In the beta subunit a much clearer distinction is seen between the mutants. In wt and βLys66Phe there is no measurable high affinity phase, whereas βLys66Tyr introduces a high affinity phase with $K_D = 3$ μM. The βLys66Tyr also had a lower K_D values for the low affinity pathway (97 μM compared to wt, (610 μM) and βLys66Phe (587 μM)).

Following ferryl reduction, the βLys66Tyr did not return exclusively to the original met form, suggesting some irreversible oxidation had occurred. This possibility was studied by HPLC. For both the wt and βLys66Tyr proteins in the absence of H_2O_2, the elution times and concentrations for the globin and heme moieties were similar (Fig. 57.2). The α-chains for both proteins eluted at the same time. The elution times for the β-chains were slightly different as expected, given the change in a charged amino acid. Following the addition of fivefold excess of H_2O_2, all peaks (heme and both globins) were noticeably decreased in absorbance, indicating oxidative damage. However, the heme peak of the mutant was decreased more in the mutant. The intensity of the peaks corresponding to the α-globins for both proteins were of similar absorbance but, that of the β-globin for the mutant was less than that of wt, indicating more damage in the particular chain containing the extra Tyr residue. Following the addition of a 50-fold excess of H_2O_2, there was no detectable heme in the βLys66Tyr mutant; a small amount remained for the wt protein.

The mutant showed a higher oxygen affinity (p50 $= 1.17 \pm 0.1$ mmHg compared to wt $= 5.21 \pm 0.17$) and a lower cooperativity (n $= 1.04 \pm 0.0$ compared to wt $= 1.3 \pm 0.07$). There was no significant difference in autoxidation rate for wild type ($t^{1/2} = 86 \pm 14.7$ min) compared to βLys66Tyr ($t^{1/2} = 98 \pm 37.6$ min). The

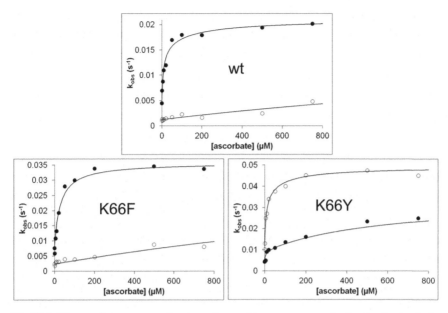

Fig. 57.1 The ascorbate concentration dependence of the rate constants for ferryl reduction in the α- and β-chains in the different mutants

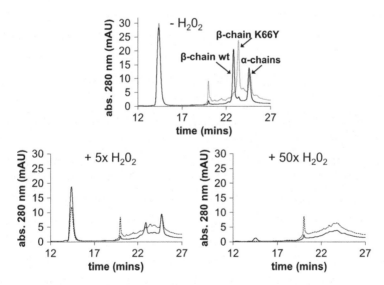

Fig. 57.2 Oxidative damage following incubation with H_2O_2 (the free heme peak elutes at 14 min and globin peaks at 23–26 min)

second order rate constant for CO binding to the heme was lower in βLys66Tyr than wt ($2.08 \pm 0.01 \times 10^5 \, M^{-1} \, s^{-1}$ compared to $3.01 \pm 0.01 \times 10^5 \, M^{-1} \, s^{-1}$), as was the CO ligand off rate ($2.98 \pm 1.84 \, s^{-1}$ compared to $6.38 \pm 1.37 \, s^{-1}$). The kinetics of cyanide binding were as expected for the wt protein, monophasic with $k = 2.5 \times 10^2$ M^{-1} s^{-1}. However, the mutant observed biphasic kinetics ($k_1 = 1.9$ and $k_2 = 0.19 \times 10^2$ M^{-1} s^{-1}) suggesting there are at least two CN^- binding species.

4 Discussion

These studies confirm that the introduction of additional tyrosine residues can enhance the detoxification of oxidizing ferryl species by physiological concentrations of plasma antioxidants, by introducing a high affinity site for electron transfer. However, in the case of βLys66Tyr, this benefit comes at the expense of protein heterogeneity, increased oxygen affinity and an enhanced tendency to oxidative damage, suggesting that βLys66Tyr is unlikely to be suitable starting material for a recombinant hemoglobin based oxygen carrier.

Acknowledgments We acknowledge financial support from BBSRC (BB/L004232/1).

Disclosure Cooper and Reeder have patents relating to the introduction of tyrosine residues into recombinant hemoglobin as a component of a blood substitute

References

1. Alayash AI (2014) Blood substitutes: why haven't we been more successful? Trends Biotechnol 32:177–185
2. Reeder BJ, Grey M, Silaghi-Dumitrescu RL et al (2008) Tyrosine residues as redox cofactors in human hemoglobin: implications for engineering non toxic blood substitutes. J Biol Chem 283:30780–30787
3. Reeder BJ, Svistunenko DA, Cooper CE et al (2012) Engineering tyrosine-based electron flow pathways in proteins: the case of aplysia myoglobin. J Am Chem Soc 134:7741–7749
4. Portoro I, Kocsis L, Herman P et al (2008) Towards a novel haemoglobin-based oxygen carrier: Euro-PEG-Hb, physico-chemical properties, vasoactivity and renal filtration. Biochim Biophys Acta 1784:1402–1409

Chapter 58
PEGylated Bovine Carboxyhemoglobin (SANGUINATE™): Results of Clinical Safety Testing and Use in Patients

A. Abuchowski

Abstract Oxygen transfer agents have long been sought as a means to treat hypoxia caused by congenital or acquired conditions. Hemoglobin-based oxygen carriers were in clinical development as blood substitutes, but development was halted due to the finding of significant vasoactivity. Rather than develop a blood substitute, a product for indications characterized by hypoxia is in development. PEGylated bovine carboxyhemoglobin (SANGUINATE™) is both a carbon monoxide releasing molecule and an oxygen transfer agent. It is comprised of three functional components that act to inhibit vasoconstriction, reduce inflammation and optimize the delivery of oxygen. SANGUINATE has the potential to reduce or prevent the effects of ischemia by inhibiting vasoconstriction and re-oxygenating tissue. Phase 1 safety trials in healthy volunteers were completed in 2013. SANGUINATE was shown to be safe and well tolerated with no serious adverse effects. Phase Ib studies have been completed in stable patients with Sickle Cell Disease. SANGUINATE has also been administered to two patients under emergency use protocols. Both patients exhibited improved status following treatment with SANGUINATE.

Keywords SANGUINATE • Clinical • Safety • Hypoxia • Ischemia

1 Introduction

The holy grail of developing a clinical substitute for blood transfusions was halted when serious adverse side effects such as, cardiac, gastrointestinal, hepatic, pancreatic, central nervous system, and renal damage with a concomitant increase in mortality were observed in some trials. The adverse events were attributed to nitric oxide (NO) scavenging and auto-regulatory vasoactivity resulting in vasoconstriction, hypertension and heme-mediated oxidative damage [1, 2]. Many of the trial

A. Abuchowski (✉)
Prolong Pharmaceuticals, South Plainfield, NJ 07080, USA
e-mail: aabuchowski@prolongpharma.com

© Springer Science+Business Media, New York 2016   461
C.E. Elwell et al. (eds.), *Oxygen Transport to Tissue XXXVII*, Advances in
Experimental Medicine and Biology 876, DOI 10.1007/978-1-4939-3023-4_58

designs for these products were hampered by its use in severely injured trauma or surgical patients, thereby making any analysis of safety and efficacy difficult.

While the development of blood substitutes has been largely abandoned by the biopharmaceutical industry, there is progress in the development of oxygen transfer agents (OTAs). These products are not blood substitutes but rather facilitate the transfer of oxygen to hypoxic tissue. OTAs are intended as therapeutics and will be developed in specific target indications that permit clearer analysis of toxicity as well as efficacy.

Despite the toxicity issues seen when used as a blood substitute, hemoglobin still remains the ideal transport molecule for oxygen. Therefore, the development has focused on optimizing the molecular design to modulate the hemoglobin affinity for oxygen (p50), reduce NO scavenging, increase the resistance to auto and chemically-induced oxidation and possess the appropriate viscosity/colloidal osmotic pressure characteristics for circulatory use.

While a number of approaches have been used to develop the optimal OTA, many attempts could not surmount the hurdles in optimizing p50 and reducing NO scavenging. For example, in patients undergoing primary hip arthroplasty, MP40X showed minimal effects on blood pressure changes, but had higher rates of adverse events, including elevation of liver enzymes and troponin levels [3]. In a Phase 2b trial in trauma patients, it failed to meet its primary endpoint [4]. A follow-on product MP4CO has been abandoned as well. This experience illustrates the importance of appropriate molecular design, choice of targeted indication and clear clinical endpoints.

SANGUINATE (PEGylated bovine carboxyhemoglobin) has been designed to avoid the problems of NO scavenging and auto-oxidation while optimizing oncotic pressure and p50. It acts both as a CO releasing molecule and an O_2 transfer agent. Its components act to prevent the toxicity and work synergistically to enhance its therapeutic potential.

SANGUINATE is produced through the polyethylene glycol modification (PEGylation) of surface lysine residues on purified bovine Hb with 5000 molecular weight PG-succinimidyl carbonate followed by carboxylation. PEGylation is a well-established drug delivery technology. Due its hydrophilic nature, PEG increases the effective molecular size of hemoglobin and thereby inhibits the interaction of haemoglobin with vascular endothelium [5]. PEG-hemoglobins have shown decreased vasoactivity as compared to acellular hemoglobin. PEG increases oncotic pressure which is known to impact blood flow dynamics and induce the production of vasodilatory NO by vascular wall shear stress activity [6]. PEGylation also has the properties of extending circulating half-life and decreasing immunogenicity of protein therapeutics [7].

Bovine hemoglobin (bHb) has several advantages over its human counterpart. The oxygen affinity (p50) of PEG modified bHb (p50 = 7–16 mmHg) is superior to PEG modified human derived hemoglobin products (p50 = 4) [8]. Since the goal is to maximize oxygen delivery to oxygen-deprived tissues, utilizing PEG-bHb, with a higher p50, offers therapeutic advantages over a non-bovine source. Also, as the p50 of PEG-bHb lies between that of the red blood cell (p50 = 26) and hypoxic tissue, it acts as a conduit to transfer the oxygen from the blood cells to the tissue. In addition, bHb requires only plasma chloride for stability, unlike human hemoglobin which is dependent upon 2,3diphosphoglycerate to remain in its active tetrameric form.

CO is an essential component to the therapeutic effect of SANGUINATE. CO reduces auto-oxidation of hemoglobin and improves shelf-life such that very little methemoglobin is formed, even after storage at room temperature for up to 18 months. More importantly, CO has therapeutic activity that complements the oxygen transfer role of hemoglobin. It is an endogenous regulatory molecule that plays a key role in regulating vascular muscle tone and reducing platelet aggregation. Delivery of exogenous CO has shown potent protection in numerous experimental models of inflammation, sepsis and hemorrhagic shock [9]. Potent anti-inflammatory activity associated with down-regulation of proinflammatory has been reported as a result of exposure to CO [10]. As such, the CO released from SANGUINATE can prevent the vasoconstriction that was observed with early hemoglobin-based products and contributes a potential therapeutic effect. In animal studies, CO was shown to be rapidly released by SANGUINATE within minutes of transfusion with a whole blood COHb peak level of 5 % that declined to baseline levels over a 2 h period [11].

To address toxicity and safety concerns of its use as a single or repeating dose therapeutic, eight toxicology preclinical studies were performed to obtain regulatory approval for the Phase I study [12]. Parameters evaluated for the assessment of toxicity included clinical observations, body weights, ophthalmic observations, food consumption, hematology, clinical chemistry, coagulation, urinalysis, and organ evaluation (44 different tissues examined including brain, heart, liver, spleen, kidney) and histopathology. Additional assessments included in some of the studies were special histopathology staining, toxicokinetics, functional observations, immunogenicity, and cardiovascular measurements. There were no adverse effects identified for any dose and, therefore, a (no) observed adverse effects level (NOEL) could not be determined even at dosage levels of 1200 mg/kg (monkey), 1600 mg/kg (pig) and 2400 mg/kg (rat). To evaluate whether the bovine protein would induce antibody formation, 6 and 9 month repeat dosing studies were performed in rats and pigs. The presence of IgG antibodies directed against SANGUINATE was measured in serum samples by the ELISA method. The data indicated that SANGUINATE did not induce an immunogenic response.

For assessment of its therapeutic effects, SANGUINATE was tested in ischemic animal models. Diabetic mouse models were also tested, as these demonstrate high oxidative stress. In animal models of stroke and myocardial infarction, SANGUINATE was effective in significantly reducing the area of infarct as compared to controls [11–13]. Studies in animal models of focal cerebral ischemia demonstrated that SANGUINATE inhibited vasoconstriction [13]. In diabetic mice, administration of SANGUINATE improved blood flow recovery and capillary density in ischemic muscle tissue after femoral artery ligation [11]. Data from an ischemic/reperfusion myocardial model in diabetic mice indicated that SANGUINATE treatment reduced oxidative stress during the early phase of reperfusion, indicating that SANGUINATE is effective in protecting against reperfusion injury [12]. SANGUINATE has also been shown to reduce neurological deficits in rat models of focal cerebral ischemia [11]. This demonstrates SANGUINATE acts to halt the hypoxic cascade and protect the cells and tissues surrounding the area of insult. Collectively, these studies support an ongoing investigation of SANGUINATE in a wide variety of indications where an OTA is needed.

SANGUINATE has been evaluated in a randomized Phase I single-blind placebo-controlled study in 24 healthy volunteers. Three cohorts of 8 [6 receiving SANGUINATE (40 mg/ml, 2 receiving saline as a placebo comparator)] in an ascending dose study received 80, 120 or 160 mg/kg. There were no clinically meaningful safety findings following the single intravenous infusion in all dose groups. In particular, there were no significant signs of systemic or pulmonary hypertension. There was an observed trend toward increased blood pressure (both systolic and diastolic) in subjects administered SANGUINATE that was not seen in placebo subjects. These changes were transient and returned to baseline by 72 h after the infusion was completed. Mean increases did not reach the level of arterial hypertension. There was no clear dose- proportionality suggesting that the increase in blood pressure was likely due to the oncotic effect of SANGUINATE. Pharmacokinetic analysis found all parameters measured (Cmax, tmax, AUC, 0-last, AUC0-inf, kel, t1/2) had a dose related response. The circulating half-life (t1/2) ranged from 7.9 to 13.8 h [14].

Due to its multiple mechanisms of action, SANGUINATE has potential use in a broad range of indications. The initial indication is for the treatment of vaso-occlusive crisis in patients with sickle cell disease (SCD). This is one of several SCD comorbidities which include stroke, leg ulcers, priapism and acute chest syndrome. The underlying pathophysiology is caused by ischemia and hemolysis resulting in extensive inflammation, reactive oxygen species formation and up-regulation of adhesion molecules.

Two patients with hemoglobinopathies that could not or would not receive blood transfusion received SANGUINATE under emergency INDs. Despite persistently low hemoglobin levels due to hemolytic manifestations of sickle cell disease, improvements in symptoms were seen in both patients with no evidence of drug-related adverse events following the administration of SANGUINATE. Conclusions regarding the efficacy of SANGUINATE cannot be drawn from these cases as they are not part of a controlled clinical trial. In each case, however, the attending physician attributed the improvement in the patient's symptoms to the treatment with SANGUINATE.

The first patient, a 61-year-old female, refused blood transfusion due to religious reasons and was admitted to the ICU with hemoglobin level of 6.5 g/dL. She was in near-comatose condition and unresponsive to voice with a pre-infusion brainoxygen saturation level of 48 %. One 500 mL bag of SANGUINATE 40 mg/mL (=290 mg/kg dose based on the recorded 69 kg mass) was administered by 2-h intravenous infusion. No acute toxicity was noted. The next morning, the patient continued to be tachypneic, but was more alert with improved brain oxygenation (EQUANOX™ Oximeter System) increasing from the high 40s to low 50s to the mid to upper 60s. The patient nodded/shook head appropriately to questions. The next day, the patient experienced acute respiratory failure and had a hemoglobin level of 2.9 and a second infusion of SANGUINATE was administered. There were no acute toxicities noted post-infusion. The patient showed poor response to therapy, with severe hypotension and negligible renal function. The family requested discontinuation of respiratory support. The site investigator credited the patient's improvement in status to the infusion with SANGUINATE, and assessed the event of death as not related to study drug [15].

The second patient was 23 years of age and was admitted to the ICU presenting with respiratory distress (PO_2 54.3, sPO_2 88.4 %, pH 7.48, PCO_2 38.2, hemoglobin 4.6) secondary to acute chest syndrome. Patient was offered transfusions and, despite risk of death, refused due to religious beliefs. A total of three doses of SANGUINATE were administered (Day 1, Day 2 and Day 8) and the patient was extubated successfully on Day 10. Over the course of treatment, the patient had improved cardiac, pulmonary and renal function. A measure of inflammation (C-reactive protein) also improved substantially. Most of the patient's laboratory values were restored to within the normal range, despite the continued severe anemia demonstrated by the extremely low hemoglobin and hematocrit levels. Despite her persistently low hemoglobin of 3.1, she reported improvement in her dyspnea upon extubation. Her initial transcranial Doppler (TCD) revealed hyperemia with high blood flow velocities that normalized following SANGUINATE infusion. Following the third administration at approximately 18 h post-dose, the TCD revealed a statistically significant ($p < 0.001$, with ANOVA and post-hoc Bonferroni tests) velocity reduction in both MCAs, which correlated with the patient's clinical improvement [16].

2   Conclusion

From the knowledge elucidated from the development of blood substitutes, the design of an OTA would require higher O_2 affinity, a viscosity similar to blood, increased oncotic pressure as compared to blood and long plasma retention [2]. The use of a single approach (i.e. PEGylation) would be insufficient to address all these issues. SANGUINATE has a mechanistic profile that addresses the issues of acellular hemoglobin toxicity and adds functionality that goes beyond oxygen transfer. Nonclinical studies have shown that SANGUINATE can play a key role in the inhibition of vasoconstriction, reduction in infarct volume and impact on oxidative stress. In vitro work has demonstrated that SANGUINATE is able to repetitively bind and transfer oxygen to deoxygenated cells (manuscript pending). The increase in cerebral oximetry readings in patients treated under eINDs without an accompanying increase in hemoglobin levels also suggests that this product is transferring oxygen.

A Human safety study has demonstrated that SANGUINATE produces no serious adverse events in humans. These results are consistent with preclinical data that no NOEL could be determined, even at the highest doses tested. A Phase Ib study in stable SCD patients has reported no serious adverse events at the lower dose and analysis of the higher dose findings will be completed shortly.

Because SANGUINATE has multiple modes of action that result in the inhibition of vasoconstriction and promotion of plasma expansion, oxygenation of hypoxic tissue, reduction of infarct volume and potential anti-inflammatory, its potential use is broad, ranging from treatment of hemoglobinopathies to stroke and chronic wounds. All these indications have a pathophysiology that results from initiation of the ischemic cascade and the ensuing damaging inflammation, apoptosis and necrosis. SANGUINATE may be able to act upon different steps in the cascade to reduce or prevent cellular death and tissue injury.

At this time, a Phase II study for the use of SANGUINATE for the reduction or prevention of delayed cerebral ischemia following subarachnoid hemorrhage has been approved by the FDA. A Phase II trial in SCD patients with vaso-occlusive crisis has been filed and other indications are under consideration. Careful selection of the initial indications and clinical endpoints for regulatory approval are the next step in the development of this product.

References

1. Tsai AG, Cabrales P, Manjula BN et al (2006) Dissociation of local nitric oxide concentration and vasoconstriction in the presence of cell-free hemoglobin oxygen carriers. Blood 108 (10):3603–3610
2. Winslow RM (2003) Current status of blood substitute research: towards a new paradigm. J Intern Med 253(5):508–517
3. Olofsson CI, Gorecki AZ, Dirksen R et al (2011) Evaluation of MP40X for prevention of perioperative hypotension in patients undergoing primary hip arthroplasty with spinal anesthesia: a randomized, double-blind, multicenter study. Anesthesiology 114(5):1048–1063
4. Carrol J (2013) Sangart goes MIA after burning through $260M-plus on R&D. Fierce Biotech. http://www.fiercebiotech.com/story/sangart-goes-mia-after-burning-through-260m-plus-rd/2013-11-07
5. Cabrales P (2013) Examining and mitigating acellular hemoglobin vasoactivity. Antioxid Redox Signal 18(17):2329–2341
6. Sriram K, Tsai AG, Cabrales P et al (2012) PEG-albumin supraplasma expansion is due to increased vessel wall shear stress induced by blood viscosity shear thinning. Am J Physiol Heart Circ Physiol 302(12):H2489–H2497
7. Veronese FM, Pasut G (2005) PEGylation, successful approach to drug delivery. Drug Discov Today 10(21):1451–1458
8. Sakai H, Masada Y, Takeoka S et al (2002) Characteristics of bovine hemoglobin as a potential source of hemoglobin-vesicles for an artificial oxygen carrier. J Biochem 131(4):611–617
9. Foresti R (2008) Use of carbon monoxide as a therapeutic agent: promises and challenges. Intensive Care Med 34:649–658
10. Klaus JA, Kibler KK, Abuchowski A et al (2010) Early treatment of transient focal cerebral ischemia with bovine PEGylated carboxy hemoglobin transfusion. Artif Cells Blood Substit Immobil Biotechnol 38(5):223–229

11. Shen X, Rosario R, Zou YS et al (2011) Improvement in angiogenesis and restoration of blood flow in diabetic mice by SANGUINATE™. FASEB J 25:1091.4
12. Ananthakrishnan R, Li Q, O'Shea KM et al (2013) Carbon monoxide form of PEGylated hemoglobin protects myocardium against ischemia/reperfusion injury in diabetic and normal mice. Artif Cells Nanomed Biotechnol 41(6):428–436
13. Zhang J, Cao S, Kwansa H et al (2012) Transfusion of hemoglobin-based oxygen carriers in the carboxy state is beneficial during transient focal cerebral ischemia. J Appl Physiol 113 (11):1709–1717
14. Misra H, Lickliter J, Kazo F et al (2014) PEGylated carboxyhemoglobin bovine (SANGUINATE™): results of a phase I clinical trial. Artif Organs 38(8):702–707
15. Parmar D (2014) A case study of SANGUINATE™ in a patient with a comorbidity due to an underlying hemoglobinopathy. J Sickle Cell Hemoglobinopath 1:2
16. Alaali Y, Mendez M, Abdelhak T et al (2014) Compassionate use of SANGUINATE™ in acute chest syndrome. J Sickle Cell Hemoglobinopath 1:1

Part X
Critical Care Neonatal

Chapter 59
Oxygen Supplementation is Effective in Attenuating Maternal Cerebral Blood Deoxygenation After Spinal Anesthesia for Cesarean Section

Noriya Hirose, Yuko Kondo, Takeshi Maeda, Takahiro Suzuki, Atsuo Yoshino, and Yoichi Katayama

Abstract The purpose of this study was to measure changes in maternal cerebral blood oxygenation using near-infrared spectroscopy (NIRS) for 15 min after spinal anesthesia performed for cesarean section, and to determine the efficacy of supplemental oxygen in maintaining maternal cerebral blood oxygenation. Thirty patients were randomly assigned to either receive 100 % oxygen via a facemask at a constant flow rate of 3 l/min throughout the study (O_2 group), or were evaluated without supplemental oxygen (Air group). Changes in cerebral blood oxygenation were evaluated using the following parameters: oxy-hemoglobin (Hb), deoxy-Hb, and total-Hb concentrations, as well as tissue oxygen index (TOI), measured over the forehead by NIRS. Mean arterial pressure (MAP) and heart rate (HR) were also recorded throughout the study. Mean oxy-Hb, total-Hb, TOI, and MAP in both groups decreased significantly from baseline values ($P < 0.05$). The reduction in oxy-Hb and TOI in the Air group was significantly greater than that in the O_2 group (oxy-Hb: -4.72 vs. -2.96 µmol/l; $P < 0.05$, TOI: -6.82 vs. -1.68 %; $P < 0.01$); however, there were no significant differences in the reduction of total-Hb and MAP between the groups. Mean deoxy-Hb in the Air group was significantly higher than that in the O_2 group (0.02 vs. -1.01 µmol/l; $P < 0.05$). The results of the present study demonstrate that oxygen supplementation attenuates cerebral blood

N. Hirose • T. Suzuki
Division of Anesthesiology, Department of Anesthesiology, Nihon University School of Medicine, 30-1, Oyaguchi-Kamicho, Itabashi-Ku, Tokyo 173-8610, Japan

Y. Kondo (✉) • T. Maeda
Division of Anesthesiology, Department of Anesthesiology, Nihon University School of Medicine, 30-1, Oyaguchi-Kamicho, Itabashi-Ku, Tokyo 173-8610, Japan

Division of Neurosurgery, Department of Neurological Surgery, Nihon University School of Medicine, 30-1, Oyaguchi-Kamicho, Itabashi-Ku, Tokyo 173-8610, Japan
e-mail: kondo.yuko@nihon-u.ac.jp

A. Yoshino • Y. Katayama
Division of Neurosurgery, Department of Neurological Surgery, Nihon University School of Medicine, 30-1, Oyaguchi-Kamicho, Itabashi-Ku, Tokyo 173-8610, Japan

© Springer Science+Business Media, New York 2016
C.E. Elwell et al. (eds.), *Oxygen Transport to Tissue XXXVII*, Advances in Experimental Medicine and Biology 876, DOI 10.1007/978-1-4939-3023-4_59

deoxygenation secondary to the reduction in cerebral blood flow following spinal anesthesia.

Keywords Spinal anesthesia • Cesarean section • Cerebral blood oxygenation changes • Near-infrared spectroscopy • Oxygen supplementation

1 Introduction

Potent sympathetic blockade after spinal anesthesia for cesarean section frequently produces maternal hypotension, which may cause subsequent maternal cerebral ischemia [1, 2]. Previous studies using near-infrared spectroscopy (NIRS) demonstrated that maternal cerebral blood oxygenation significantly decreases after spinal anesthesia and depends on arterial pressure [2, 3]. Hence, oxygen supplementation is routinely given to patients undergoing cesarean section under spinal anesthesia in clinical settings. However, the efficacy of supplemental oxygen in maintaining maternal cerebral blood oxygenation has not been determined. We therefore used NIRS to compare the changes in maternal cerebral blood flow (CBF) and oxygenation (CBO) after spinal anesthesia with and without supplemental oxygen.

2 Methods

This study was approved by the Hospital Ethics Committee on Human Rights Clinical Trials and Research, and written informed consent was obtained from all patients included in this prospective, randomized control study. We studied 30 ASA (American Society of Anesthesiologists) physical status I-II female patients scheduled for elective cesarean section under spinal anesthesia. Patients were randomly assigned to receive 100 % oxygen via a facemask at a constant flow rate of 3 l/min throughout the study (O_2 group) or were not given supplemental oxygen (Air group).

2.1 Monitoring and Anesthetic Procedure

The patients were monitored with automatic non-invasive blood pressure measurements on the right arm, pulse oximetry, and electrocardiography. We measured CBF and CBO changes over the forehead using NIRS (NIRO Pulse™, Hamamatsu Photonics, Hamamatsu City, Japan). The NIR light from three laser diodes (775, 810, 850 nm) was directed at the forehead through a fiberoptic bundle, and the reflected light was transmitted to a multi-segment photodiode detector array. The NIRS system measures the concentrations of oxy-hemoglobin (Hb), deoxy-Hb, total-Hb (total-Hb: oxy-Hb + deoxy-Hb), and the tissue oxygen index (TOI). Measurements of the NIRS parameters, mean arterial pressure (MAP), and heart rate (HR) were commenced before the anesthetic procedure (baseline), and were repeated every 1 min for 15 min after administration of spinal anesthesia.

All patients fasted overnight and did not receive any premedication. On arrival at the operating room, each patient was placed in the supine position and received intravenous colloids (6 % hydroxyethyl starch: HES) at a rate of 20 ml/kg/h. After recording baseline values of all the parameters, the patients in the O_2 group received 100 % oxygen via a facemask at a constant flow rate of 3 l/min, oxygenation being continued for 15 min after administration of spinal anesthesia, while the patients in the Air group rested without oxygen supplementation. Three minutes later, patients were placed in the right lateral decubitus position to receive spinal and epidural anesthesia. An epidural catheter was inserted into the L1/2 interspace and lumbar puncture was performed at the L3/4 interspace. Hyperbaric 0.5 % bupivacaine was used for spinal anesthesia in all the patients. The patients were returned to the supine position immediately after intrathecal injection of bupivacaine. If hypotension (MAP < 80 % of baseline values or systolic blood pressure < 90 mmHg) was observed, the patient was treated with ephedrine and/or phenylephrine (IV bolus) and left lateral uterine displacement, as required.

2.2   Data Analysis

All results are expressed as mean ± SD, or as median and range. The unpaired Student's t-test or Welch's t-test followed by F-test was used for comparisons of demographic and clinical data between the O_2 and Air groups. The incidence of adverse symptoms between the groups was compared using Fisher's exact probability test. Intra- and inter-group comparisons of MAP, HR, and mean changes in NIRS variables during the 15 min period following administration of spinal anesthesia were performed using repeated-measures ANOVA and Tukey-Kramer multiple comparison test for post-hoc analysis. A P-value of < 0.05 was considered statistically significant.

3   Results

We were able to obtain data from 12 to 13 patients in the O_2 and Air groups, respectively. Data of five patients were poor due to adhesive unstability of the optical sensor of NIRS and therefore excluded from this study. There were no significant differences in demographic and clinical data between the groups (Table 59.1). Changes in MAP, HR, and the values measured using NIRS in each group are shown in Fig. 59.1A–F. Repeated measures analyses indicated significant differences in MAP, oxy-Hb, deoxy-Hb, total-Hb, and TOI with time (MAP and

Table 59.1 Comparison of demographic and clinical data between the air group and O_2 group

	Air group (n = 13)	O_2 group (n = 12)
Age (years)	34.5 (6.4)	32.6 (5.2)
Height (cm)	156.2 (5.2)	158.4 (6.4)
Weight (kg)	58.3 (7.3)	61.7 (7.8)
Gestational age (weeks)	37.2 (0.8)	37.8 (0.7)
Dose of bupivacaine (mg)	11.1 (1.4)	11.1 (0.7)
Block height, dermatome (median)	Th 3 (Th 2–6)	Th 4 (Th 2–7)
Fluids (ml)		
Before spinal anesthesia	415.3 (117.3)	455.8 (93.5)
During 15 min after spinal anesthesia	295.7 (36.4)	318.8 (34.4)
Adverse symptoms		
Incidence, n	5	1
Variety (n)	Nausea (4), dizziness (1)	Nausea (1)

Data are presented as mean (SD) or median (range)

oxy-Hb; $P < 0.05$, total-Hb and TOI; $P < 0.01$) and by group (oxy-Hb and deoxy-Hb; $P < 0.05$, TOI; $P < 0.01$). Mean values of oxy-Hb, total-Hb, TOI, and MAP in both groups decreased significantly from baseline values ($P < 0.05$). In the comparison between groups, the reduction of oxy-Hb and TOI in the Air group was significantly greater than that in the O_2 group ($P < 0.05$), while there were no significant differences in the reduction of MAP and total-Hb between the groups. Mean deoxy-Hb in the Air group was significantly higher than that in the O_2 group, although there were no significant changes in these parameters from baseline values in both groups ($P < 0.05$).

4 Discussion

The results of the present study demonstrated that CBF estimated from oxy-Hb and total-Hb [4], CBO estimated from TOI [5], and MAP significantly decreased after spinal anesthesia when compared with baseline in both groups. These results are consistent with previous studies using NIRS [2, 3] and suggest that maternal CBF and CBO decrease secondary to the hypotension that occurs after spinal anesthesia. The present study also demonstrated that oxy-Hb and TOI were significantly decreased and deoxy-Hb was significantly increased in the Air group compared to the O_2 group, whereas the decrease in total-Hb was similar between the groups. These results suggest that even a low concentration of supplemental oxygen, at an inspired oxygen fraction (FiO_2) of approximately 0.24–0.3, as was administered in this study, is effective in attenuating the maternal cerebral blood deoxygenation associated with a decrease in CBF after spinal anesthesia.

Oxygen supplementation for parturients undergoing cesarean section under spinal anesthesia is traditionally recommended for the prevention of both fetal

Fig. 59.1 Changes in NIRS parameters and hemodynamics from baseline for 15 min after administration of spinal anesthesia in the Air and O_2 groups. Graphs: (**A**) Mean arterial pressure; (**B**) Heart rate; (**C**) oxy-Hb; (**D**) deoxy-Hb; (**E**) total-Hb; (**F**) Tissue oxygen index (TOI). Time points: *A*; baseline, *B*; 3 min after the beginning of supplemental oxygen (O_2 group), *C*; immediately after spinal anesthesia. *$P < 0.05$ between groups

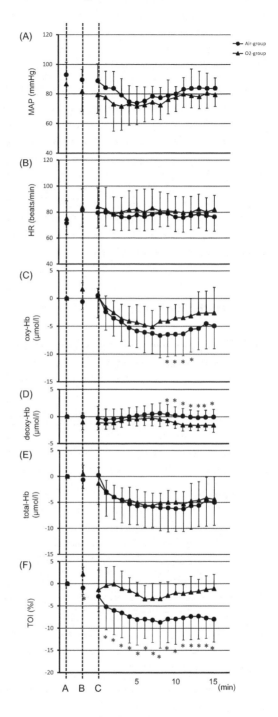

acidemia and maternal cerebral ischemia [1, 6]. However, Khaw et al. demonstrated that a high concentration of supplemental oxygen (FiO_2 0.6) had no impact on umbilical pH, although it modestly increased maternal arterial and umbilical venous blood oxygenation [7]. Further, breathing a FiO_2 of more than 0.3 reportedly markedly increased oxygen free radical activity in both mother and fetus [8]. Thus, they suggested that empirical use of oxygen supplementation in parturients may not be necessary. However, in the present study, there were four cases in the Air group whose TOI decreased transiently by more than 13 % from baseline, which raises the suspicion of severe ischemia based on evaluation by NIRS [9], although the mean decrease in TOI remained within the safe range. Additionally, the incidence of side effects such as nausea and dizziness, which could be related to cerebral ischemia [1], was higher in the Air group, although there were no statistically significant differences between the groups. Based on these results, supplemental oxygen for parturients may be desirable in terms of prevention of maternal cerebral blood deoxygenation after spinal anesthesia for cesarean section.

5 Conclusion

Oxygen supplementation for parturients undergoing cesarean section under spinal anesthesia is effective in attenuating maternal cerebral blood deoxygenation after spinal anesthesia-induced reduction in cerebral blood flow.

Acknowledgments The authors thank Kaoru Sakatani MD, PhD, (NEWCAT Institute, Department of Electronic Engineering, Nihon University College of Engineering, Koriyama, Japan) for assistance with data analysis.

References

1. Datta S, Alper MH, Ostheimer GW et al (1982) Method of ephedrine administration and nausea and hypotension during spinal anesthesia for cesarean section. Anesthesiology 56:68–70
2. Berlac PA, Rasmussen YH (2005) Per-operative cerebral near-infrared spectroscopy (NIRS) predicts maternal hypotension during elective caesarean delivery in spinal anaesthesia. Int J Obstet Anesth 14:26–31
3. Kondo Y, Sakatani K, Hirose N et al (2013) Effect of spinal anesthesia for elective cesarean section on cerebral blood oxygenation changes: comparison of hyperbaric and isobaric bupivacaine. Adv Exp Med Biol 765:109–114
4. Pryds O, Greisen G, Skov LL et al (1990) Carbon dioxide-related changes in cerebral blood volume and cerebral blood flow in mechanically ventilated preterm neonates: comparison of near infrared spectrophotometry and 133 Xenon clearance. Pediatr Res 27:445–449
5. Terborg C, Birkner T, Schack B et al (2003) Noninvasive monitoring of cerebral oxygenation during vasomotor reactivity tests by a new near-infrared spectroscopy device. Cerebrovasc Dis 16:36–41
6. Corke BC, Datta S, Ostheimer GW et al (1982) Spinal anesthesia for caesarean section. The influence of hypotension on neonatal outcome. Anaesthesia 37:658–662

7. Khaw KS, Wang CC, Ngan Kee WD et al (2002) Effects of high inspired oxygen fraction during elective caesarean section under spinal anesthesia on maternal and fetal oxygenation and lipid peroxidation. Br J Anaesth 88:18–23
8. Khaw KS, Ngan Kee WD, Chu CY et al (2010) Effects of different inspired oxygen fraction on lipid peroxidation during general anaesthesia for elective caesarean section. Br J Anaesth 105:355–360
9. Al-Rawi PG, Kirkpatrick PJ (2006) Tissue oxygen index: thresholds for cerebral ischemia using near-infrared spectroscopy. Stroke 37:2720–2725

Chapter 60
Changes in Cerebral Blood Flow and Oxygenation During Induction of General Anesthesia with Sevoflurane Versus Propofol

Yuko Kondo, Noriya Hirose, Takeshi Maeda, Takahiro Suzuki, Atsuo Yoshino, and Yoichi Katayama

Abstract Sevoflurane and propofol are widely used for induction and maintenance of general anesthesia. Although the effects of sevoflurane and propofol on cerebral hemodynamics during maintenance of general anesthesia have been demonstrated, the effects during induction of general anesthesia have still not been clarified. We therefore compared changes in cerebral blood flow (CBF) and oxygenation (CBO) during induction of anesthesia using sevoflurane (group S: $n = 9$) or propofol (group P: $n = 9$). CBF and CBO were evaluated using the following variables: oxy-, deoxy-, and total-hemoglobin (Hb) concentrations and tissue oxygen index (TOI), measured on the forehead by near-infrared spectroscopy. The variables were recorded immediately before administration of sevoflurane or propofol and at every 10 s for 4 min after administration of the induction agent. Patients received 8 % sevoflurane in 100 % oxygen via an anesthesia mask in group S, and an IV bolus of 2 mg/kg of propofol during oxygenation in group P. We found that oxy-Hb, total-Hb, and TOI were significantly higher in group S than in group P ($P > 0.05$). Changes in deoxy-Hb, MBP, and HR did not differ between the groups. The results of the present study demonstrated that sevoflurane increases CBF and CBO during induction of general anesthesia.

Y. Kondo • T. Maeda
Division of Neurosurgery, Department of Neurological Surgery, Nihon University School of Medicine, 30-1,Oyaguchi-Kamicho, Itabashi-ku, Tokyo 173-8610, Japan

Division of Anesthesiology, Department of Anesthesiology, Nihon University School of Medicine, 30-1,Oyaguchi-Kamicho, Itabashi-ku, Tokyo 173-8610, Japan

N. Hirose (✉) • T. Suzuki
Division of Anesthesiology, Department of Anesthesiology, Nihon University School of Medicine, 30-1,Oyaguchi-Kamicho, Itabashi-ku, Tokyo 173-8610, Japan
e-mail: hirose.noriya@nihon-u.ac.jp

A. Yoshino • Y. Katayama
Division of Neurosurgery, Department of Neurological Surgery, Nihon University School of Medicine, 30-1,Oyaguchi-Kamicho, Itabashi-ku, Tokyo 173-8610, Japan

© Springer Science+Business Media, New York 2016   479
C.E. Elwell et al. (eds.), *Oxygen Transport to Tissue XXXVII*, Advances in Experimental Medicine and Biology 876, DOI 10.1007/978-1-4939-3023-4_60

Keywords Near-infrared spectroscopy • General anesthesia • Sevoflurane, propofol • Cerebral blood flow • Oxygenation

1 Introduction

Sevoflurane and propofol are widely used for induction and maintenance of general anesthesia. The effects of sevoflurane and propofol on cerebral hemodynamics during maintenance of general anesthesia have been previously demonstrated [1, 2]. However, the effects during induction of general anesthesia, when hemodynamic changes are usually greater than during maintenance of anesthesia, have still not been clarified. We therefore compared changes in cerebral blood flow (CBF) and oxygenation (CBO) during induction of general anesthesia with sevoflurane or propofol using near infrared spectroscopy (NIRS).

2 Methods

After obtaining approval from our hospital ethics committee, we recruited and obtained written informed consent from 18 American Society of Anesthesiologists (ASA) I patients scheduled for elective surgical procedures under general anesthesia. The patients were divided into two groups according to whether they were anesthetized with sevoflurane (Group S) or propofol (Group P).

2.1 Monitoring

Intraoperative monitoring included automatic non-invasive blood pressure (NIBP) on the right arm, pulse oximetry, and electrocardiography. We evaluated the changes in CBF and CBO on the forehead using NIRS (NIRO-200NX®, Hamamatsu Photonics, Hamamatsu City, Japan). The NIR light from three laser diodes (735, 810, and 850 nm) was directed at the forehead, and the reflected light was transmitted to a multisegment photodiode detector array. The source-detector distance of the optode is 4 cm. With this, we continuously measured the changes in oxy-Hb, deoxy-Hb, total-Hb concentrations, and TOI (tissue oxygen index). Baseline data were recorded using NIRS immediately before administration of sevoflurane or propofol, and the time course of the variables was evaluated every 10 s for 4 min after administration of sevoflurane or propofol. Mean blood pressure (MBP) and heart rate (HR) were recorded every 1 min for 4 min after administration of sevoflurane or propofol.

2.2   Anesthetic Procedure

Eighteen patients were randomly allocated to group S (age, 40.0 ± 16.2 years) or group P (age, 44.6 ± 10.4 years). None of the patients received premedication. An intravenous line was inserted while the patient was still in the ward. In the operating room, the patients were placed in the supine position and received 6 l/min of oxygen through a face mask. After oxygenation for 3 min, patients were given 2 μg/kg fentanyl IV. One minute later, the patients who were assigned to group S received 8 % sevoflurane in 100 % oxygen via an anesthesia mask, while those in group P received an IV bolus of 2 mg/kg of propofol during oxygenation. All patients were administered 1 mg/kg rocuronium for muscle relaxation after loss of consciousness and were mechanically ventilated via an anesthesia mask, with maintenance of end-tidal CO_2 within 35–40 mmHg for 4 min after administration of sevoflurane or propofol. Tracheal intubation was performed after the study measurements.

2.3   Data Analysis

All results are expressed as mean $\pm$ SD. Patient characteristics were compared between the groups using the unpaired Student's t-test. Changes in MBP, HR, and NIRS variables during induction of anesthesia were analyzed using repeated-measures ANOVA and the Tukey-Kramer multiple comparison test for post-hoc analysis. $P < 0.05$ was considered to represent a statistically significant difference.

3   Results

There were no significant differences in patient characteristics between the groups (Table 60.1). Figure 60.1 shows that there were no significant decreases in MBP and HR in both groups. As shown in Fig. 60.2, oxy-Hb (1.93 μmol/l), total-Hb (1.73 μmol/l), and TOI (2.93 %) increased from the baseline values and reached plateau within about 2 min in group S. When compared with the values observed in group P, the increases showed statistically significant differences. Changes in deoxy-Hb, MBP, and HR did not differ between the groups. In group P, oxy-Hb and TOI transiently decreased within 1 min after an injection of propofol, however there were no statistically significant changes from the baseline and shortly thereafter recovered to baseline level (Fig. 60.2).

Table 60.1 Characteristics of the patients in the two groups

	Group S (n = 9)	Group P (n = 9)
Age (years)	40.0 ± 16.2	44.6 ± 10.4
Height (cm)	166.9 ± 6.1	164.6 ± 6.3
Weight (kg)	64.2 ± 8.5	64.8 ± 17.4

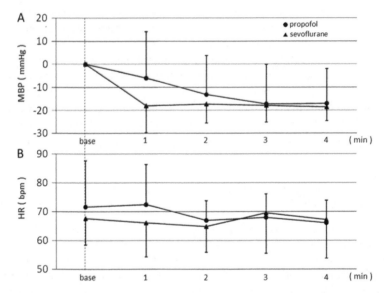

Fig. 60.1 MBP (**A**) and HR (**B**) versus time in the sevoflurane and propofol groups (data are expressed as mean ± SD). This figure demonstrates the amount of variation in MBP from baseline in the two groups

4 Discussion

This is the first study to use NIRS to evaluate the changes in CBF and CBO during induction of general anesthesia with sevoflurane or propofol.

The present study demonstrated that oxy-Hb, total-Hb, and TOI increased during induction of general anesthesia with sevoflurane, while no time course change was observed after a bolus of propofol. It is well known that changes in total-Hb on NIRS reflect changes in cerebral blood volume (CBV) and correlate with changes in CBF [3, 4], while TOI reflects CBO [5]. Based on these facts, the results of the present study suggest that sevoflurane inhalation during induction of general anesthesia increases CBF and CBO, while propofol does not influence these variables.

It has been suggested that 1.0 minimum alveolar concentration (MAC) sevoflurane decreased CBF due to reduction in cerebral metabolism, whereas more than 1.5 MAC of sevoflurane dose-dependently increased CBF caused by its cerebral vasodilating effect [6–9]. Thus, 8 % (4.6 MAC) sevoflurane used in the present study may give rise to luxury perfusion secondary to uncoupling of flow and metabolism, and cause an impairment of cerebral autoregulation [1]. Therefore, a

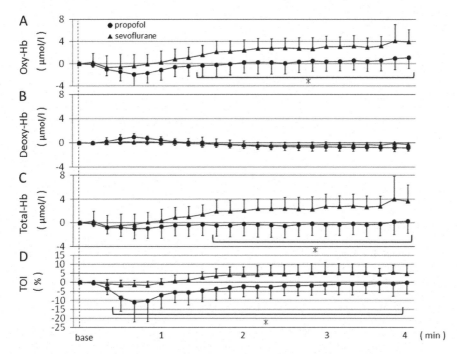

Fig. 60.2 Changes in Oxy-Hb (**A**), Deoxy-Hb (**B**), Total-Hb (**C**), and TOI (**D**) versus time in the sevoflurane and propofol groups (data are expressed as mean ± SD). *Asterisk*: Inter-group comparisons indicated that oxy-Hb, total-Hb and TOI in group S were significantly higher than in group P ($P < 0.05$)

high concentration of sevoflurane may be unsuitable for patients with increased intracranial pressure, as occurs with intracranial space-occupying lesions, although it may be suitable for patients with carotid artery stenosis and cerebral ischemia.

The present study showed that oxy-Hb and TOI in group P transiently decreased within 1 min after an injection of propofol. It has been demonstrated that reduction in cerebral metabolism caused by propofol decreases CBF [1]. However, no concomitant change in total-Hb and deoxy-Hb suggests that CBF and cerebral metabolism remained stable. A further study to explain the phenomenon will be needed.

NIRS has the advantage of enabling evaluation of CBO using the ratio between oxy-Hb and total-Hb. When both cerebral metabolism and hemodynamics reduce concurrently during induction of general anesthesia, NIRS is specifically useful in evaluating both CBO and CBF.

5 Conclusion

Sevoflurane increases CBO and CBF during induction of general anesthesia, while propofol has no effect on these variables during anesthesia induction.

Acknowledgments The authors thank Kaoru Sakatani MD, PhD, (NEWCAT Institute, Department of Electronic Engineering, Nihon University College of Engineering, Koriyama, Japan) for assistance with data analysis.

References

1. Conti A, Iacopino DG, Fodale V et al (2006) Cerebral haemodynamic changes during propofol-remifentanil or sevoflurane anaesthesia: transcranial Doppler study under bispectral index monitoring. Br J Anaesth 97:333–339
2. Kaisti KK, Metsahonkala L, Teras M et al (2002) Effects of surgical levels of propofol and sevoflurane anesthesia on cerebral blood flow in healthy subjects studied with positron emission tomography. Anesthesiology 96:1358–1370
3. Ferrari M, Wilson DA, Hanley DF et al (1992) Effects of graded hypotension on cerebral blood flow, blood volume, and mean transit time in dogs. Am J Physiol 262:1908–1914
4. Pryds O, Greisen G, Skov LL et al (1990) Carbon dioxide-related changes in cerebral blood volume and cerebral blood flow in mechanically ventilated preterm neonates: comparison of near infrared spectrophotometry and 133 Xenon clearance. Pediatr Res 27:445–449
5. Yagi T, Nagao K, Sakatani K et al (2013) Changes of oxygen metabolism and hemodynamics during ECPR with hypothermia measured by near-infrared spectroscopy: a pilot study. Adv Exp Med Biol 789:121–128
6. Matta BF, Hearth KJ, Tipping K et al (1999) Direct cerebral vasodilatory effects of sevoflurane and isoflurane. Anesthesiology 91:677–680
7. Bundgaard H, von Oettingen G, Larsen KM et al (1998) Effects of sevoflurane on intracranial pressure, cerebral blood flow and cerebral metabolism. A dose-response study in patients subjected to craniotomy for cerebral tumours. Acta Anaesthesiol Scand 42:621–627
8. Berkowitz RA, Hoffman WE, Cunningham F et al (1996) Changes in cerebral blood flow velocity in children during sevoflurane and halothane anesthesia. J Neurosurg Anesthesiol 8: 194–198
9. Rhodali O, Juhel S, Mathew S et al (2014) Impact of sevoflurane anesthesia on brain oxygenation in children younger than 2 years. Paediatr Anaesth 24:734–740

Chapter 61
Neurovascular Interactions in the Neurologically Compromised Neonatal Brain

H. Singh, R. Cooper, C.W. Lee, L. Dempsey, S. Brigadoi, A. Edwards,
D. Airantzis, N. Everdell, A. Michell, D. Holder, T. Austin, and J. Hebden

Abstract Neurological brain injuries such as hypoxic ischaemic encephalopathy
(HIE) and associated conditions such as seizures have been associated with poor
developmental outcome in neonates. Our limited knowledge of the neurological
and cerebrovascular processes underlying seizures limits their diagnosis and timely
treatment. Diffuse optical tomography (DOT) provides haemodynamic information
in the form of changes in concentration of de/oxygenated haemoglobin, which can
improve our understanding of seizures and the relationship between neural and
vascular processes. Using simultaneous EEG-DOT, we observed distinct
haemodynamic changes which are temporally correlated with electrographic seizures. Here, we present DOT-EEG data from two neonates clinically diagnosed as
HIE. Our results highlight the wealth of mutually-informative data that can be
obtained using DOT-EEG techniques to understand neurovascular coupling in HIE
neonates.

Keywords Neurovascular coupling • Diffuse optical tomography • Neonatal
seizures • Burst suppression • Hypoxic–ischaemic encephalopathy

1 Introduction

Brain injury in the term infant caused by a shortage of oxygen and blood flow
(hypoxia-ischaemia) is a major cause of death and lifelong neurodisability [1]. Seizures are common in these infants and can exacerbate existing brain injury

H. Singh (✉) • R. Cooper • L. Dempsey • S. Brigadoi • N. Everdell • J. Hebden
neoLAB, The Evelyn Perinatal Imaging Centre, Rosie Hospital, Cambridge, UK

BORL, Department of Medical Physics and Bioengineering, UCL, London, UK
e-mail: harsimrat.singh@ucl.ac.uk

C.W. Lee • A. Edwards • D. Airantzis • A. Michell • D. Holder • T. Austin
neoLAB, The Evelyn Perinatal Imaging Centre, Rosie Hospital, Cambridge, UK

Neonatal Unit, Rosie Hospital, Cambridge University Hospitals, Cambridge, UK

© Springer Science+Business Media, New York 2016   485
C.E. Elwell et al. (eds.), *Oxygen Transport to Tissue XXXVII*, Advances in
Experimental Medicine and Biology 876, DOI 10.1007/978-1-4939-3023-4_61

[2]. However, the clinical manifestation of seizures can be subtle or even absent. Whilst video-EEG is the standard approach to monitoring seizures [3], it has limited spatial resolution and provides little information on the potential mechanisms by which seizures could exacerbate brain damage. Neurophysiologists rely on surges in electrical activity to mark seizure occurrence, which can be challenging in neonates, given such events are often unaccompanied by behavioural correlates [4, 5]; as a result many seizures go unnoticed [6]. There is a paucity of data on the metabolic and haemodynamic changes associated with seizures [7].

As well as seizures, another common feature in neonates with hypoxic-ischaemic brain injury is the electrographic pattern of 'burst-suppression': intermittent fluctuating brain states with profound inactivity or sudden surges. This phenomenon is commonly observed in a variety of pathological conditions. Neurophysiologic model of burst-suppression suggests metabolic processes [8] associated with it, the haemodynamic response to such phenomena is largely unknown.

Diffuse optical tomography (DOT) and electroencephalography (EEG) are two techniques that are able to record changes in cerebral blood oxygenation and electrical activity respectively. The portable and non-invasive nature of these techniques makes them ideally suited to cot-side measurements in the neonatal intensive care unit (NICU). Concurrent changes in these signals can be attributed to dynamic brain mechanisms by which changes in neuronal activity effects changes in dilation or constriction of cerebral blood vessels, known as neurovascular coupling [9]. The haemodynamic parameters measured using optical techniques are considered to be surrogate measures of oxygen transport in the brain, and can be used to study the complex, incompletely understood neurovascular mechanisms.

This paper presents two sets of DOT-EEG data which exemplify the wealth of information about the neurovascular interactions during two common yet complex conditions in HIE neonates (a) seizures (b) burst-suppressions.

2 Methods

Data Acquisition A group (n = 13) of term age infants diagnosed with hypoxic ischaemic encephalopathy (HIE), treated with therapeutic hypothermia (a standard clinical treatment for HIE in the newborn) were recruited into the study (REC reference 09/H0308/125). Simultaneous EEG-DOT acquisition was performed during three stages of therapeutic hypothermia, namely during cooling, rewarming and post-rewarming phases. Optodes (16 detectors, 16 emitters) and EEG (13) electrodes were coupled onto the baby's head using an integrated EEG-DOT head cap. This arrangement provides 58 DOT channels and a standard 11-channel EEG neonatal montage. The sampling frequency of the EEG (Micromed/BrainProducts) and DOT (UCL topography system) were 256 and 10 Hz, respectively.

Clinical History The EEG-DOT data presented are from two infants, both with severe HIE. In subject 1 (Female, Gestational Age 40 weeks), a 60-min recording acquired during post-warming is presented in the following sections. In subject 2, (Male, Gestational Age 41 weeks) a 75-min recording during the cooling phase is presented. The datasets are referred to as dataset 1 and 2, respectively, in the following sections.

Data Analysis A clinical inspection of the Video-EEG was performed by two clinical neurophysiologists. Seven subtle seizures in dataset 1 and 9 suppressed periods in dataset 2 were identified in the EEG recordings along with their respective onset and duration. DOT and EEG data were temporally aligned using signals recorded from a custom-built random pulse generator. The EEG data were pre-processed and band-pass filtered (0.5–70 Hz) using FIELDTRIP [10]. Initial processing of the DOT data was performed using the HOMER 2 toolbox [11]. Individual channels were inspected for motion artifacts and spline-corrected [12, 13]. The DOT data were low pass filtered at 1 Hz to remove high frequency noise and cardiac effects. Four channels from dataset 1 and 15 channels from dataset 2 exhibited a low signal-to-noise ratio (mean intensity values of channels outside a specified range $[-0.0005 \ 1]$) and were excluded from further analysis. Channel-wise raw data in the form of intensity values for each wavelength were converted to changes in optical density relative to the mean of each channel across the entire acquisition period. Changes in optical density were converted to changes in concentration of oxy-haemoglobin (HbO), deoxy-haemoglobin (HbR), and total haemoglobin (HbT = HbO + HbR) using the modified Beer-Lambert law with an estimated differential pathlength factor of 4.9 [14]. The average concentration changes across channels were calculated for dataset 1 to obtain a mean haemodynamic signal to help identify the dominant features of the global haemodynamic data.

Whole Head Image Reconstruction An age-matched voxelized tissue mask (identifying scalp, skull, cerebrospinal fluid, grey and white matter), taken from UCL's 4D neonatal head model package [15] was spatially registered to an up-sampled clinical MRI image for subject 1, while an un-registered, age-matched atlas was used for subject 2. Volumetric whole head and a grey matter surface meshes were then produced with the iso2mesh package [16]. A forward model was computed using TOAST [17] and multispectral linear reconstruction was performed to compute volumetric images of HbO and HbR. The reconstructed volumetric images were then projected on to the grey matter surface mesh.

3   Results

Neurovascular Interactions in Neonatal Seizures Dataset 1 exhibited a discontinuous EEG, and seven electrographic seizures were identified with very subtle clinical signs. Ictal EEG showed high amplitude at onset, including some 1 Hz slow activity. These features evolved into semi-rhythmic generalised slow activity at 2–5 Hz, with loss of the background EEG discontinuity. There was no significant difference between hemispheres seen in the EEG data. Inter-ictal EEG highlighted the discontinuity with short periods of bursts interspersed with suppression EEG. Electrographic seizure duration ranged between 30 and 90 s. Most of the DOT channels exhibited a large amplitude haemodynamic change concomitant to electrographic seizures. The mean haemodynamic signal across all channels highlights the large yet remarkably consistent response for all seven seizures (Fig. 61.1, top panel). This abnormally large haemodynamic response exhibits an increasing trend prior to the electrographic onset but reaches maximum just after the onset, followed by an extended decrease and slow recovery to a steady state. This is seen in the enlarged version of seizure 3 in Fig. 61.1, where the averaged EEG signal across channels is shown along with the HbO, HbR and HbT changes over the period beginning ~350s before and ~350s after the electrographic onset. Blood volume changes are shown in the whole head images in the bottom panel, demonstrating spatial dynamics with large areas of blood volume decrease but also an increase in some areas.

Neurovascular Interactions in Burst-Suppression The EEG of dataset 2 showed several periods of very low amplitude activity lasting 10–20 s; this discontinuous EEG evolved back into continuity soon after these periods terminated. Time-frequency decompositions show very low EEG power during these suppressed segments. There is a remarkably large response in the form of blood volume decrease to these suppression segments in the EEG. This decrease in HbT has a general spread over the cortex. Interestingly, there is pattern of build up to this HbT decrease in time after the onset of the EEG suppression. It originates from the right anterior frontal areas spreading to left central sulcus and right parietal areas (Fig. 61.2) reaching a maximum level of decrease, after which it recovers back to a steady-state. The duration of this blood volume change is significantly longer than the suppression period in the EEG.

4   Discussion and Conclusions

The electrographic seizure events observed in subject 1 show that distinct changes in DOT measurements are temporally correlated with abnormal electrographic activity. Although the haemodynamic response is much longer than the electrographic assessment of seizure duration, it is remarkably consistent across seizure events, predominantly consisting of an initial increase, then prolonged decrease of blood volume. The reconstructed DOT images exhibit complex spatial dynamics reminiscent of a number of animal studies, which have shown an 'inhibitory surround' phenomenon associated with epileptic foci [15–17].

Burst-suppression is defined as an "EEG pattern where high-voltage activity alternates with isoelectric quiescence" [8]. The global HbT decrease, correlated in time with the suppression segments in Fig. 61.2, is indicative of a vasculature response to the brain's neural inactivity. It is also notable that the irregular

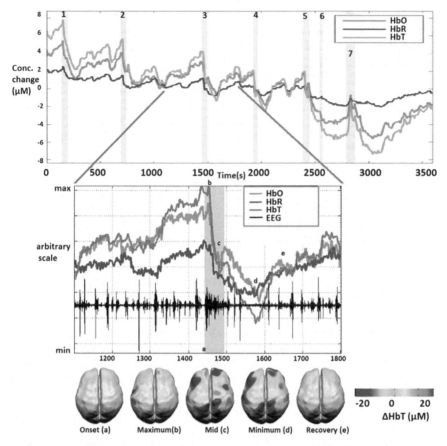

Fig. 61.1 Seven seizure events are indicated on the time course of the global mean hemodynamic concentration changes for subject 1. Seizure events are depicted by *grey shaded areas*. A typical response with reconstructed images (HbT) for seizure 3 at five distinct time points highlights the response profile

suppression periods presented here occurred during the cooling phase and may be associated with, or exacerbated by, therapeutic hypothermia. Both seizures and burst suppressions are associated with profound regional changes in haemodynamics.

Though only part of a series of DOT-EEG recordings performed in a cohort of HIE babies, the two datasets presented here highlight the complex underlying interactions between the neuronal and the vascular processes in the pathological neonatal brain. The neurovascular interactions during neonatal seizures do not follow the expected increase in HbO and decrease in HbR pattern that has been observed in previous functional studies in newborns [18–20] children and adults [21, 22], but similar large abnormal biphasic haemodynamic responses were previously reported by our group in 2011 [7] but in the absence of an electrographic

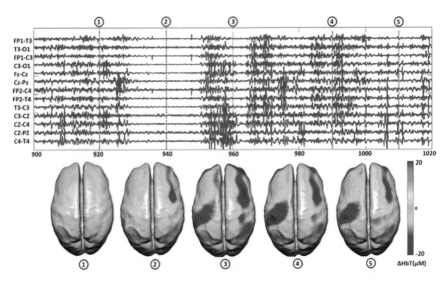

Fig. 61.2 Reconstructed images of the changes in HbT associated with the suppression periods in the EEG in dataset2. Changes in HbT are relative to a baseline defined as the initial 30s

correlate. Recently neurophysiological models of burst-suppression in animals and adults [8, 23] have been reported. To our knowledge, the data presented here is the first study to show haemodynamic response to burst-suppression in newborns. The decrease in blood volume in response to suppression in EEG is noticeably large and temporally correlated with the suppression periods. Furthermore, this response has been represented with a localised spatial resolution in the form of whole head images (Fig. 61.2).

The results show that haemodynamic response to both electrographic seizures and suppression is slower than the neuronal changes. Similar conclusions are also reported in the neurovascular models studies using fMRI-EEG [24, 25], where the neuronal signals (MUA, LFP, ERP) are said to be transient and the BOLD response is much prolonged [26]. But in comparison to fMRI, DOT allows measurement of both HbO and HbR and in turn can be used to reflect blood volume changes [27]. For multimodal acquisition with EEG, DOT and fMRI provide localised spatial resolution and complement the neuronal information of the EEG. Though fMRI has a better depth resolution, DOT measured HbO [28] and HbR [29] are closely correlated with BOLD signal interpretation. The comparative studies with DOT-EEG, report a superior contrast-to-noise ratio than fMRI when accounted for confounding factors (optical properties of heterogeneous tissue layers, partial volume errors and optical sensitivity) [28]. There are practical advantages of DOT-EEG in terms of non-invasive, portability, cost, innocuous to routine clinical care (DOT-EEG can be done at cot/bed-side) and freedom of research in more natural environments. fMRI-EEG studies can be cumbersome with technicalities such as immobility of subject, inherent noise issues in the fMRI scanner and overall lack of easy access to fMRI, costs and paraphernalia required for it.

The data presented here, though novel, is limited by patient numbers to draw general conclusions about neurovascular coupling in neonates. The clinical significance of these observations is unclear, and further work is required to understand the prevalence and mechanisms of these neurovascular changes in the injured neonatal brain.

Acknowledgments The authors are thankful to the parents and staff on the Neonatal Intensive Care Unit at the Rosie Hospital, Cambridge, UK; funders Action Medical Research (AMR-1945) and EPSRC (EP/J021318/1).

References

1. Back S (2014) Cerebral white and gray matter injury in newborns: new insights into pathophysiology and management. Clin Perinatol 41:1–24. doi:10.1016/j.clp.2013.11.001
2. Glass HC, Kan J, Bonifacio SL, Ferriero DM (2012) Neonatal seizures: treatment practices among term and preterm infants. Pediatr Neurol 46:111–115. doi:10.1016/j.pediatrneurol. 2011.11.006
3. Cilio MR (2009) EEG and the newborn. J Pediatr Neurol 7:25–43. doi:10.3233/JPN-2009-0272
4. Shellhaas R, Barks A (2012) Impact of amplitude-integrated electroencephalograms on clinical care for neonates with seizures. Pediatr Neurol 46:32–35. doi:10.1016/j.pediatrneurol. 2011.11.004
5. Van Rooij LGM, Hellström-Westas L, de Vries LS (2013) Treatment of neonatal seizures. Semin Fetal Neonatal Med 18:209–215. doi:10.1016/j.siny.2013.01.001
6. Shah DK, Boylan GB, Rennie JM (2012) Monitoring of seizures in the newborn. Arch Dis Child Fetal Neonatal Ed 97:F65–F69. doi:10.1136/adc.2009.169508
7. Cooper RJ, Hebden JC, O'Reilly H et al (2011) Transient haemodynamic events in neurologically compromised infants: a simultaneous EEG and diffuse optical imaging study. Neuroimage 55:1610–1616. doi:10.1016/j.neuroimage.2011.01.022
8. Ching S, Purdon PL, Vijayan S et al (2012) A neurophysiological-metabolic model for burst suppression. Proc Natl Acad Sci U S A 109:3095–3100. doi:10.1073/pnas.1121461109
9. Leithner C, Royl G (2014) The oxygen paradox of neurovascular coupling. J Cereb Blood Flow Metab 34:19–29. doi:10.1038/jcbfm.2013.181
10. Oostenveld R, Fries P, Maris E, Schoffelen J-M (2011) FieldTrip: open source software for advanced analysis of MEG, EEG, and invasive electrophysiological data. Comput Intell Neurosci 2011:156869. doi:10.1155/2011/156869
11. Huppert TJ, Diamond SG, Franceschini MA, Boas DA (2009) HomER: a review of time-series analysis methods for near-infrared spectroscopy of the brain. Appl Opt 48:D280–D298
12. Cooper RJ, Selb J, Gagnon L et al (2012) A systematic comparison of motion artifact correction techniques for functional near-infrared spectroscopy. Front Neurosci 6:147. doi:10.3389/fnins.2012.00147
13. Scholkmann F, Spichtig S, Muehlemann T, Wolf M (2010) How to detect and reduce movement artifacts in near-infrared imaging using moving standard deviation and spline interpolation. Physiol Meas 31:649–662. doi:10.1088/0967-3334/31/5/004
14. Scholkmann F, Wolf M (2013) General equation for the differential pathlength factor of the frontal human head depending on wavelength and age. J Biomed Opt 18:105004. doi:10.1117/1.JBO.18.10.105004

15. Brigadoi S, Aljabar P, Kuklisova-Murgasova M et al (2014) A 4D neonatal head model for diffuse optical imaging of pre-term to term infants. Neuroimage. doi:10.1016/j.neuroimage. 2014.06.028

16. Fang Q, Boas D (2009) Tetrahedral mesh generation from volumetric binary and gray-scale images. In: Proceedings of IEEE International Symposium Biomedical Imaging ISBI'09, Boston, pp 1142–1145

17. Schweiger M, Arridge S (2014) The Toast++ software suite for forward and inverse modeling in optical tomography. J Biomed Opt 19:40801

18. Roche-Labarbe N, Fenoglio A, Radhakrishnan H et al (2014) Somatosensory evoked changes in cerebral oxygen consumption measured non-invasively in premature neonates. Neuroimage 85(Part 1):279–286. doi:10.1016/j.neuroimage.2013.01.035

19. Gibson AP, Austin T, Everdell NL et al (2006) Three-dimensional whole-head optical tomography of passive motor evoked responses in the neonate. Neuroimage 30:521–528

20. Liao SM, Gregg NM, White BR et al (2010) Neonatal hemodynamic response to visual cortex activity: high-density near-infrared spectroscopy study. J Biomed Opt 15:026010–026019

21. Obrig H, Wenzel R, Kohl M et al (2000) Near-infrared spectroscopy: does it function in functional activation studies of the adult brain? Int J Psychophysiol 35:125–142. doi:10.1016/ S0167-8760(99)00048-3

22. Lloyd-Fox S, Blasi A, Elwell CE (2010) Illuminating the developing brain: the past, present and future of functional near infrared spectroscopy. Neurosci Biobehav Rev 34:269–284. doi:10.1016/j.neubiorev.2009.07.008

23. Liu X, Zhu X-H, Zhang Y, Chen W (2011) Neural origin of spontaneous hemodynamic fluctuations in rats under burst-suppression anesthesia condition. Cereb Cortex 21:374–384. doi:10.1093/cercor/bhq105

24. Friston K (2008) Neurophysiology: the brain at work. Curr Biol 18:R418–R420. doi:10.1016/j. cub.2008.03.042

25. Rosa MJ, Kilner JM, Penny WD (2011) Bayesian comparison of neurovascular coupling models using EEG-fMRI. PLoS Comput Biol 7:e1002070. doi:10.1371/journal.pcbi.1002070

26. Goense JBM, Logothetis NK (2008) Neurophysiology of the BOLD fMRI signal in awake monkeys. Curr Biol 18:631–640. doi:10.1016/j.cub.2008.03.054

27. Wallois F, Mahmoudzadeh M, Patil A, Grebe R (2012) Usefulness of simultaneous EEG-NIRS recording in language studies. Brain Lang 121:110–123. doi:10.1016/j.bandl.2011.03.010

28. Strangman G, Culver JP, Thompson JH, Boas DA (2002) A quantitative comparison of simultaneous BOLD fMRI and NIRS recordings during functional brain activation. Neuroimage 17:719–731. doi:10.1006/nimg.2002.1227

29. Huppert TJ, Hoge RD, Diamond SG et al (2006) A temporal comparison of BOLD, ASL, and NIRS hemodynamic responses to motor stimuli in adult humans. Neuroimage 29:368–382. doi:10.1016/j.neuroimage.2005.08.065

Chapter 62
In Vivo Measurement of Cerebral Mitochondrial Metabolism Using Broadband Near Infrared Spectroscopy Following Neonatal Stroke

Subhabrata Mitra, Gemma Bale, Judith Meek, Sean Mathieson, Cristina Uria, Giles Kendall, Nicola J. Robertson, and Ilias Tachtsidis

Abstract Neonatal stroke presents with features of encephalopathy and can result in significant morbidity and mortality. We investigated the cerebral metabolic and haemodynamic changes following neonatal stroke in a term infant at 24 h of life. Changes in oxidation state of cytochrome-c-oxidase (oxCCO) concentration were monitored along with changes in oxy- and deoxy- haemoglobin using a new broadband near-infrared spectroscopy (NIRS) system. Repeated transient changes in cerebral haemodynamics and metabolism were noted over a 3-h study period with decrease in oxyhaemoglobin (HbO_2), deoxy haemoglobin (HHb) and oxCCO in both cerebral hemispheres without significant changes in systemic observations. A clear asymmetry was noted in the degree of change between the two cerebral hemispheres. Changes in cerebral oxygenation (measured as $HbDiff = HbO_2 - HHb$) and cerebral metabolism (measured as oxCCO) were highly coupled on the injured side of the brain.

Keywords Broadband NIRS • Cytochrome c oxidase • Mitochondrial metabolism • Cerebral hemodynamics • Neonatal stroke

This chapter was originally published under a CC BY-NC 4.0 license, but has now been made available under a CC BY 4.0 license. An erratum to this chapter can be found at DOI 10.1007/978-1-4939-3023-4_66.

S. Mitra (✉) • J. Meek • S. Mathieson • C. Uria • G. Kendall • N.J. Robertson
Institute for Women's Health, University College London and Neonatal Unit, University College London Hospitals Trust, London, UK
e-mail: subhabratamitra@hotmail.com

G. Bale • I. Tachtsidis
Biomedical Optics Research Laboratory, Department of Medical Physics and Biomedical Engineering, University College London, London, UK

1 Introduction

Perinatal stroke commonly presents with features of encephalopathy, seizures, or neurologic deficit during the early neonatal period. It can result in significant morbidity and severe long-term neurologic and cognitive deficits, including cerebral palsy, epilepsy and behavioural disorders. The incidence is high and has been estimated at 1 in 1600–5000 live births with estimated annual mortality rate of 3.49 per 100,000 live births [1]. Although seizures can be monitored with cerebral function monitor (CFM) or electroencephalography (EEG), the diagnosis of cerebral injury is typically confirmed on brain magnetic resonance imaging (MRI) once the infant becomes clinically stable [2, 3].

In contrast to adult stroke, the initial presentation of stroke in neonates can be subtle and non-specific. Neonates can present with lethargy, poor feeding, apnoea and hypotonia. This often delays the diagnosis and can influence the outcome. Any improvement in bedside non-invasive monitoring to aid early diagnosis and management would greatly benefit this group of infants.

Near-infrared spectroscopy (NIRS) is a non-invasive tool that has been widely used for continuous bedside monitoring of cerebral oxygenation and haemodynamic changes. NIRS can measure the concentration changes of oxygenated ($\Delta[HbO_2]$) and deoxygenated haemoglobin ($\Delta[HHb]$) which in turn can be used to derive changes in total haemoglobin ($\Delta[HbT] = \Delta[HbO_2] + \Delta[HHb]$) and haemoglobin difference ($\Delta[HbDiff] = \Delta[HbO_2] - \Delta[HHb]$). HbT and HbDiff are indicative of cerebral blood volume and brain oxygenation, respectively. These measurements have been widely used to assess the haemodynamic changes in the cerebral tissue, but a clear assessment of cerebral metabolism during the same period is absolutely essential for a better understanding of the pathophysiology of cerebral injury and its management.

Cytochrome-c-oxidase (CCO) is the terminal electron acceptor in the mitochondrial electron transport chain (ETC). It plays a crucial role in mitochondrial oxidative metabolism and ATP synthesis and is responsible for more than 95 % of oxygen metabolism in the body [4]. CCO contains four redox centres, one of which—copper A (CuA)—has a broad absorption peak in the near-infrared (NIR) spectrum, which changes depending on its redox state [5]. As the total concentration of CCO is assumed constant, the changes in the NIRS-measured oxCCO concentration are indicative of the changes in CCO redox state in cerebral tissue, representing the status of cerebral mitochondrial oxidative metabolism.

Our group has recently demonstrated that brain mitochondrial oxidative metabolism measured by $\Delta[oxCCO]$ using broadband NIRS system during and after cerebral hypoxia-ischemia correlates well with simultaneous phosphorus magnetic resonance spectroscopy parameters of cerebral energetics in a preclinical model [6].

We have recently developed a new broadband NIRS system which is capable of absolute measurements of optical absorption and scattering to quantify $\Delta[oxCCO]$ as well as $\Delta[HbO_2]$ and $\Delta[HHb]$ in neonatal brain [7]. In this study, we present the haemodynamic and metabolic changes following neonatal stroke. Our aim was to compare the haemodynamic and metabolic responses between the injured and non-injured side of the brain following neonatal stroke, using broadband NIRS measurement of changes in oxCCO.

2   Methods

Ethical approval for the Baby Brain Study at University College London Hospitals NHS Foundation Trust (UCLH) was obtained from the North West Research Ethics Centre (REC reference: 13/LO/0106). We studied a term (40 weeks 6 days) newborn infant (birth weight 3370 g), admitted with clinical seizures. Seizures were first noted at 9 h of age and stopped at 17 h of age after treatment with multiple anticonvulsants (phenobarbitone, phenytoin, midazolam and paraldehyde). Seizures initially involved only the right upper and lower limbs. EEG recordings revealed repeated seizure episodes originating from the left hemisphere.

NIRS monitoring was commenced at 24 h of age. One NIRS channel was placed on either side of the forehead and data were collected at 1 Hz. Four detector optodes were placed horizontally against each source optode on either side with source-detector separations of 1.0, 1.5, 2.0 and 2.5 cm for multi-distance measurements. The longest optode source-detector distance of 2.5 cm was chosen to ensure a better depth penetration [8]. Differential path length (DPF) was chosen as 4.99 [9].

A program was created in LabVIEW 2011 (National Instruments, USA) to control the charge-coupled device (CCD), collect the raw data and calculate the corresponding concentrations. The changes in chromophore concentrations were calculated from the measured changes in broadband NIR light attenuation using the modified Beer-Lambert law as applied with the UCLn algorithm [10] across 136 wavelengths (770–906 nm). Systemic data from the Intellivue Monitors (Philips Healthcare, UK) were collected using ixTrend software (ixellence GmbH, Germany). Systemic and EEG data were synchronised with the NIRS data. Electroencephalography (EEG) data was collected using a Nicolet EEG monitor (Natus Medical, Incorporated, USA). Brain magnetic resonance imaging (MRI) and venography were performed on day 5 using a 3T Philips MRI scanner (Philips Healthcare, UK) on day 5. T1 and T2 weighted images with an apparent diffusion coefficient (ADC) map were obtained on MRI.

3   Data Analysis

Initial data analysis was carried out in MATLAB R2013a (Mathworks, USA). NIRS data were visually checked and were processed with an automatic wavelet de-noising function, which reduces the high frequency noise but maintains the trend information. Systemic data were down-sampled and interpolated to the NIRS data timeframe (1 Hz). Artefacts from movement or changes in external lighting were removed using the method suggested by Scholkmann et al. [11]. This method also corrects shifts in the baseline due to artefact. All statistical analysis was performed using GraphPad Prism 6 (GraphPad Software, USA).

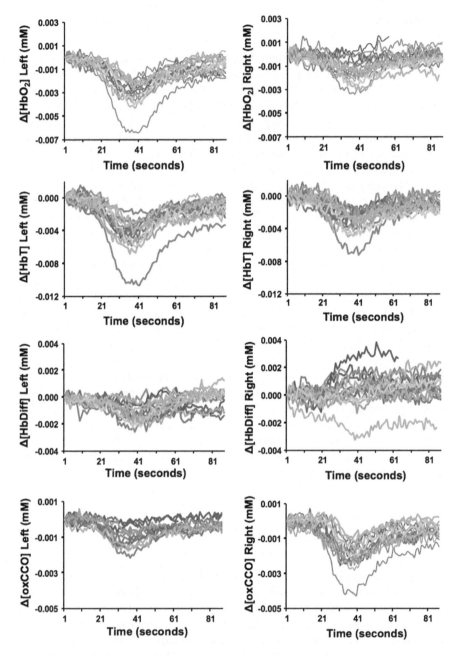

Fig. 62.1 NIRSsignals from each side of the brain during all events (*each coloured line* represents a single event). $\Delta[HbO_2]$, $\Delta[HbT]$, and $\Delta[HbDiff]$ reflect higher changes on the *left side*, but $\Delta[oxCCO]$ revealed minimal change on the injured left side compared to the right side

4   Results

NIRS data were collected over a 3-h period without any clinical or electrographical seizure noted during this period. Synchronous and repeated transient changes in $\Delta[HbO_2]$, $\Delta[HHb]$ and $\Delta[oxCCO]$ were noted on both sides (Fig. 62.1). Following an acute drop in these parameters, signals returned slowly towards baseline. These changes were noted over an average duration of 90 s. A change in $\Delta[HbT]$ of more than 2 μM was considered a significant event and 16 similar events were identified and analysed during the study.

A significant difference was noted between right and left sides in both cerebral metabolism, oxygenation and their relationship. $\Delta[HbO_2]$, $\Delta[HbT]$ and $\Delta[HbDiff]$ were higher on the left (injured) side. However changes in [oxCCO] were more prominent on the right side during the events (Fig. 62.1). During the events, maximum concentration changes (fall) in $\Delta[HbO_2]$, $\Delta[HbT]$, $\Delta[HbDiff]$ and $\Delta[oxCCO]$ were significantly different between the two sides (Table 62.1) but $\Delta[HHb]$ did not show any significant difference between the sides. $\Delta[oxCCO]$ responded differently to changes in $\Delta[HbDiff]$ between the left side (slope 0.64, r^2 0.5) and right side (slope -0.21, r^2 0.05) (Fig. 62.2).

MRI of brain on day 5 revealed low signal intensity on TI weighted images and high signal intensity on T2 weighted images in the left parieto-occipital region

Table 62.1   Differences in the maximum change between the left and right sides. Mean ± standard deviations of changes on both sides are presented with two-tailed p values

	Left	Right	p value
$\Delta[HbO_2]$ (mmolar)	-0.0032 ± 0.0002	-0.0018 ± 0.0002	0.0002
$\Delta[HHb]$ (mmolar)	-0.0020 ± 0.0001	-0.0020 ± 0.0002	0.9933
$\Delta[HbT]$ (mmolar)	-0.0049 ± 0.0004	-0.0036 ± 0.0003	0.0315
$\Delta[HbDiff]$ (mmolar)	-0.0016 ± 0.0001	-0.0008 ± 0.0001	0.0012
$\Delta[oxCCO]$ (mmolar)	-0.0011 ± 0.0001	-0.0021 ± 0.0001	0.0003

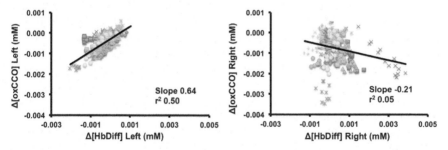

Fig. 62.2   Linear regression analysis between $\Delta[oxCCO]$ with $\Delta[HbDiff]$ on both sides on day 1. Each coloured and different shaped point represents an event

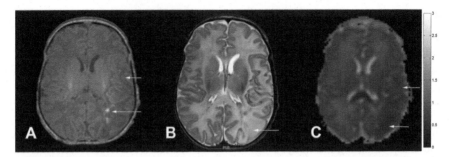

Fig. 62.3 MRI scan taken at 3T on day 5. (**a**) T1 weighted axial image demonstrating generalised low signal intensity in the left parieto-occipital region with T1 shortening, (**b**) T2 weighted axial image demonstrating high signal intensity in the affected region with loss of cortical ribbon, (**c**) Apparent diffusion coefficient (ADC) map showing restricted diffusion in the affected area

indicating a left sided neonatal stroke. Apparent diffusion coefficient (ADC) map demonstrated restricted diffusion in the same area on the left side (Fig. 62.3).

5   Discussion

Spontaneous transient changes in NIRS parameters were recorded repeatedly from both cerebral hemispheres; a clear asymmetry was evident in these spontaneous haemodynamic and metabolic changes between the injured left side and the right side. The origin of these events is unclear. Similar events have been described previously following seizures using a different optical system [12]. We did not find any significant changes in systemic observations and electrical activity on EEG during the events in our study. Absolute band power in EEG was suppressed on the injured left side when compared to the right side during the study. It is possible that neuronal metabolic changes following seizures were driving the haemodynamic changes. Cerebral oxygenation (measured as HbDiff) and cerebral metabolism (measured as oxCCO) were tightly coupled on the injured side (left).

Following stroke, a persistent reduction in blood flow leads to a decrease in both substrate supply and oxygenation on the injured side [13]. These changes have opposite effects on Δ[oxCCO]. A decrease in substrate supply would lead to a change in redox state towards oxidation whereas a decrease in oxygenation will lead to a reduced redox state. These changes in redox state in opposite directions may explain why the Δ[oxCCO] response on the left side during these events was attenuated compared to the right side. The oxygenation and haemodynamic responses were however more exaggerated on the injured side. This restricted oxCCO change on the injured side of the brain is likely to reflect a persistent abnormal mitochondrial metabolism following unilateral seizures and reduced ATP turnover. An asymmetry in the cerebral energy state has been described with ^{31}P MRS recorded from right and left cerebral hemispheres after seizures in a newborn baby [14]. This persisting abnormal cerebral metabolism may be due to the increased energy demand that occurs during persistent seizures; this is known to lead to unpredictable changes in the redox states of ETC metabolites [13].

In summary, we identified asymmetric cerebral oxidative and metabolic responses following neonatal seizures on day 1 using broadband NIRS measurement in a newborn infant. Although we were able to make an earlier predictive assessment, compared to the current standard clinical assessment tool (MRI) in this case study, a generalisation should be avoided at this point.

Acknowledgments We thank all the families and neonatal staff in UCLH for their support. This project was supported by EPSRC (EP/G037256/1) and The Wellcome Trust (088429/Z/09/Z). We thank Alan Bainbridge for his help in preparing the figures.

References

1. Lynch JK (2009) Epidemiology and classification of perinatal stroke. Semin Fetal Neonatal Med 14(5):245–249
2. Lequin MH, Dudink J, Tong KA et al (2009) Magnetic resonance imaging in neonatal stroke. Semin Fetal Neonatal Med 14(5):299–310
3. Govaert P (2009) Sonographic stroke templates. Semin Fetal Neonatal Med 14(5):284–298
4. Richter OM, Ludwig B (2003) Cytochrome c oxidase—structure, function, and physiology of a redox-driven molecular machine. Rev Physiol Biochem Pharmacol 147:47–74
5. Jöbsis FF (1977) Noninvasive, infrared monitoring of cerebral and myocardial oxygen sufficiency and circulatory parameters. Science 198(4323):1264–1267
6. Bainbridge A, Tachtsidis I, Faulkner SD et al (2013) Brain mitochondrial oxidative metabolism during and after cerebral hypoxia-ischemia studied by simultaneous phosphorus magnetic-resonance and broadband near-infrared spectroscopy. Neuroimage pii: S1053-8119 (13)00870-7
7. Bale G, Mitra S, Meek J et al (2014) A new broadband near-infrared spectroscopy system for in-vivo measurements of cerebral cytochrome-c-oxidase changes in neonatal brain Injury. Biomedical Optics, OSA Technical Digest, paper BS3A.39
8. Grant PE, Roche-Labarbe N, Surova A et al (2009) Increased cerebral blood volume and oxygen consumption in neonatal brain injury. J Cereb Blood Flow Metab 29(10):1704–1713
9. Duncan A, Meek JH, Clemence M et al (1996) Measurement of cranial optical path length as a function of age using phase resolved near infrared spectroscopy. Pediatr Res 39:889–894
10. Matcher S, Elwell C, Cooper C (1995) Performance comparison of several published tissue near-infrared spectroscopy algorithms. Anal Biochem 227(1):54–68
11. Scholkmann F, Spichtig S, Muehlemann T et al (2010) How to detect and reduce movement artifacts in near-infrared imaging using moving standard deviation and spline interpolation. Physiol Meas 31(5):649–662

12. Cooper RJ, Hebden JC, O'Reilly H et al (2011) Transient haemodynamic events in neurologically compromised infants: a simultaneous EEG and diffuse optical imaging study. Neuroimage 55(4):1610–1616
13. Banaji M (2006) A generic model of electron transport in mitochondria. J Theor Biol 243 (4):501–516
14. Younkin DP, Delivoria-Papadopoulos M, Maris J et al (1986) Cerebral metabolic effects of neonatal seizures measured with in vivo 31P NMR spectroscopy. Ann Neurol 20(4):513–519

Chapter 63
A New Framework for the Assessment of Cerebral Hemodynamics Regulation in Neonates Using NIRS

Alexander Caicedo, Thomas Alderliesten, Gunnar Naulaers, Petra Lemmers, Frank van Bel, and Sabine Van Huffel

Abstract We present a new framework for the assessment of cerebral hemodynamics regulation (CHR) in neonates using near-infrared spectroscopy (NIRS). In premature infants, NIRS measurements have been used as surrogate variables for cerebral blood flow (CBF) in the assessment of cerebral autoregulation (CA). However, NIRS measurements only reflect changes in CBF under constant changes in arterial oxygen saturation (SaO_2). This condition is unlikely to be met at the bedside in the NICU. Additionally, CA is just one of the different highly coupled mechanisms that regulate brain hemodynamics. Traditional methods for the assessment of CA do not take into account the multivariate nature of CHR, producing inconclusive results. In this study we propose a newly developed multivariate methodology for the assessment of CHR. This method is able to effectively decouple the influences of SaO_2 from the NIRS measurements, and at the same time, produces scores indicating the strength of the coupling between the systemic variables and NIRS recordings. We explore the use of this method, and its derived scores, for the monitoring of CHR using data from premature infants who developed a grade III-IV intra-ventricular hemorrhage during the first 3 days of life.

Keywords Cerebral autoregulation • NIRS • Multivariable • Premature infants

A. Caicedo (✉) • S. Van Huffel
Department of Electrical Engineering (ESAT), STADIUS Center for Dynamical Systems, Signal Processing, and Data Analytics, KU Leuven, Leuven, Belgium

iMinds Medical IT, Leuven, Belgium
e-mail: acaicedo@esat.kuleuven.be

T. Alderliesten • P. Lemmers • F. van Bel
Department of Neonatology, University Medical Center, Wilhelmina Children's Hospital, Utrecht, The Netherlands

G. Naulaers
Neonatal Intensive Care Unit, University Hospitals Leuven, KU Leuven, Leuven, Belgium

© Springer Science+Business Media, New York 2016   501
C.E. Elwell et al. (eds.), *Oxygen Transport to Tissue XXXVII*, Advances in
Experimental Medicine and Biology 876, DOI 10.1007/978-1-4939-3023-4_63

1 Introduction

Cerebral hemodynamics regulation (CHR) comprises a set of mechanisms that interact in order to maintain brain homeostasis [1]. These mechanisms are highly coupled and act upon sudden changes in systemic variables mitigating its effect on brain hemodynamics. In premature infants these mechanisms are likely to be impaired [2, 3]. Due to the fragility of the cerebral vascular bed in this population, monitoring CHR is of paramount importance in order to avoid brain damage. Near-infrared spectroscopy (NIRS) signals can be used in the continuous monitoring of CHR as markers for brain oxygenation. In addition, NIRS can also be used as surrogate variables for cerebral blood flow (CBF) and cerebral blood volume (CBV) [2, 3]. However, NIRS measurements are highly sensitive to changes in arterial oxygen saturation (SaO_2), which hinders their interpretation as a surrogate variable for both CBF and CBV. In the framework of CHR monitoring, this problem is avoided by performing the analysis in segments with relatively constant SaO_2. However, this condition is hardly met at the bedside in the neonatal intensive care units (NICU), hampering the introduction of CHR monitoring in clinical practice.

Additionally, most of the interest in this research field has been given to the monitoring of cerebral autoregulation (CA). CA is the mechanism that tries to maintain a proper CBF in the presence of changes in mean arterial blood pressure (MABP). This mechanism is described as a plateau region in the static relationship between MABP-CBF. Normal values of MABP are located inside the flat region of the CA curve. However, in the presence of hypotension/hypertension CBF might become pressure passive, which exposes the brain to damage due to ischemia or hemorrhage. In neonates, the upper and lower MABP thresholds of the CA plateau are unknown. In addition, these infants are likely to present low values of blood pressure. For this reason, these infants normally receive treatment for hypotension in order to avoid low cerebral perfusion. However, CA is not an isolated mechanism, there are several protection mechanisms that interact and affect the auto-regulation curve. Metabolic influences caused by changes in blood gases concentration, as well as neurogenic activity, play an important role in the regulation of brain hemodynamics [1]. For this reason a multivariate approach for CHR monitoring is needed.

In this context, an appropriate CHR monitor should include measurements of MABP, partial pressure of CO_2 ($PaCO_2$) and a surrogate marker for neurogenic activity. For the latter, measurements of heart rate (HR) might be of particular interest, since they carry information about the sympathetic mediated and vagal outflow [4, 5]. In addition, measurements of SaO_2 should also be included in order to decouple its dynamics from NIRS measurements. The main goal in this paper is to highlight the importance of the use of multivariate models for the assessment of CHR, as well as the importance of a proper decoupling of physiological noise from the NIRS measurements.

2   Methods

Data For this study we used measurements from nine premature infants with a gestational age less than 32 weeks. From this population five subjects presented III-IV grade of intra-ventricular hemorrhage (IVH). The remaining four patients were matched controls. Concomitant measurements of HR, MABP, SaO_2 and regional cerebral oxygen saturation ($rScO_2$) measured by the INVOS 4100 (Somanetics, Troy, Michigan, USA) were obtained during their first 3 days of life. All data were recorded and stored in a personal computer using Poly 5 (Inspector Research Systems, Amsterdam, the Netherlands) with a sampling frequency of 1 Hz. Additionally, cranial ultrasound was performed daily in order to determine the presence of cerebral injury (e.g. white matter damage, IVH). These data are part of a large data collection presented in [6], and have also been used for a separate analysis in [7]. Figure 63.1 presents measurements from one representative subject.

Mathematical Tools We propose the use of a newly developed decomposition algorithm based on wavelet regression and oblique subspace projections (WR-OBP). In short, WR-OBP is a hybrid between wavelet regression and subspace system identification that uses input–output observations of the system in order to produce a mathematical model that can explain the measured output. Furthermore, WR-OBP is able to decouple the linked dynamics between the different underlying subsystems in order to decompose the observed output in terms of the partial contributions of each input variable. This is of considerable

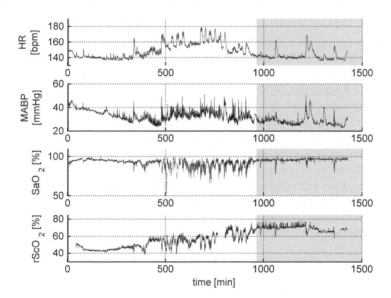

Fig. 63.1 Concomitant recordings of HR, MABP, SaO_2 and $rScO_2$ from a representative subject. The *shaded area* represents the interval of time between the last clinical evaluation with good outcome and the detection of a grade III IVH

interest in the framework of CHR monitoring, since it allows not only to decouple the influence of SaO_2 from the NIRS measurements, but also provides a set of signals that represent the time course relationship between each input variable and the respective output. This set of signals can be used to define scores for the assessment of the coupling between systemic and brain hemodynamic variables. A preliminary study using an early version of WR-OBP can be found in [8].

WR-OBP uses HR, MABP and SaO_2 as input variables and $rScO_2$ as the output of a mean average filter model of the form $y = Ax$. In order to construct the regression matrix A, the input variables are denoised using wavelet denoising and a block Hankel structured is formed with the denoised variables. The block Hankel matrix A, then, contains a set of delayed versions of the denoised input variables. The optimal number of delays is found by means of tenfold cross-validation. Once the linear regression has been solved, we used oblique subspace projections in order to find the contribution of each input variable in the $rScO_2$ changes. This can be interpreted as projecting the $rScO_2$ onto the signals subspaces of the input variables. In order to construct the oblique projectors, the columns of the regression matrix A that correspond to one of the input variables is used as the target subspace, while the rest of the columns are used as the reference subspace. The oblique projector is constructed using $P_{i.(i)} = A_i (A^T_i Q_{(i)} A_i)^{-1} A^T_i Q_{(i)}$, where $P_{i.(i)}$ is the oblique projection onto the input variable i along the reference subspace indicated by all the remaining variables, A_i is a partition of A containing only the columns related to the input variable i, $Q_{(i)} = I - P_{(i)}$, with $P_{(i)} = A_{(i)} (A^T_{(i)} A_{(i)})^{-1} A^T_{(i)}$ is the orthogonal projector onto the null space of the $A_{(i)}$, and $A_{(i)}$ the partition of A that remains after removing A_i. Finally, the contribution of the input variable I on the output is computed using $y_i = P_{i.(i)} y$.

Signal Processing We divided the signals in consecutive overlapping segments with a duration of 1000 s, and an overlapping of 500 s. For each segment we used MABP, HR and SaO_2 as inputs and $rScO_2$ as output in the WR-OBP decomposition algorithm. As a result, WR-OBP decomposed $rScO_2$ in three different signals, each one corresponding to one input variable, in such a way that $rScO_2 = (rScO_2)_{MABP} + (rScO_2)_{HR} + (rScO_2)_{SaO2} + \varepsilon$, with $(rScO_2)_i$ being the partial contribution of the variable i on the measured $rScO_2$ and ε an error component representing measurement noise and/or not modeled dynamics. The coupling coefficient between the output and a respective input variable was computed by dividing the power of the partial contribution of the respective input by the power of the output, e.g. the coupling coefficient between MABP-$rScO_2$ is computed as the ratio $Power[(rScO_2)_{MABP}]/Power[rScO_2]$. The decomposition of a representative segment of the measurements is shown in Fig. 63.2.

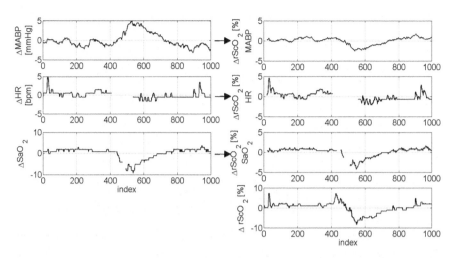

Fig. 63.2 Decomposition of a representative segment of the data using WR-OBP, on the *left* the input variables are displayed, from top to bottom MABP, HR and SaO_2. On the *right* the respective partial contribution of each input variable, as well as the measured output can be found. From *top* to *bottom* the contributions from MABP, HR, and SaO_2, the *last plot* represents the measured output, $rScO_2$

3   Results and Discussion

Figure 63.3 shows the results from WR-OBP applied to the measurements from the patient whose data are shown in Fig. 63.1. Since WR-OBP uses oblique projectors, and oblique projectors do not preserve the Euclidean norm, the coupling coefficients can be larger than 1. As a first approach to differentiate between control and IVH patients, in Fig. 63.4 we explored the dispersion of the scores using a scatter plot, where the x-axis represents the coupling coefficients between $MABP$-$rScO_2$ and the y-axis represents the coupling coefficients between HR-$rScO_2$. Furthermore, we divided the scores from the IVH-group in two subgroups, namely the pre-IVH region and the IVH region. We hypothesized that the scores of the pre-IVH region should behave similarly to the ones from the control group, while scores in the IVH region indicating the presence of an IVH episode should reflect the change in the coupling between systemic and hemodynamics variables. Since most of the data points for the control subjects and the pre-IVH sub-group were located in regions of low $MABP$-$rScO_2$ and/or HR-$rScO_2$ coupling, we propose to use the product of both coupling coefficients as a marker for IVH.

Figure 63.5 shows the time series with the product of the coupling scores for each subject who suffered an IVH. Interestingly, in all the patients, high coupling scores were found inside the shaded area, which represents the time interval between the last clinical evaluation with positive outcome and when the IVH was detected. We might hypothesize that the location in time of the high scores could be related to the onset of the IVH episode. However, we do not have the exact time

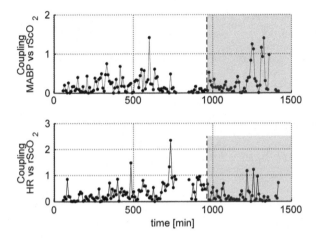

Fig. 63.3 Coupling coefficients between MABP-rScO$_2$ and HR-rScO$_2$ computed for a representative subject. The *shaded area* represents the time between the last clinical evaluation with positive outcome and the time when a grade III IVH was detected

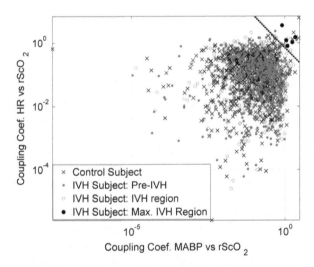

Fig. 63.4 Scatter plot of the coupling scores for CHR monitoring for all the subjects. The x-axis as we'll as the y-axis is in logarithmic scale. The *dashed line* indicate the limit of the boundary where the product of the coupling coefficients is lower than a selected threshold; this region is assumed to represent segments with low values of coupling between systemic and hemodynamics variables. The scores for the control subjects are plotted using 'x'. The scores for the IVH subjects are divided in two: pre-IVH region plotted using *grey dots*, and the region where the IVH occurred using *grey circles*. Finally, the *solid black circles* represent the segments that describe the maximum peaks in the IVH region

location of the episodes in order to prove this argument. Assuming this hypothesis as correct, just for illustrative purposes, we arbitrarily defined a threshold of 0.7 to identify the occurrence of an IVH event. Using this threshold, in Fig. 63.4 we plotted the area for which the product of the coupling coefficients was lower than this specified threshold. Additionally, we also marked in Fig. 63.4 the coupling coefficients corresponding to the segments with the maximum peaks in the IVH region found in Fig. 63.5. For subject 1 we took two peaks, since the maximum peak contains two data points.

The dispersion of the data in Fig. 63.4 indicates that, in the presence of a high coupling between systemic and cerebral hemodynamics, for some segments MABP dominates the changes in $rScO_2$, while in other segments HR drives the changes in $rScO_2$. This might indicate that the pathophysiology leading to the IVH could have differed between the patients. Therefore, looking only at one of the mechanisms involved in CHR might lead to inconclusive results. Additionally, in Fig. 63.4 some coupling coefficients corresponding to segments for the control group and the pre-IVH region can be found outside the normal (shaded) region. These segments with an apparent high coupling between systemic and hemodynamic variables produce false detections that might have been caused by the presence of artefacts, inaccuracies in the used algorithm, or unlabeled episodes of hemodynamic instability.

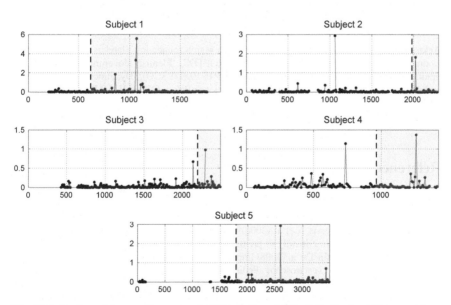

Fig. 63.5 Product of the coupling scores MABP-$rScO_2$ and HR-$rScO_2$ for the different subjects that suffered a grade III-IV IVH. The *shaded area* represents the time between the last clinical evaluation with positive outcome and the time when a grade III-IV IVH was detected. The y-axis represents the magnitude of the product of both scores, while the x-axis represents the time in minutes

We want to stress that the main goal of this paper is not to introduce a new marker for IVH, but to point out the importance of a multivariate framework for CHR monitoring. With the case study presented here we have found a promising relation between the results provided by our multivariate analysis and the possible detection of the onset of an IVH episode. However, several limitations should be taken into account. First, because the population included in the study is small, no strong conclusion can be drawn. Second, we only illustrate how a multivariate analysis can be used in the framework of CHR monitoring and how the results from this framework might be used to define markers for disease; however, these markers were not optimized, mainly due to the small population size and because the exact time of the IVH episode is unknown. The exact time of the IVH episode needs to be known in order to validate the proposed marker. However, despite the limitations of the study, these preliminary results are promising and highlight the importance of a multivariate framework for the monitoring of CHR using NIRS.

To conclude, we have presented a new framework for the assessment of CHR. This approach requires the use of multivariate models that are able to assess the influence of each measured systemic variable on brainhemodynamics. More importantly, this new framework suggests the need to study the information provided by the combined scores rather than studying their influences separately, to cope with inter-patient differences as well as differences in pathophysiology. Additionally, in a preliminary study, we showed that scores for CHR monitoring, obtained by combining the information of coupling between different systemic variables on brain hemodynamics, might be useful to detect the onset of IVH.

Acknowledgments Research supported by the Research foundation Flanders (FWO); Research Council KUL: GOA MaNet, CoE PFV/10/002 (OPTEC). Flemish Government: FWO: travel grant. Belgian Federal Science Policy Office: IUAP P719/(DYSCO), 'Dynamical systems, control and optimization', 2012–2017. Belgian Federal Science Policy Office: IUAP P7/19 DYSCO. EU HIP Trial FP7-HEALTH/2007–2013 (n° 260777).

References

1. Peng T, Rowley A, Plessis A et al (2000) Multivariate system identification for cerebral autoregulation. Ann Biomed Eng 36(2):308–320
2. Tsuji M, Saul J, du Plessis A et al (2000) Cerebral intravascular oxygenation correlates with mean arterial pressure in critically ill premature infants. Pediatr Rev 106(4):625–632
3. Wong F, Leung T, Austin T et al (2008) Impaired autoregulation in preterm infants identified by using spatially resolved spectroscopy. Pediatrics 121:604–611
4. Purkayastha S, Saxena A et al (2012) α_1-Adrenergic receptor control of the cerebral vasculature in humans at rest and during exercise. Exp Physiol. doi:10.1113/expphysiol.2012.066118
5. Schaffer L, Burkhardt T et al (2008) Cardiac autonomic balance in small-for-gestational-age neonates. Am J Physiol Heart Circ Physiol 294:H884–H890
6. Alderliesten T, Lemmers P et al (2013) Cerebral oxygenation, extraction, and autoregulation in very premature infants who develop peri-intraventricular hemorrhage. J Pediatr 162(4):698–704

7. Caicedo A, Varon C et al (2014) Differences in the cerebral hemoynamics regulation mechanisms of premature infants with intra-ventricular hemorrhage assessed by means of phase rectified signal averaging. In: Proceedings of the IEEE EMBC 2014, Chicago, 4208–4211 pp
8. Caicedo A, Tachtsidis I et al (2012) Decoupling the influence of systemic variables in the peripheral and cerebral haemodynamics during ECMO procedure by means of oblique and orthogonal subspace projections. In: 2012 Annual international conference of the IEEE Engineering in Medicine and Biology Society (EMBC), 6153–6156 pp

Chapter 64
Characterizing Fluctuations of Arterial and Cerebral Tissue Oxygenation in Preterm Neonates by Means of Data Analysis Techniques for Nonlinear Dynamical Systems

Stefan Kleiser, Marcin Pastewski, Tharindi Hapuarachchi, Cornelia Hagmann, Jean-Claude Fauchère, Ilias Tachtsidis, Martin Wolf, and Felix Scholkmann

Abstract The cerebral autoregulatory state as well as fluctuations in arterial (SpO_2) and cerebral tissue oxygen saturation (StO_2) are potentially new relevant clinical parameters in preterm neonates. The aim of the present study was to test the investigative capabilities of data analysis techniques for nonlinear dynamical systems, looking at fluctuations and their interdependence. StO_2, SpO_2 and the heart rate (HR) were measured on four preterm neonates for several hours. The fractional tissue oxygenation extraction (FTOE) was calculated. To characterize the fluctuations in StO_2, SpO_2, FTOE and HR, two methods were employed: (1) phase-space modeling and application of the recurrence quantification analysis (RQA), and (2) maximum entropy spectral analysis (MESA). The correlation between StO_2 and SpO_2 as well as FTOE and HR was quantified by (1) nonparametric nonlinear regression based on the alternating conditional expectation (ACE) algorithm, and (2) the maximal information-based nonparametric exploration (MINE) technique. We found that (1) each neonate showed individual characteristics, (2) a ~60 min oscillation was observed in all of the signals, (3) the nonlinear correlation strength between StO_2 and SpO_2 as well as FTOE and HR was specific for each neonate and showed a high value for a neonate with a reduced health status, possibly indicating an impaired cerebral

---

This chapter was originally published under a CC BY-NC 4.0 license, but has now been made available under a CC BY 4.0 license. An erratum to this chapter can be found at DOI 10.1007/978-1-4939-3023-4_66.

S. Kleiser • M. Pastewski • M. Wolf • F. Scholkmann (✉)
Biomedical Optics Research Laboratory, Department of Neonatology, University Hospital Zurich, University of Zurich, 8091 Zurich, Switzerland
e-mail: Felix.Scholkmann@usz.ch

T. Hapuarachchi • I. Tachtsidis
Department of Medical Physics and Bioengineering, University College London, London, UK

C. Hagmann • J.-C. Fauchère
Division of Neonatology, University of Zurich, 8091 Zurich, Switzerland

511
C.E. Elwell et al. (eds.), *Oxygen Transport to Tissue XXXVII*, Advances in Experimental Medicine and Biology 876, DOI 10.1007/978-1-4939-3023-4_64

autoregulation. In conclusion, our data analysis framework enabled novel insights into the characteristics of hemodynamic and oxygenation changes in preterm infants. To the best of our knowledge, this is the first application of RQA, MESA, ACE and MINE to human StO_2 data measured with near-infrared spectroscopy (NIRS).

Keywords Long term measurements • Autoregulation • Near infrared spectroscopy • Correlation analysis • Spontaneous fluctuations

1   Introduction

Preterm infants exhibit an immature regulation of respiration as well as systemic and cerebral blood circulation (i.e. cerebral autoregulation, CO_2 vasoreactivity), leading to an increased incidence of hypoxic and hyperoxic episodes due to (1) large fluctuations in cerebral hemodynamics, and (2) impaired coupling between cerebral blood flow (CBF) and metabolic demand [1]. Episodes of intermittent hypoxemia occur in 74 % of preterm infants, compared to 62 % of term infants [2]. Hyperoxemia or hypoxemia may lead to an increase in mortality and neurological morbidity with long-term effects in later adult life. Greater variability in arterial oxygen saturation (SpO_2) [3] correlates with an increased incidence of retinopathy of prematurity (ROP). Thus, the assessment of the dynamics of SpO_2 and cerebral tissue oxygen saturation (StO_2) in preterm neonates may be of high clinical relevance. Due to continuous advancement in biomedical optics [4, 5], a reliable noninvasive long-term measurement of StO_2 in preterm neonates is in principle feasible [6, 7].

   The aim of the present study was to analyze long-term measurements of StO_2 (conducted by multi-distance near-infrared spectroscopy, MD-NIRS) and SpO_2, heart rate (HR) and the fractional tissue extraction (FTOE) in preterm infants by means of data analysis techniques for nonlinear dynamical systems in order to investigate the characteristics of cerebral and systemic hemodynamic fluctuations and their interdependence.

2   Material and Methods

2.1   Subjects, Instrumentation and Experimental Protocol

A total of 20 clinically stable preterm neonates were enrolled. The study was approved by the ethics committee, and written informed consent was obtained from the parents before the study. Four neonates were selected for the present analysis, namely those with long continuous signals and the highest signal-to-noise ratio (SNR) (Table 64.1). SpO_2 and HR were determined by a standard patient monitor (Infinity Delta XL, Dräger, Germany) and StO_2 by an internally developed MD-NIRS device (OxyPrem, 4×3 [760, 805, 870 nm] light sources, two source-detector distances, i.e. 1.5, 2.5 cm [8]). OxyPrem uses the self-calibrating approach

Table 64.1 Description of the study sample

Characteristics	Neonate #1	Neonate #2	Neonate #3	Neonate #4
GA at birth (weeks)	33.4	26.4	29.4	26.8
GA at measurement (weeks)	34.7	28.5	29.9	30.7
Weight at measurement (g)	2220	1280	1090	1440
Apgar (1, 5, 10)	8, 8, 9	5, 4, 5	8, 8, 8	5, 8, 8
Respiration	Spontaneous	SIMV	CPAP	Spontaneous
FiO_2 (%), Hct (%), Hb (g/dL)	21, 50.6, 16.6	25, 40.9, 13.4	21, 49.5, 16.1	21, 36, 11.7
PDA	No	No	Yes	No
Length of analyzed data (min)	111	271	145	308

GA gestational age, *FiO_2* fraction of inspired oxygen, *SIMV* synchronized intermittent mandatory ventilation, *CPAP* continuous positive airway pressure, *Hct* hematocrit, *Hb* hemoglobin, *PDA* persistent ductus arteriosus

[9] which ensures a robust and high-precision measurement of absolute StO_2 values [10]. The NIRS optode was positioned over the left prefrontal cortex (PFC).

Measurements were performed continuously during the night (from ~10 pm till ~6 am), i.e. NIRS measured the resting-state activity of cerebral hemodynamics.

2.2 Signal Processing and Data Analysis

From the SpO_2 and StO_2 we calculated the fractional tissue oxygenation extraction ($FTOE = (SpO_2\text{-}StO_2)/SpO_2) \times 100$ [%]). FTOE quantifies the balance between oxygen delivery and oxygen consumption and correlates significantly with the invasively measured oxygen extraction fraction [11]. All signals (SpO_2, StO_2, FTOE and HR) were downsampled to 0.05 Hz to increase the SNR and since only low frequencies were of interest. For each of the four datasets, an interval was chosen for the subsequent analysis which contains data without any signal distortion. The lengths of the data are given in Table 64.1. To characterize the fluctuations in StO_2, SpO_2, FTOE and HR, two different methods were applied:

- *Phase-space modeling and application of the recurrence quantification analysis (RQA)* [12, 13]. Each signal (StO_2, SpO_2, FTOE and HR) was embedded into a phase space with the dimension m and time delay τ. The optimal values for m and τ were determined by finding the first minimum of the false nearest neighbors function depending on m, and the autocorrelation function depending on τ, respectively. In a subsequent step, the phase space trajectories were characterized by the RQA. In particular, the determinism (*DET*, i.e. the predictability of the system), entropy of the diagonal length (*ENT*, i.e. the complexity of the system's deterministic dynamics), and laminarity (*LAM*, i.e. the amount of intermittency of the system's dynamics) were calculated.
- *Maximum entropy spectral analysis (MESA)* [14]. This method enables a high-precision spectral analysis based on the principle of maximum entropy.

To prevent spurious peaks, the order of the MESA-based periodogram was set at one third of the number of samples [15].

The correlation between StO_2 and SpO_2 as well as FTOE and HR were quantified by two nonparametric methods:

- *Nonparametric nonlinear regression based on the alternating conditional expectation (ACE) algorithm* [16]. This technique finds the optimal transformations for the dependent and independent variables in order to maximize the correlation. The correlation strength is quantified by the maximal correlation coefficient, r_{ACE}.
- *Maximal information-based nonparametric exploration (MINE) technique* [17]. MINE enables the characterization of dependencies between variables. We calculated the maximal information coefficient (*MIC*) (relationship strength) and maximum asymmetry score (*MAS*) (departure from monotonicity).

In addition, each signal was characterized by calculating the median, and variability index 1 (VI_1, quantified as the mean of the modulus of the first derivation). In addition, the relationship of the fluctuation strength of StO_2 vs. SpO_2 was determined by the ratio of their standard deviations (variability index 2, VI_2).

3 Results, Discussion, Conclusion and Outlook

Figure 64.1a–d shows the time courses of SpO_2, StO_2, HR and FTOE. In Fig. 64.1b the normalized (i.e. subtraction of the mean value) SpO_2 and StO_2 were plotted to increase the visibility of similar dynamics. Figure 64.1d shows FTOE and HR after normalization (z-score) and smoothing (Kolmogorov-Zurbenko filter, window length: 180 s, iterations: 2) which increases the visibility of the similar long-term variability of both signals. The ACE correlation plots as well as the RQA, ACE and MINE results are visualized in Fig. 64.2. All signals show subject-specific dynamics:

- RQA: Noticeable low values for *DET* and *LAM* in neonate #4, high values for neonate #1. *ENT* (for StO_2 and SpO_2) has low values for neonate #3.
- MESA: (1) Neonate #1 exhibited a large oscillation with a period length (T) of 60 min in StO_2 and SpO_2 as well as HR and FTOE. (2) A large oscillation with $T \approx 30$ min is present in neonate #2 for StO_2 and SpO_2, followed by a second strongest oscillation with $T \approx 60$ min. In FTOE the predominant oscillation was at $T \approx 60$ min. The spectra of StO_2 and SpO_2 have a remarkably similar fine structure of oscillatory peaks indicating a large degree of similarity in the dynamics. In addition, an oscillation with $T \approx 15$ min can be seen in StO_2, SpO_2, HR and FTOE. (3) Neonate #3 shows an oscillation peak with $T \approx 60$ min in StO_2, SpO_2, HR and FTOE, whereas for FTOE a larger oscillation with $T \approx 30$ min is present. The spectra of StO_2 and SpO_2 have different fine

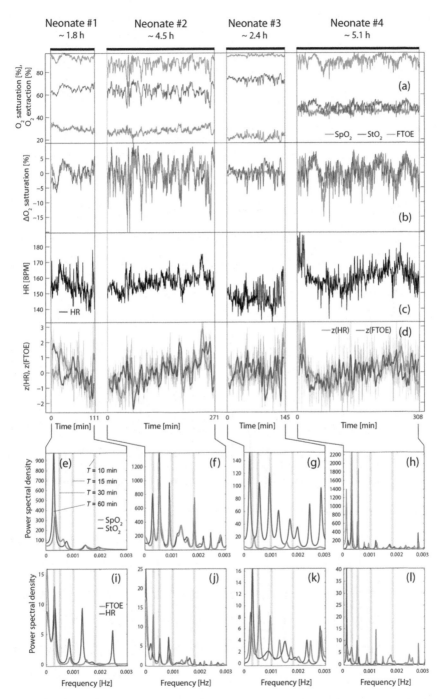

Fig. 64.1 (**a–d**) Visualization of the analyzed signals (StO_2, SpO_2, FTOE and HR). (**e–l**) Frequency spectra obtained by MESA

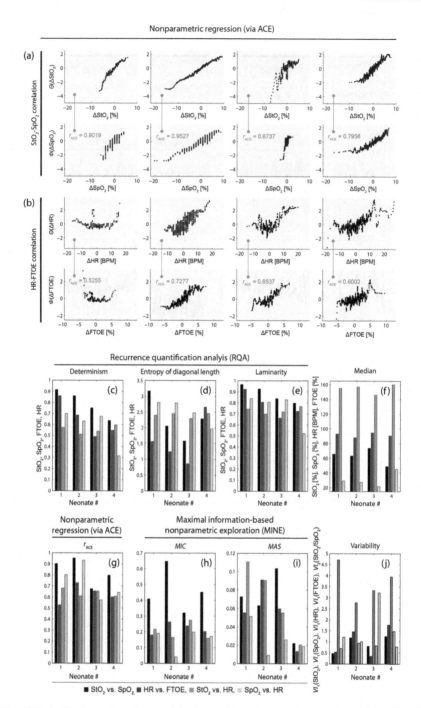

Fig. 64.2 (**a, b**) Correlation diagrams based on ACE nonparametric nonlinear regression. (**c–j**) Parameters obtained by RQA, MESA, ACE, MINE as well as the values for the median and variability

structures, which is also true for the spectra of HR and FTOE. (4) Neonate #4 shows a strong peak with $T \approx 60$ and 30 min in StO_2 and SpO_2.

- ACE and MINE: Concerning the relationship between StO_2 and SpO_2, the largest r_{ACE} and *MIC* value was found for neonate #2, the lowest for neonate #3; *MAS* was highest in neonate #3. Concerning HR vs. FTOE, neonate #2 had the largest r_{ACE}, *MIC* and *MAS* values. Neonate #4 showed significantly low *MAS* values for all four conditions (StO_2 vs. SpO_2 or HR, HR vs. FTOE or SpO_2).

To interpret the results it is helpful to discuss the similarities and differences of the signal characteristics with respects to the four neonates:

- *Similarities*: (1) The values for *DET*, *ENT* and *LAM* were all higher for StO_2 compared to SpO_2 (except for *ENT* of neonate #4), indicating more complex signal characteristics of StO_2 than of SpO_2. (2) The correlation between StO_2 and SpO_2 is higher than that observed between HR and FTOE ($r_{ACE}(StO_2$, $SpO_2) = 0.8310 \pm 0.1236$, $MIC(StO_2$, $SpO_2) = 0.4570 \pm 0.1385$; $r_{ACE}(HR$, $FTOE) = 0.6268 \pm 0.0854$, $MIC(HR$, $FTOE) = 0.2216 \pm 0.0371$). (3) All neonates showed an oscillation in StO_2, SpO_2, FTOE and HR with a period of $T \approx 60$ min, whereas the amplitude was specific for each neonate.
- *Differences*: (1) *DET*, *LAM* and *ENT* of SpO_2 were highest for neonate #3. (2) The correlation (r_{ACE}, *MIC*) between StO_2 and SpO_2 as well as HR and FTOE was highest for neonate #2; neonate #3 showed the lowest StO_2/SpO_2 correlation. (3) The smallest *MAS* values for all four correlations were found for neonate #4 (except for StO_2 vs. HR, neonate #2). (4) In neonate #3 we observed the highest median values for StO_2 and SpO_2 as well as the lowest ones for HR and FTOE. In neonate #4 the lowest StO_2 and the highest FTOE value were measured. (5) High VI_1 values in StO_2 and SpO_2 were present in neonates #2 and #4. Neonate #3 showed the largest VI_2 value for StO_2/SpO_2. (6) The mean StO_2 values correlated inversely with the individual Hct and GA (at birth) ($r = -0.927$ and $r = -0.982$, respectively; $p < 0.05$).

The physiological interpretation of these findings is not straightforward since all patient-specific characteristics have an influence on the analyzed parameters. In particular, the general health state (e.g. PDA, microbleeds, ischemia: yes/no), the type of respiration (ventilatory support: yes/no, type of support), and the GA (at birth/measurement) could potentially have a strong impact on the parameters. The following observations were made based on our analysis: (1) The general inverse correlation observed between StO_2 and Hct was also noticed by other studies (e.g. [18]). (2) Neonate #3 exhibited a large VI_2, e.g. the fluctuations in StO_2 were much stronger than in SpO_2 (especially the decreases), a pattern that is observed by neonates with a PDA—indeed, neonate #3 had a PDA (which was however classified as not hemodynamically relevant). The low StO_2/SpO_2 correlation (r_{ACE}, *MIC*) and the different frequency spectra (StO_2 vs. SpO_2, HR vs. FTOE) point also to a specific state of the systemic-cerebral hemodynamic coupling. The observation that neonate #3 had the highest median values for StO_2 and SpO_2 as well as the lowest ones for HR and FTOE is surprising since one would expect an increased FTOE and decrease StO_2 in case of a PDA [19]. (3) The oscillations in the data with $T \approx 60$ and 30 min could originate from sleep phases. A sleep-wake cycling (with a quiet sleep phase with $T \approx 20$ min) is known [20] in term newborns with $T \approx 50$–60 min and an increase in total hemoglobin and HR during active sleep (compared to quite sleep) has previously been observed [21, 22]. (4) The two neonates with the lowest GA at birth (#2, #4) had the largest variability of StO_2, SpO_2 and FTOE which could indicate an immature functioning of cerebral hemodynamic regulation.

In conclusion, using four case studies, we demonstrated the possibility of realizing long-term measurements in preterm neonates with MD-NIRS and we presented a novel framework for investigating the characteristics of cerebral and systemic hemodynamic fluctuations and their interdependence. A follow-up study, investigating the signal characteristics in healthy and ill preterm neonates using the same framework would be the next step. Focusing on the fluctuation characteristics of the signals may offer novel insights into systemic and cerebral hemodynamics that are not assessed routinely only using traditional analyses (i.e. based on moments and linear correlations). To the best of our knowledge, this is the first application of RQA, MESA, ACE and MINE to human NIRS data.

Acknowledgments The work was supported by the Wellcome Trust. We gratefully acknowledge funding by Nano-Tera (NeoSense, ObeSense), the Swiss National Science Foundation, the Neuroscience Center Zurich (UCL-Zurich Collaboration), The Danish Council for Strategic Resarch (SafeBoosC). We thank Caroline Guyer MD (Children's Hospital Zurich) for her valuable input, and Rachel Scholkmann for proofreading the manuscript.

References

1. Fyfe KL et al (2013) The development of cardiovascular and cerebral vascular control in preterm infants. Sleep Med Rev 18:299–310
2. Hunt CE et al (2011) Longitudinal assessment of hemoglobin oxygen saturation in preterm and term infants in the first six months of life. J Pediatr 159(3):377–383
3. Di Fiore JM et al (2012) The relationship between patterns of intermittent hypoxia and retinopathy of prematurity in preterm infants. Pediatr Res 72(6):606–612

4. Scholkmann F et al (2014) A review on continuous wave functional near-infrared spectroscopy and imaging instrumentation and methodology. Neuroimage 85(Part 1):6–27
5. Wolf M, Ferrari M, Quaresima V (2007) Progress of near-infrared spectroscopy and topography for brain and muscle clinical applications. J Biomed Opt 12(6):062104
6. Wolf M, Greisen G (2009) Advances in near-infrared spectroscopy to study the brain of the preterm and term neonate. Clin Perinatol 36(4):807–834
7. Wolf M et al (2012) A review of near infrared spectroscopy for term and preterm newborns. J Near Infrared Spectrosc 20(1):43–55
8. Hyttel-Sorensen S et al (2013) Calibration of a prototype NIRS oximeter against two commercial devices on a blood-lipid phantom. Biomed Opt Express 4(9):1662–1672
9. Hueber DM et al (1999) New optical probe designs for absolute (self-calibrating) NIR tissue hemoglobin measurements. Opt Tomogr Spectrosc Tissue III Proc SPIE 3597:618–631
10. Scholkmann F, Metz AJ, Wolf M (2014) Measuring tissue hemodynamics and oxygenation by continuous-wave functional near-infrared spectroscopy – how robust are the different calculation methods against movement artifacts? Physiol Meas 35(4):717–734
11. Naulaers G et al (2007) Use of tissue oxygenation index and fractional tissue oxygen extraction as non-invasive parameters for cerebral oxygenation. A validation study in piglets. Neonatology 92(2):120–126
12. Webber CL Jr, Zbilut JP (1985) Dynamical assessment of physiological systems and states using recurrence plot strategies. J Appl Physiol 76(2):965–973
13. Marwan N et al (2002) Recurrence plot based measures of complexity and its application to heart rate variability data. Phys Rev E 66(2):026702
14. Ulrych TJ, Bishop TN (1975) Maximum entropy spectral analysis and autoregressive decomposition. Rev Geophys 13(1):183–200
15. Kay SM (1979) The effects of noise on the autoregressive spectral estimator. IEEE Trans Acoust Speech Signal Process 27(5):478–485
16. Breiman L, Friedman JH (1985) Estimating optimal transformations for multiple regression and correlation. J Am Stat Assoc 80(391):580–598
17. Reshef DN et al (2011) Detecting novel associations in large data sets. Science 334 (6062):1518–1524
18. Youkin DP et al (1987) The effect of hematocrit and systolic blood pressure on cerebral blood flow in newborn infants. J Cereb Blood Flow Metab 7(3):295–299
19. Vanderhaegen J et al (2008) Surgical closure of the patent ductus arteriosus and its effect on the cerebral tissue oxygenation. Acta Paediatr 97(12):1640–1644
20. Osredkar D et al (2005) Sleep-wake cycling on amplitude-integrated electroencephalography in term newborns with hypoxic-ischemic encephalopathy. Pediatrics 115(2):327–332
21. Münger DM, Bucher HU, Duc G (1998) Sleep state changes associated with cerebral blood volume changes in healthy term newborn infants. Early Hum Dev 52(1):27–42
22. Stéphan-Blanchard E et al (2013) Heart rate variability in sleeping preterm neonates exposed to cool and warm thermal conditions. PLoS One 8(7):e68211

Chapter 65
Can the Assessment of Spontaneous Oscillations by Near Infrared Spectrophotometry Predict Neurological Outcome of Preterm Infants?

André Stammwitz, Kurt von Siebenthal, Hans U. Bucher, and Martin Wolf

Abstract The aim was to assess the correlation between cerebral autoregulation and outcome. Included were 31 preterm infants, gestational age 26 1/7 to 32 2/7 and <24 h life. Coherence between cerebral total haemoglobin (tHb) or oxygenation index (OI) measured by near-infrared spectrophotometry (NIRS) and systemic heart rate (HR) or arterial blood pressure (MAP) was calculated as a measure of autoregulation. In contrast to previous studies, low coherences in the first 24 h were significantly associated with intraventricular haemorrhage, death or abnormal neurodevelopmental outcome at 18 months or later. We suggest that our results can be explained by the concept of a multi-oscillatory-functions-order.

Keywords Autoregulation • Near infrared spectroscopy • Neurological outcome • Preterm infants • Spontaneous oscillations

1 Introduction

In the last two decades mortality for preterm infants has decreased continuously in industrialised countries, but morbidity and long-term cerebral outcome remained stable [1] due to the unchanged incidence of germinal matrix-intraventricular haemorrhage (GMH-IVH) or white brain matter injury. Fluctuations in cerebral blood flow (CBF) are a main factor in the pathogenesis of GMH-IVH [2]. They may be caused by impaired cerebral autoregulation, i.e. changes in blood pressure transmitted to the brain [3, 4]. The coherence between cerebral (tHb or OI measured by NIRS) and systemic (HR or MAP) spontaneous oscillations, is a novel approach to assess autoregulation [5] (Fig. 65.1). A high coherence between cerebral and systemic parameters was interpreted as impaired autoregulation and associated with increased incidence of GMH-IVH [6, 7] or subsequent death [8]. In contrast,

A. Stammwitz • K. von Siebenthal • H.U. Bucher • M. Wolf (✉)
Division of Neonatology, University Hospital Zurich, Zurich, Switzerland
e-mail: martin.wolf@usz.ch

© Springer Science+Business Media, New York 2016   521
C.E. Elwell et al. (eds.), *Oxygen Transport to Tissue XXXVII*, Advances in
Experimental Medicine and Biology 876, DOI 10.1007/978-1-4939-3023-4_65

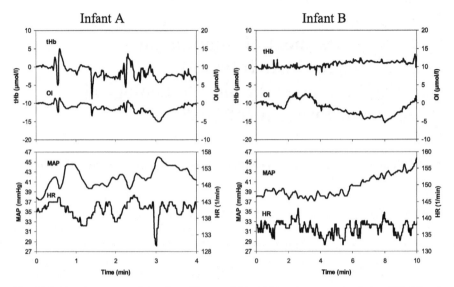

Fig. 65.1 Simultaneous tracings of total haemoglobin (tHb) and oxygenation index (OI) (*above*) and mean arterial blood pressure (MAP) and heart rate (HR) (*below*). *Left*: Example of coherent tracings of 1-day-old premature neonate (1940 g), who had normal MDIs (89/100; 96/100) and normal ultrasounds. Coherence scores were between 0.9 and 0.99. *Right*: Incoherent tracings of a 1-day-old premature neonate (1050 g), who had pathologic ultrasound scans and an abnormal MDI at 18 months (62/100). Coherence scores were between 0.14 and 0.39

Doppler sonography studies showed that CBF is independent of MAP [9, 10]. To clarify these conflicting results, the aim was to determine in very preterm infant the relation between autoregulation during the first 3 days of life and their neurological outcome in the first 10 years.

2 Methods

After approval by the ethical committee and parental consent, we enrolled 31 pre-term neonates with normal capillary refilling, MAP, HR and without arrhythmias. Included were infants with an intraarterial catheter. Excluded were patients with blood transfusions in the first 6 h, severe sepsis with low blood pressure, severe malformations or severe asphyxia (Sarnat III). Gestational age ranged from 26 1/7 to 32 2/7 weeks (median: 27 2/7), birth weight from 690 to 2440 g (median: 1030) and head circumference from 22.5 to 35.5 (median: 25.60); 14 infants were boys, 17 girls. The median socio-economic index, based on mother's education and father's occupation was 6 (from 12), suggesting a normal distribution of the popu-lation. 28 infants were delivered by caesarean section and 3 spontaneously. The median APGAR score 5 min postpartum was 7.5 (3–9). The median pH value in the umbilical artery was 7.25 (7.16–7.33). All infants had lung disease diagnosed by clinical examinations and chest radiography, consistent with a respiratory distress syndrome. Five neonates received nasal CPAP, the others were mechanically ventilated for a median time of 7 days. All infants received supplemental oxygen with a mean inspired oxygen fraction of 0.42 ± 0.15 %. Twenty infants received pethidine as analgesic and five diazepam for sedation. In six infants a patent ductus arteriosus was diagnosed and treated with indomethacine after day 2. The following parameters were determined at the time of measurement: Haemoglobin 16.6 ± 2.7 g/100 ml, blood glucose 4.9 ± 2.2 mM, SaO_2 91.2 ± 2.3 % and $tcpCO_2$ 5.6 ± 1.1 kPa. One infant received dopamine for low MAP.

In the neonatal brainNIRS [11] measures concentration changes in cerebral oxy-, deoxy-, total haemoglobin (O_2Hb, HHb, tHb in µM) and oxygen index (OI = (O_2Hb-HHb)/2) [12]. The neonatal sensor of the Critikon Cerebral Oxygenation Monitor 2001 was placed fronto-parietally. It emitted light at 774, 815, 867, 904 nm. The interoptode distance was 35 mm, differential pathlength factor 4.4 [13], and sampling time 0.56 s. NIRS was measured at <6 h for 2 h, at 12–16 h, at 24–28 h and at 68–76 h (if the infant still had additional O_2 and invasive MAP). Transcutaneous pO_2 ($tcpO_2$, Hellige Oxymonitor), pCO_2 ($tcpCO_2$, Hellige Kapnomonitor), SaO_2 (Nellcor 200 or Radiometer Oximeter), HR in (1/min) and MAP (in mm Hg from an umbilical artery catheter (Hellige Vicom-SM)) were recorded together with the NIRS data. All were kept within normal ranges.

A cranial ultrasound examination (Acuson 128XP, 7 MHz transducer) was performed at <24 h, 3, 7, 14 days of life and every 2 weeks until discharge. Findings were classified into grade of intraventricular haemorrhages (IVH): 0 = none, 1 = subependymal, 2 = intraventricular, 3 = intraventricular with dilatation, 4 = parenchymal [14]. Grade of parenchymal lesions (PVL): 0 = none, 1 = echodensities persisting more than 10 days, 2 = localized periventricular cysts, 3 = cystic periventricular leukomalacia, and 4 = cystic subcortical leukomalacia [15]. To compare with previous studies two ultrasound groups were formed: (1) Poor ultrasound: IVH grades 3, 4 and/or any PVL and (2) Normal ultrasound: IVH grade 0, 1 or 2 and no PVL.

Developmental follow-up was examined at 9 and 18 months age corrected for prematurity using a standardised neurological test and the Bayley II scales of infant development: psycho-motor-developmental index (PDI) and mental-developmental index (MDI) ≥ 84 were considered normal and < 84 abnormal [16]. Further neurological examinations were performed at 3 years (3 infants) and 10 years (10 infants). Neurological findings were classified abnormal for markedly increased muscle tone (score ≥ 2 on Ashworth scale [17]). Correlation between MDI, PDI or neurological status and coherences were estimated on the basis of the worst outcome in any of the assessments. Neonates who died in the first days of life were classified as abnormal Bayley indices.

All measurements were screened and periods of ≥ 12 min without artefacts or changes in SaO_2 (<5 %) were selected. Groups of 9 samples were aggregated to obtain a sample rate of ~5 s. Coherence spectra were calculated by MATLAB's cohere function (parameters: nfft = 144, numoverlap = 143, subtraction of 'mean') [18]. The coherence was calculated between the tHb or OI and MAP or HR

[6, 19]. We calculated the mean coherence for the band between 0 and 0.01 Hz (ultra-low frequency = ULF) and 0–0.1 Hz (low frequency = LF).

We used Spearman's rho for continuous variables and to compare groups.

3 Results

Since analysis of variance did not show a significant difference between the first three measurement (6, 12 and 24 h), for each infant the coherences of the first 24 h were averaged. The different coherence indices showed a significant linear correlation (p < 0.00001, Fig. 65.2). The grade of correlation was higher for LF compared to ULF. LF and ULF were also highly linearly correlated (p < 0.005). The different coherence indices correlated significantly with the clinical parameters (Table 65.1, Fig. 65.3). Low coherence during the first 24 h of life was consistently associated with unfavourable outcome.

Nine infants died between 2 and 34 days (median 3 days) due to grade 3/4 IVH (N = 5) and respiratory failure (N = 4). In 12 of 22 surviving infants an abnormal neurology was found at discharge (mostly increased muscle tonus). The number decreased with age: at 9 months 10 of 20 infants, at 18 months 3 of 18 infants and at 10 years 3 of 13 infants. MDI at 18 months was normal in 13 of 19 infants, suspect (83–68) in two and abnormal (<68) in four infants. The respective values for the PDI were: 11, 5 and 3 infants. Neonates with high MDI and/or normal neurological outcome had significantly higher coherences than neonates with low MDI and/or abnormal neurology.

Ultrasound scans showed no lesion in 5 of 31 infants. IVH grade 1/2 was found in 9, and grade 3/4 in 5 infants. PVL grade 1 was observed in 11, grade 3 in 1 infant.

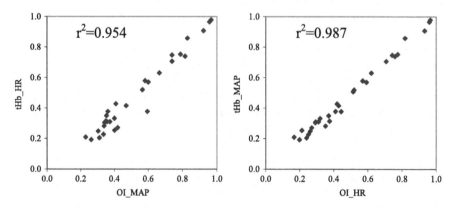

Fig. 65.2 Examples of correlation of two coherences between cerebral and systemic parameters: tHb_HR versus OI_MAP (*left*) and tHb_MAP versus OI_HR (*right*). *Each dot* represents one infant. All correlations were highly significant (p < 0.00001, Spearman), which indicates that the systemic variables HR and MAP are interchangeable to a high degree. The same is true for the cerebral variables tHb and OI

Table 65.1 Correlation between coherences and outcome. Contrary to former studies high coherences (bold) are correlated with good outcome

Coherence	Variables	Mean	SD	N	Variables	Mean	SD	N	Significance
OI_MAP	Normal neurology	0.636	0.234	10	Abnormal neurology	0.477	0.208	19	0.073
	Normal MDI	0.654	0.231	13	Abnormal MDI	0.441	0.175	15	**0.010**
	Survivors	0.567	0.226	22	Dead	0.417	0.192	8	0.107
	No PVL	0.496	0.217	14	PVL	0.554	0.235	16	0.488
	No IVH	0.561	0.234	24	IVH	0.389	0.114	6	**0.019**
OI_HR	Normal neurology	0.614	0.251	10	Abnormal neurology	0.434	0.22	20	0.054
	Normal MDI	0.633	0.242	13	Abnormal MDI	0.397	0.188	16	**0.006**
	Survivors	0.535	0.245	22	Dead	0.372	0.197	9	0.088
	No PVL	0.464	0.225	15	PVL	0.510	0.260	16	0.600
	No IVH	0.525	0.257	24	IVH	0.359	0.11	7	**0.020**
tHb_MAP	Normal neurology	0.635	0.23	10	Abnormal neurology	0.458	0.213	19	**0.048**
	Normal MDI	0.645	0.235	13	Abnormal MDI	0.424	0.178	15	**0.009**
	Survivors	0.559	0.228	22	Dead	0.394	0.191	8	0.079
	No PVL	0.486	0.214	14	PVL	0.541	0.243	16	0.514
	No IVH	0.555	0.235	24	IVH	0.359	0.106	6	**0.007**
tHb_HR	Normal neurology	0.612	0.254	10	Abnormal neurology	0.424	0.222	20	0.095
	Normal MDI	0.621	0.258	13	Abnormal MDI	0.395	0.184	16	**0.011**
	Survivors	0.522	0.257	22	Dead	0.379	0.184	9	0.142
	No PVL	0.446	0.227	15	PVL	0.513	0.262	16	0.457
	No IVH	0.523	0.259	24	IVH	0.359	0.106	7	**0.008**

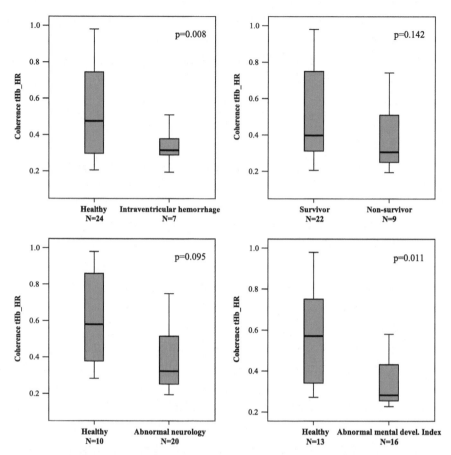

Fig. 65.3 Correlation of the different coherence scores with clinical outcome. All four outcome variables intraventricular hemorrhage, survival, neurology and mental developmental show an association of unfavourable outcome with a low coherence value. The p-values are indicated

Again low coherence was significantly associated with unfavourable IVH grade ≥ 3 (Fig. 65.3). PVL did not correlate with the coherences.

To summarise, in our population of preterm infants a low coherence during the first 24 h of life was consistently associated with unfavourable outcome.

4 Discussion

In this study we demonstrated that low coherences between MAP or HR and OI or tHb in the first 24 h of life in very preterm infants are associated with unfavourable outcome. The *coherence-indices* were high in 20 of 31 neonates. They showed no significant *variability* over time, indicating no linear tendency towards lower values during the first

24 h as found by Menke et al. [20]. Nevertheless, in some infants coherence changed during the first 24 h. This may be due to individually different time course of entrainment with environmental rhythms (discussed below). Remarkable is the high correlation between the types of coherences: HR or MAP and OI or tHb yield very similar results.

It was postulated [6] that coherence scores reflect cerebral autoregulation and therefore represent a predictive value for neurological outcome. Previous studies found high ULF coherences (>0.5), i.e. impaired autoregulation, in 17 of 31 neonates, which was associated with a higher incidence of IVH/PVL [6–8, 21, 22]. These results are in contrast to our findings: We found a significant correlation of *high* coherence values with favourable outcome, not only for IVH but also for later outcome assessments. Our data are consistent, since all outcome parameters showed the same tendency. We were not able to identify differences between the previous and our study that could explain this discrepancy.

In the following the concept of autoregulation has to be discussed in order to possibly explain our results. Traditionally, autoregulation was seen as a *myogenic concept*, as a cerebral protective mechanism leading to a constriction/dilatation of arteries in response to an excessive increase/decrease of MAP [23]. On the one hand assessing cerebral autoregulation in neonates is difficult, since it is "very easy to obtain completely misleading results" [23]. Autoregulation may vary considerably over short periods of time within the same subject [24]. On the other hand even in adults, the text book concept of autoregulation (i.e. the CBF stable from 50 to 150 mmHg of MAP) is obsolete: today autoregulation merely stabilizes CBF in a range of only 10 mmHg [25]. Recently this led to a more complex view, the *metabolic concept*, i.e. autoregulation as a response to metabolic demands [24, 26–28]. Also this concept postulates that a high coherence is a sign of impaired autoregulation and thus neither approach is able to explain our results.

Spontaneous oscillations in our frequency range (0–0.1 Hz) may not reflect autoregulation. Since our results indicate that a high coherence is associated with beneficial outcome, let us consider an example where a high coherence was beneficial [29, 30]. The coordination or coherence of different central oscillators, e.g. of neurons in the medulla oblongata, is of great importance for a healthy functioning of the respiratory and sucking functions. A loss of such coherence leads to aspiration pneumonia [31]. Analogously we hypothesize that high coherences indicate a coordination of physiological sub-systems and thus are a sign of health. This concept called *multi-oscillatory-functions-order* (MOFO) was postulated already in 1980 [32]. It is applicable to a wide spectrum of rhythmically changing physiological processes. The *coordination of interacting oscillations* may lead to system stabilisation [33, 34]. The following examples illustrate the scope, productivity and relevance of this MOFO approach.

Fractured relations (i.e. 3.55:1) of oscillations in reaction time were a sign of immaturity [32] and were followed by integer relations (i.e. 3:1) considered as mature. This constitutes evidence of von Holst's "principle of absolute and relative coordination", i.e. the maturation from fractured to integer relations [35]. This experiment was expanded to other physiologic parameters, such as HR, respiratory rate, short-term memory, auditory-motor reaction time or those recorded by electrooculogram. Sinz defined a *coupling ratio* i.e., the ratio between realised and possible frequency synchronisations. This ratio was shown to have diagnostic value for infants with congenital heart disease, patients with neurosis, office workers and students listening to music [32]. Similar efforts were continued by others [36–41]. All these findings indicate that synchronization is actually a sign of health and maturity. In addition, high short-term coherence values between HR, respiration and MAP were reduced by severe brain disorders [41, 42]. This can be explained by an impaired central autonomic coordination. Thus a reduced coherence, (or 'uncoupling' [37] or 'decomplexification' [38]) indicates a pathological process. It is expected that the more the coupling decreases, the poorer the outcome, because uncoupling between lungs, heart and the vascular system decreases their functional performance.

Summarizing our findings, all outcome parameters showed the same tendency: Favourable outcome was associated with high coherence. According to the MOFO approach, the high coherences in our study indicate a strong coordination of physiologic sub-systems, which is a sign of stability and explains the favourable short and long-term outcome. This MOFO evolves in the first months of life [29, 30, 43]. The autoregulation concept is not necessarily erroneous. Probably the concept of autoregulation does not apply to oscillations in our frequency range of up to 0.1 Hz. They are understandable by a MOFO concept. Future studies should investigate the areas of validity of autoregulation and MOFO approaches.

Almost all of our neonates were mechanically ventilated. Possibly this affected our coherence values by entrainment [39, 40]. A methodological problem of calculating the coherence is the intermittently low amplitude of spontaneous MAP variations. In a previous study MAP variations of 25 % were observed, but the cerebral spectral power was <2.5 % in a third of the subjects [27]. It was suggested that variations in MAP of >10 % are necessary to reduce noise to an acceptable level [23]. Therefore, in our recordings, we selected periods fulfilling this criterion.

Many studies demonstrate both the productivity of a MOFO-like approach and the lack of knowledge concerning coupling phenomena. Our results are internally consistent and support the MOFO hypothesis. They demonstrate that it is necessary to overcome the concept of autoregulation for the coherence analysis of spontaneous oscillations in the frequency range of 0–0.1 Hz.

The first ultrasound examination in our study was performed <24 h and IVH usually appeared later. Thus low coherence precedes brain damage. This is of potentially high clinical value, because there may be time to prevent IVH.

5 Conclusion

In contrast to previous studies, low coherences between systemic and cerebral parameters in the first 24 h of life were associated with IVH, death, abnormal MDI and abnormal neurological outcome. Similar predictive results were obtained with coherences using OI and tHb as cerebral variables and MAP and HR as systemic variables. The time of measurement in the first 24 h of life did not influence coherences significantly. According to a MOFO approach, high coherences reflect a high coordination of all physiological cycles. We propose that high coherences are an indicator of maturation and integrity.

Acknowledgements We thank Dr. Vera Dietz and Matthias Keel for their help during the measurements and Dr. Daniel Haensse for help during data analysis.

References

1. Ruegger C, Hegglin M, Adams M et al (2012) Population based trends in mortality, morbidity and treatment for very preterm- and very low birth weight infants over 12 years. BMC Pediatr 12:17
2. Wigglesworth JS (1979) Prevention of cerebral hemorrhage in preterm infants. Lancet 2:256
3. Lou HC, Lassen NA, Friis-Hansen B (1979) Impaired autoregulation of cerebral blood flow in the distressed newborn infant. J Pediatr 94:118–121
4. Pryds O (1991) Control of cerebral circulation in the high-risk neonate. Ann Neurol 30: 321–329
5. De Smet D, Vanderhaegen J, Naulaers G et al (2009) New measurements for assessment of impaired cerebral autoregulation using near-infrared spectroscopy. Adv Exp Med Biol 645: 273–278
6. Tsuji M, Saul JP, du Plessis A et al (2000) Cerebral intravascular oxygenation correlates with mean arterial pressure in critically ill premature infants. Pediatrics 106:625–632
7. Alderliesten T, Lemmers PM, Smarius JJ et al (2013) Cerebral oxygenation, extraction, and autoregulation in very preterm infants who develop peri-intraventricular hemorrhage. J Pediatr 162:698–704 e2
8. Wong FY, Leung TS, Austin T et al (2008) Impaired autoregulation in preterm infants identified by using spatially resolved spectroscopy. Pediatrics 121:e604–e611
9. Pellicer A, Valverde E, Gaya F et al (2001) Postnatal adaptation of brain circulation in preterm infants. Pediatr Neurol 24:103–109
10. Tyszczuk L, Meek J, Elwell C et al (1998) Cerebral blood flow is independent of mean arterial blood pressure in preterm infants undergoing intensive care. Pediatrics 102:337–341
11. Wray S, Cope M, Delpy DT et al (1988) Characterization of the near infrared absorption spectra of cytochrome aa3 and haemoglobin for the non-invasive monitoring of cerebral oxygenation. Biochim Biophys Acta 933:184–192
12. Fukui Y, Ajichi Y, Okada E (2003) Monte Carlo prediction of near-infrared light propagation in realistic adult and neonatal head models. Appl Optics 42:2881–2887
13. Wyatt JS, Cope M, Delpy DT et al (1990) Measurement of optical path length for cerebral - near-infrared spectroscopy in newborn infants. Dev Neurosci 12:140–144
14. Papile LA, Burstein J, Burstein R et al (1978) Incidence and evolution of subependymal and intraventricular hemorrhage: a study of infants with birth weights less than 1,500 gm. J Pediatr 92:529–534
15. de Vries LS, Eken P, Dubowitz LM (1992) The spectrum of leukomalacia using cranial ultrasound. Behav Brain Res 49:1–6
16. Maas YG, Mirmiran M, Hart AA et al (2000) Predictive value of neonatal neurological tests for developmental outcome of preterm infants. J Pediatr 137:100–106

17. Damiano DL, Quinlivan JM, Owen BF et al (2002) What does the Ashworth scale really measure and are instrumented measures more valid and precise? Dev Med Child Neurol 44: 112–118
18. Morren G, Lemmerling P, van Huffel S et al (2001) Detection of the autoregulation in the brain using a novel subspace-based technique. In: Proceedings of the 23rd annual international conference of the IEEE Engineering in Medicine and Biology Society (EMBC 2001), Istanbul
19. von Siebenthal K, Beran J, Wolf M et al (1999) Cyclical fluctuations in blood pressure, heart rate and cerebral blood volume in preterm infants. Brain Dev 21:529–534
20. Menke J, Michel E, Hillebrand S et al (1997) Cross-spectral analysis of cerebral autoregulation dynamics in high risk preterm infants during the perinatal period. Pediatr Res 42:690–699
21. Volpe JJ (2001) Neurobiology of periventricular leukomalacia in the premature infant. Pediatr Res 50:553–562
22. O'Leary H, Gregas MC, Limperopoulos C et al (2009) Elevated cerebral pressure passivity is associated with prematurity-related intracranial hemorrhage. Pediatrics 124:302–309
23. Panerai RB (1998) Assessment of cerebral pressure autoregulation in humans--a review of measurement methods. Physiol Meas 19:305–338
24. Panerai RB, Eames PJ, Potter JF (2003) Variability of time-domain indices of dynamic cerebral autoregulation. Physiol Meas 24:367–381
25. Willie CK, Tzeng YC, Fisher JA et al (2014) Integrative regulation of human brain blood flow. J Physiol 592(5):841–859
26. Panerai RB, Carey BJ, Potter JF (2003) Short-term variability of cerebral blood flow velocity responses to arterial blood pressure transients. Ultrasound Med Biol 29:31–38
27. Giller CA, Mueller M (2003) Linearity and non-linearity in cerebral hemodynamics. Med Eng Phys 25:633–646
28. Caicedo A, Naulaers G, Wolf M et al (2012) Assessment of the myogenic and metabolic mechanism influence in cerebral autoregulation using near-infrared spectroscopy. Adv Exp Med Biol 737:37–44
29. Lohr B, Siegmund R (1999) Ultradian and circadian rhythms of sleep-wake and food-intake behavior during early infancy. Chronobiol Int 16:129–148
30. Korte J, Wulff K, Oppe C et al (2001) Ultradian and circadian activity-rest rhythms of preterm neonates compared to full-term neonates using actigraphic monitoring. Chronobiol Int 18:697–708
31. Hellbrugge T (1967) Chronophysiology of the child. Verh Dtsch Ges Inn Med 73:895–921
32. Sinz R (1980) Chronopsychophysiologie. Akademie, Berlin
33. Gonze D, Halloy J, Goldbeter A (2002) Robustness of circadian rhythms with respect to molecular noise. Proc Natl Acad Sci U S A 99:673–678
34. Goldbeter A (2002) Computational approaches to cellular rhythms. Nature 420:238–245
35. von Holst E (1939) Die relative Koordination als Phänomen und als Methode zentralnervöser Funktionsanalyse. Ergeb Physiol 42:228–306
36. Biswal B, Hudetz AG, Yetkin FZ et al (1997) Hypercapnia reversibly suppresses low-frequency fluctuations in the human motor cortex during rest using echo-planar MRI. J Cereb Blood Flow Metab 17:301–308
37. Godin PJ, Buchman TG (1996) Uncoupling of biological oscillators: a complementary hypothesis concerning the pathogenesis of multiple organ dysfunction syndrome. Crit Care Med 24:1107–1116
38. Goldstein B, Fiser DH, Kelly MM et al (1998) Decomplexification in critical illness and injury: relationship between heart rate variability, severity of illness, and outcome. Crit Care Med 26: 352–357
39. Larsen PD, Galletly DC (2001) Cardioventilatory coupling in heart rate variability: the value of standard analytical techniques. Br J Anaesth 87:819–826
40. Censi F, Calcagnini G, Cerutti S (2002) Coupling patterns between spontaneous rhythms and respiration in cardiovascular variability signals. Comput Methods Programs Biomed 68:37–47

41. Zwiener U, Schelenz C, Bramer S et al (2003) Short-term dynamics of coherence between respiratory movements, heart rate, and arterial pressure fluctuations in severe acute brain disorders. Physiol Res 52:517–524

42. Arnold M, Miltner WH, Witte H et al (1998) Adaptive AR modeling of nonstationary time series by means of Kalman filtering. IEEE Trans Biomed Eng 45:553–562

43. Mirmiran M, Baldwin RB, Ariagno RL (2003) Circadian and sleep development in preterm infants occurs independently from the influences of environmental lighting. Pediatr Res 53: 933–938

Erratum to: Oxygen Transport to Tissue XXXVII

Clare E. Elwell, Terence S. Leung, and David K. Harrison

Erratum to:
Chapter 14 in: C.E. Elwell et al. (eds.), *Oxygen Transport to Tissue XXXVII*, DOI 10.1007/978-1-4939-3023-4_14

Chapter 17 in: C.E. Elwell et al. (eds.), *Oxygen Transport to Tissue XXXVII*, DOI 10.1007/978-1-4939-3023-4_17

Chapter 29 in: C.E. Elwell et al. (eds.), *Oxygen Transport to Tissue XXXVII*, DOI 10.1007/978-1-4939-3023-4_29

Chapter 62 in: C.E. Elwell et al. (eds.), *Oxygen Transport to Tissue XXXVII*, DOI 10.1007/978-1-4939-3023-4_62

Chapter 64 in: C.E. Elwell et al. (eds.), *Oxygen Transport to Tissue XXXVII*, DOI 10.1007/978-1-4939-3023-4_64

C.E. Elwell
Department of Medical Physics and Biomedical Engineering, University College London, London, UK
e-mail: c.elwell@ucl.ac.uk

T.S. Leung
Department of Medical Physics and Biomedical Engineering, University College London, London, UK
e-mail: t.leung@ucl.ac.uk

D.K. Harrison
Microvascular Measurements, St. Lorenzen, Italy
e-mail: Harrison.David.K@gmail.com

© The Author(s) 2016
C.E. Elwell et al. (eds.), *Oxygen Transport to Tissue XXXVII*, Advances in Experimental Medicine and Biology 876, DOI 10.1007/978-1-4939-3023-4_66

Chapters 14, 17, 29, 62 and 64 were originally published under a CC BY-NC 4.0 license, but have now been made available under a CC BY 4.0 license.

Updated versions of the original chapters can be found at
http://dx.doi.org/10.1007/978-1-4939-3023-4_14
http://dx.doi.org/10.1007/978-1-4939-3023-4_17
http://dx.doi.org/10.1007/978-1-4939-3023-4_29
http://dx.doi.org/10.1007/978-1-4939-3023-4_62
http://dx.doi.org/10.1007/978-1-4939-3023-4_64

Index

© Springer Science+Business Media, New York 2016
C.E. Elwell et al. (eds.), *Oxygen Transport to Tissue XXXVII*, Advances in
Experimental Medicine and Biology 876, DOI 10.1007/978-1-4939-3023-4

Printed by Printforce, the Netherlands